FLAVONOIDS AND RELATED COMPOUNDS
Bioavailability and Function

OXIDATIVE STRESS AND DISEASE

Series Editors

LESTER PACKER, PhD
ENRIQUE CADENAS, MD, PhD

UNIVERSITY OF SOUTHERN CALIFORNIA SCHOOL OF PHARMACY
LOS ANGELES, CALIFORNIA

FLAVONOIDS AND RELATED COMPOUNDS
Bioavailability and Function

Edited by

JEREMY P. E. SPENCER • ALAN CROZIER

CRC Press
Taylor & Francis Group
Boca Raton   London   New York

CRC Press is an imprint of the
Taylor & Francis Group, an **informa** business

CRC Press
Taylor & Francis Group
6000 Broken Sound Parkway NW, Suite 300
Boca Raton, FL 33487-2742

First issued in paperback 2016

ISBN 13: 978-1-138-19941-5 (pbk)
ISBN 13: 978-1-4398-4826-5 (hbk)

Library of Congress Cataloging-in-Publication Data

Flavonoids and related compounds : bioavailability and functions / editors, Jeremy P.E.
    Spencer, Alan Crozier.
        p. ; cm. -- (Oxidative stress and disease ; 29)
    Includes bibliographical references and index.
    ISBN 978-1-4398-4826-5 (hardcover : alk. paper)
    I. Spencer, Jeremy P.E. II. Crozier, Alan. III. Series: Oxidative stress and disease ; 29.
    [DNLM: 1. Flavonoids. 2. Biological Availability. W1 OX626 v.29 2012 / QU 220]

572'.2--dc23   2012005189

Visit the Taylor & Francis Web site at
http://www.taylorandfrancis.com

and the CRC Press Web site at
http://www.crcpress.com

In memory of James A. Joseph (1944–2010): unconventional thinker, pioneer in nutritional neuroscience, inspirational speaker, caring mentor, generous scientist, outdoor adventurer, entertaining friend to many, and a shining example to us all.

Contents

Series Preface

Through evolution, oxygen—itself a free radical—was chosen as the terminal electron acceptor for respiration; hence, the formation of oxygen-derived free radicals is a consequence of aerobic metabolism. These oxygen-derived radicals are involved in oxidative damage to cell components inherent in several pathophysiological situations. Conversely, cells convene antioxidant mechanisms to counteract the effects of oxidants by either a highly specific manner (e.g., superoxide dismutases) or in a less specific manner (e.g., through small molecules, such as glutathione, vitamin E, and vitamin C). Oxidative stress—as classically defined—entails an imbalance between oxidants and antioxidants. However, the same free radicals that are generated during oxidative stress are produced during normal metabolism and, as a corollary, are involved in both human health and disease by virtue of their involvement in the regulation of signal transduction and gene expression, activation of receptors and nuclear transcription factors, antimicrobial and cytotoxic actions of immune system cells, as well as in aging and age-related degenerative diseases.

In recent years, the research disciplines interested in oxidative stress have increased our knowledge of the importance of the cell redox status and the recognition of oxidative stress as a process with implications for many pathophysiological states. From this multi- and interdisciplinary interest in oxidative stress emerges a concept that attests to the vast consequences of the complex and dynamic interplay of oxidants and antioxidants in cellular and tissue settings. Consequently, our view of oxidative stress is growing in scope and new future directions. Likewise, the term *reactive oxygen species*, adopted at some stage to highlight nonradical/radical oxidants, now fails to reflect the rich variety of other species in free radical biology and medicine, encompassing nitrogen-, sulfur-, oxygen-, and carbon-centered radicals. These reactive species are involved in the redox regulation of cell functions and, as a corollary, oxidative stress is increasingly viewed as a major upstream component in cell signaling cascades involved in inflammatory responses, stimulation of cell adhesion molecules, and chemoattractant production and as an early component in age-related neurodegenerative disorders, such as Alzheimer's, Parkinson's, and Huntington's diseases, and amyotrophic lateral sclerosis. Hydrogen peroxide is probably the most important redox signaling molecule that, among others, can activate NFκB, Nrf2, and other universal transcription factors and is involved in the redox regulation of insulin- and MAPK-signaling. These pleiotropic effects of hydrogen peroxide are largely accounted for by changes in the thiol/disulfide status of the cell, an important determinant of the redox status of the cell with clear involvement in adaptation, proliferation, differentiation, apoptosis, and necrosis.

The identification of oxidants in regulation of redox cell signaling and gene expression was a significant breakthrough in the field of oxidative stress: the classical definition of oxidative stress as an imbalance between the production of oxidants and the occurrence of antioxidant defenses now seems to provide a limited depiction of oxidative stress, but it emphasizes the significance of cell redox status. Because individual signaling and control events occur through discrete redox pathways rather than through global balances, a new definition of oxidative stress was advanced by Dean P. Jones as a disruption of redox signaling and control that recognizes the occurrence of compartmentalized cellular redox circuits. These concepts are anticipated to serve as platforms for the development of tissue-specific therapeutics tailored to discrete, compartmentalized redox circuits. This, in essence, dictates principles of drug development–guided knowledge of mechanisms of oxidative stress. Hence, successful interventions will take advantage of new knowledge of compartmentalized redox control and free radical scavenging.

Virtually all diseases thus far examined involve free radicals. In most cases, free radicals are secondary to the disease process, but in some instances causality is established by free radicals. Thus, there is a delicate balance between oxidants and antioxidants in health and diseases. Their proper balance is essential for ensuring healthy aging. Compelling support for the involvement of free radicals in disease development originates from epidemiological studies showing that enhanced antioxidant status is associated with reduced risk of several diseases. Of great significance is the role played by micronutrients in the modulation of cell signaling. This establishes a strong linking of diet and health and disease centered on the abilities of micronutrients to regulate redox cell signaling and modify gene expression.

Oxidative stress is an underlying factor in health and disease. In this series of books, the importance of oxidative stress and diseases associated with organ systems is highlighted by exploring the scientific evidence and clinical applications of this knowledge. This series is intended for researchers in the basic biomedical sciences and clinicians. The potential of such knowledge for healthy aging and disease prevention warrants further knowledge about how oxidants and antioxidants modulate cell and tissue function.

Flavonoids and Related Compounds: Bioavailability and Function edited by Alan Crozier (University of Glasgow) and Jeremy P.E. Spencer (University of Reading) is an authoritative treatise that reports updated information on the bioavailability, absorption, and metabolism of several flavonoids and polyphenols, their effect on cell signaling pathways, and their role in vascular function, neurodegeneration (Alzheimer's and Parkinson's disease), and cancer. The number of flavonoids and polyphenols recognized to date is staggering, and the editors have focused on those with promising functions in health and disease in light of the current knowledge concerning their bioavailability and basic biological mechanisms of action. Alan Crozier and Jeremy Spencer, internationally recognized leaders in the field of dietary flavonoids, are congratulated for this excellent and timely book.

Lester Packer
Enrique Cadenas

Preface

Representing one of the most important lifestyle factors, diet can strongly influence the incidence and onset of cardiovascular disease and neurodegenerative disorders. Recent dietary intervention studies in several mammalian species, including humans, with flavonoid-rich foods, in particular *Vitis vinifera* (grape), *Camellia sinensis* (tea), *Theobroma cacao* (cocoa), and *Vaccinium* spp. (blueberry), have indicated an ability of these dietary components to improve memory and learning. While these foods and beverages differ greatly in chemical composition, macro- and micronutrient content and caloric load per serving, they have in common that they are among the major dietary sources for a group of phytochemicals called flavonoids and related phenolic compounds. There is now a wealth of information to suggest that these compounds exert a multiplicity of biological effects in humans, including beneficial actions on the cardiovascular system, various effects on the brain, and a range of activities against cancer development. However, even though there is extensive evidence for their beneficial effects, there are still question marks over the extent of their absorption, the degree of their metabolism, and their precise mechanisms of action *in vivo*.

The book begins by examining the current knowledge regarding the absorption, metabolism, and bioavailability of individual flavonoids and phenolic subgroups. Individual chapters summarize the current thinking with regard to the biotransformation and conjugation of individual compounds in the gastrointestinal tract, liver, large intestine, and cells. In particular, the extent to which dietary phenolics components undergo metabolism in the large intestine, which has been largely ignored to date, is highlighted as is the generation of potentially novel bacterially derived metabolites. These individual chapters highlight which metabolites enter the circulatory system and likely mediate protective actions against the various human diseases.

Historically, the biological actions of flavonoids and related (poly)phenolic compounds were attributed to their ability to exert antioxidant actions. However, it is now thought highly unlikely that this classical hydrogen-donating antioxidant activity accounts for the bioactivity of these compounds *in vivo*. Instead, evidence has accumulated that the cellular effects are mediated by interactions with specific proteins central to intracellular signaling cascades. Several chapters of the book examine the latest evidence for the beneficial actions of flavonoids against various human pathological conditions, including cardiovascular disease, neurodegeneration, and cancer, and strive to provide logical and scientifically valid augments for how such protective effects are mediated.

Overall, the book provides an excellent overview for anyone interested in the bioavailability and biological function of a range of flavonoids relevant to a wide array of plant-based foods.

Editors

Jeremy P. E. Spencer, PhD, received his doctorate from King's College London in 1997 and is currently professor of nutritional medicine at the University of Reading. His initial work focused on the cellular and molecular mechanisms underlying neuronal death in Parkinson's and Alzheimer's diseases. His recent interests concern how flavonoids influence brain health through their interactions with specific cellular signaling pathways pivotal in protecting against neurotoxins, in preventing neuroinflammation and in controlling memory, learning, and neurocognitive performance.

Alan Crozier, PhD, graduated from the University of Durham in the United Kingdom and after completing postgraduate studies at the University of London, he moved to a postdoctoral position at the University of Calgary in Alberta. He then lectured at the University of Canterbury in Christchurch, New Zealand, before transferring to the University of Glasgow where until recently he was professor of plant biochemistry and human nutrition. He is currently a senior research fellow and has published more than 250 papers and edited eight books. He has carried out research on plant hormones and purine alkaloids, but the focus of his activities is now in the field of dietary flavonoids and phenolics. His research group has extensive national and international collaborations, with especially strong links to colleagues in Japan, Italy, France, the United States, and Malaysia. Their research is focused principally on teas, coffee, fruit juices, and wines, and the absorption and metabolism of a diversity of potentially protective polyphenolic compounds in the body following the ingestion of these beverages by humans—topics that are covered in depth in the first 11 chapters of this book.

Contributors

Tamar Amit, PhD
Department of Pharmacology
Technion Faculty of Medicine
Haifa, Israel

Anna-Marja Aura, DSc
VTT Technical Research Centre
  of Finland
Espoo, Finland

Denis Barron, PhD
Nestlé Institute of Health Sciences
Lausanne, Switzerland

Aalt Bast, PhD
Department of Toxicology
Faculty of Health,
  Medicine and Life Sciences
Maastricht University
Maastricht, the Netherlands

Aedín Cassidy, PhD
Department of Nutrition
Norwich Medical School
University of East Anglia
Norwich, United Kingdom

Michael N. Clifford, PhD
Centre for Nutrition and
  Food Safety
Faculty of Health and
  Medical Sciences
University of Surrey
Guildford, United Kingdom

Alan Crozier, PhD
School of Medicine
College of Medical,
  Veterinary and Life Sciences
University of Glasgow
Glasgow, United Kingdom

Daniele Del Rio, PhD
The φ^2 Laboratory of
  Phytochemicals in Physiology
Human Nutrition Unit
Department of Public Health
University of Parma
Parma, Italy

Juan C. Espín, PhD
CEBAS-CSIC
Espinardo (Murcia), Spain

Cesar G. Fraga, PhD
Physical Chemistry-PRALIB
School of Pharmacy and Biochemistry
University of Buenos Aires-CONICET
Buenos Aires, Argentina
and
Department of Nutrition
University of California
Davis, California

René Fumeaux, PhD
Nestlé Research Center
Lausanne, Switzerland

María T. García-Conesa, PhD
CEBAS-CSIC
Espinardo (Murcia), Spain

Peter Hollman, PhD
Division of Nutrition
RIKILT
Wageningen, the Netherlands

Annett Klinder, PhD
Hugh Sinclair Unit of Human Nutrition
Department of Food and
  Nutritional Sciences
University of Reading
Reading, United Kingdom

Johanna W. Lampe, PhD, RD
Cancer Prevention Research
  Program
Public Health Sciences Division
Fred Hutchinson Cancer
  Research Center
Seattle, Washington

Mar Larrosa, PhD
CEBAS-CSIC
Espinardo (Murcia), Spain

Claudine Manach, PhD
INRA UMR1019
Auvergne University
Clermont-Ferrand, France

Silvia Mandel, PhD
Department of Pharmacology
Technion Faculty of Medicine
Haifa, Israel

Christine Morand, PhD
INRA UMR1019
Auvergne University
Clermont-Ferrand, France

Kaeko Murota, PhD
Department of Life Science
Faculty of Science and
  Engineering
Kinki University
Higashi-Osaka, Japan

Patricia I. Oteiza, PhD
Department of Nutrition and
  Department of Environmental
  Toxicology
University of California
Davis, California

José Peñalvo, PhD
Department of Epidemiology
CNIC
Madrid, Spain

Mariusz Konrad Piskula, PhD
Division of Food Science
Institute of Animal Reproduction and
  Food Research
Polish Academy of Sciences
Olsztyn, Poland

Shibu M. Poulose, PhD
USDA-ARS
Jean Mayer Human Nutrition
  Research Center on Aging
Tufts University
Boston, Massachusetts

Ronald L. Prior, PhD (Retired)
Searcy, Arkansas

Elke Richling, PhD
Department of Chemistry
Food Chemistry and Toxicology
University of Kaiserslautern
Kaiserslautern, Germany

Ana Rodriguez-Mateos, PhD
Molecular Nutrition Group
Department of Food and
  Nutritional Sciences
School of Chemistry, Food and Pharmacy
University of Reading
Reading, United Kingdom

Joseph Rothwell, PhD
INRA UMR1019
Auvergne University
Clermont-Ferrand, France

Ian Rowland, PhD
Hugh Sinclair Unit of Human Nutrition
Department of Food and
  Nutritional Sciences
University of Reading
Reading, United Kingdom

Stefano Sforza, PhD
Department of Organic and
  Industrial Chemistry
University of Parma
Parma, Italy

Barbara Shukitt-Hale, PhD
USDA-ARS
Neuroscience Laboratory
Jean Mayer Human Nutrition
  Research Center on Aging
Tufts University
Boston, Massachusetts

Piyawan Sitthiphong, PhD
Hugh Sinclair Unit of Human
  Nutrition
Department of Food and
  Nutritional Sciences
University of Reading
Reading, United Kingdom

Candice Smarrito-Menozz, PhD
Nestlé Research Center
Lausanne, Switzerland

Jeremy P.E. Spencer, PhD
Molecular Nutrition Group
School of Chemistry, Food and
  Pharmacy
University of Reading
Reading, United Kingdom

Angelique Stalmach, PhD
College of Medical,
  Veterinary and Life Sciences
University of Glasgow
Glasgow, United Kingdom

Junji Terao, PhD
Department of Food Science
Graduate School of Nutrition and
  Bioscience
University of Tokushima
Tokushima, Japan

Francisco A. Tomás-Barberán, PhD
CEBAS-CSIC
Espinardo (Murcia), Spain

Mireia Urpi-Sarda, PhD
Department of Internal Medicine
Hospital Clinic
August Pi i Sunyer Biomedical
  Research Institute (IDIBAPS)
Barcelona, Spain

and

CIBER 06/03: Fisiopatología de la
  Obesidad y la Nutrición
Instituto de Salud Carlos III
Madrid, Spain

Katerina Vafeiadou, PhD
Hugh Sinclair Unit of Human
  Nutrition
Department of Food and
  Nutritional Sciences
University of Reading
Reading, United Kingdom

David Vauzour, PhD
Faculty of Medicine and Health
  Sciences
Norwich Medical School
University of East Anglia
Norwich, United Kingdom

Paola Vitaglione, PhD
Department of Food Science
University of Napoli "Federico II,"
Portici, Italy

Florian Viton, PhD
Nestlé Research Center
Lausanne, Switzerland

Thomas Walle, PhD
Department of Pharmacology
Medical University of South Carolina
Charleston, South Carolina

Orly Weinreb, PhD
Department of Pharmacology
Technion Faculty of Medicine
Haifa, Israel

Antje R. Weseler, PhD
Department of Toxicology
Faculty of Health,
  Medicine and Life Sciences
Maastricht University
Maastricht, the Netherlands

Gary Williamson, PhD
Department of Food Science and
  Nutrition
University of Leeds
Leeds, United Kingdom

Moussa B.H. Youdim, PhD
Department of Pharmacology
Technion Faculty of Medicine
Haifa, Israel

1 Bioavailability of Flavanones

*Mireia Urpi-Sarda, Joseph Rothwell,
Christine Morand, and Claudine Manach*

CONTENTS

1.1 INTRODUCTION

Flavanones are a class of flavonoids found mainly in citrus fruits, although minor amounts have also been detected in herbs, red wine, and tomatoes (Neveu et al. 2010). Flavanones are widely consumed in Western countries. In the adult Spanish population, for instance, intake may reach 50 mg/day, or around 17% of the estimated total flavonoid intake, which ranks them as the most consumed flavonoid sub-class after proanthocyanidins (Zamora-Ros et al. 2010). Flavanone intakes of 14.4, 20.4, 22, 33.5, and 34.7 mg/day have also been reported in the United States, United Kingdom, Finland, Greece, and Italy, respectively (Zamora-Ros et al. 2010).

Epidemiological and clinical studies have associated the consumption of citrus fruits or juices with a lower risk of ischemic stroke and acute coronary events and with an improvement of vascular function (Joshipura et al. 1999; Morand et al. 2011). These beneficial effects of citrus products appear to be linked to their flavanone content, even if they are also important sources of other dietary bioactives such as vitamin C, vitamin B9, carotenoids, and organic acids. Convincing data from numerous animal studies suggest the involvement of dietary flavanones in lowering blood lipids, reducing plasma markers of endothelial dysfunction, reducing atherosclerosis plaque progression, and improving insulin sensitivity (Choe et al. 2001; Lee et al. 2001; Akiyama et al. 2009; Mulvihill et al. 2009, 2010; Chanet et al. 2011).

Flavanones are based on a diphenylpropane skeleton, two benzene rings (A and B) connected by a saturated three-carbon chain forming a closed pyran ring with the benzene A ring. An epoxide group is present at the C4 position (Figure 1.1). Hesperetin and naringenin are the most common flavanones in fruits, and they are usually conjugated to a glucose-rhamnose disaccharide at the 7-position, typically rutinose or neohesperidose. Flavanone rutinosides are tasteless, whereas flavanone neohesperidoside conjugates such as hesperetin-7-*O*-neohesperidoside (neohesperidin) from bitter orange (*Citrus aurantium*) and naringenin-7-*O*-neohesperidoside (naringin) from grapefruit (*Citrus paradisi*) are intensely bitter (Tomás-Barberán and Clifford 2000). By virtue of their abundance in citrus fruit, hesperetin and naringenin conjugates are the most studied flavanones with regard to metabolism and bioavailability.

This chapter provides an overview of the *in vivo* bioavailability of flavanones with a particular focus on the flavanone metabolites identified in the biofluids of humans and animals. A comprehensive understanding of the absorption, metabolism, and circulating forms of these polyphenolic compounds will be crucial to assess the mechanisms by which they exert bioactivity and will also open up the possibility

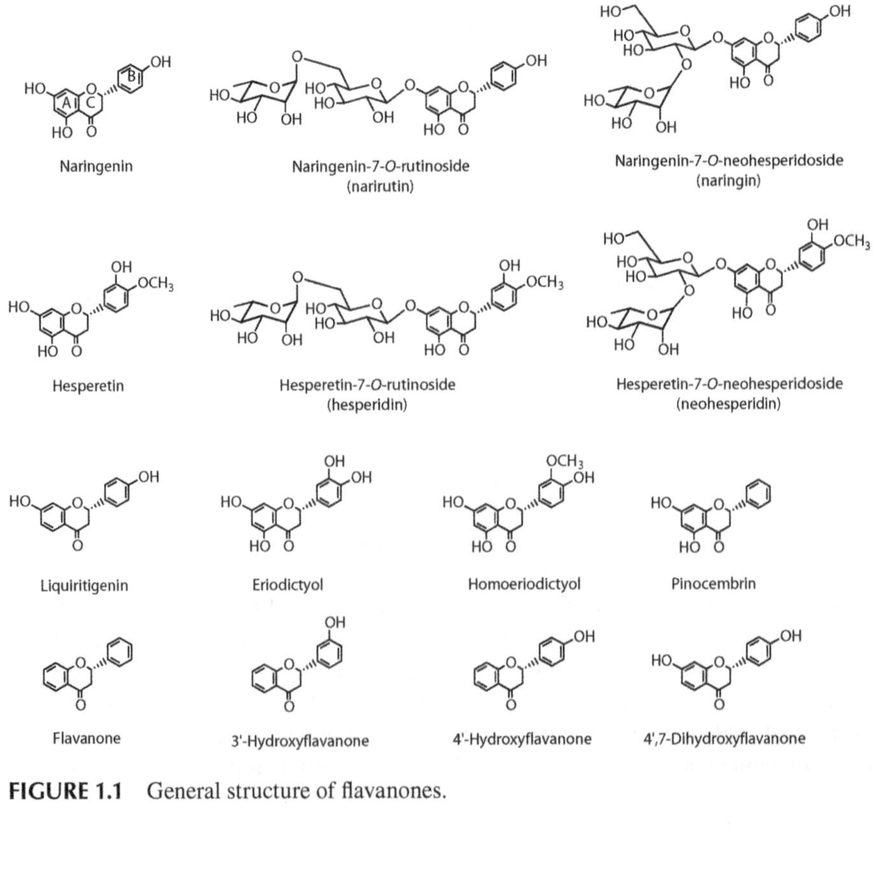

FIGURE 1.1 General structure of flavanones.

of optimizing the nature and quantity of flavanone doses for possible protection against disease.

1.2 BIOAVAILABILITY OF FLAVANONES

1.2.1 ABSORPTION OF FLAVANONES

Flavanones are mainly present in foods as diglycosides. It has long been known that such glycosides cannot be absorbed in their native form in the small intestine but must by hydrolyzed by intestinal microflora before absorption of their aglycone moieties in the colon. The nature of the attached sugar moiety has been shown to be an important determinant of the mode of absorption. The kinetics and efficiency of absorption of isolated naringenin, naringenin-7-O-glucoside, and naringenin-7-O-neohesperidoside were compared in rats either after a single flavanone-containing meal or after adaptation for 14 days to a supplemented diet (Felgines et al. 2000). Similar kinetics and levels of absorption were reported for naringenin and naringenin-7-O-glucoside, but the time of the peak plasma concentration (T_{max}) was markedly delayed in the case of naringenin-7-O-neohesperidoside, reflecting an absorption in more distal parts of the intestine. In humans, the absorption of aglycone and glycosides has never been compared using pure compounds. However, several intervention studies with citrus fruit juices, which mainly contain rutinosides and hesperidosides, have consistently reported plasma T_{max} times to be between 4.5 and 7 h, indicative of absorption in the colon (Erlund et al. 2001; Manach et al. 2003; Gardana et al. 2007; Mullen et al. 2008a; Brett et al. 2009; Bredsdorff et al. 2010; Cao et al. 2010; Vallejo et al. 2010).

Enzymes present in the small intestine are able to hydrolyze some flavonoid glucosides, but flavonoid rhamnoglucosides, such as naringenin-7-O-neohesperidoside, are not hydrolyzed by cell-free extracts from the human small intestine (Day et al. 1998). Griffiths and Barrow (1972) revealed that the gut microbiota play a crucial role in the release of flavanone aglycones from their glycosides. They showed that naringenin-7-O-neohesperidoside and hesperetin-7-O-rutinoside (hesperidin) when administered to germ-free rats were recovered intact in feces, whereas low recoveries were obtained from rats with a normal microflora. In support, a recent study found that a 6-day pretreatment of rats with antibiotics markedly lowered the absorption of hesperetin-7-O-rutinoside (Jin et al. 2011). The key enzymes required for hydrolysis and subsequent absorption of flavanone glycosides are α-L-rhamnosidases, and several bacterial strains present in the human colon have been reported to produce α-L-rhamnosidases able to cleave flavonoid rutinosides and neohesperidosides (Bokkenheuser et al. 1987; Yadav et al. 2010). In contrast to rutinosides and hesperidosides, flavanone glucosides are quite rare in foods. According to the web database Phenol-Explorer, naringenin-7-O-glucoside (prunin) is found in tomatoes and almonds, and eriodictyol-7-O-glucoside is present in peppermint. Moreover, a number of patented processes for the debittering of citrus fruit juices are based on the cleavage of naringenin-7-O-neohesperidoside by α-L-rhamnosidases to produce the less bitter glucoside naringenin-7-O-glucoside (Yadav et al. 2010). Such a process has been used to compare the bioavailability in humans of flavanones from a natural orange juice and an orange juice treated with hesperidinase to yield hesperetin-7-O-glucoside (Nielsen et al. 2006). In a double-blind randomized cross-over study on 16 volunteers, the conversion of the rutinose to a glucose group markedly improved the bioavailability of hesperetin through the change of absorption site from the colon to the small intestine (Nielsen et al. 2006). Plasma T_{max} decreased from 7.0 ± 3.0 h to 0.6 ± 0.1 h, and correspondingly the peak plasma concentration (C_{max}), the area under the plasma concentration versus time curve (AUC), and urinary excretion of hesperetin increased more than 3-fold. Similarly, Bredsdorff et al. (2010) compared the bioavailability of naringenin from untreated orange juice, naturally rich in naringenin-7-O-rutinoside, to an orange juice treated with α-rhamnosidase, abundant in naringenin-7-O-glucoside. Again, the α-rhamnosidase treatment of the orange juice considerably increased the plasma AUC and C_{max} of naringenin (4- and 5.4-fold, respectively), whereas the T_{max} fell from 311 to 92 min. The urinary excretion of naringenin increased from 7 to 47% of intake after the α-rhamnosidase treatment. It is notable that the ingestion of flavanones as glucosides instead of rutinosides increases bioavailability without changing the profile of phase II metabolites (Bredsdorff et al. 2010). Because of the striking effectiveness of converting natural diglycoside forms into glucosides to improve flavanone bioavailability, interest has been recently arisen in developing industrial applications for the fermentation of food sources using *Lactobacillus* strains, which express rhamnosidases (Avila et al. 2009; Beekwilder et al. 2009).

The mechanisms involved in intestinal absorption of flavanones have been investigated using various *in vitro* systems and animal models. Several authors have studied the metabolism and transport of flavanones using cultured Caco-2 cells as an *in vitro* model for the intestinal epithelium. This cell line exhibits many morphological and functional similarities to the normal human intestinal epithelial cells when grown as polarized cells. Tourniaire et al. (2005) observed a very poor absorption of naringenin-7-O-neohesperidoside by Caco-2 cells and suggested an involvement of the P-glycoprotein (P-gp) transporter in effluxing the neohesperidoside back to the apical side. Similarly, the permeation rate of hesperetin-7-O-rutinoside across the Caco-2 cell monolayer was shown to be very low, and transport occurred via a paracellular route (Kim et al. 1999; Kobayashi et al. 2008; Serra et al. 2008). In contrast, Kobayashi et al. (2008) demonstrated that hesperetin was efficiently absorbed, with a permeation rate 400-fold higher than for hesperitin-7-O-rutinoside. The aglycone was reported to be absorbed via a transcellular route, by means of a proton-coupled active transport Na^+-independent transporter as well as passive diffusion, made possible by the small size and relatively high lipophilicity of hesperetin (a log P of 2.55 predicted by ALOGPS computational method.). Londoño-Londoño et al. (2010) reported that hesperetin had much stronger molecular interactions with lipophilic membranes than hesperitin-7-O-rutinoside, in part due to the ability of the aglycone to adopt a more planar conformation. Using the Caco-2 cell model, Brand et al. (2008) investigated the metabolism of the aglycone hesperetin and the role of ATP-binding cassette (ABC) transporters in the efflux of hesperetin and its metabolites. Hesperetin was extensively metabolized in the Caco-2 cells to its 7-O-glucuronide (86% of the total metabolites) and 7-O-sulfate conjugates, which were predominantly transported to the apical side but also, to a lesser extent, to the basolateral side. Co-administration experiments with inhibitors of several ABC transporters indicated that this efflux of hesperetin metabolites to the apical side mainly involved the breast cancer-resistant protein (BCRP) transporter. However, involvement of the multidrug resistant transporter MRP2, as previously described for other flavonoids, was not ruled out.

Kobayashi and Konishi (2008) also studied the transport of naringenin and eriodictyol through Caco-2 cells. Both flavanones were absorbed through a proton-driven active transport as previously described for hesperetin. Similarly, Chabane et al. (2009) found that naringenin was partially absorbed by transcellular passive diffusion but also transported by an active ATP-dependent system mediated by MRP1, which is expressed at the basolateral side of the intestinal cells. Naringenin was also shown to be secreted to the apical side via active P-gp and MRP2 efflux transporters. The role of BCRP, involved in hesperetin efflux, was not investigated. Although all the transporters involved may not yet be identified, it is clear that the intestinal absorption of flavanone aglycones occurs via both passive transcellular diffusion and active transport.

There is a general consensus in the flavanone literature that, demonstrated most notably by the Caco-2 model, the deglycoslyation of naringenin-7-O-neohesperido-side and hesperetin-7-O-rutinoside by intestinal microbiota is necessary for effective intestinal absorption. Moreover, once absorbed into enterocytes, flavanones appear to be conjugated and efficiently transported back into the gut lumen by active transporters. This process may be responsible for the limited access of flavanones to systemic circulation.

1.2.2 METABOLISM OF FLAVANONES *IN VIVO*

There is an extensive amount of literature available concerning the *in vitro* metabolism and bioactivity of flavanones. Despite the value of *in vitro* data, *in vivo* studies on metabolism and bioavailability are the most crucial for knowledge of the flavanone forms to which tissues are exposed and the magnitude and time scales of this exposure.

All phase II metabolites of flavanones reported to be formed *in vivo* in animal and human studies are shown in Table 1.1. In all instances, mass spectrometry was used for the identification and quantification of metabolites. Few standards are commercially available for polyphenol metabolites. Analytical methods should, therefore, not only be sensitive, selective, and robust, but data must be analyzed meticulously. Advancing rapidly, mass spectrometry has emerged as the technique that best meets these needs and has, thus, become the preferred means of characterization of metabolites from biofluids and tissues.

Most of the literature on flavanone metabolism has treated the biotransformations of hesperetin, naringenin, and their respective glycosides. The majority of metabolites identified have been glucuronide and sulfate conjugates (Table 1.1). Some sulfoglucuronide and diglucuronide conjugates have also been described, but in lower concentrations (Zhang and Brodbelt 2004; Mullen et al. 2008a; Brett et al. 2009; Vallejo et al. 2010). In humans, the main sites of O-glucuronidation of naringenin are the 7- and 4′-hydroxyl groups (Mullen et al. 2008a; Brett et al. 2009; Bredsdorff et al. 2010; Vallejo et al. 2010). Glucuronidation at the 5-position has additionally

TABLE 1.1

Phase II Metabolites of Flavanones

Study Group	Flavanone Source (Dose)	Metabolite Identified *in vivo*	Presence in Plasma	Presence in Urine	Biofluid in Which Identified	Reference
		Human Studies				
10 humans	350 mL Polyphenol-rich juice (hesperetin-7-*O*-rutinoside: 45 μmol; naringenin-7-*O*-neohesperidoside: 5.9 μmol)	Hesperetin-7-*O*-glucuronide	–	Minor	U	Borges et al. 2010
		Hesperetin-*O*-glucuronides (2)	Major	Major	P,U	
		Hesperetin-*O*-glucuronide-*O*-sulfate	–	Major +	U	
16 humans	i) Orange juice (0.83 mg/kg bw naringenin-7-*O*-rutinoside)	Hesperetin-3'-*O*-glucuronide	–	–	U	Bredsdorff et al. 2010
		Hesperetin-7-*O*-glucuronide			U	
		Hesperetin-5,7-*O*-diglucuronide			U	
	ii) α-Rhamnosidase-treated orange juice (0.52 mg/kg bw naringenin-7-*O*-glucoside)	Hesperetin-3',7-*O*-diglucuronide			U	
		Hesperetin-3'-*O*-sulfate			U	
		Hesperetin-*O*-sulfo-*O*-glucuronide			U	
		Naringenin-7-*O*-glucuronide			U	
		Naringenin-4'-*O*-glucuronide			U	
20 humans	300 g Orange juice or 150 g orange fruit (11.8 mg naringenin-7-*O*-rutinoside and 79.7 mg hesperetin-7-*O*-rutinoside)	Hesperetin-7-*O*-glucuronide	–	–	P,U	Brett et al. 2009
		Hesperetin-3'-*O*-glucuronide			P,U	
		Naringenin-7-*O*-glucuronide			P,U	
		Naringenin-4'-*O*-glucuronide			P,U	
		Hesperetin-*O*-diglucuronides (2)			U	
		Hesperetin-*O*-sulfo-*O*-glucuronide			U	
		Naringenin-*O*-diglucuronide			U	
16 humans	10 capsules of almond extract (total of 4 g) (0.12 mg/g of flavanones: naringenin, naringenin-7-*O*-glucoside and eriodictyol)	Naringenin-*O*-glucuronide (2)	–	–	U	Garrido et al. 2010

Subjects	Treatment	Metabolite			P/U	Reference
6 humans	500 mg Naringenin-7-O-neohesperidoside powder	Naringenin-O-glucuronide	—	Major	U	Ishii et al. 2000
		Naringenin-7-O-neohesperidoside		Minor		
5 men	0.5 or 1L Orange juice (110 mg and 220 mg, respectively, of hesperetin-7-O-rutinoside)	Naringenin-O-glucuronide	—	Major	P	Manach et al. 2003
		Naringenin-O-sulfo-O-glucuronide		Minor	P	
8 humans	Orange juice fortified with 131 μmol of hesperetin-7-O-rutinoside with and without 150 mL of natural yogurt	Naringenin-7-O-glucuronide	—	Major	P,U	Mullen et al. 2008a
		Naringenin-4'-O-glucuronide		Major	P,U	
		Hesperetin-7-O-glucuronide		Minor	P,U	
		Hesperetin-O-diglucuronide	n.d.	Minor	U	
		Naringenin-O-diglucuronide		Minor	U	
		Hesperetin-O-glucuronide-O-sulfates		Medium	U	
10 humans	500 mL Fermented rooibos tea (FRT) (84 μmol flavonoids [23.1 μmol of eriodictyol-C-glucosides)]	Eriodictyol-O-sulfate	—	Major	U	Stalmach et al. 2009
		Eriodictyol-O-glucuronide	—	Traces	U	
	500 mL unfermented rooibos tea (URT) (159 μmol flavonoids [5.5 μmol of eriodictyol-C-glucosides)]	Eriodictyol-O-glucuronide-O-sulfate				
2 humans	10 capsules of almond extract (total 4 g) (0.12 mg/g flavanones: naringenin, naringenin-7-O-glucoside and eriodictyol)	Naringenin-O-glucuronide (2)	Main	Main	P,U	Urpi-Sarda et al. 2009
10 humans	Orange juices (commercial and experimental) (total flavonoids: from 117 to 441 mg)	Naringenin-7-O-glucuronide	Main	—	P,U	Vallejo et al. 2010
		Naringenin-4'-O-glucuronide	Major	Major	P,U	
		Naringenin-O-sulfate	Major	Major	P,U	
		Naringenin-O-sulfo-O-glucuronide	Detected		P	
		Naringenin-O-diglucuronide	Minor		P	
		Hesperetin-3'-O-glucuronide	Major	Major	P,U	
		Hesperetin-7-O-glucuronide	Major	Major	P,U	

(Continued)

TABLE 1.1 (CONTINUED)
Phase II Metabolites of Flavanones

Study Group	Flavanone Source (Dose)	Metabolite Identified *in vivo*	Presence in Plasma	Presence in Urine	Biofluid in Which Identified	Reference
3 humans	900 mL Grapefruit juice (two brands, A or B) (Brand A: naringenin-7-O-neohesperidoside: 343 µg/mL; naringenin-7-O-rutinoside: 126 µg/mL; brand B: naringenin-7-O-neohesperidoside: 328 µg/mL; naringenin-7-O-rutinoside: 88 µg/mL)	Hesperetin-O-sulfate (2 in urine)	Major	—	P,U	Zhang et al. 2004
		Hesperetin-O-sulfo-O-glucuronide	Detected		P	
		Hesperetin-O-diglucuronide	Minor		P	
		Naringenin-7-O-glucuronide (t)	—	—	U	
		Naringenin-4'-O-glucuronide (t)			U	
		Naringenin-7-sulfate (t)			U	
		Naringenin-4'-sulfate (t)			U	
		Naringenin-5-sulfate (t)			U	
		Naringenin-O-glucuronide-O-sulfate (t)			U	
		Naringenin-O-diglucuronide			U	
Animal Studies						
3 pigs	Extract of *Cyclopia genistoides*: 75 g/day for 11 days (1 mg hesperetin-7-O-rutinoside /kg/day)	Hesperetin-O-glucuronide	n.d.	Main	U	Bock et al. 2008
		Eriodictyol-O-glucuronide	n.d.	Minor	U	
Male Sprague-Dawley rats	Naringenin (50 mg/kg bw) orally	Naringenin-5-O-β-glucuronide	—	—	P	El Mohsen et al. 2004
		Naringenin-7-O-β-glucuronide			P	
		Naringenin-O-glucuronide	Main	Main	P,U	
6 male Sprague-Dawley rats	746.7 mg/kg bw naringenin-7-O-neohesperidoside	Naringenin-O-glucuronide	Main	—	P	Fang et al. 2006

Model	Treatment	Metabolite				Reference
Wistar rats	Enriched orange juice (33 mg/g hesperetin-7-O-rutinoside) or hesperetin-7-O-glucoside (0.25 and 0.5% w/w)	Hesperetin-7-O-glucuronide	Major	–	P	Habauzit et al. 2009
		Hesperetin-O-sulfate (2)	Minor		P	
	Orange juice enriched with hesperetin-7-O-rutinoside (0.25% w/w) (33 mg/g hesperetin-7-O-rutinoside)	Homoeriodictyol-4'-O-glucuronide	Major		P	
		Homoeriodictyol-7-O-glucuronide			P	
		Hesperetin-3'-O-glucuronide	–		P	
		Homoeriodictyol	Minor		P	
10 Beagle dogs	Two capsules containing 200 mg grapefruit extract (70 mg flavanones)	Naringenin-O-glucuronide	Major	–	P	Mata-Bilbao et al. 2007
Wistar rats	50 or 100 mg/kg hesperetin-7-O-rutinoside in 10 mL of 0.5% carboxymethyl cellulose	Hesperetin-7-β-D-glucuronide	4h: major	–	P,U	Matsumoto et al. 2004
		Hesperetin-3'-O-β-D-glucuronide	4h: major		P,U	
		Hesperetin-O-sulfate	nd		P	
		Homoeriodictyol-O-glucuronide	6h: major		P	
		Homoeriodictyol-O-sulfate	nd		P	
Sprague-Dawley rats	Eriodictyol-7-O-rutinoside (50 and 75 μmol/kg bw) by gavage in distilled water	Hesperetin-O-sulfate	–	–	P,U	Miyake et al. 2000
		Hesperetin-O-glucuronide			P,U	
		Eriodictyol-O-glucuronide			P,U	
		Eriodictyol-O-sulfate			P,U	
		Homoeriodictyol-O-glucuronide			P,U	
		Homoeriodictyol-O-sulfate			P,U	
6 Healthy and 6 tumor-bearing rats	Semi-synthetic diet supplemented with 0.5% naringenin-7-O-neohesperidoside (w/w) (7 days)	Naringenin-7-O-glucuronide	–	–	U	Silberberg et al. 2006a
		Naringenin-4'-O-glucuronide			U,PL,K	
		Hesperetin-7-O-glucuronide			U,L,K	
		Isosakuranetin-O-glucuronides (2)			U,PL,K	
		Naringenin-O-sulfates (2)			U,PL,K	
		Hesperetin-O-sulfate			U,L,K	
		Hesperetin-O-sulfates (2)			P	

P: plasma; U: urine; L: liver tissue; K: kidney tissue; bw: body weight; n.d.: not detected; t: tentative identification.

been observed in rats fed naringenin aglycone, by reference to a synthesized standard mixture of naringenin 5- and 7-*O*-glucuronide (El Mohsen et al. 2004). Sulfation of naringenin appears to be less prevalent than glucuronidation. Of the 12 studies (eight in humans), which characterized naringenin metabolites, only three (two in humans and one in rats) identified sulfated forms of naringenin in either plasma or urine (Zhang and Brodbelt 2004; Silberberg et al. 2006a; Vallejo et al. 2010). The *O*-sulfation of naringenin could occur at the 7-, 4′-, or 5-hydroxyl groups (Zhang and Brodbelt 2004), but the predominant position of sulfation remains unclear (Zhang et al. 2004; Silberberg et al. 2006a; Vallejo et al. 2010) (Figure 1.2).

Glucuronidation also appears to be the principal biotransformation of hesperetin. In both humans and rats, the main positions of hesperetin glucuronidation are the 7- and 3′-hydroxyl groups (Mullen et al. 2008a; Brett et al. 2009; Bredsdorff et al. 2010; Vallejo et al. 2010). In humans, the 5,7-*O*-diglucuronide and the 3′,7-*O*-diglucuronide were also identified in urine after orange juice consumption (Bredsdorff et al. 2010) (Figure 1.2). The same study identified a 3′-sulfate of hesperetin, but in lower

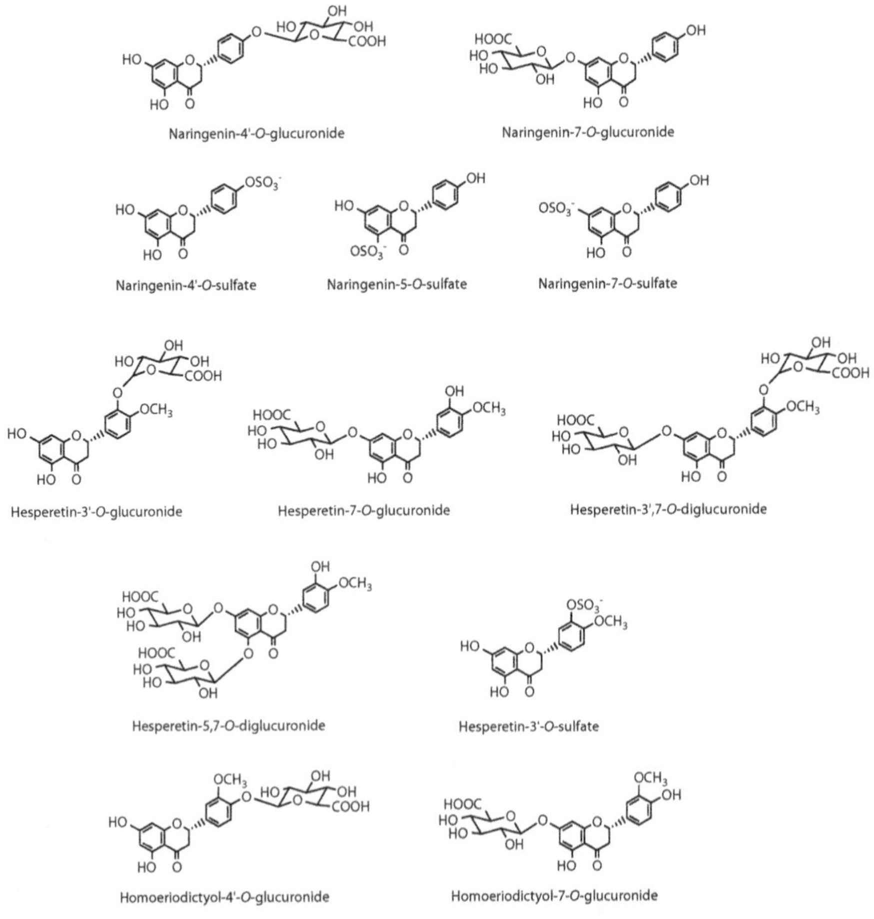

FIGURE 1.2   Structures of phase II metabolites of flavanones.

concentrations. The position of conjugation was elucidated through spectrophotometry at different pH.

Sulfoglucuronides are also major urinary metabolites of hesperetin although they are minor in or absent from plasma (Mullen et al. 2008a; Borges et al. 2010). Sulfate conjugates were identified in half the studies ($n = 10$) that detected hesperetin metabolites, and these came mainly from animal data ($n = 8$). The presence of high amounts of sulfoglucuronide and sulfate conjugates of hesperetin in urine with respect to those of naringenin could be a consequence of a different specificity of the sulfotransferase for hesperetin and naringenin (Mullen et al. 2008a).

The physiopathologic state does not appear to qualitatively affect the metabolism of naringenin-7-O-neohesperidoside, as evidenced by the similarities of the profiles of circulating metabolites of naringenin between healthy and tumor-bearing rats fed a 0.5% naringenin-7-O-neohesperidoside diet for 7 days (Silberberg et al. 2006a). However, the total plasma concentrations of naringenin were reduced by almost 40% in tumor-bearing rats. This decreased bioavailability of naringenin could result from a higher efflux of naringenin metabolites by the MRPs expressed at the apical sites of the intestinal cells, and for which an increased activity has been reported in cancer (Sesink et al. 2005).

Free flavanone aglycones, such as naringenin and hesperetin, are detected, if at all, in only trace amounts after the consumption of flavanone-rich products (Felgines et al. 2000; Bugianesi et al. 2002; Manach et al. 2003; Wang et al. 2006; Brett et al. 2009). To a degree, this appears to be dose related as Ma et al. (2006) found that 3–17% of the total naringenin-based compounds present in rat plasma were the aglycone after administration of increasing amounts of naringenin.

Intact flavanone glycosides are generally not recovered in plasma or urine, since they are thought to be too polar to be absorbed passively from the gastrointestinal tract and would also be susceptible to deglycosylation and subsequent metabolism in the intestinal mucosa or liver. Nevertheless, one publication has reported the presence of intact naringenin-7-O-neohesperidoside in urine. After oral administration of pure naringenin-7-O-neohesperidoside to volunteers, the disaccharide was found at 0.5% of the concentration of naringenin-O-glucuronides (Ishii et al. 2000). It is notable that the dose administered was particularly high (500 mg), suggesting that passive absorption of native glycosides through the small intestine enterocytes might have occurred to a limited extent.

Until recently, phase II metabolites of flavanones were not commercially available, and putative identification of the circulating metabolites was performed by analysis of ^{1}H NMR data after isolation, differential pH spectrophotometry, and when possible by comparison with synthetic standards. In addition, metabolites were quantified indirectly by the comparison of free forms before and after specific enzymatic hydrolyses using mass spectrometry by reference to the calibration curves of respective aglycones. Recently, Brett et al. (2009) developed an innovative technique to determine the location of the glucuronic acid moieties in naringenin and hesperetin using LC-MS^n and metal complexation with the Co^{2+} ion and an auxiliary ligand, 4,7-diphenyl-1,10-phenanthroline. The chemical synthesis of some flavanone glucuronides, such as 7,4′-di-O-methyleriodictyol-3′-O-β-D-glucuronide, naringenin 4′-, and 7-O-β-D-glucuronide and hesperetin-3′- and 7-O-β-D-glucuronide has been published (Boumendjel et al. 2009; Khan et al. 2010). In the future, the commercial availability (Cayman Chemical, Michigan; Toronto Research Chemicals, Ontario, Canada) of naringenin-7- and 4′-O-glucuronide conjugates will allow researchers obtain greater accuracy in their quantification of these metabolites in biological fluids. It will also allow the investigation of biological effects of physiological metabolites in cellular models.

The intestine is known to participate in the phase II conjugation of flavanone aglycones during first-pass metabolism. The hydrophobic aglycones are able to passively diffuse through the permeable gut mucosa and are conjugated within mucosal cells (Brand et al. 2010a). However, absorption into systemic circulation is limited because of the efflux of these conjugates back into the intestinal lumen by specific transporters (Liu and Hu 2007; Brand et al. 2008). Flavanones that do enter systemic circulation may also undergo metabolism in the liver as substrates of UDP-glucuronosyl-transferase (UGT) and sulfotransferase (SULT) enzymes to form glucuronidated and sulfated metabolites (Silberberg et al. 2006b). Flavanones may be substrates for many isoforms of these enzymes. A total of 22 different UGT and 10 different SULT proteins have been detected in human tissues (Mackenzie et al. 2005; Riches et al. 2009). These enzymes have different efficiencies, kinetics, and specificities for the conjugation of hesperetin, as demonstrated by Brand et al. (2010a) with 12 individual UGTs and 12 individual SULTs from rats and humans. Three UGT enzymes (UGT1A3, UGT1A6, and UGT2B4) were reported to exclusively produce hesperetin-7-O-glucuronide, whereas UGT1A7 mainly produced the 3′-O-glucuronide. The remainder (UGT1A1, UGT1A8, UGT1A9, UGT1A10, UGT2B7, and UGT2B15) produced conjugates at both the 3′- and 7- positions. UGT1A9, UGT1A1, UGT1A7, UGT1A8, and UGT1A3 were found to be the most efficient at catalyzing hesperetin glucuronidation. Incubation of hesperetin with human or rat microsomal fractions also resulted in the formation of hesperetin-3′- and 7-O-glucuronides (Brand et al. 2010a). Conjugation at the C5 position was not observed, with isolated enzymes, rat or human microsomes, or extracts from small intestine, colon, and liver.

Sulfotransferases also showed marked regioselectivity for hesperetin conjugation (Brand et al. 2010a). Human cytosolic fractions predominantly produced hesperetin-3′-O-sulfate (80–95%). Considering the catalytic efficiency as well as the abundance of the various SULT isoforms in the intestine and the liver, it was concluded that SULT1A1 is involved in the sulfation of hesperetin in the liver, whereas SULT1B1 and SULT1A3 are mainly responsible for the sulfonation of hesperetin in the intestine. Hesperetin-7-O-sulfate is probably produced by SULT1C4 in the rat liver (Brand et al. 2010a).

Hesperetin is chiral and exists as two enantiomeric forms in nature. The 2S-hesperidin configuration is the predominant form in orange juice with a ratio of 92:8 in favor of the S-epimer (Aturki et al. 2004; Si-Ahmed et al. 2010). With improving analytical techniques, there is growing interest in distinguishing and measuring the two enantiomers, although pure enantiomers are not yet commercially available. Until recently, only the pharmacokinetics of racemic flavanones had been determined, but some pharmacokinetic studies have been carried out with flavanone R- and S-enantiomers. *In vitro* studies have explored the differences in the metabolism of *S*- and *R*-hesperetin by UGT and SULT, as well as their transport by Caco-2 cells. Although Brand et al. (2010b) observed a 5.2-fold higher efficiency in the glucuronidation of *S*-(–)-hesperetin compared to the glucuronidation of *R*-(+)-hesperetin when incubated with human small intestine microsomes, the overall differences in intestinal metabolism and transport were small.

Using a novel high-performance liquid chromatography method to distinguish flavanone enantiomers, Yañez et al. (2008) investigated the stereoselective pharmacokinetics of flavanones after intravenous administration of 20 mg/kg body weight (bw) racemic hesperetin, naringenin, and eriodictyol to rats. While *S*-(–)-naringenin and eriodictyol were excreted in greater amounts than their *R*-(+) enantiomers (10% and 50% higher, respectively), 50% more *R*-(+)-hesperetin was excreted than *S*-(–)-hesperetin. Yañez and Davies (2005) also determined naringenin enantiomers in the urine of a human volunteer after consumption of tomato juice containing racemic naringenin and naringenin-7-*O*-neohesperidoside. The cumulative urinary excretion of the *S*-(–)-enantiomer was 40% higher than that of the *R*-(+)-enantiomer.

Hesperetin enantiomers were also measured in both human and rat urine following administration of orange juice, which contained 6-fold higher *S*-(–)- than *R*-(+)-hesperidin and 10-fold higher *S*-(–)- than *R*-(+)-hesperetin, after administration of oral racemic hesperetin-7-*O*-rutinoside (200 mg/kg) (Yanez et al. 2005). In both experiments, the relative excretion of the *R*-(+)-hesperetin was higher than that of the *S*-(–)-enantiomer. Si-Ahmed et al. (2010) measured hesperetin enantiomers in the urine of male volunteers who had consumed 1L of commercial blood orange juice. A total of 22 mg of *S*-(–)-hesperetin was excreted in urine, compared to less than 8 mg for *R*(+)-hesperetin. However, relative urinary excretion was 6.4% for *R*-(+)-hesperetin when compared to 3.5% for the *S*-(–)-enantiomer, since the intake of the latter was much greater (Si-Ahmed et al. 2010). These data suggested therefore that the *R*-(+)-enantiomer is more efficiently absorbed. Further work may be required to assess to what degree *in vitro* studies investigating the biological effects of flavanones using racemic compounds represent the *in vivo* situation.

There is limited evidence to suggest that flavanones may also be subjected to phase I metabolism reactions *in vivo*. For example, hesperetin and naringenin were converted to eriodictyol by rat liver microsomes (Nielsen et al. 1998). It has been suggested that, in the rat, eriodictyol could be an intermediate metabolite that is very rapidly remethylated to hesperetin and homoeriodictyol in the liver (Miyake et al. 2000; Matsumoto et al. 2004). However, eriodictyol has not been detected in biofluids after hesperetin administration to humans. Only one *in vitro* study has demonstrated other metabolic routes governed by the cytochrome P450 enzyme system. Nikolic and van Breemen (2004) investigated the *in vitro* metabolism of flavanone, 3′- and 4′-hydroxyflavanone, pinocembrin, naringenin, and 7,4′-dihydroxyflavanone (Figure 1.1) in rat liver microsomes and described several metabolic routes such as oxidation, formation of flavones by the loss of two hydrogen atoms, B-ring cleavage of flavanones, and formation of chromone compounds and reduction of the C4 carbonyl group (Nikolic and van Breemen 2004). To our knowledge, this is the only paper that has suggested these metabolic routes *in vitro*, and no *in vivo* data are available at present. Further studies using appropriate analytical technology are needed to assess the importance of such pathways.

1.2.3 Microbial Metabolism

The vast majority of ingested flavanone derivatives are not absorbed in the small intestine but are carried to the colon, where they are degraded by the microflora. *In vitro* and *in vivo* animal studies have demonstrated the colonic breakdown of flavonoid aglycones to phenolic acids and ring fission products, which may subsequently be absorbed into the systemic circulation. Phenolic acid breakdown products of flavanones include propionic, hydroxyphenylacetic, hydroxycinnamic, and hydroxybenzoic acid derivatives (Table 1.2 and Figure 1.3). *In vivo*, such compounds are often found conjugated as well as free acids (Felgines et al. 2000; Miyake et al. 2000; El Mohsen et al. 2004). The availability of commercial standards for many of these compounds has facilitated their identification and quantification in *in vivo* and *in vitro* studies, and mass spectrometry facilitated the characterization of a number of phenolic acids (Roowi et al. 2009) and their conjugates (Vallejo et al. 2010). A complete list of identified metabolites produced by the microflora from hesperetin and naringenin is given in Table 1.2.

At present, only one study has investigated the profile of phenolic acids excreted after consumption of a flavanone-rich food by humans. After administration of orange juice, an increase was noted in the excretion of five phenolic acids: 3-hydroxyphenylacetic acid, 3-(3′-hydroxyphenyl) hydracrylic acid, dihydroferulic acid, 3-(3′-methoxy-4′-hydroxyphenyl) hydracrylic acid, and 3′-hydroxyhippuric acid in urine (Figure 1.3) (Roowi et al. 2009). Excretion of these products was observed between 10 and 24 h after administration, corresponding to the time needed for the transit and degradation of hesperetin-7-*O*-rutinoside by the colonic microflora.

Data from rat studies and *in vitro* studies, which have incubated pure compounds in the presence of human colonic microflora, have enabled the degradation pathway of flavanones by microflora to be elucidated. Following deglycosylation by microbial β-glucosidases, flavanones undergo cleavage of the heterocyclic C-ring (Figure 1.3) and dehydrogenation of the C-ring to form 3-(3′-methoxy-4′-hydroxyphenyl) hydracrylic acid or 3-(3′-hydroxyphenyl)propionic acid in the case of hesperetin (Labib et al. 2004; Roowi et al. 2009), 3-(4′-hydroxyphenyl)propionic acid in the case of naringenin or 3-(3′,4′-dihydroxyphenyl)propionic acid in the case of eriodictyol (Fuhr and Kummert 1995; Felgines et al. 2000; Miyake et al. 2000; Labib et al. 2004; Rechner et al. 2004; Possemiers et al. 2011) (Figure 1.3). Phloroglucinol (1,3,5-trihydroxybenzene) is probably also produced from the B ring of naringenin and hesperetin (Labib et al. 2004; Possemiers et al. 2011). The presence of 3-(3′-hydroxyphenyl) hydracrylic acid has also been observed, possibly produced from C-ring fission of hesperetin and subsequent demethylation, or alternatively from the *O*-demethylation of 3-(3′-methoxy-4′-hydroxyphenyl) hydracrylic acid. The resulting compounds can be further degraded, oxidized, or conjugated to glycine. Dihydroferulic acid, 3′-hydroxyhippuric acid, 3′-methoxy-4′-hydroxyphenyl acetic acid, and 3′-hydroxyphenylacetic acid are thus produced from hesperetin, whereas hippuric acid, 4′-hydroxyhippuric acid, 4′-hydroxyphenylacetic acid, 3-(4′-hydroxyphenyl)propionic acid, *p*-hydroxybenzoic acid, and *p*-coumaric acid were observed after fermentation of naringenin (Roowi et al. 2009) (Figure 1.3). The presence of hippuric and hydroxyhippuric acids in human urine indicates that

TABLE 1.2
Microbial Metabolites of Flavanones

Study Group	Flavanone Source (Dose)	Metabolites Identified *in vivo*	Presence in Plasma	Presence in Urine	Reference
		Human Studies			
10 humans	Orange juice, 250 mL fortified with 131 μmol of hesperetin-7-*O*-rutinoside with and without 150 mL of natural yoghurt	3′-Hydroxyphenylacetic acid	–	Medium	Roowi et al. (2009)
		3′-Hydroxyphenylhydracrylic acid		Medium	
		Dihydroferulic acid		Medium	
		3′-Methoxy-4′-hydroxyphenylhydracrylic acid		Medium high	
		3′-Hydroxyhippuric acid		Minor	
		4-Hydroxybenzoic acid		Minor	
		4′-Hydroxyphenylacetic acid		Major	
		3′-Methoxy-4′-hydroxyphenylacetic acid		Medium	
		Hippuric acid		Major	
10 healthy humans	Orange juices (commercial and experimental) (total flavonoids from 117 to 441 mg)	4′-Hydroxyphenylpropionic acid glucuronide	i.d.	i.d.	Vallejo et al. (2010)
		p-Coumaric acid glucuronide			
		4-Hydroxybenzoic acid glucuronide			
		3′-Methoxy-4′-hydroxyphenylacetic acid glucuronide			
		Hippuric acid glucuronide			
		3′-Hydroxyphenylacetic acid glucuronide			
		4′-Hydroxyphenylacetic acid glucuronide			

(Continued)

TABLE 1.2 (CONTINUED)
Microbial Metabolites of Flavanones

Study Group	Flavanone Source (Dose)	Metabolites Identified *in vivo*	Presence in Plasma	Presence in Urine	Reference
		Animal Studies			
Sprague-Dawley rats	Naringenin, 50 mg/kg orally	3(4'-Hydroxyphenyl)propionic acid	–	Major	El Mohsen et al. (2004)
Male Wistar rats	Control diet supplemented with 0.25% naringenin or with 0.5% naringenin-7-*O*-neohesperidoside	3-(4'-Hydroxyphenyl)propionic acid *p*-Coumaric acid *p*-Hydroxybenzoic acid	–	Major Medium Minor	Felgines et al. (2000)
Sprague-Dawley rats	Eriocitrin, 50 and 75 μmol/kg by gavage	3',4'-Dihydroxyhydrocinnamic acid	Main	Major	Miyake et al. (2000)

i.d.: identified only.

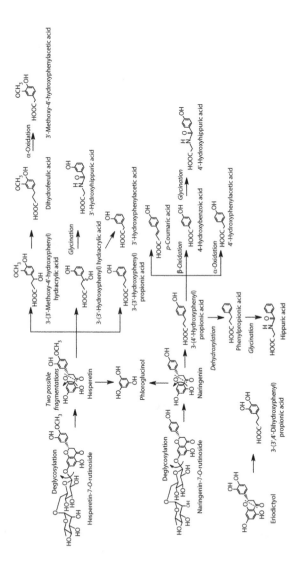

FIGURE 1.3 Proposed metabolic pathway for the catabolism of flavanones by intestinal microbiota. (Based on data of Felgines, C. et al., *Am. J. Physiol. Gastrointest. Liver Physiol.*, 279, G1148–G1154, 2000; Miyake, Y. et al., *J. Agric. Food Chem.*, 48, 3217–3224, 2000; El Mohsen, M.A. et al., *Free Radic. Res.*, 38, 1329–1340, 2004; Labib, S. et al., *Mol. Nutr. Food Res.*, 48, 326–332, 2004; Rechner, A.R. et al., *Free Radic. Biol. Med.*, 36, 212–225, 2004; Roowi, S. et al., *Mol. Nutr. Food Res.*, 53, S68–S75, 2009; Vallejo, F. et al., *J. Agric. Food Chem.*, 58, 6516–6524, 2010 and Possemiers, S. et al., *Fitoterapia*, 82, 53–66, 2011.)

these compounds undergo glycination in the liver (Rechner et al. 2004; Possemiers et al. 2011).

Limited data are available on the relative excretion of microbial and phase II metabolites of flavanones. In rats fed pure naringenin, the concentration of the microbial metabolite 3-(4′-hydroxyphenyl)propionic acid in urine was found to be 5.6-fold lower than that of free and conjugated naringenin 18 h after administration (El Mohsen et al. 2004). In addition, the urinary concentration of the three microbial metabolites determined by Felgines et al. (2000) in rat urine was also 3- and 2-fold lower than conjugated naringenin after the single or repeated administration of naringenin to rats, respectively (Table 1.2). However, after administration of naringenin-7-O-rutinoside, microbial metabolite concentrations were slightly higher than conjugated naringenin concentrations. It is noteworthy that in a human study reported by Roowi et al. (2009), the five phenolic acids recovered in urine after orange juice intake accounted for 37% of the ingested flavanones, which is considerably higher than the proportion usually excreted as hesperetin conjugates. It is clear that the microbial metabolites of flavanones warrant further study, especially with regard to potential bioactivity.

1.2.4   Tissue Distribution

Knowledge of the extent and duration of exposure of tissues to circulating flavonoids, and the forms to which tissues are exposed, is essential for understanding and predicting bioactivity at target sites. However, as for most polyphenols, data on the tissue distribution of flavanones in animals are still scarce, and no data are available for humans at present. Naringenin aglycone has been the main compound used for studies of the tissue distribution of flavanones in animals or cell cultures.

After absorption, flavanones circulate in the bloodstream. Determination of the dissociation equilibrium constant for the binding of naringenin to human serum albumin, using docking simulation and *in vitro* incubation experiments, has indicated strong binding to albumin in plasma (Bolli et al. 2010). However, the relevance of this finding is questionable as the aglycone was studied rather than a phase II metabolite. The effects of glucuronidation and sulfation upon binding have not been studied for the flavanones, but in the same study experiments with isoflavones showed that the sulfation of daidzein did not impair binding relative to the aglycone. The degree of binding to albumin may determine the extent of delivery to cells and tissues, as well as influence plasma clearance. The classic view is that cellular uptake is proportional to the unbound concentration of metabolites. However, binding to albumin is reversible, and conformational changes occurring in the vicinity of the membrane may lead to the dissociation of the ligand-albumin complex (Kragh-Hansen et al. 2002). Although the binding to albumin has been shown to affect the biological activity of many drugs, its impact on that of flavonoid metabolites is not yet documented.

The distribution of naringenin in heart, brain, lungs, spleen, liver, and kidneys has been investigated after gastric gavage of [³H]-naringenin (10 and 50 mg/kg bw) to rats (El Mohsen et al. 2004). Some radioactivity was detected in plasma and tissues 2 h after gavage, but much higher levels were observed after 18 h. While high levels of radioactivity were detected in the urine, implying efficient excretion, plasma and tissue radioactivity levels at 18 h postgavage were not reduced compared to those at 2 h. The apparent extent of this absorption, indicated by the exceptionally high levels of radioactivity incorporated into the tissues, is striking, given that a much lower absorption would be expected from studies with unlabeled flavonoids. These data should be viewed with caution, as the radioactivity detected may well correspond to tritium-labeled water rather than to tritiated naringenin, due to exchange of the label.

El Mohsen et al. (2004) also examined the concentration and nature of naringenin metabolites present in tissues after gavage with unlabeled naringenin (50 mg/kg bw). At 2 h postgavage, monoglucuronides were the major metabolites of naringenin in plasma (98% of total metabolites) and tissues (about 25–80%). After 18 h, the aglycone became the predominant form and was the only form detected in the liver and heart. These findings are consistent with data reported for other flavonoids indicating that higher proportions of aglycone are generally present in tissues compared to plasma (Chang et al. 2000). However, it must be noted that adequate quality controls for assessing artifacts such as hydrolysis of conjugates during sample preparation are not always provided. In addition to conjugated metabolites detected after 18 h, it is noteworthy that the microbial metabolite 3-(4′-hydroxyphenyl)propionic acid was also identified as a major metabolite in the urine. Total metabolites detected after 18 h were only 1–5% of the levels detected after 2 h in most tissues. However, the brain and lungs retained 27 and 20%, respectively, of the total metabolites detected at 2 h. This suggests that the kinetics and/or the level of exposure can differ substantially between organs.

Other *in vivo* and *in vitro* studies have indicated that the flavanones hesperetin and naringenin as well as their relevant *in vivo* metabolites are able to cross the blood–brain barrier (BBB). Ten minutes after intravenous administration of 20 mg/kg bw naringenin to rats, the cerebral cortex concentrations of aglycone and total metabolites were 1.6 ± 0.2 and 2.1 ± 0.4 µg/g, respectively (Peng et al. 1998). In comparison, the total plasma concentration was approximately 2-fold higher (4.8 ± 0.3 µg/g) but the aglycone concentration was considerably lower (0.7 ± 0.1 µg/g). The profile of tissue metabolites may be different from that of plasma metabolites because of the specific uptake or elimination of some metabolites or because of intracellular metabolism. Using brain endothelial cell lines from mouse (b.END5) and rat (RBE4), Youdim et al. (2003) showed that the hesperetin and naringenin were taken up efficiently, as would be expected, given their substantial lipophilicities. This uptake increased with time and as a function of concentration. Flavanone monoglucuronides, obtained from enzymatic glucuronidation of aglycones and subsequent purification, were also able to enter cultured brain endothelial cells, but only after a prolonged period of exposure and to a much lower extent than their corresponding aglycones. After exposure to flavanone glucuronides, free aglycones were detected in cell extracts and the incubation media, indicating a deglucuronidation process. Further studies are required to identify the mechanism by which the anionic glucuronides can cross the membranes, as their more polar structure means transcellular passive diffusion will be limited.

ECV304 cell monolayers cocultured with C6 glioma cells represent a useful *in vitro* model for assessing and ranking passive permeability of compounds across the BBB. The co-cultured cell model mimics the BBB, which is formed by the endothelium of brain microvessels, under the inductive influence of associated cells, especially astrocytes. In this model, flavanone aglycones were reported to have a high apparent permeability (P_{app}) (Youdim et al. 2003). No glucuronidation was detected during their monolayer permeation. Naringenin had a higher permeability (P_{app} = 350 nm/s) than did hesperetin (P_{app} = 290 nm/s). Their respective glucuronides had 2-fold lower apparent permeabilities (113–182 nm/s). It is notable that, in the same study, anthocyanins had similar permeabilities to flavanone glucuronides and that (–)-epicatechin and its methylated and glucuronidated metabolites were unable to cross the monolayer model.

In another *in vivo* model of BBB permeability, consisting of an *in situ* perfusion of the [^{3}H]-labeled compound in rat brain, naringenin was shown to enter all brain regions studied (right hemisphere homogenate, cerebellum, hypothalamus, hippocampus, superior colliculus, striatum, medulla, cortex, and left hemisphere homogenate), with distribution volume values ranging from 89 to 288 µL/g (Youdim et al. 2004). In addition to lipophilicity, the ability of flavonoids to enter the brain depends on their interactions with specific efflux transporters expressed in the BBB, such as P-gp, which plays an important role in brain uptake. Pretreatment with inhibitors of efflux transporters such as P-gp did not significantly increase the distribution volume of naringenin in brain in the rat *in situ* perfusion model, suggesting that naringenin is not a substrate for the efflux systems studied. Conversely, in two cell lines MDCK-MDR1 and RBE-4, which functionally express P-gp, naringenin uptake was increased in the presence of P-gp inhibitors (Youdim et al. 2004). Further work is needed to investigate the role of transporters in flavanone uptake and efflux in brain endothelial cells. However, the ability of flavanones to cross the BBB is already of considerable interest since a number of studies have suggested a potential role of flavanones in the maintenance of cognitive function and prevention of neurodegenerative diseases (Zbarsky et al. 2005; Vauzour et al. 2007; Kumar et al. 2010; Spencer 2010).

The cellular uptake of flavanones has also been studied in dermal fibroblasts (Spencer et al. 2004). The uptake of hesperetin and naringenin was 20- and 10-fold higher than that of quercetin, respectively, under the same conditions, demonstrating the high efficiency of passive diffusion for flavanones. In contrast, naringenin glucuronide and hesperetin glucuronide did not penetrate the cells. More studies with a range of cell types are needed to investigate the possibility that flavanone glucuronides cross cell membranes without prior deconjugation. Data on the cellular uptake of flavanone sulfates, which have never been studied, are also required. Particular care must be taken when interpreting the data from immortalized cell lines, as they may have lost key membrane carriers.

In conclusion, flavanones are bound to albumin in blood and are rapidly distributed to tissues such as brain, lung, heart, liver, spleen, and kidney after absorption. Shortly after ingestion, glucuronides are the main metabolites in plasma and tissues, whereas later after ingestion, the aglycone predominates, although some breakdown products of flavanones could also be present in tissues. Various *in vitro* and *in vivo* studies have demonstrated that flavanones are able to cross the BBB. However, there is a lack of descriptive data on the distribution of dietary flavanones in many tissues, as well as on their metabolism and kinetics at target sites.

1.3 PHARMACOKINETICS OF FLAVANONES

The pharmacokinetic parameters of the flavanones hesperetin, naringenin, and liquiritigenin have been studied in human volunteers (Table 1.3) and in animal models (Table 1.4) after the intake of citrus fruits and juices, other flavanone-rich foods or extracts, and pure compounds. Tables 1.3 and 1.4 give key pharmacokinetics parameters, including C_{max}, T_{max}, AUC, elimination half-life ($T_{1/2}$), as well as percentage of the ingested dose excreted in urine. Flavanones are mainly present in citrus foods and juices, and therefore, the majority studies have administered these food sources. A few studies have investigated flavanone pharmacokinetics after ingestion of less common sources such as Zhi Zhu Wan (BZZW) Chinese medical formulations or capsules containing pure compounds. Although these sources were less nutritionally relevant, the investigations were able to provide complementary information on flavanone bioavailability (Table 1.3).

1.3.1 STUDIES IN HUMANS

Maximum plasma concentrations of hesperetin and naringenin metabolites are usually reached 4–7 h after flavanone intake from food sources (Figure 1.4). As discussed earlier, this is consistent with absorption in the distal part of the intestine. When flavanones are consumed in the more easily absorbable aglycone form, total metabolites attain a C_{max} in 1–2 h (Bugianesi et al. 2002, 2004; Nielsen et al. 2006; Bredsdorff et al. 2010). The T_{max} duration was shorter for naringenin metabolites than for hesperetin metabolites in all intervention studies that used flavanone sources providing both hesperetin and naringenin glycosides (Erlund et al. 2001; Manach et al. 2003; Gardana et al. 2007; Brett et al. 2009; Krogholm et al. 2010; Vallejo et al. 2010). It is not clear whether this could be explained by differences in the amount ingested. Pharmacokinetic data for hesperetin and naringenin after administration of the pure flavanones are needed to clarify any differences in their absorption. In studies where the flavanone sources were citrus juices or fruits, plasma C_{max} values for total naringenin ranged from 0.04 µM after consumption of 400 mL commercial orange juice produced from concentrate to 0.64 µM after consumption of orange juice (8 mL/kg bw providing 23 ± 2 mg naringenin equivalents), and 6.0 µM after consumption of grapefruit juice (8 mL/kg bw providing 199 ± 42 mg naringenin equivalents) (Erlund et al. 2001; Vallejo et al. 2010) (Table 1.3). The differences observed between studies are primarily explained by the size of the flavanone doses ingested. For hesperetin, C_{max} values of total metabolites ranged from 0.09 µM after consumption of whole oranges (providing 79.7 ± 17.7 mg hesperetin eq.) to 2.2 µM after consumption of orange juice (8 mL/kg bw providing 126 ± 26 mg naringenin equivalents) (Erlund et al. 2001; Brett et al. 2009). After administration of BZZW Chinese medicine providing 0.89 mg/kg bw naringenin-7-O-neohesperidoside, and 2.58 mg/kg bw hesperetin-7-O-rutinoside, C_{max} values as high as 11.6 ± 8.1 µM and 10.5 ± 7.3 µM were measured for naringenin and hesperetin, respectively (Cao et al.

TABLE 1.3

Pharmacokinetic Studies of Flavanones and Their Metabolites in Humans

Species	Flavanone Source (Dose)	Compound	C_{max} (μmol/L)	T_{max} (h)	$T_{1/2}$ (h)	$AUC_{0-\infty}$ (μmol·h/L)	Urinary Excretion (%)	Reference
16 humans	Orange juice (0.83 mg/kg bw naringenin-7-O-rutinoside)	Total naring	0.12 ± 0.14	5.18 ± 3.05	–	0.30 ± 0.20	7 ± 3	Bredsdorff et al. 2010
15 humans	α-Rhamnosidase-treated orange juice (0.52 mg/kg bw naringenin-7-O-glucoside)	Total naring	0.77 ± 0.35	1.55 ± 2.77	–	1.17 ± 0.48	47 ± 17	Bredsdorff et al. 2010
20 humans	150 g orange fruit (79.7 mg hesperetin-7-O-rutinoside)	Total hesp	0.095 ± 0.12	7.00 ± 4.23	–	1.25 ± 1.68	4.53 ± 3.44	Brett et al. 2009
	300 g orange juice (71.8 mg hesperetin-7-O-rutinoside)	Total hesp	0.10 ± 0.13	6.20 ± 2.04	–	1.15 ± 1.20	4.63 ± 3.05	
129 humans	300 g orange juice (71.8 mg hesperetin-7-O-rutinoside)	Total hesp	–	–	–	–	3.9 ± 3.9	Brett et al. 2009
20 humans	150 g orange fruit (11.8 mg naringenin-7-O-rutinoside)	Total naring	0.085 ± 0.12	5.88 ± 1.83	–	0.85 ± 1.04	12.5 ± 10.6	Brett et al. 2009
	300 g orange juice (9.4 mg naringenin-7-O-rutinoside)	Total naring	0.053 ± 0.053	4.46 ± 2.56	–	0.65 ± 0.63	10.2 ± 6.8	
129 humans	300 g orange juice (9.4 mg naringenin-7-O-rutinoside)	Total naring	–	–	–	–	14.5 ± 11.9	Brett et al. 2009
8 humans	Orange juice (8 mL/kg) (417 μmol hesperetin-7-O-rutinoside)	Total hesp	2.20 ± 1.58	5.4 ± 1.6	2.2 ± 0.8	AUC_{0-24}: 10.3 ± 8.2	5.3 ± 3.1	Erlund et al. 2001
	Orange juice (8 mL/kg) (85 μmol naringenin-7-O-neohesperidoside)	Total naring	0.64 ± 0.40	5.5 ± 2.9	1.3 ± 0.6	AUC_{0-24}: 2.6 ± 1.6	1.1 ± 0.8	

5 humans	Grapefruit juice (8 mL/kg) (731 μmol naringenin-7-O-neohesperidoside)	Total naring	5.99 ± 5.36	4.8 ± 1.1	2.2 ± 0.1	AUC$_{0-24}$: 27.7 ± 26.3	30.2 ± 25.5	Erlund et al. 2001
6 humans	20 mL/kg bw grapefruit juice (621 μmol/L naringenin-7-O-neohesperidoside))	Naring	–	–	–	–	0.09 (0.01–0.37)[d]	Fuhr et al. 1995
	20 mL/kg bw grapefruit juice (621 μmol/L naringenin-7-O-neohesperidoside))	Naring glcAs	–	–	–	–	8.8 (5.2–58.5)[d]	
7 women	150 or 300 mL blood orange juice (51 and 102 mg hesperetin)	Total hesperetin	150 mL: 0.14 ± 0.11; 300 mL: 0.26 ± 0.20	150 mL: 5.3 ± 0.8; 300 mL: 5.1 ± 0.7	–	150 mL: 0.65 ± 0.77; 300 mL: 1.31 ± 1.56	–	Gardana et al. 2007
	150 or 300 mL blood orange juice (6 and 12 mg naringenin)	Total naringenin	150 mL: 0.06 ± 0.04; 300 mL: 0.13 ± 0.08	150 mL: 5.0 ± 0.6; 300 mL: 5.0 ± 0.0	–	150 mL: 0.20 ± 0.21; 300 ml: 0.35 ± 0.25	–	
5 men	0.5 or 1L orange juice (110 mg and 220 mg, respectively, of hesperetin)	Total hesperetin	After 0.5L: 0.46 ± 0.07; after 1L: 1.28 ± 0.13	After 0.5L: 5.40 ± 0.40; after 1L: 5.8 ± 0.37	–	AUC$_{0-24}$: After 0.5L: 4.19 ± 1.11; after 1L: 9.28 ± 1.95	After 0.5L: 4.13 ± 1.18; after 1L: 6.41 ± 1.32	Manach et al. 2003
	0.5 or 1L orange juice (22.6 mg and 45.2 mg, respectively, of naringenin)	Total naringenin	After 0.5L: 0.06 ± 0.02; after 1L: 0.20 ± 0.04	After 0.5L: 4.6 ± .60; after 1L: 5.0 ± 0.45	–	AUC$_{0-24}$: After 0.5L: 0.43 ± 0.17; after 1L: 1.29 ± 0.33	After 0.5L: 7.11 ± 1.86; after 1L: 7.87 ± 1.69	
8 humans	250 mL orange juice (12 μmol of naringenin-7-O-rutinoside)	Total naringenin	–	–	–	–	17.7 ± 3.9[c]	Mullen et al. 2008a
	250 mL orange juice (12 μmol of naringenin-7-O-rutinoside) with 150 mL yogurt	Total naringenin	–	–	–	–	15.7 ± 3.4[c]	

(Continued)

TABLE 1.3 (CONTINUED)

Pharmacokinetic Studies of Flavanones and Their Metabolites in Humans

Species	Flavanone Source (Dose)	Compound	C_{max} (μmol/L)	T_{max} (h)	$T_{1/2}$ (h)	$AUC_{0-\infty}$ (μmol·h/L)	Urinary Excretion (%)	Reference
	250 mL orange juice (168 μmol hesperetin-7-O-rutinoside)	–	Hesp-glcAs: 0.92 ± 0.22[a]	Hesp-glcAs: 4.4 ± 0.5[a]	Hesp-glcAs: 3.6 ± 1.3[a]	Hesp-glcAs: 4.1 ± 2.9[a]	Total Hesp: 6.3 ± 2.0[b]	Mullen et al. 2008a
	250 mL orange juice (168 μmol hesperetin-7-O-rutinoside) with 150 mL yogurt	–	Hesp-glcAs: 0.66 ± 0.17[a]	Hesp-glcAs: 5.1 ± 0.4[a]	Hesp-glcAs: 3.8 ± 0.8[a]	Hesp-glcAs: 3.0 ± 2.6[a]	Total Hesp: 6.4 ± 2.0[b]	
Humans	Orange juice (61 mg hesperetin-7-O-rutinoside)	Total hesperetin	0.48 ± 0.27	7 ± 3	–	AUC_{0-10}: 1.16 ± 0.52	4.06 ± 1.77	Nielsen et al. 2006
	Hesperidinase-treated orange juice (80 mg hesperetin-7-O-glucoside)	Total hesperetin	2.60 ± 1.07	0.6 ± 0.1	–	AUC_{0-10}: 3.45 ± 1.27	14.40 ± 6.75	
	Orange juice supplemented with hesperetin-7-O-rutinoside (192 ± 30 mg hesperetin-7-O-rutinoside)	Total hesperetin	1.05 ± 0.25	7.4 ± 2.0	–	AUC_{0-10}: 4.16 ± 1.50	8.90 ± 3.83	
10 humans	400 mL commercial orange juice produced by concentrate (TF: 116.8 mg)	Total hesp	0.32 ± 0.07	4.6 ± 0.7	–	1.18 ± 0.28	5.4 ± 1.2	Vallejo et al. 2010
	400 mL commercial orange juice produced by concentrate (TF: 215.2 mg)	Total hesp	0.37 ± 0.07	6.4 ± 0.7	–	0.95 ± 0.23	1.7 ± 0.4	

400 mL commercial pulp-enriched orange juice (TF: 281.2 mg)	Total hesp	0.15 ± 0.03	6.0 ± 0.7	–	0.50 ± 0.07	1.0 ± 0.5
400 mL orange flavonoid extract enriched juice B (TF: 207.2 mg)	Total hesp	1.16 ± 0.19	7.3 ± 0.2	–	4.43 ± 0.91	4.6 ± 1.0
400 mL orange flavonoid extract diluted in water (TF: 440.8 mg)	Total hesp	1.48 ± 0.42	6.3 ± 0.6	–	4.15 ± 0.97	8.9 ± 2.9
400 mL commercial orange juice produced by concentrate (TF: 116.8 mg)	Total naring	0.04 ± 0.01	4.7 ± 1.1	–	0.07 ± 0.027	2.6 ± 0.5
400 mL commercial orange juice produced by concentrate (TF: 215.2mg)	Total naring	0.44 ± 0.11	5.7 ± 0.7	–	0.74 ± 0.18	0.7 ± 0.2
400 mL commercial pulp-enriched orange juice (TF: 281.2 mg)	Total naring	0.07 ± 0.007	5.0 ± 0.6	–	0.20 ± 0.03	0.5 ± 0.2
400 mL orange flavonoid extract enriched juice (TF: 207.2 mg)	Total naring	0.54 ± 0.13	5.9 ± 0.6	–	1.46 ± 0.38	2.5 ± 0.5
400 mL orange flavonoid extract diluted in water (TF: 440.8 mg)	Total naring	0.44 ± 0.11	6.4 ± 1.1	–	0.96 ± 0.19	5.4 ± 1.5

(Continued)

TABLE 1.3 (CONTINUED)
Pharmacokinetic Studies of Flavanones and Their Metabolites in Humans

Species	Flavanone Source (Dose)	Compound	C_{max} (μmol/L)	T_{max} (h)	$T_{1/2}$ (h)	$AUC_{0-\infty}$ (μmol·h/L)	Urinary Excretion (%)	Reference
1 man	1250 mL orange juice (37.6 ± 2.2 mg/L naringenin-7-O-rutinoside; 65 ± 3.3 hesperetin-7-O-rutinoside) and 1250 mL grapefruit juice (373.1 ± 22.1 mg/L naringenin-7-O-neohesperidoside ; 141 ± 9.5 mg/L naringenin-7-O-rutinoside; 6.5 ± 0.4 mg/L hesperetin-7-O-rutinoside) (divided in 5 doses during 48 h)	Total naring	–	–	–	–	6.81	Ameer et al. 1996
		Total hesp	–	–	–	–	24.43	
	500 mg naringenin-7-O-neohesperidoside and 500 mg hesperetin-7-O-rutinoside	Total naring	–	–	–	–	4.89	
	500 mg naringenin-7-O-neohesperidoside and 500 mg hesperetin-7-O-rutinoside	Hesp	–	–	–	–	2.97	
	500 mg naringenin-7-O-neohesperidoside	Naring	–	–	–	–	4.84	

			Hesp-O-glcA (2): 0.17 ± 0.05	Hesp-O-glcA (2): 3.7 ± 0.2	Hesp O-glcA (2): 1.3 ± 0.5	Hesp O-glcA (2): 0.56 ± 0.16	Total hesp: 12.0	
10 humans	350 mL polyphenols-rich juice (45 μmol hesperetin-7-O-rutinoside + 5.9 μmol naringenin-7-O-neohesperidoside)	—	0.17 ± 0.05	3.7 ± 0.2	1.3 ± 0.5	0.56 ± 0.16	12.0	Borges et al. 2010
10 men	Juice mix (28 mg/L naringenin)	Naring	0.25 ± 0.13	3.6 ± .6	—	AUC$_{0-48}$: 2.64 ± 1.95	22.6 ± 11.5 (0–48h)	Krogholm et al. 2010
	Juice mix (32 mg/L hesperetin)	Hesp	0.18 ± 0.13	4.9 ± 1.4	—	AUC$_{0-48}$: 2.13 ± 1.59	14.2 ± 9.1 (0–48h)	
5 humans	500 g cooked cherry tomatoes (17 mg naringenin-7-O-neohesperidoside)	Naring	0.06 ± 0.02	2	—	—	—	Bugianesi et al. 2004
5 men	150 g cooked commercial tomato paste (3.8 mg naringenin-7-O-neohesperidoside)	Naring	0.12 ± 0.03	2	—	—	—	Bugianesi et al. 2002
10 humans	Blue-labeled Zhi Zhu Wan (0.08 g/kg) (1.16 mg hesperetin-7-O-rutinoside and 8.10 mg/g hesperetin-7-O-neohesperidoside)	Hesp	8.51 ± 5.07	3.71 ± 2.22	5.49 ± 3.93	115.53 ± 81.62	—	Cao et al. 2010
	Red-labeled ZZW (0.08 g/kg) (32.3 mg/g hesperetin-7-O-rutinoside and 0.33 mg/g hesperetin-7-O-neohesperidoside)	Hespe	10.46 ± 7.28	8.52 ± 3.52	5.10 ± 2.41	146.72 ± 74.57	—	
	BZZW (0.08 g/kg) (11.18 mg/g naringenin-7-O-neohesperidoside)	Naring	11.62 ± 8.09	8.52 ± 3.52	5.10 ± 2.41	162.90 ± 82.79	—	

(Continued)

TABLE 1.3 (CONTINUED)

Pharmacokinetic Studies of Flavanones and Their Metabolites in Humans

Species	Flavanone Source (Dose)	Compound	C_{max} (μmol/L)	T_{max} (h)	$T_{1/2}$ (h)	$AUC_{0-\infty}$ (μmol·h/L)	Urinary Excretion (%)	Reference
6 humans	Solid dispersion capsules (135 mg hesperetin)	Hesp	2.73 ± 1.36	3.67 ± 0.52	3.05 ± 0.91	16.05 ± 5.55	3.26 ± 0.44	Kanaze et al. 2007
	Solid dispersion capsules (135 mg naringenin)	Naring	7.39 ± 2.83	3.67 ± 0.82	2.31 ± 0.40	34.65 ± 10.88	5.81 ± 0.81	
10 men	4.5 g of FG, wrapped with oblat, with 500 mL water (2.27 mmol eriodictyol-7-O-rutinoside)	—	Eriodictyol: ~2; homoeriodictyol: ~0.5; hesperetin: 2	Eriodictyol: ~4; homoeriodictyol: ~1; hesperetin: ~1	—	AUC_{0-4}: Eriodictyol: 5.2 ± 0.9; homoeriodictyol: 1.5 ± 0.3; hesperetin: 5.6 ± 0.6	—	Miyake et al. 2006
	3.7 g of FA, wrapped with oblat, with 500 mL water (82.2% aglycones prepared from FG) (1.89 mmol eriodictyol)	—	Eriodictyol: ~7.5; homoeriodictyol: ~4; hesperetin: ~4.5	Eriodictyol: ~1; Homoeriodictyol: ~1; hesperetin: ~1	—	AUC_{0-4}: Eriodictyol: 21.9 ± 1.8; homoeriodictyol: 13.4 ± 1.7; hesperetin: 10.3 ± 1.4	—	
4 men	4.5 g FG with 500 mL lemon juice (2.27 mmol eriodictyol-7-O-rutinoside)	—	Eriodictyol: ~2; homoeriodictyol: ~0.5; hesperetin: ~2	Eriodictyol: ~4; Homoeriodictyol: ~4; hesperetin: ~4	—	AUC_{0-4}: Eriodictyol: 1.5 ± 0.3; homoeriodictyol: 0.5 ± 0.2; hesperetin: 2.5 ± 0.4	—	Miyake et al. 2006

| 7 men | 3.7 g FA with 50 mL vodka and 450 mL water (1.89 mmol eriodictyol) | – | Eriodictyol: ~11.5 ; homoeriodictyol: ~4; hesperetin: ~5 | Eriodictyol: ~0.5 Homoeriodictyol: ~0.5; hesperetin: ~0.5 | – | AUC_{0-t}: Eriodictyol: 25.1 ± 5.3; homoeriodictyol: 8.3 ± 1.8; hesperetin: 13.1 ± 2.5 | – | Miyake et al. 2006 |

Values are mean ± SD unless otherwise stated.

IV: intravenous; bw: body weight; Hesp: hesperetin; Naring: naringenin; glcA: glucuronide; Hesp-glcAs: sum of hesperetin-7-O-glucuronide and unidentified hesperetin-O-glucuronide; Hesp O-glcA (2): hesperetin-O-glucuronide (two isomers); FG: flavanone glycosides (30% eriodictyol-7-O-rutinoside. 0.1% 6,8-C-diglucosyldiosmetin, 0.05% hesperetin-7-O-rutinoside); FA: flavanone aglycones (contained 14.7% eriodictyol, 3.0% hesperetin, 0.3% eriodictyol-7-O-rutinoside); (α): half-life in the distribution phase; (β): half-life in the elimination phase; TF: total flavonoids; Erio: Eriodictyol. All studies were performed after oral intake of aglycone or food.

a   Mean ± SE.

b   Expressed as percentage of hesperetin-7-O-rutinoside intake.

c   Expressed as percentage of naringenin-7-O-rutinoside intake.

d   Median and range.

TABLE 1.4

Pharmacokinetic Studies of Flavanones and Their Metabolites in Animal Models

Species	Source (Dose)	Administration	C_{max} (μmol/L)	T_{max} (h)	$T_{1/2}$ (h)	$AUC_{0-\infty}$ or AUC_{0-t} (μmol·h/L)	Urinary Excretion (%)	Reference
Male Wistar rats	Control diet supplemented with 0.25% naringenin (9.2 mmol/kg diet of aglycone equivalents)	Oral	Naringenin: 128 ± 2	–	–	–	66.2 ± 3.1	Felgines et al. 2000
	Control diet supplemented with 0.25% naringenin (9.2 mmol/kg diet of aglycone equivalents) (14 days)	Oral	Naringenin: ~135	–	–	–	58.7 ± 5.7	
	Control diet supplemented with 0.38% naringenin-7-O-glucoside (9.2 mmol/kg diet of aglycone equivalents)	Oral	Naringenin: 144 ± 8	–	–	–	–	
	Control diet supplemented with 0.5% naringenin-7-O-neohesperidoside (9.2 mmol/kg diet of aglycone equivalents)	Oral	Naringenin: 139 ± 15	–	–	–	32.0 ± 5.4	
	Control diet supplemented with 0.5% naringenin-7-O-neohesperidoside (9.2 mmol/kg diet of aglycone equivalents) (14 days)	Oral	Naringenin: ~142	–	–	–	23.2 ± 2.1	

6 rabbits	Naringenin (25 mg/kg)	IV	—	(α): 0.07 ± 0.02; (β): 0.28 ± 0.07	Free[a]: 27.80 ± 4.39; Conjugates[a]: 7.27± 2.38	—	Hsiu et al. 2002
	Naringenin (25 mg/kg)	Oral	Naringenin: ~2.7; Naringenin conjugates: ~2.3	Free and conjugates: ~0.17	Free[a]: 0.94 ± 0.32; Conjugates[a]: 1.75± 0.56	—	
5 rabbits	Naringenin-7-O-neohesperidoside (225 mg/kg)	Oral	Naringenin: ~0.5; Naringenin conjugates: ~2.6	Free and conjugates: ~1.5	Free[a]: 1.86 ± 0.48; Conjugates[a]: 8.55 ± 2.65	—	Hsiu et al. 2002
10 Wistar rats	Naringenin (30 mg/kg)	Gastric gavage	Free:10.7; Total: 62.4	Free: 0.08; Total: 0.5	Free: 3.5; Total: 114	—	Ma et al. 2006
	Naringenin (90 mg/kg)	Gastric gavage	Free:13.7; Total:103	Free: 0.25; Total: 2.0	Free: 66; Total: 489	—	
	Naringenin (270 mg/kg)	Gastric gavage	Free:16.2; Total:161	Free: 0.08; Total: 2.0	Free: 80; Total: 1666	—	
8 male Sprague-Dawley rats	Naringenin (37 μmol/kg)	IV	—	Naringenin: 0.66 ± 0.49; GlcAs: 1.19 ± 0.32; Sulf: 4.26 ± 0.81	AUC_{0-12}: 5.56 ± 0.49; AUC_{0-12}: 7.58 ± 2.18; AUC_{0-12}: 200 ± 32	—	Wang et al. 2006
	Naringenin (184 μmol/kg)	Oral	GlcAs[a] 7.4 ± 1.3; Sulf[a] 11.7 ± 1.6	GlcAs[a] 0.66 ± 0.3; Sulf[a] 1.57 ± 0.81	AUC_{0-24}: GlcAs[a]: 16.8 ± 4.7; Sulf[a]: 49.3 ± 6.9	—	
	Naringenin-7-O-neohesperidoside (184 μmol/kg)	Oral	GlcAs[a] 0.6 ± 0.1; Sulf[a] 3.6 ± 0.2	GlcAs[a] 2.16 ± 0.77; Sulf[a] 2.25 ± 0.55	AUC_{0-96}: GlcAs[a]: 5.0 ± 0.5; Sulf[a]: 83.1 ± 6.2	—	
7 male Sprague-Dawley rats	Naringenin-7-O-neohesperidoside (367 μmol/kg)	Oral	GlcAs[a] 1.5 ± 0.4; Sulf[a] 5.8 ± 0.9	GlcAs[a] 4.67± 1.21; Sulf[a] 6.92 ± 1.08	AUC_{0-96}: GlcAs[a]: 5.0 ± 1.2; Sulf[a]: 106 ± 11	—	Wang et al. 2006

(Continued)

TABLE 1.4 (CONTINUED)
Pharmacokinetic Studies of Flavanones and Their Metabolites in Animal Models

Species	Source (Dose)	Administration	C_{max} (μmol/L)	T_{max} (h)	$T_{1/2}$ (h)	$AUC_{0-\infty}$ or AUC_{0-t} (μmol·h/L)	Urinary Excretion (%)	Reference
6 male Sprague-Dawley rats	Naringenin-7-O-neohesperidoside (367 μmol/kg twice daily for 9 days)	Oral	GlcAs:[a] 2.8 ± 0.5; Sulf:[a] 6.4 ± 1.2	GlcAs:[a] 2.17 ± 1.21; Sulf:[a] 3.46 ± 1.94	—	AUC_{0-12}: GlcAs:[a]: 21.0 ± 5.5; Sulf:[a] 113 ± 5	—	Wang et al. 2006
	Racemic naringenin (20 mg/kg)	IV	—	—	R(+)-Naring:[a] 3.40 ± 0.27; S(−)-Naring:[a] 3.65 ± 0.21	R(+)-Naring:[a] 42.8 ± 2.0; S(−)-Naring:[a] 49.3 ± 1.7	R(+)-Naring:[a] 8.02 ± 0.97; S(−)-Naring:[a] 8.83 ± 0.92	Yanez et al. 2008
	Racemic hesperetin (20 mg/kg)	IV	—	—	R(+)-hesp:[a] 6.99 ± 0.26; S(−)-hesp:[a]: 3.73 ± 0.47	R(+)-hesp:[a] 80.6 ± 4.9; S(−)-hesp:[a] 24.6 ± 1.9	R(+)-hesp:[a] 3.77 ± 0.55; S(−)-hesp:[a] 2.44 ± 0.39	Yanez et al. 2008
	Racemic eriodictyol (20 mg/kg)	IV	—	—	R(+)-Erio:[a] 4.16 ± 0.36; S(−)-Erio:[a] 3.68 ± 0.25	R(+)-Erio:[a] 42.9 ± 3.2; S(−)-Erio:[a] 38.6 ± 2.3	R(+)-Erio:[a] 5.07 ± 0.99; S(−)-Erio:[a] 7.32 ± 1.75	Yanez et al. 2008
6 male healthy SD rats	*Rhizoma drynariae* extract (700 mg/kg naringenin-7-O-neohesperidoside)	Oral	Naringenin-7-O-neohesperidoside: 9.41 ± 2.83; Naring: 0.82 ± 0.17	Naringenin-7-O-neohesperidoside: 0.67 ± 0.20; Naring: 8.0 ± 1.3	Naringenin-7-O-neohesperidoside: 4.1 ± 0.8	Naringenin-7-O-neohesperidoside: 5.67 ± 0.93; Naring: 6.58 ± 1.58	—	Xiao-Hong et al. 2010

10 beagles	Grapefruit extract (400 mg of which 140 mg flavanones: 84.2 mg naringenin-7-O-neohesperidoside, 48 mg naringenin-7-O-rutinoside, 8.6 mg naringenin)	Oral	Naringenin-7-O-neohesperidoside: 0.24; Naringenin: 0.02; Naringenin glcA: 0.09	Naringenin-7-O-neohesperidoside: 1.3; Naringenin: 0.3; Naring gluc: 0.5	–	AUC_{0-24}; Naringenin-7-O-neohesperidoside: 0.39; Naringenin: 0.03; Naringenin glcA: 0.13	–	(Mata-Bilbao et al. 2007)
3 male mice	Liquiritigenin (20 mg/kg)	IV	–	–	Free: 0.08	Free: 5.9; GlcA 1: 9.2; GlcA 2: 43.1	–	Kang et al. 2009b
	Liquiritigenin (50 mg/kg)	IV	–	–	Free: 0.09	$AUC_{0.2}$; Free: 18.2; GlcA 1: 20.9; GlcA 2: 14.	–	
	Liquiritigenin (50 mg/kg)	Oral	Free: 6.52	Free: 0.1	–	Free: 2.74; AUC_{0-6}: GlcA 1: 37.8; $AUC_{0.8}$: GlcA 2: 196	–	
	Liquiritigenin (100 mg/kg)	Oral	Free: 20.86	Free: 0.1	–	Free: 4.9; AUC_{0-6}: GlcA 1: 63.1; $AUC_{0.8}$: GlcA 2: 358	–	
7 male Sprague–Dawley rats	Liquiritigenin (5 mg/kg)	IV	–	–	Free: 0.08 ± 0.02	Free: 6.5 ± 2.3; $AUC_{0.8}$: GlcA 1: 8.8 ± 1.3; GlcA 2: 6.7 ± 2.1	Free: 22.3± 5.95	Kang et al. 2009b
	Liquiritigenin (10 mg/kg)	IV	–	–	Free: 0.12 ± 0.01	Free: 11.07 ± 3.72; $AUC_{0.8}$: GlcA 1: 24.0 ± 6.8; Gluc 2: 19.4 ± 7.7	Free: 19.7 ± 4.49	
	Liquiritigenin (20 mg/kg)	IV	–	–	Free: 0.13 ± 0.02	Free: 26.8 ± 6.6; $AUC_{0.8}$: GlcA 1: 52.3 ± 14.5; GlcA 2: 49.1 ± 10.9	Free: 24.7 ± 5.76	

(Continued)

TABLE 1.4 (CONTINUED)

Pharmacokinetic Studies of Flavanones and Their Metabolites in Animal Models

Species	Source (Dose)	Administration	C_{max} (μmol/L)	T_{max} (h)	$T_{1/2}$ (h)	$AUC_{0-\infty}$ or AUC_{0-t} (μmol·h/L)	Urinary Excretion (%)	Reference
6 male Sprague-Dawley rats	Liquiritigenin (20 mg/kg)	Oral	Free: 2.8 ± 1.2	Free: 0.12	Free: 0.37 ± 0.19	Free: 0.83 ± 0.34; AUC_{0-12}: GlcA 1: 56.2 ± 12.6; GlcA 2: 56.0 ± 14.7	Free: 17.4 ± 8.06	Kang et al. 2009b
7 male Sprague-Dawley rats	Liquiritigenin (50 mg/kg)	IV	–	–	Free: 0.13 ± 0.02	Free: 65.1 ± 8.0; AUC_{0-8}: GlcA 1: 113 ± 14.6; GlcA 2: 108 ± 18.3	Free: 18.3 ± 5.80	Kang et al. 2009b
6 male Sprague-Dawley rats	Liquiritigenin (50 mg/kg)	Oral	Free: 6.9 ± 3.1	Free: 0.25	Free: 0.25 ± 0.18	Free: 2.8 ± 1.2; AUC_{0-12}: GlcA 1: 96.3 ± 17.8; GlcA 2: 111 ± 28.6	Free: 14.3 ± 8.95	Kang et al. 2009b
	Liquiritigenin (100 mg/kg)	Oral	Free: 34.9 ± 16.4	Free: 0.12	Free: 0.25 ± 0.15	Free: 6.43 ± 3.34; AUC_{0-12}: GlcA 1: 206 ± 54; GlcA 2: 197 ± 62	Free: 16.2 ± 7.92	
6 male white rabbits	Liquiritigenin (20 mg/kg)	IV	–	–	Free: 0.40 ± 0.04	Free: 35.5 ± 5.0; GlcA 1: 6.9 ± 3.3; GlcA 2: 51.5 ± 10.3	Free: 0.62 ± 0.24	Kang et al. 2009b
	Liquiritigenin (50 mg/kg)	IV	–	–	Free: 0.36 ± 0.05	Free: 98.3 ± 21.4; Gluc 1: 19.60 ± 7.2; GlcA 2: 144 ± 54	Free: 0.93 ± 1.03	

5 male Beagle dogs	Liquiritigenin (10 mg/kg)	IV	–	–	Free: 0.13 ± 0.06	Free: 15.8 ± 2.2; GlcA 1: 41.2 ± 16.0; GlcA 2: 40.3 ± 8.6	Free: 10.4 ± 3.81	Kang et al. 2009b
	Liquiritigenin (20 mg/kg)	Oral	Free: 0.10 ± 0.03	Free: 1	Free: 0.88 ± 0.29	Free: 0.17 ± 0.03; GlcA 1: 25.5 ± 8.0; GlcA 2: 26.2 ± 8.7	Free: 3.77 ± 2.40	
	Liquiritigenin (20 mg/kg)	IV	–	–	Free: 0.13 ± 0.01	Free: 32.0 ± 2.8; GlcA 1: 105 ± 30; GlcA 2: 95.7 ± 39.2	Free: 8.82 ± 3.94	
	Liquiritigenin (50 mg/kg)	Oral	Free: 0.20 ± 0.13	Free: 1.5	Free: 0.38 ± 0.21	Free: 0.18 ± 8.01; GlcA 1: 47.1 ± 10.6; GlcA 2: 78.8 ± 26.4	Free: 2.00 ± 1.04	
Rats	Liquiritigenin (20 mg/kg)	IV	–	4'-O-GlcA: 0.08; 7-O-GlcA: 0.08	Free: 0.09 ± 0.01; 4'-O-GlcA: 3.3 ± 2.6; 7-O-GlcA: 3.5 ± 1.8	Free: 19.4 ± 2.3; 4'-O-GlcA: 33.7 ± 6.1; 7-O-GlcA: 25.3 ± 9.0	Free: 15.0 ± 6.1; 4'-O-GlcA: 10.1 ± 9.6; 7-O-GlcA: 1.57 ± 2.03	Kang et al. 2009a
	Liquiritigenin (20 mg/kg)	Oral	Free: 5.2 ± 2.3; 4'-O-GlcA: 20.8 ± 13.1; 7-O-GlcA: 33.1 ± 17.1	Free: 0.12; 4'-O-GlcA: 0.25; 7-O-GlcA: 0.25	Free: 0.12 ± 0.08; 4'-O-GlcA: 6.7 ± 2.5; 7-O-GlcA: 5.4 ± 2.5	Free: 1.30 ± 0.51; 4'-O-GlcA: 59.0 ± 14.4; 7-O-GlcA: 50.5 ± 8.2	Free: 10.9 ± 7.8; 4'-O-GlcA: 10.6 ± 3.2; 7-O-GlcA: 11.3 ± 6.7	

Values are mean ± SD unless otherwise stated.

IV: intravenous; Naring: naringenin; glcA: glucuronide; sulf: sulfate; (α): half-life in the distribution phase; (β): half-life in the elimination phase; bw: body weight; TF: total flavonoids; Erio: Eriodictyol.

[a] Mean ± SE.

[b] Expressed as percentage of naringenin-7-O-rutinoside intake.

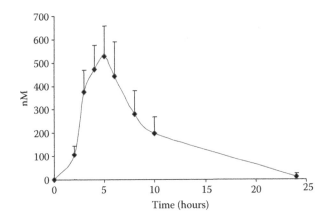

FIGURE 1.4 Combined concentrations of hesperetin-7-O-glucuronide and an unassigned hesperetin-O-glucuronide in the plasma of eight human subjects 0–24 h after ingesting 250 mL of orange juice. Data expressed as nM are presented as mean values ± standard errors depicted by vertical bars ($n = 8$). (After Mullen, W. et al., *J. Agric. Food Chem.*, **56**, 11157–11164, 2008a.)

2010). Other constituents of the extract may have enhanced the absorption of the flavanones. It is not yet known whether such levels of flavanone metabolites are achievable with normal dietary sources.

Flavanone concentrations measured in plasma after relevant dietary doses are generally much lower than those that have been used to induce biological effects in most *in vitro* studies available in the literature. This does not exclude potential protective actions *in vivo*, but future studies must aim to better simulate realistic plasma flavanone levels in order to characterize these actions and their targets.

The elimination of flavonoids generally occurs through both biliary and urinary pathways. Excretion via bile is usual for compounds with molecular weights higher than 500 Da and has been reported to be important for some polyphenol metabolites such as genistein-O-glucuronides (Ma et al. 2006). Although data for humans are not available, several animal studies support the involvement of significant biliary excretion and enterohepatic recycling for flavanones. For example, an *in situ* perfusion model of the rat small intestine with cannulation of the bile duct showed that eriodictyol was eliminated through bile in high proportions, as conjugated derivatives (Crespy et al. 2003). After oral administration of naringenin to Wistar rats, a later second peak in the concentration of total plasma metabolites was observed, indicating intestinal reabsorption of metabolites excreted in bile (Ma et al. 2006). This peak was not present in the plasma time courses of rats whose bile ducts had been ligated. In support, kinetic data obtained after the intravenous administration of naringenin to rats indicated the occurrence of enterohepatic recycling (Yanez et al. 2008).

Urinary excretion of flavanones is more widely documented. The relative urinary excretion (urinary recovery as a proportion of the ingested dose) is considered to give a minimum measure of absorption, since when compounds are extensively eliminated through the bile or degraded into nondetectable derivatives, absorption can be underestimated. Estimates of the relative urinary excretion of total naringenin in the literature are highly variable. Most studies have calculated recoveries between 7 and 23%, although one study reported a recovery of only 1.1% after consumption of orange juice and a recovery of 30.2% after consumption of grapefruit juice (Erlund et al. 2001) (Table 1.3). In general, recovery of hesperetin was lower than that of naringenin, with estimates ranging from 3.9 to 14.2% (Table 1.3). For both flavanones, the wide variability observed between studies is probably accounted for by differences in the aglycone/glycoside profile of the respective food sources.

The elimination of flavanones from plasma is rapid with mean $T_{1/2}$ times of 1.3–2.2 h being reported (Erlund et al. 2001; Mullen et al. 2008a). Furthermore, flavanone derivatives were not usually recovered in plasma 24 h after intake. As flavanone storage in tissues is unlikely (see Section 1.2.4), the presence of flavanone derivatives in systemic circulation is transitory.

Flavanones are thought to be among the most absorbable polyphenols. When hesperetin, naringenin, and quercetin were administered to humans via the same fruit juice mix, the relative urinary excretion of hesperetin and naringenin was 9- to 15-fold higher than that of quercetin (Krogholm et al. 2010). This is in keeping with data obtained in a compilation of pharmacokinetics data from 97 studies on polyphenol bioavailability which revealed that C_{max}, AUC, and relative urinary excretion values for flavanones were higher than those for most other classes of polyphenols with the exception of the isoflavones daidzein and genistein (Manach et al. 2005) (see also Chapter 5).

Most human intervention studies have indicated that naringenin is more bioavailable than hesperetin (Mullen et al. 2008a; Brett et al. 2009) (Table 1.3). The reasons for this difference are not known, although the lower bioavailability of hesperetin corresponds to other data indicating that for other classes of flavonoids such as anthocyanins and flavonols, absorption of 4′-hydroxylated compounds is greater than that of their corresponding 3′,4′-dihydroxy analogues (Bonetti et al. 2007; Mullen et al. 2008b). Naringenin has a free 4′-hydroxyl group, and hesperetin has free 3′-hydroxyl and 4′-methoxy groups, and the presence of these groups is likely to influence absorption and degradation (Mullen et al. 2008a).

1.3.1.1 Effect of Food Processing and Matrix

Some studies have examined the bioavailability of flavanones from the same food consumed after different processing conditions (e.g., cooking, squeezing). For instance, the bioavailability of flavanones from a 150 g single portion of orange fruit was compared with that from 300 g of orange juice in healthy volunteers (Table 1.3). No significant differences in the plasma pharmacokinetic parameters and urinary excretion of naringenin and hesperetin were observed on a dose-adjusted basis (Brett et al. 2009). After consumption of a polyphenol-rich juice, hesperetin metabolites were excreted in higher proportions (12%) than after administration of orange juice (~6%). This could be due to the higher content of hesperetin-7-O-rutinoside in orange juice than in the polyphenol-rich juice (Borges et al. 2010). Bugianesi et al. (2004) compared the bioavailability of naringenin after either a single dose of fresh tomatoes or a cooked tomato meal. After an intake of 17 mg naringenin from the cooked cherry tomatoes, naringenin was quantifiable (C_{max} of 0.06 ± 0.02 µM after 2 h) but was not detected after an intake of 19 mg naringenin from fresh tomatoes. Thus, the impact of heat treatment and matrix disruption upon bioaccessibility and overall bioavailability of flavanones may require further investigation.

Vallejo et al. (2010) investigated the effects of flavanone solubility on their bioavailability from orange beverages. The enrichment of orange juice with a flavanone extract led to higher concentrations in plasma, whereas enrichment with pulp decreased flavanone concentrations (Table 1.3). Moreover, flavanone excretion and plasma C_{max} correlated well with the soluble flavanone content of the juice, but not with the total flavanone content. Therefore, the solubility of the flavanones could be a key determinant of bioavailability. On a dose-adjusted basis, the C_{max}, AUC, and urinary recovery were higher with a flavanone extract diluted in water compared to an orange juice enriched with the same flavanone extract, indicating that the orange juice matrix has an effect on the absorption of flavanones. This study also reported a higher AUC/soluble ratio for hesperetin than for naringenin, suggesting that the juice matrix can cause different flavanones to be absorbed to different extents (Vallejo et al. 2010).

The bioavailability of flavanones may be influenced by interactions with other foods or dietary components. For example, the effect of full-fat yogurt on the bioavailability of flavanones from orange juice has been studied in humans (Mullen et al. 2008a). When orange juice was consumed with yogurt, a short delay in the urinary excretion of phase II metabolites was noted, although the plasma levels of hesperetin phase II metabolites were not affected (Mullen et al. 2008a). Similarly, the microbial catabolism of hesperetin-7-O-rutinoside to ring fission products was studied in humans after acute supplementation of orange juice with or without yogurt. The urinary excretion of five phenolic acids (Table 1.2) was significantly reduced by the concomitant intake of yogurt with orange juice (Roowi et al. 2009). Overall urinary excretion was found to be only 9.3 ± 4.4 µmol after the consumption of orange juice with yogurt, compared to 62 ± 18 µmol after the consumption or orange juice alone—a decrease of ~85%.

1.3.2 STUDIES IN ANIMALS

Table 1.4 details pharmacokinetic studies in animal models. C_{max} values after oral administration ranged from 5.0 to 161 µM. In rats, plasma AUC was found to increase linearly with increasing doses of naringenin (Ma et al. 2006).

Wang et al. (2006) investigated the pharmacokinetics of naringenin-7-O-neohesperidoside and naringenin in Sprague-Dawley rats after the administration of an oral or intravenous single dose of naringenin or single or repeated oral doses of naringenin-7-O-neohesperidoside (twice per day for 9 days). After the repeated administration of naringenin-7-O-neohesperidoside, total plasma metabolites gradually increased, reaching a steady state on day 5.

Most publications on pharmacokinetics have treated naringenin and hesperetin as the main flavanones. Only a few studies have investigated other flavanones, such as liquiritigenin or eriodictyol, in animals. The most complete study for liquiritigenin is that of Kang et al. (2009a) who studied its pharmacokinetics in several animal models (mice, rats, rabbits, and dogs) after oral and intravenous administration. The same group studied the pharmacokinetics of liquiritigenin in rats after oral and intravenous administration (Kang et al. 2009b). Liquiritigenin was found to be poorly bioavailable because of extensive gastrointestinal first-pass effects. Urinary excretion was lower than that of naringenin (Table 1.4) although only free liquiritigenin was measured (Kang et al. 2009a). When glucuronide conjugates were additionally included, excretion increased and was similar to that of naringenin after administration of naringenin-7-O-neohesperidoside (Felgines et al. 2000; Kang et al. 2009b).

1.4   CONCLUSIONS

Despite the importance of flavanone consumption from citrus fruits, many aspects of *in vivo* fate after ingestion are not clearly understood. The nature and extent of flavanone absorption has been the subject of numerous detailed studies, but there is a paucity of data, particularly in humans, relating to their subsequent distribution to tissues. More attention must be paid to tissue distribution because any protective effects against disease will be determined by bioactivity elicited at target sites. In addition, absorption and metabolism data often show considerable variability, not only according to the profile of flavanone derivatives ingested and the matrix of administration but also to interindividual differences. Parameters under genetic control, such as enzyme expression, may be as important as well as established factors such as the composition of gut microflora and the influence of administered food matrices. It is hoped that alongside the advance of sensitive analytical techniques and improving availability of commercial metabolite standards, fresh experimental approaches will allow these unknowns to be addressed.

ACKNOWLEDGMENTS

Mireia Urpi-Sarda was supported by the postdoctoral program Sara Borrell CD09/00134 from the Spanish Ministry of Science and Innovation.

REFERENCES

Akiyama, S., Katsumata, S., Suzuki, K. et al. (2009). Hypoglycemic and hypolipidemic effects of hesperidin and cyclodextrin-clathrated hesperetin in Goto-Kakizaki rats with type 2 diabetes. *Biosci. Biotechnol. Biochem.*, **73**, 2779–2782.

Ameer, B., Weintraub, R.A., Johnson, J.V. et al. (1996). Flavanone absorption after naringin, hesperidin, and citrus administration. *Clin. Pharmacol. Ther.*, **60**, 34–40.

Aturki, Z., Brandi, V., and Sinibaldi, M. (2004). Separation of flavanone-7-O-glycoside diastereomers and analysis in citrus juices by multidimensional liquid chromatography coupled with mass spectrometry. *J. Agric. Food Chem.*, **52**, 5303–5308.

Avila, M., Jaquet, M., Moine, D. et al. (2009). Physiological and biochemical characterization of the two α-L-rhamnosidases of *Lactobacillus plantarum* NCC245. *Microbiology.*, **155**, 2739–2749.

Beekwilder, J., Marcozzi, D., Vecchi, S. et al. (2009). Characterization of Rhamnosidases from *Lactobacillus plantarum* and *Lactobacillus acidophilus*. *Appl. Environ. Microbiol.*, **75**, 3447–3454.

Bock, C., Waldmann, K.H., and Ternes, W. (2008). Mangiferin and hesperidin metabolites are absorbed from the gastrointestinal tract of pigs after oral ingestion of a *Cyclopia genistoides* (honeybush tea) extract. *Nutr. Res.*, **28**, 879–891.

Bokkenheuser, V.D., Shackleton, C.H., and Winter, J. (1987). Hydrolysis of dietary flavonoid glycosides by strains of intestinal Bacteroides from humans. *Biochem. J.*, **248**, 953–956.

Bolli, A., Marino, M., Rimbach, G. et al. (2010). Flavonoid binding to human serum albumin. *Biochem. Biophys. Res. Commun.*, **398**, 444–449.

Bonetti, A., Marotti, I., and Dinelli, G. (2007). Urinary excretion of kaempferol from common beans (*Phaseolus vulgaris* L.) in humans. *Int. J. Food Sci. Nutr.*, **58**, 261–269.

Borges, G., Mullen, W., Mullan, A. et al. (2010). Bioavailability of multiple components following acute ingestion of a polyphenol-rich juice drink. *Mol. Nutr. Food Res.*, **54**, S268–S277.

Boumendjel, A., Blanc, M., Williamson, G. et al. (2009). Efficient synthesis of flavanone glucuronides. *J. Agric. Food Chem.*, **57**, 7264–7267.

Brand, W., van der Wel, P.A., Rein, M.J. et al. (2008). Metabolism and transport of the citrus flavonoid hesperetin in Caco-2 cell monolayers. *Drug Metab. Dispos.*, **36**, 1794–1802.

Brand, W., Boersma, M.G., Bik, H. et al. (2010a). Phase II metabolism of hesperetin by individual UDP-glucuronosyltransferases and sulfotransferases and rat and human tissue samples. *Drug Metab. Dispos.*, **38**, 617–625.

Brand, W., Shao, J., Hoek-van den Hil, E.F. et al. (2010b). Stereoselective conjugation, transport and bioactivity of *S*- and *R*-hesperetin enantiomers *in vitro*. *J. Agric. Food Chem.*, **58**, 6119–6125.

Bredsdorff, L., Nielsen, I.L., Rasmussen, S.E. et al. (2010). Absorption, conjugation and excretion of the flavanones, naringenin and hesperetin from α-rhamnosidase-treated orange juice in human subjects. *Br. J. Nutr.*, **103**, 1602–1609.

Brett, G.M., Hollands, W., Needs, P.W. et al. (2009). Absorption, metabolism and excretion of flavanones from single portions of orange fruit and juice and effects of anthropometric variables and contraceptive pill use on flavanone excretion. *Br. J. Nutr.*, **101**, 664–675.

Bugianesi, R., Catasta, G., Spigno, P. et al. (2002). Naringenin from cooked tomato paste is bioavailable in men. *J. Nutr.*, **132**, 3349–3352.

Bugianesi, R., Salucci, M., Leonardi, C. et al. (2004). Effect of domestic cooking on human bioavailability of naringenin, chlorogenic acid, lycopene and beta-carotene in cherry tomatoes. *Eur. J. Nutr.*, **43**, 360–366.

Cao, H., Chen, X., Sun, H. et al. (2010). Pharmacokinetics-based elucidation on disparity in clinical effectiveness between varieties of Zhi Zhu Wan, a traditional Chinese Medical formula. *J. Ethnopharmacol.*, **128**, 606–610.

Chabane, M., Al Ahmad, A., Peluso, J. et al. (2009). Quercetin and naringenin transport across human intestinal Caco-2 cells. *J. Pharm. Pharmacol.*, **61**, 1473–1483.

Chanet, A., Milenkovic, D., Deval, D. et al. (2011). Naringin, the major grapefruit flavonoid, specifically affects atherosclerosis development in diet-induced hypercholesterolemia in mice. *J. Nutr. Biochem.*, in press.

Chang, H.C., Churchwell, M.I., Delclos, K.B. et al. (2000). Mass spectrometric determination of genistein tissue distribution in diet-exposed Sprague-Dawley rats. *J. Nutr.*, **130**, 1963–1970.

Choe, S.C., Kim, H.S., and Jeong, T.S. (2001). Naringin has an antiatherogenic effect with the inhibition of intercellular adhesion molecule-1 in hypercholesterolemic rabbits. *J. Cardiovasc. Pharmacol.*, **38**, 947–955.

Crespy, V., Morand, C., Besson, C. et al. (2003). The splanchnic metabolism of flavonoids highly differed according to the nature of the compound. *Am. J. Physiol. Gastrointest. Liver Physiol.*, **284**, G980–G988.

Day, A.J., DuPont, M.S., Ridley, S. et al. (1998). Deglycosylation of flavonoid and isoflavonoid glycosides by human small intestine and liver beta-glucosidase activity. *FEBS Lett.*, **436**, 71–75.

El Mohsen, M.A., Marks, J., Kuhnle, G. et al. (2004). The differential tissue distribution of the citrus flavanone naringenin following gastric instillation. *Free Radic. Res.*, **38**, 1329–1340.

Erlund, I., Meririnne, E., Alfthan, G. et al. (2001). Plasma kinetics and urinary excretion of the flavanones naringenin and hesperetin in humans after ingestion of orange juice and grapefruit juice. *J. Nutr.*, **131**, 235–241.

Fang, T., Wang, Y., Ma, Y. et al. (2006). A rapid LC/MS/MS quantitation assay for naringin and its two metabolites in rats plasma. *J. Pharm. Biomed. Anal.*, **40**, 454–459.

Felgines, C., Texier, O., Morand, C. et al. (2000). Bioavailability of the flavanone naringenin and its glycosides in rats. *Am. J. Physiol. Gastrointest. Liver Physiol.*, **279**, G1148–G1154.

Fuhr, U. and Kummert, A.L. (1995). The fate of naringin in humans: A key to grapefruit juice-drug interactions? *Clin. Pharmacol. Ther.*, **58**, 365–373.

Gardana, C., Guarnieri, S., Riso, P. et al. (2007). Flavanone plasma pharmacokinetics from blood orange juice in human subjects. *Br. J. Nutr.*, **98**, 165–172.

Garrido, I., Urpi-Sarda, M., Monagas, M. et al. (2010). Targeted analysis of conjugated and microbial-derived phenolic metabolites in human urine after consumption of an almond skin phenolic extract. *J. Nutr.*, **140**, 1799–1807.

Habauzit, V., Nielsen, I.L., Gil-Izquierdo, A. et al. (2009). Increased bioavailability of hesperetin-7-glucoside compared with hesperidin results in more efficient prevention of bone loss in adult ovariectomised rats. *Br. J. Nutr.*, **102**, 976–984.

Hsiu, S.L., Huang, T.Y., Hou, Y.C. et al. (2002). Comparison of metabolic pharmacokinetics of naringin and naringenin in rabbits. *Life Sci.*, **70**, 1481–1489.

Griffiths, L.A. and Barrow, A. (1972). Metabolism of flavonoid compounds in germ-free rats. *Biochem. J.*, **130**, 1161–1162.

Ishii, K., Furuta, T., and Kasuya, Y. (2000). Mass spectrometric identification and high-performance liquid chromatographic determination of a flavonoid glycoside naringin in human urine. *J. Agric. Food Chem.*, **48**, 56–59.

Jin, M.J., Kim, U., Kim, I.S. et al. (2011). Effects of gut microflora on pharmacokinetics of hesperidin: A study on non-antibiotic and pseudo-germ-free rats. *J. Toxicol. Environ. Health A.*, **73**, 1441–1450.

Joshipura, K.J., Ascherio, A., Manson, J.E. et al. (1999). Fruit and vegetable intake in relation to risk of ischemic stroke. *JAMA*, **282**, 1233–1239.

Kang, H.E., Jung, H.Y., Cho, Y.K. et al. (2009a). Pharmacokinetics of liquiritigenin in mice, rats, rabbits, and dogs, and animal scale-up. *J. Pharm. Sci.*, **98**, 4327–4342.

Kang, H.E., Cho, Y.K., Jung, H.Y. et al. (2009b). Pharmacokinetics and first-pass effects of liquiritigenin in rats: Low bioavailability is primarily due to extensive gastrointestinal first-pass effect. *Xenobiotica*, **39**, 465–475.

Khan, M.K., Rakotomanomana, N., Loonis, M. et al. (2010). Chemical synthesis of citrus flavanone glucuronides. *J. Agric. Food Chem.*, **58**, 8437–8443.

Kim, M., Kometani, T., Okada, S. et al. (1999). Permeation of hesperidin glycosides across Caco-2 cell monolayers via the paracellular pathway. *Biosci. Biotechnol. Biochem.*, **63**, 2183–2188.

Kobayashi, S. and Konishi, Y. (2008). Transepithelial transport of flavanone in intestinal Caco-2 cell monolayers. *Biochem. Biophys. Res. Commun.*, **368**, 23–29.

Kobayashi, S., Tanabe, S., Sugiyama, M. et al. (2008). Transepithelial transport of hesperetin and hesperidin in intestinal Caco-2 cell monolayers. *Biochim. Biophys. Acta*, **1778**, 33–41.

Kragh-Hansen, U., Chuang, V.T., and Otagiri, M. (2002). Practical aspects of the ligand-binding and enzymatic properties of human serum albumin. *Biol. Pharm. Bull.*, **25**, 695–704.

Krogholm, K.S., Bredsdorff, L., Knuthsen, P. et al. (2010). Relative bioavailability of the flavonoids quercetin, hesperetin and naringenin given simultaneously through diet. *Eur. J. Clin. Nutr.*, **64**, 432–435.

Kumar, A., Dogra, S., and Prakash, A. (2010). Protective effect of naringin, a citrus flavonoid, against colchicine-induced cognitive dysfunction and oxidative damage in rats. *J. Med. Food*, **13**, 976–984.

Labib, S., Erb, A., Kraus, M. et al. (2004). The pig caecum model: A suitable tool to study the intestinal metabolism of flavonoids. *Mol. Nutr. Food Res.*, **48**, 326–332.

Lee, C.H., Jeong, T.S., Choi, Y.K. et al. (2001). Anti-atherogenic effect of citrus flavonoids, naringin and naringenin, associated with hepatic ACAT and aortic VCAM-1 and MCP-1 in high cholesterol-fed rabbits. *Biochem. Biophys. Res. Commun.*, **284**, 681–688.

Liu, Z. and Hu, M. (2007). Natural polyphenol disposition via coupled metabolic pathways. *Expert. Opin. Drug Metab. Toxicol.*, **3**, 389–406.

Londoño-Londoño, J., Lima, V.R., Jaramillo, C. et al. (2010). Hesperidin and hesperetin membrane interaction: Understanding the role of 7-*O*-glycoside moiety in flavonoids. *Arch. Biochem. Biophys.*, **499**, 6–16.

Ma, Y., Li, P., Chen, D. et al. (2006). LC/MS/MS quantitation assay for pharmacokinetics of naringenin and double peaks phenomenon in rats plasma. *Int. J. Pharm.*, **307**, 292–299.

Mackenzie, P.I., Bock, K.W., Burchell, B. et al. (2005). Nomenclature update for the mammalian UDP glycosyltransferase (UGT) gene superfamily. *Pharmacogenet. Genomics*, **15**, 677–685.

Manach, C., Morand, C., Gil-Izquierdo, A. et al. (2003). Bioavailability in humans of the flavanones hesperidin and narirutin after the ingestion of two doses of orange juice. *Eur. J. Clin. Nutr.*, **57**, 235–242.

Manach, C., Williamson, G., Morand, C. et al. (2005). Bioavailability and bioefficacy of polyphenols in humans. I. Review of 97 bioavailability studies. *Am. J. Clin. Nutr.*, **81**, 230S–242S.

Mata-Bilbao, M., Andres-Lacueva, C., Roura, E. et al. (2007). Absorption and pharmacokinetics of grapefruit flavanones in beagles. *Br. J. Nutr.*, **98**, 86–92.

Matsumoto, H., Ikoma, Y., Sugiura, M. et al. (2004). Identification and quantification of the conjugated metabolites derived from orally administered hesperidin in rat plasma. *J. Agric. Food Chem.*, **52**, 6653–6659.

Miyake, Y., Shimoi, K., Kumazawa, S. et al. (2000). Identification and antioxidant activity of flavonoid metabolites in plasma and urine of eriocitrin-treated rats. *J. Agric. Food Chem.*, **48**, 3217–3224.

Miyake, Y., Sakurai, C., Usuda, M. et al. (2006). Difference in plasma metabolite concentration after ingestion of lemon flavonoids and their aglycones in humans. *J. Nutr. Sci. Vitaminol. (Tokyo)*. **52**, 54–60.

Morand, C., Dubray, C., Milenkovic, D. et al. (2011). Hesperidin contributes to the vascular protective effects of orange juice: A randomized crossover study in healthy volunteers. *Am. J. Clin. Nutr.*, **93**, 73–80.

Mullen, W., Archeveque, M.A., Edwards, C.A. et al. (2008a). Bioavailability and metabolism of orange juice flavanones in humans: Impact of a full-fat yogurt. *J. Agric. Food Chem.*, **56**, 11157–11164.

Mullen, W., Edwards, C.A., Serafini, M. et al. (2008b). Bioavailability of pelargonidin-3-*O*-glucoside and its metabolites in humans following the ingestion of strawberries with and without cream. *J. Agric. Food Chem.*, **56**, 713–719.

Mulvihill, E.E., Allister, E.M., Sutherland, B.G. et al. (2009). Naringenin prevents dyslipidemia, apolipoprotein B overproduction, and hyperinsulinemia in LDL receptor-null mice with diet-induced insulin resistance. *Diabetes*, **58**, 2198–2210.

Mulvihill, E.E., Assini, J.M., Sutherland, B.G. et al. (2010). Naringenin decreases progression of atherosclerosis by improving dyslipidemia in high-fat-fed low-density lipoprotein receptor-null mice. *Arterioscler. Thromb. Vasc. Biol.*, **30**, 742–748.

Neveu, V., Perez-Jimenez, J., Vos, F. et al. (2010). Phenol-Explorer: An online comprehensive database on polyphenol contents in foods. *Database (Oxford)*, bap024.

Nielsen, I.L.F., Chee, W.S.S., Poulsen, L. et al. (2006). Bioavailability is improved by enzymatic modification of the citrus flavonoid hesperidin in humans: A randomized, double-blind, crossover trial. *J. Nutr.*, **136**, 404–408.

Nielsen, S.E., Breinholt, V., Justesen, U. et al. (1998). *In vitro* biotransformation of flavonoids by rat liver microsomes. *Xenobiotica*, **28**, 389–401.

Nikolic, D. and van Breemen, R.B. (2004). New metabolic pathways for flavanones catalyzed by rat liver microsomes. *Drug Metab. Dispos.*, **32**, 387–397.

Peng, H.W., Cheng, F.C., Huang, Y.T. et al. (1998). Determination of naringenin and its glucuronide conjugate in rat plasma and brain tissue by high-performance liquid chromatography. *J. Chromatogr. B Biomed. Sci. Appl.*, **714**, 369–374.

Possemiers, S., Bolca, S., Verstraete, W. et al. (2011). The intestinal microbiome: A separate organ inside the body with the metabolic potential to influence the bioactivity of botanicals. *Fitoterapia*, **82**, 53–66.

Rechner, A.R., Smith, M.A., Kuhnle, G. et al. (2004). Colonic metabolism of dietary polyphenols: Influence of structure on microbial fermentation products. *Free Radic Biol Med.*, **36**, 212–225.

Riches, Z., Stanley, E.L., Bloomer, J.C. et al. (2009). Quantitative evaluation of the expression and activity of five major sulfotransferases (SULTs) in human tissues: The SULT "pie." *Drug Metab. Dispos.*, **37**, 2255–2261.

Roowi, S., Mullen, W., Edwards, C.A. et al. (2009). Yoghurt impacts on the excretion of phenolic acids derived from colonic breakdown of orange juice flavanones in humans. *Mol. Nutr. Food Res.*, **53**, S68–S75.

Serra, H., Mendes, T., Bronze, M.R. et al. (2008). Prediction of intestinal absorption and metabolism of pharmacologically active flavones and flavanones. *Bioorg. Med. Chem.*, **16**, 4009–4018.

Sesink, A.L., Arts, I.C., de Boer, V.C. et al. (2005). Breast cancer resistance protein (Bcrp1/ Abcg2) limits net intestinal uptake of quercetin in rats by facilitating apical efflux of glucuronides. *Mol. Pharmacol.*, **67**, 1999–2006.

Si-Ahmed, K., Tazerouti, F., Badjah-Hadj-Ahmed, A.Y. et al. (2010). Analysis of hesperetin enantiomers in human urine after ingestion of blood orange juice by using nano-liquid chromatography. *J. Pharm. Biomed. Anal.*, **51**, 225–229.

Silberberg, M., Gil-Izquierdo, A., Combaret, L. et al. (2006a). Flavanone metabolism in healthy and tumor-bearing rats. *Biomed. Pharmacother.*, **60**, 529–535.

Silberberg, M., Morand, C., Mathevon, T. et al. (2006b). The bioavailability of polyphenols is highly governed by the capacity of the intestine and of the liver to secrete conjugated metabolites. *Eur. J. Nutr.*, **45**, 88–96.

Spencer, J.P. (2010). The impact of fruit flavonoids on memory and cognition. *Br. J. Nutr.*, **104**, S40–S47.

Spencer, J.P., El-Mohsen, M.A., and Rice-Evans, C. (2004). Cellular uptake and metabolism of flavonoids and their metabolites: Implications for their bioactivity. *Arch. Biochem. Biophys.*, **423**, 148–161.

Stalmach, A., Mullen, W., Pecorari, M. et al. (2009). Bioavailability of C-linked dihydrochalcone and flavanone glucosides in humans following ingestion of unfermented and fermented rooibos teas. *J. Agric. Food Chem.*, **57**, 7104–7111.

Tomás-Barberán, F. and Clifford, M. (2000). Flavanones, chalcones and dihydrochalcones—nature, occurrence and dietary burden. *J. Sci. Food Agric.*, **80**, 1073–1080.

Tourniaire, F., Hassan, M., Andre, M. et al. (2005). Molecular mechanisms of the naringin low uptake by intestinal Caco-2 cells. *Mol. Nutr. Food Res.*, **49**, 957–962.

Urpi-Sarda, M., Garrido, I., Monagas, M. et al. (2009). Profile of plasma and urine metabolites after the intake of almond [*Prunus dulcis* (Mill.) D.A. Webb] polyphenols in humans. *J. Agric. Food Chem.*, **57**, 10134–10142.

Tourniaire, F., Hassan, M., Andre, M. et al. (2005). Molecular mechanisms of the naringin low uptake by intestinal Caco-2 cells. *Mol. Nutr. Food Res.*, **49**, 957–962.

Vallejo, F., Larrosa, M., Escudero, E. et al. (2010). Concentration and solubility of flavanones in orange beverages affect their bioavailability in humans. *J. Agric. Food Chem.*, **58**, 6516–6524.

Vauzour, D., Vafeiadou, K., Rice-Evans, C. et al. (2007). Activation of pro-survival Akt and ERK1/2 signalling pathways underlie the anti-apoptotic effects of flavanones in cortical neurons. *J Neurochem.*, **103**, 1355–1367.

Wang, M.J., Chao, P.D.L., Hou, Y.C. et al. (2006). Pharmacokinetics and conjugation metabolism of naringin and naringenin in rats after single dose and multiple dose administrations. *J. Food Drug Anal.*, **14**, 247–253.

Yadav, V., Yadav, P.K., Yadav, S. et al. (2010). α-L-Rhamnosidase: A review. *Process Biochem.*, **45**, 1226–1235.

Yañez, J.A. and Davies, N.M. (2005). Stereospecific high-performance liquid chromatographic analysis of naringenin in urine. *J. Pharm. Biomed. Anal.*, **39**, 164–169.

Yañez, J.A., Teng, X.W., Roupe, K.A. et al. (2005). Stereospecific high-performance liquid chromatographic analysis of hesperetin in biological matrices. *J. Pharm. Biomed. Anal.*, **37**, 591–595.

Yañez, J.A., Remsberg, C.M., Miranda, N.D. et al. (2008). Pharmacokinetics of selected chiral flavonoids: Hesperetin, naringenin and eriodictyol in rats and their content in fruit juices. *Biopharm. Drug Dispos.*, **29**, 63–82.

Youdim, K.A., Dobbie, M.S., Kuhnle, G. et al. (2003). Interaction between flavonoids and the blood-brain barrier: *In vitro* studies. *J. Neurochem.*, **85**, 180–192.

Youdim, K.A., Qaiser, M.Z., Begley, D.J. et al. (2004). Flavonoid permeability across an *in situ* model of the blood-brain barrier. *Free Radic Biol Med.*, **36**, 592–604.

Zamora-Ros, R., Andres-Lacueva, C., Lamuela-Raventos, R.M. et al. (2010). Estimation of dietary sources and flavonoid intake in a Spanish adult population (EPIC-Spain). *J. Am. Diet. Assoc.*, **110**, 390–398.

Zbarsky, V., Datla, K.P., Parkar, S. et al. (2005). Neuroprotective properties of the natural phenolic antioxidants curcumin and naringenin but not quercetin and fisetin in a 6-OHDA model of Parkinson's disease. *Free Radic Res.*, **39**, 1119–1125.

Zhang, J. and Brodbelt, J.S. (2004). Screening flavonoid metabolites of naringin and narirutin in urine after human consumption of grapefruit juice by LC-MS and LC-MS/MS. *Analyst*, **129**, 1227–1233.

2 Bioavailability of Dietary Monomeric and Polymeric Flavan-3-ols

Alan Crozier, Michael N. Clifford,
and Daniele Del Rio

CONTENTS

2.1 INTRODUCTION

Flavan-3-ols are the most complex subclass of flavonoids, ranging from the simple monomers to the oligomeric and polymeric proanthocyanidins, which are also known as condensed tannins. The two chiral centers at C2 and C3 of the monomeric flavan-3-ol produce four isomers for each level of B-ring hydroxylation, two of which, (+)-catechin and (−)-epicatechin, are widespread in nature, whereas others such as (−)-epiafzelechin have a more limited distribution (Figure 2.1) (Crozier et al. 2006; Aron and Kennedy 2008). Pairs of enantiomers are not resolved on the commonly used reversed phase HPLC columns and, as a consequence, are easily overlooked. It is interesting to note that humans fed (−)-epicatechin excrete some (+)-epicatechin, indicating ring opening and racemization, possibly in the gastrointestinal tract (Yang et al. 2000). Transformation can also occur during food processing (Seto et al. 1997).

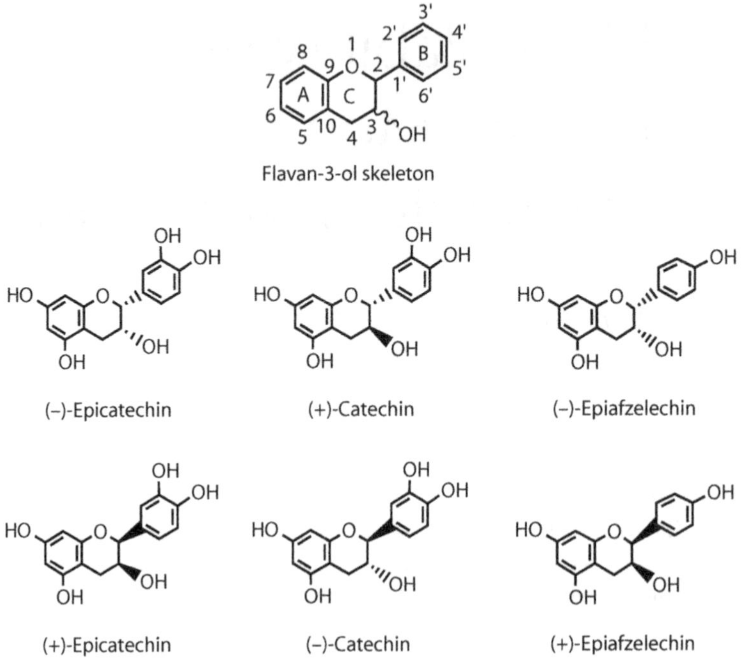

Flavan-3-ol skeleton

(−)-Epicatechin   (+)-Catechin   (−)-Epiafzelechin

(+)-Epicatechin   (−)-Catechin   (+)-Epiafzelechin

FIGURE 2.1   Flavan-3-ol monomer stereoisomers.

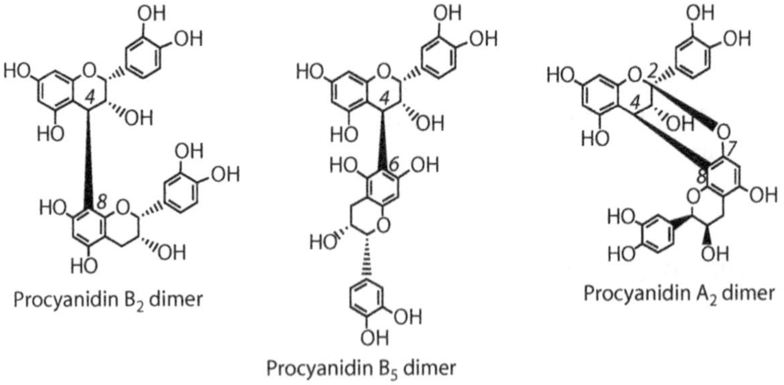

Procyanidin B₂ dimer   Procyanidin A₂ dimer

Procyanidin B₅ dimer

FIGURE 2.2   Proanthocyanidin type A and type B dimers.

Oligomeric and polymeric proanthocyanidins have an additional chiral center at C4 in the "upper" and "lower" units. Type B proanthocyanidins are formed by oxidative coupling between the C-4 of the upper monomer and the C-6 or C-8 of the adjacent lower or "extension" unit to create oligomers or polymers. Type A proanthocyanidins have an additional ether bond between C-2 in the C-ring of one monomer and C-7 in the A-ring of the other monomer (Figure 2.2). Proanthocyanidins can occur as polymers of up to 50 units.

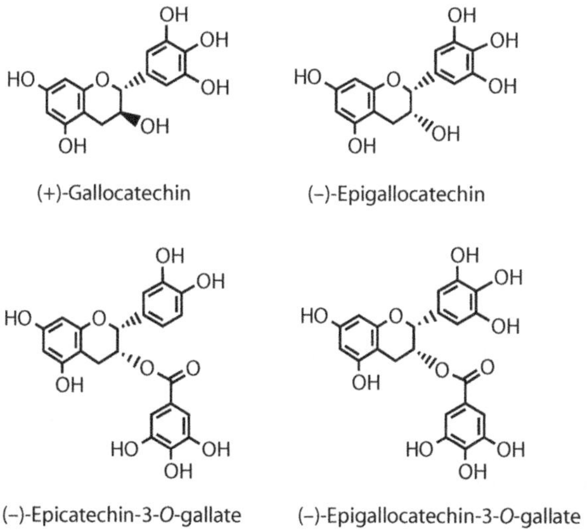

(+)-Gallocatechin

(−)-Epigallocatechin

(−)-Epicatechin-3-O-gallate

(−)-Epigallocatechin-3-O-gallate

FIGURE 2.3 Galloyl- and 3-O-gallated flavan-3-ol derivatives which occur principally in green tea.

Monomeric flavan-3-ols also occur as C-3 esters, commonly with gallic acid but sometimes with other aromatic acids, and occasionally as digallates. Oligomeric and polymeric proanthocyanidins may also contain one or more gallate esters (Figure 2.3). Proanthocyanidins that consist exclusively of (epi)catechin units are called procyanidins and are the most abundant type of proanthocyanidins in plants. The less common proanthocyanidins containing (epi)afzelechin or (epi)gallocatechin subunits are called propelargonidins and prodelphinidins, respectively. Mixed oligomers and polymers are also known. The terms propelargonidins, procyanidins and prodelphinidins reflect the tendency of proanthocyanidins to undergo acid catalyzed depolymerization, often incorrectly described as a hydrolysis, yielding the relevant anthocyanidin from the upper and extension unit(s) but not from the lower terminal unit. Although rarely addressed, strictly, this causes some nomenclature problems with mixed oligomers and polymers. For example, a dimer with an upper (epi)afzelchin unit and a lower (epi)catechin unit is a propelargonidin. A dimer with an upper (epi)catechin and a lower (epi)afzelchin unit is a procyanidin, etc. Similarly, with oligomers containing at least two different upper and extension monomers, more than one anthocyanidin will be produced. For example, a trimer with an upper (epi)afzelchin unit, a middle (epi)catechin unit, and a lower (epi)gallocatechin unit would yield pelargonidin and cyanidin in approximately equal quantities. Occasionally, the terminal (bottom) monomer is a flavan-3,4-diol unit.

Red wines contain oligomeric procyanidins and prodelphinidins originating mainly from the seeds of grapes (Auger et al. 2004), while procyanidins, derived from the roasted seeds of cocoa (*Theobroma cacao*), occur in dark chocolate (Hammerstone et al. 1999) along with (−)-epicatechin and (−)-catechin, the latter being formed from (−)-epicatechin during processing (Gu et al. 2004). Monomeric and polymeric flavan-3-ols are found widely in fruits and vegetables (Crozier et al. 2006, 2009). Nut skins are a rich source of proanthocyanidins with varying degrees of polymerization (Bartolomè et al. 2010), and hazelnut skins are probably among the richest edible source of this class of compounds (Del Rio et al. 2011).

Green tea (*Camellia sinensis*) contains high levels of flavan-3-ol monomers with the main components being (–)-epigallocatechin, (–)-epigallocatechin-3-*O*-gallate, and (–)-epicatechin-3-*O*-gallate (Figure 2.4). The levels of these flavan-3-ols decline during fermentation of the tea leaves, principally as a result of the action of polyphenol oxidase, and there is a concomitant accumulation of theaflavins and thearubigins (Del Rio et al. 2004). For a review of the low molecular mass polyphenols in black tea, see Drynan et al. (2010).

Theaflavin, theaflavin-3-*O*-gallate, theaflavin-3′-*O*-gallate, and theaflavin-3,3′-*O*-digallate are dimer-like structures (Figure 2.4) that contribute to the quality of the black tea beverage. The brownish, water-soluble, high molecular weight thearubigins are the major phenolic fraction in black tea. A significant breakthrough in the characterization of thearubigins was reported in 2010 as the result of collaborative

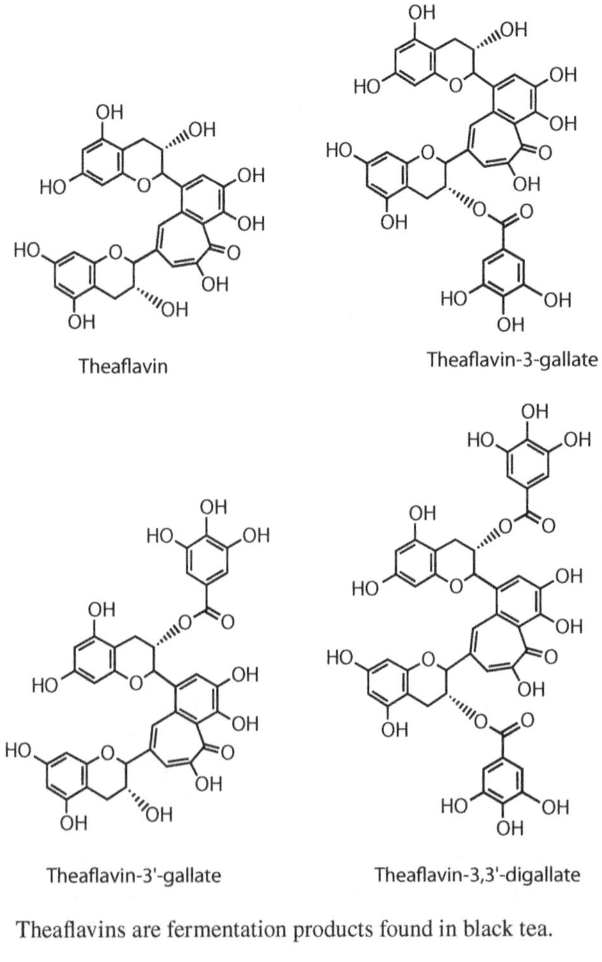

Theaflavin

Theaflavin-3-gallate

Theaflavin-3′-gallate

Theaflavin-3,3′-digallate

FIGURE 2.4 Theaflavins are fermentation products found in black tea.

studies involving researchers at the University of Surrey in the UK and Jacobs University Bremen in Germany that made use of modern mass spectrometric procedures, in particular ion trap–MS$^{(n)}$ and Fourier transform ion cyclotron resonance mass spectrometry (FTICR-MS) (Kuhnert et al. 2010a, 2010b). These studies of the thearubigins from 15 commercial black teas, selected to represent the major types of black teas produced world-wide, detected on average 5000 thearubigin components in the mass range between m/z 1000 and 2100. FTICR-MS data revealed the presence of a maximum of 9428 peaks in the mass range of 300 to 1000 m/z. This vast number of products easily explains why the thearubigins do not resolve during chromatography, appearing as a streak during paper or thin-layer chromatography and as a Gaussian-like hump when analysed by reverse phase HPLC. A typical cup of black tea contains approximately 100 mg of thearubigins (Gosnay et al. 2002; Woods et al. 2003), suggesting that very few if any individual thearubigins will exceed 100 µg per cup.

The accurate high-resolution mass data obtained by FTICR-MS allowed molecular formulas to be assigned to 1517 thearubigins; this facilitated the development of the "oxidative cascade" hypothesis to explain the immense number of compounds formed primarily by the action of a single enzyme, polyphenol oxidase, and a handful of flavan-3-ol substrates (Kuhnert et al. 2010a). The hypothesis is summarized elsewhere by Clifford and Crozier (2011).

2.2   BIOAVAILABILITY—ABSORPTION, DISTRIBUTION, METABOLISM, AND EXCRETION

Dietary flavan-3-ols, unlike most other flavonoids, exist *in planta* predominantly as aglycones rather than glycoside conjugates. Following their ingestion, absorption of some but not all components into the circulatory system occurs in the small intestine, although limited absorption may also occur in the stomach (Crozier et al. 2009). Movement of flavan-3-ols monomers into the wall of the proximal gastrointestinal tract appears to be by diffusion. Prior to passage into the bloodstream, they undergo metabolism forming sulfate, glucuronide, and/or methylated metabolites through the respective action of sulfotransferases (SULT), uridine-5′-diphosphate glucuronosyltransferases (UGT) and catechol-O-methyltransferases (COMT). There is also efflux of at least some of the metabolites back into the lumen of the small intestine; this is thought to involve members of the adenosine triphosphate (ATP)-binding cassette (ABC) family of transporters, including multidrug resistance protein (MRP) and P-glycoprotein (P-gp). Once in the bloodstream, metabolites can be subjected to further phase II metabolism with additional conversions occurring in the liver, where enterohepatic transport via the bile duct may result in some recycling to the small intestine (Donovan et al. 2006b). Flavan-3-ols and their metabolites not absorbed in the small intestine can be absorbed in the large intestine where they will also be subjected to the action of the colonic microflora, which cleave conjugating moieties, allowing the released aglycones to undergo ring fission, thus leading to the production of phenolic acids and hydroxycinnamates. These compounds can be absorbed and may be subjected to some degree of phase II metabolism in the colonic epithelium and/or the liver before being excreted in the urine in substantial quantities that, in most instances, are well in excess of the flavan-3-ol metabolites that enter the circulatory system via the small intestine (Roowi et al. 2010; Williamson and Clifford 2010).

2.2.1 ANALYSIS OF FLAVAN-3-OL METABOLITES

A detailed review on the bioavailability of polyphenols in humans was published by Manach et al. (2005). Much of the research covered involved feeding volunteers a single supplement and monitoring the levels of flavonoids in plasma and urine over a 24-h period. As flavonoid metabolites were, and indeed still are, rarely available commercially, these analyses almost invariably involved treatment of samples with mollusc glucuronidase/sulfatase preparations and subsequent quantification of the released aglycones by HPLC using absorbance, fluorescence, or electrochemical detection. Some recent bioavailability studies have analyzed samples directly by HPLC with tandem mass spectrometric (MS^2) detection without recourse to enzyme hydrolysis. The availability of reference compounds enables specific metabolites to be identified by HPLC-MS (Day et al. 2001). In the absence of standards, it is not always possible to distinguish between isomers and ascertain the position of conjugating groups on the flavonoid skeleton. A further complication associated with the analysis of flavan-3-ol metabolites is that without reference compounds and chiral chromatography (Donovan et al. 2006a), reverse phase HPLC is unable to distinguish between the four possible enantiomers of any (epi)catechin or any (epi)gallocatechin derivative. Thus, what in reality is (−)-epicatechin-3'-*O*-glucuronide (Figure 2.5) can be only partially identified, even with MS^3 detection, as an (epi)catechin-*O*-glucuronide. Nonetheless, the use of MS in this way represents a powerful HPLC detection system, as with low nanogram quantities of sample it provides structural information on analytes of interest that is not obtained with other detectors.

Quantification of partially identified metabolites by MS using consecutive reaction monitoring (CRM) or selected ion monitoring (SIM) is, of necessity, based on a calibration curve of a related compound, which in the instance cited above could be (−)-epicatechin as it is readily available from commercial sources. In such circumstances, as the slopes of the aglycone and glucuronide SIM dose–response curves are unlikely to be identical, this approach introduces a potential source of error in the quantitative estimates, and there is a view that quantitative estimates

(−)-Epicatechin-3'-O-glucuronide   (−)-Epicatechin-3'-O-sulfate

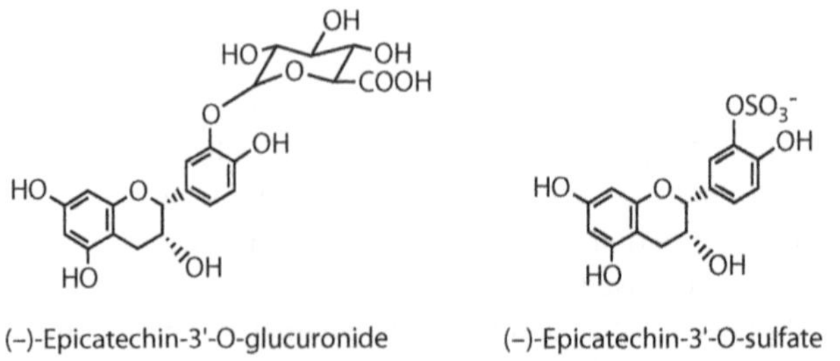

FIGURE 2.5 (−)-Epicatechin-3'-*O*-glucuronide and (−)-epicatechin-3'-*O*-sulfate.

based on enzyme hydrolysis are, therefore, much more accurate. However, enzyme hydrolysis makes use of glucuronidase/sulfatase preparations that contain a mixture of enzyme activities, and there can be substantial batch-to-batch variation in their specificity (Donovan et al. 2006b). As yet there are no reports of flavonoid bioavailability studies using glucuronidase/sulfatase preparations where information on the identity, number, and quantity of the individual sulfate and glucuronide conjugates in the samples of interest has been obtained. As a consequence, there are no direct data on the efficiency with which the enzymes hydrolyse the individual metabolites and release the aglycone and methylated derivatives. This introduces a varying, unmeasured error factor. The accuracy of quantitative estimates based on the use of glucuronidase/sulfatase preparations are, therefore, probably no better, and possibly much worse, than those based on HPLC-CRM/SIM. The fact that enzyme hydrolysis results in very reproducible data is not relevant as reproducibility is a measure of precision, not accuracy (Reeve and Crozier 1980). These shortcomings of analyses based on enzyme hydrolysis apply to bioavailability studies with all dietary flavonoids, and it is interesting to note that the one publication on the subject to date reports that the use of enzyme hydrolysis results in an underestimation of isoflavone metabolites (Gu et al. 2005).

2.2.2 BIOAVAILABILITY OF FLAVAN-3-OL ENANTIOMERS

Donovan et al. (2006) reported that (−)-catechin is absorbed less readily than (+)-catechin. In a more recent study, Ottaviani et al. (2011) investigated the bioavailability of different enantiomeric forms of flavan-3-ol monomers in an acute study in which adult human males consumed equal quantities of (−)-epicatechin, (−)-catechin, (+)-epicatechin, and (+)-catechin (see Figure 2.1) in a cocoa drink. Based on plasma concentrations and urinary excretion, the bioavailability of the stereoisomers was ranked as (−)-epicatechin > (+)-epicatechin = (+)-catechin > (−)-catechin. There were also differences in the metabolic fate of the catechin and epicatechin epimers as reflected in the ratios of their 3′- and 4′-O-methylated metabolites. In addition, the levels of nonmethylated metabolites of (−)- and (+)-epicatechin in plasma and urine differed, indicating that flavan-3-ol stereochemistry also affects metabolic pathways other than O-methylation. As samples were analyzed after glucuronidase/sulfatase treatment, it was not possible to determine to what degree this impacted on the production of glucuronide and sulfate metabolites, and this remains a topic for further investigation. As the individual flavan-3-ol stereoisomers in cocoa products used in feeding studies are usually not fully determined, this finding raises the possibility that varying stereochemical ratios could be a contributing factor in the different (epi) catechin metabolite profiles reported in the literature, with the stereochemical variation reflecting variation in cocoa processing.

2.2.3 COCOA FLAVAN-3-OL MONOMERS

Because it is now recognized that the stereochemistry of the flavan-3-ols consumed in cocoa, and especially the stereochemistry of the flavan-3-ol metabolites observed in plasma and urine, has rarely been characterized to enantiomer level, this uncertainty will be made clear in the discussion that follows by using, for example, (epi)catechin rather than the apparently more precise term(s) used in the original publications.

Early human studies on the post-ingestion fate of cocoa flavan-3-ols treated plasma and urine samples with β-glucuronidase/sulfatase prior to the analysis of the released (epi)catechin monomers by reverse phase HPLC, typically with fluorescence (Ho et al. 1995; Richelle et al. 1999) or electrochemical detection (Rein et al. 2000; Wang et al. 2000). With this methodology, Richelle et al. (1999) showed that following the consumption of 40 g of dark chocolate containing 282 µmol of (−)-epicatechin, the (epi)catechin levels rose rapidly and reached a peak plasma concentration (C_{max}) of 355 nM after 2.0 h (T_{max}), indicative of absorption in the upper gastrointestinal tract rather than the large intestine. With double the chocolate intake, the C_{max} increased to 676 nM, whereas the T_{max} was extended to 2.6 h, which was attributed to the *ad libitum* consumption of bread by the volunteers rather than the increased intake of chocolate. Wang et al. (2000) also carried out a dose study in which varying amounts of chocolate were served with 40 g of bread. The data, which are summarized in Table 2.1, show a positive relationship between intake and (epi)catechin plasma concentration.

In a further study, Baba et al. (2000) fed a chocolate containing 760 µmol of (−)-epicatechin and 214 µmol of catechin, most probably the (−)-isomer, to human subjects and collected plasma and urine over the ensuing 24 h period. By selective incubation of samples with glucuronidase and sulfatase, it was possible to distinguish among glucuronide, sulfate, and sulfoglucuronide metabolites of the released (epi)catechin and methyl-(epi)catechin aglycones. The main metabolites in plasma were sulfates and sulfoglucuronide of (epi)catechin and methyl-(epi)catechin with lower levels (epi)catechin-glucuronide and (epi)catechin. The combined C_{max} of the metabolites was 4.8 ± 0.9 µM and the T_{max} was 2 h. The same metabolites were present in urine with most being excreted in the initial 8-h period after consumption of

TABLE 2.1

Concentration of (Epi)Catechin Metabolites in Plasma of Volunteers After the Ingestion of Chocolate Containing 159, 312, and 417 µmol of (−)-Epicatechin[a]

(−)-Epicatechin Intake (µmol)	0 h	2 h	6 h
0	1 ± 1[b]	19 ± 14[b]	1 ± 1[b]
159	2 ± 2[b]	133 ± 27[b]	26 ± 8[c]
312	4 ± 2[b]	258 ± 29[b]	66 ± 8[c]
471	4 ± 3[b]	355 ± 49[b]	103 ± 16[c]

Source: After Wang, J.F. et al., *J. Nutr.*, **130**, 2115S–2119S, 2000.

Mean values with a different superscript are significantly different ($P < 0.05$).

*Data expressed as mean values in nM ± standard error ($n = 9$–13).

TABLE 2.2

Total (Epi)Catechin Metabolites Excreted in Urine by Volunteers 0–24 h After the Consumption of Chocolate Containing 760 µmol of (–)-Epicatechin[a]

Excretion Period	Excretion (µmol)	Excretion (% of Intake)
0–8 h	188 ± 33	24.7 ± 4.3
8–24 h	39 ± 19	5.1 ± 2.5
Total (0–24 h)	**227 ± 39**	**29.9 ± 5.2**

Source: After Baba, S. et al., *Free Radic. Res.*, **33**, 635–641, 2000.

[a] Data expressed as mean values ± standard error ($n = 5$).

the chocolate. The total 0–24 h urinary excretion of the (epi)catechin and methyl-(epi)catechin metabolites was 227 ± 39 µmol, which corresponds to 29.9% of the ingested (–)-epicatechin (Table 2.2). Baba et al. (2000) obtained broadly similar data when the same dose of flavan-3-ols was consumed as a cocoa drink rather than as chocolate.

More recent cocoa flavan-3-ol bioavailability studies have analyzed plasma and urine samples using HPLC-MS[2] methodology. In one such study, volunteers drank a beverage made with 10 g commercial cocoa powder and 250 ml of hot water (Mullen et al. 2009). The drink contained 22.3 µmol of catechin, almost all of it as the less bioavailable (–)-isomer, and 23.0 µmol of (–)-epicatechin along with 70 mg of procyanidins. Two flavan-3-ol metabolites were detected in plasma, an O-methyl-(epi)catechin-O-sulfate and an (epi)catechin-O-sulfate, possibly (-)-epicatechin-3'-O-sulfate (Figure 2.5). Both had a C_{max} below 100 nM and a T_{max} of < 1.5 h, and after 8 h, only trace levels remained in the circulatory system (Figure 2.6). The two sulfated flavan-3-ols were also the main metabolites in urine which, in addition, contained smaller quantities of an (epi)catechin-O-glucuronide, probably (–)-epicatechin-3'-O-glucuronide (Figure 2.5), and an additional (epi)catechin-O-sulfate. The amount of flavan-3-ol metabolites excreted in urine over the 0–24 h collection period was 7.32 ± 0.82 µmol, which is equivalent to 16.3 ± 1.8% of intake. Considering that almost half of the flavan-3-ol monomer content of the cocoa was (–)-catechin, which has reduced bioavailability, the real figure for (–)-epicatechin absorption was probably nearer 32% and as such is comparable with urinary (epi)catechin excretion levels obtained by Baba et al. (2000) and presented in Table 2.2.

It should be noted that rather different (epi)catechin metabolite profiles have been obtained by different research groups. In the Mullen et al. (2009) investigation, an (epi)catechin-O-sulfate and an O-methyl-(epi)catechin-O-sulfate were detected in plasma. These sulfates were also the main metabolites in urine, which also contained an additional (epi)catechin-O-sulfate and an (epi)catechin-O-glucuronide (Table 2.3). Stalmach et al. (2009, 2010) detected a similar spectrum of (epi)catechin metabolites in plasma and urine collected after ingestion of green tea (see Section 2.2.4). These findings are also in keeping with the data of Baba et al. (2000),

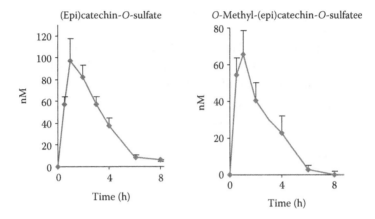

FIGURE 2.6   Concentration of an (epi)catechin-O-sulfate and an O-methyl-(epi)catechin-O-sulfate in plasma 0–8 h after the ingestion of a cocoa, prepared with water containing 45 μmol of (epi)catechins, by volunteers. No flavan-3-ol metabolites were detected in plasma collected 24 h after cocoa intake. Data presented as means values ± standard error ($n = 8$). (After Mullen, W. et al., *Am. J Clin. Nutr.*, **89**, 1784–1791, 2009.)

TABLE 2.3

Quantities of Flavan-3-ol Metabolites Excreted in Urine by Human Subjects 0–24 h After the Consumption of 250 mL of a Water-Based Cocoa Drink, Containing 45 μmol of Flavan-3-ol Monomers[a]

Flavan-3-ol Metabolites	0–2 h	2–5 h	5–8 h	8–24 h	Total (0–24 h)
(−)-Epicatechin-O-glucuronide	0.41 ± 0.04	0.25 ± 0.04	0.08 ± 0.02	0.02 ± 0.01	0.76 ± 0.07
(Epi)catechin-O-sulfate	0.93 ± 0.11	0.54 ± 0.06	0.16 ± 0.02	0.13 ± 0.06	1.76 ± 0.12
(Epi)catechin-O-sulfate	1.13 ± 0.20	0.92 ± 0.19	0.17 ± 0.04	0.09 ± 0.05	2.31 ± 0.30
O-Methyl-(epi) catechin-O-sulfate	1.15 ± 0.23	0.90 ± 0.17	0.23 ± 0.04	0.21 ± 0.15	2.49 ± 0.42
Total	**3.62 ± 0.54**	**2.61 ± 0.41**	**0.64 ± 0.11**	**0.45 ± 0.17**	**7.32 ± 0.82**
	(49.5%)	*(35.7%)*	*(8.7%)*	*(6.1%)*	*(100%)*

Source: After Mullen, W. et al., *Am. J Clin. Nutr.,* **89**, 1784–1791, 2009.

[a]   Data expressed as mean values in μmol ± standard error ($n = 9$). Figures in bold, italicized parentheses represent flavan-3-ol metabolites excreted in each collection period as a percentage of the total 0–24 h excretion.

discussed earlier, in which postenzyme hydrolysis indicated, albeit indirectly, the presence of sulfate and sulfo-glucuronides of (epi)catechin and methyl-(epi)catechin as the main metabolites in both plasma and urine. However, in other cocoa studies in which plasma and urine samples were analyzed by HPLC-MS[2], an (epi)catechin-O-glucuronide was the main metabolite and sulfates were either absent or minor components (Roura et al. 2005, 2007a, 2007b, 2008; Tomás-Barberán et al. 2007). Initially, the reason for these varying metabolite profiles, especially the absence of sulfated metabolites in plasma in some studies and not others, was unclear, although it was thought that the amount of (–)-epicatechin ingested could be a factor (Mullen et al. 2009b). However, it is more likely that one of the causes of the discrepancy is losses of (epi)catechin-O-sulfates, which are known to be unstable (Day and Morgan 2003) and are therefore likely to break down during processing of plasma and urine prior to analysis.

The body appears to treat (epi)catechin metabolites as xenobiotics and, as a consequence, urinary excretion in µmol amounts corresponding to ~30% of intake appears to provide a more realistic guide to bioavailability than either the nM concentrations in plasma (see Figure 2.6) or calculations based on plasma area-under-the curve (AUC) values. Urinary excretion figures, however, do not include the possibility of metabolites being sequestered in body tissues, and so this is also theoretically an underestimate of absorption, but the degree to which this occurs remains to be determined. Nor does it include the portion lost in feces following elimination in bile or direct from the enterocyte and not subsequently reabsorbed. The fact that tissue sequestration has yet to be convincingly demonstrated suggests that it can only be at low levels, if at all.

2.2.3.1 Matrix Effects

Neilson et al. (2009) investigated the influence of the chocolate matrix on the bioavailability of cocoa (epi)catechins. In a cross-over study, six volunteers ingested a dark chocolate and a similar chocolate containing either high sucrose or milk protein. In addition, two cocoa drinks containing milk protein and either sucrose at the same level as the dark chocolate, or an artificial sweetener, were consumed. All the matrices contained 94 µmol of (–)-epicatechin and 32 µmol of catechin, most of which was presumably the (–)-enantiomer. Plasma collected over a 6-h post-ingestion period was then subjected to enzyme hydrolysis before analysis of (epi)catechin by HPLC. Pharmacokinetic analysis revealed C_{max} values ranging from 24.7 to 42.6 nM and AUC figures of 100.7–142.7 nmol/L/h. Significantly higher C_{max} and AUC values were obtained with the two cocoa drinks compared with the three chocolate products. Although the T_{max} values were not statistically different, those with the cocoa drink were 0.9 and 1.1 h compared with 2.3, 1.8, and 2.3 h with the chocolates (Table 2.4). Thus, with different cocoa matrices, there appear to be differences in (epi)catechin bioavailability, but they are relatively small. It is unfortunate that urinary excretion of (epi)catechin metabolites was not investigated as it would have been substantial, as discussed in Section 2.2.3, and would almost certainly have provided evidence of more clear-cut differences in (epi)catechin bioavailability with the five cocoa products than the comparison of plasma pharmacokinetics.

In a further investigation into the influence of the chocolate matrix on the bioavailability of cocoa (epi)catechins with rats, Neilson et al. (2010) showed that plasma concentrations of O-methyl-(epi)catechin-O-glucuronides and (epi)catechin-O-glucuronides, analyzed by HPLC-MS, were highest with a high sucrose chocolate and lowest with milk chocolate, while a reference dark chocolate yielded intermediate concentrations.

TABLE 2.4
(Epi)Catechin Metabolite Pharmacokinetic Parameters in Plasma of Volunteers 0–6 h After the Consumption of Different Chocolate Formulations Containing 94 μmol of (−)-Epicatechin and 32 μmol of Catechin[a]

Product	Composition	C_{max} (nmol/L)	T_{max} (h)	AUC (nmol/L/h)
Chocolate	sucrose (6.6 g)	31.6 ± 2.8[bc]	2.3 ± 0.8[a]	121.1 ± 12.6[ab]
	sucrose (14.6 g)	34.0 ± 3.3[ab]	1.8 ± 0.5[a]	128.1 ± 13.1[ab]
	sucrose (6.6 g), milk protein (6.0 g)	24.7 ± 1.9[c]	2.3 ± 0.8[a]	107 ± 11.1[b]
Cocoa beverage	sucrose (6.6 g), milk protein (6.1 g)	42.5 ± 4.3[a]	0.9 ± 0.1[a]	132.1 ± 14.8[a]
	artificial sweetener, milk protein (6.1 g)	41.6 ± 2.1[a]	1.1 ± 0.3[a]	± 9.3[a]

Source: After Neilson, A.P. et al., *J. Agric. Food. Chem.*, **57**, 9418–9426, 2009.

[a] Data expressed as mean values ± standard error ($n = 6$).

Common superscripts in the same column indicate no significant difference ($P > 0.05$) between formulations.

The impact of cocoa flavan-3-ols on the physico-chemical properties of milk gels has been investigated (Vega and Grover 2011), and there is much interest in whether or not milk reduces the bioavailability of cocoa flavan-3-ols. Polyphenols generally, and especially procyanidin oligomers, (−)-epigallocatechin-3-O-gallate, and theaflavins, bind strongly to proteins, especially those rich in proline such as the caseins of milk. The precise nature of the interaction is greatly influenced by the protein–polyphenols ratio and by other components of food (Luck et al. 1994; Haslam 1998).

A study with male Sprague–Dawley rats showed that whole milk and heavy cream, and to a lesser extent skimmed milk, lowered absorption of cocoa flavan-3-ol monomers in direct proportion to their fats, but not their protein content (Gossai and Lau-Cam 2009). In an earlier study with volunteers, Serafini et al. (2003) found that although consumption of 100 g of dark chocolate brought about an increase in plasma antioxidant capacity, this effect was substantially reduced when the chocolate was ingested with 200 mL of milk, and no increase in antioxidant capacity was observed after eating milk chocolate. They also showed that the absorption of (−)-epicatechin from chocolate was reduced when consumed with milk or as milk chocolate. It was hypothesized that proteins in the milk bind to the flavan-3-ols and limit their absorption from the gastrointestinal tract. This report generated much controversy with subsequent studies producing conflicting data on the impact of milk on cocoa flavan-3-ol absorption.

In the Mullen et al. (2009) investigation referred to in Section 2.2.3, volunteers drank 250 mL of a commercial cocoa made not just with hot water but also with hot milk. Both drinks contained 45 μmol of flavan-3-ol monomers. Milk did not have a significant effect on either the plasma C_{max} or T_{max} of sulfated and methylated (epi)catechin metabolites, but did bring about a significant reduction in (epi)catechin metabolites excreted 0–2 h and 2–5 h after ingestion with the overall amounts being excreted over a 24-h period declining from 16.3 to 10.4% of intake. This was not due to the effects of milk on either gastric emptying or on the time for the head of the meal to reach the colon, ruling out the possibility that milk slowed the rate of transport of the meal through the gastrointestinal tract. The reduced excretion of the flavan-3-ol metabolites was, therefore, probably a consequence of components in the milk that either bind directly to flavan-3-ols or interfere with the mechanism involved in their transport across the wall of the small intestine into the portal vein. The findings of this study contrast with reports that milk does not affect the absorption of flavan-3-ols. These include a study by Roura et al. (2007a) who monitored flavan-3-ol metabolites in plasma collected 2 h after drinking cocoa containing 128 μmol of flavan-3-ol monomers, a threefold higher quantity than the 45 μmol ingested in the Mullen et al. (2009) study. It is, however, interesting to note that although not statistically significant, urinary excretion in the study by Roura et al. (2007b) was 20% lower with cocoa-milk compared with cocoa-water. Keogh et al. (2007), who analyzed plasma 0–8 h after the consumption of a flavan-3-ol rich cocoa drink, also reported that milk had no effect on the absorption of catechin and epicatechin. In this instance, the ingested dose of flavan-3-ol monomers was 2374 μmol, 53-fold higher than in the Mullen et al. study. This high dose was reflected in a C_{max} of ~12 μM compared with ~150 nM in the Mullen et al. (2009) study. Schroeter et al. (2003) also reported that milk did not influence plasma epicatechin after consumption of a cocoa beverage, which in this instance was consumed at a dose of 1314 μmol of flavan-3-ol monomers for a 70-kg human.

There is an explanation for these seemingly contradictory reports. It would appear that with high flavan-3-ol cocoas, which are principally research products, the factors in milk that reduce absorption have a minimal overall impact. With drinks with a lower flavan-3-ol content, such as the one used in the Mullen et al. (2009) study, which is typical of many commercial cocoas that are on supermarket shelves (Miller et al. 2008) and available to the general public, milk does have the capacity to interfere with absorption.

2.2.4 Green Tea Flavan-3-ol Monomers

Green tea contains high concentrations of flavan-3-ol monomers. As well as (–)-epicatechin and (+)-catechin, (epi)gallocatechins, and 3-O-galloylated flavan-3-ols are present, components that do not occur in cocoa. Typically, (–)-epigallocatechin-3-O-gallate, (–)-epigallocatechin, and (–)-epicatechin predominate (Figures 2.1 and 2.3).

In an acute feeding study, healthy human subjects consumed 500 mL of Choladi, a commercial bottled green tea containing 648 μmol of flavan-3-ols, after which plasma and urine were collected over a 24-h period and analyzed by HPLC-MS^2 (Stalmach et al. 2009). The plasma contained a total of 12 metabolites, in the form of O-methylated, sulfated, and glucuronide conjugates of (epi)catechin and (epi)gallocatechin along with the native green tea flavan-3-ols (–)-epigallocatechin-3-O-gallate and (–)-epicatechin-3-O-gallate. An analysis of the pharmacokinetic profiles of these compounds illustrated in Figure 2.7 is presented in Table 2.5. None of the flavan-3-ols were present in the circulatory system at 0 h, but they did appear in detectable quantities 30 min after green tea consumption. The main component which accumulated was an (epi)gallocatechin-O-glucuronide, with a C_{max} of 126 nmol/L and a T_{max} of 2.2 h, while an (epi)catechin-O-glucuronide, probably the 3'-O-conjugate, attained a C_{max} of 29 nmol/L with a 1.7-h T_{max}. The unmetabolized flavan-3-ols (−)-epigallocatechin-3-O-gallate and (−)-epicatechin-3-O-gallate had C_{max} values of 55 and 25 nmol/L after 1.6 and 2.3 h, respectively. The T_{max} durations ranged from 1.6 to 2.3 h (Table 2.5); all the flavan-3-ols and their metabolites were present in only trace quantities after 8 h (Figure 2.7) and were not detected in the 24 h plasma. These T_{max} values and the pharmacokinetic profiles are indicative of absorption in the small intestine. The appearance of unmetabolized flavonoids in plasma is unusual. The passage of (−)-epicatechin-3-O-gallate and (−)-epigallocatechin-3-O-gallate through

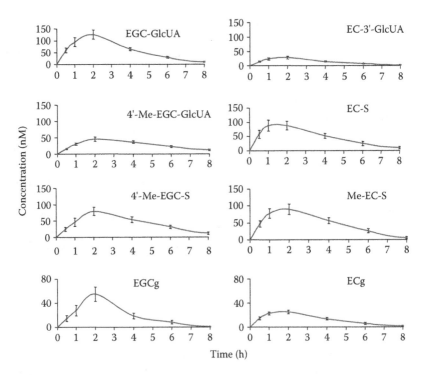

FIGURE 2.7 Concentrations of (epi)gallocatechin-O-glucuronide (EGC-GlcUA), 4'-O-methyl-(epi)gallocatechin-O-glucuronide (4'-Me-EGC-GlcUA), 4'-O-methyl-(epi)gallocatechin-O-sulfates (4'-Me-EGC-S), (−)-epicatechin-3'-O-glucuronide (EC-3'-GlcUA), (epi)catechin-O-sulfate (EC-S), 3'- and 4'-O-methyl-(epi)catechin-O-sulfates (Me-EC-S), (−)-epigallocatechin-3-gallate (EGCg), and (−)-epicatechin-3-gallate (ECg) in the plasma of human subjects 0–8 h after the ingestion of 500 ml of green tea. Data expressed as mean values with their standard errors ($n = 10$) depicted by vertical bars. Note that no flavan-3-ols or their metabolites were detected in plasma collected 24 h after ingestion of the green tea. (After Stalmach, A. et al., *Mol. Nutr. Food Res.*, **53**, S44–S53, 2009.)

TABLE 2.5
Pharmacokinetic Analysis of Flavan-3-ols and Their Metabolites Detected in Plasma of Volunteers Following the Ingestion of 500 mL of Green Tea[a]

Flavan-3-ols (Number of Isomers)	C_{max} (nM)	T_{max} (h)
(Epi)catechin-O-glucuronide (1)	29 ± 4.7	1.7 ± 0.2
(Epi)catechin-O-sulfates (2)	89 ± 15	1.6 ± 0.2
O-Methyl-(epi)catechin-O-sulfates (5)	90 ± 15	1.7 ± 0.2
(Epi)gallocatechin-O-glucuronide (1)	126 ± 19	2.2 ± 0.2
4′-O-Methyl-(epi)gallocatechin-O-glucuronide (1)	46 ± 6.3	2.3 ± 0.3
4′-O-Methyl-(epi)gallocatechin-O-sulfates (2)	79 ± 12	2.2 ± 0.2
(−)-Epigallocatechin-3-O-gallate (1)	55 ± 12	1.9 ± 0.1
(−)-Epicatechin-3-O-gallate (1)	25 ± 3.0	1.6 ± 0.2

Source: After Stalmach, A. et al., *Mol. Nutr. Food Res.*, **53**, S44–S53, 2009.
[a] Data expressed as mean values ± SE ($n = 10$).

the wall of the small intestine into the circulatory system without metabolism could be a consequence of the presence of the 3-O-galloyl moiety interfering with phase II metabolism. However, gallic acid per se is readily absorbed with a reported urinary excretion of 37% of intake (Shahrzad and Bitsch 1998; Shahrzad et al. 2001), and the gallate ester might also exhibit improved absorption.

Urine collected 0–24 h after green tea ingestion contained an array of flavan-3-ol metabolites similar to those detected in plasma, except for the presence of minor amounts of three additional (epi)gallocatechin-O-sulfates and the absence of (−)-epicatechin-3-O-gallate and (−)-epigallocatechin-3-O-gallate (Table 2.6) (Stalmach et al. 2009). This indicates that the flavan-3-ols do not undergo extensive further phase II metabolism prior to removal from the circulatory system and excretion in the urine. In total, 52.4 μmol of metabolites were excreted, which was equivalent to 8.1% of the ingested green tea flavan-3-ols. When the urinary (epi)gallocatechin and (epi)catechin metabolites were considered separately, a somewhat different picture emerged. The 33.3 μmol excretion of (epi)gallocatechin metabolites was 11.4% of the ingested (−)-epigallocatechin and (+)-gallocatechin, while the 19.1 ± 2.2 μmol recovery of (epi)catechin represented 28.5% of intake (Table 2.6). This is in keeping with high urinary recoveries of (epi)catechin metabolites obtained in the studies with cocoa products (see Section 2.2.3), indicating that (−)-epicatechin, in particular, is highly bioavailable and is absorbed and excreted to a much greater extent that other flavonoids, with the possible exception of isoflavones (Clifford and Crozier, 2011).

The absence of detectable amounts of (−)-epigallocatechin-3-O-gallate in urine, despite its presence in plasma, an event observed by several investigators (Unno et al. 1996; Chow et al. 2001; Henning et al. 2005), is difficult to explain. It is possible that the kidneys are unable to remove (−)-epigallocatechin-3-O-gallate from the

TABLE 2.6

Quantification of the Major Groups of Flavan-3-ol Metabolites Excreted in Urine 0–24 h After the Ingestion of 500 mL of Green Tea by Volunteers[a]

Flavan-3-ol Metabolites *(Number of Isomers)*	0–24 h Excretion (µmol)
(Epi)gallocatechin-O-glucuronide *(1)*	6.5 ± 1.2
4′-O-Methyl-(epi)gallocatechin-O-glucuronide *(1)*	4.4 ± 1.5
4′-O-Methyl-(epi)gallocatechin-O-sulfates *(2)*	19.8 ± 0.3
(Epi)gallocatechin-O-sulfates *(3)*	2.6 ± 3.0
Total (epi)gallocatechin metabolites	*33.3 (11.4%)*
(Epi)catechin-O-glucuronide *(1)*	1.5 ± 0.3
(Epi)catechin-O-sulfates *(2)*	6.7 ± 0.7
O-Methyl-(epi)catechin-O-sulfates *(5)*	10.9 ± 1.2
Total (epi)catechin metabolites	*19.1 (28.5%)*
Total flavan-3-ol metabolites	**52.4 (8.1%)**

Source: After Stalmach, A. et al., *Mol. Nutr. Food Res.*, **53**, S44–S53, 2009.

[a] Data expressed as mean value ± standard error (*n* = 10). Italicized figures in parentheses indicate amount excreted as a percentage of intake.

bloodstream, but if this is the case, there must be other mechanisms that result in its rapid decline after reaching C_{max}. Studies with rats have led to the speculation that (−)-epigallocatechin-3-O-gallate may be removed from the bloodstream in the liver and returned to the small intestine in the bile (Kida et al. 2000; Kohri et al. 2001). To what extent enterohepatic recirculation of (−)-epigallocatechin-3-O-gallate, and also (−)-epicatechin-3-O-gallate, occurs in humans remains to be established. Quite possibly these bile-excreted flavan-3-ols would be degallated by the gut microflora and, if not more extensively degraded, would be excreted in urine as (epi)catechin and (epi)gallocatechin metabolites. It is of note that feeding studies with [2-^{14}C]resveratrol have provided evidence that that stilbene metabolites do undergo enterohepatic recycling in humans (Walle et al. 2004).

2.2.4.1   Dose and Matrix Effects

Auger et al. (2008) fed ileostomists, volunteers who had had their colon removed surgically, increasing doses of Polyphenon E, a green tea extract containing a characteristic array of flavan-3-ols; urinary excretion of metabolites was used as a measure of absorption in the small intestine. The data obtained with (epi)gallocatechin and (epi)catechin metabolites are summarized in Table 2.7. At a dose of 22 µmol, the 0–24 h excretion of (epi)gallocatechin metabolites was 5.7 ± 1.9 µmol, and this figure did not increase significantly with intakes of 55 and 165 µmol. There is, therefore, a strict limit on the extent to which (epi)gallocatechins can be absorbed. Following the ingestion of 77 µmol of (epi)catechins, 36 ± 9 µmol were excreted, and with doses of 192 and 577 µmol urinary excretion increased significantly to

TABLE 2.7

Urinary Excretion of Flavan-3-ol Metabolites After the Ingestion of Increasing Doses of (–)-Epicatechin and (–)-Epigallocatechin in Poly phenon E by Humans with an Ileostomy. Metabolites Excreted Over a 24-h Period After Ingestion Expressed as μmol and Italicized Figures in Parentheses Represent the Percentage of Glucuronide, Sulfate, and Methylated Metabolites[a]

	Dose		
(Epi)Gallocatechin Metabolites	**22 mmol**	**55 mmol**	**165 mmol**
Glucuronides	1.8 *(17%)*	1.0 *(18%)*	1.7 *(19%)*
Sulfates	3.9 *(37%)*	2.0 *(36%)*	3.6 *(40%)*
Methylated	4.8 *(46%)*	2.6 *(46%)*	4.7 *(51%)*
Total (epi)gallocatechin metabolites	**5.7 ± 1.9**[a]	**3.0 ± 0.8**[a]	**5.3 ± 1.2**[a]
	Dose		
(Epi)Catechin Metabolites	**77 mmol**	**192 mmol**	**577 mmol**
Glucuronides	3.4 *(7%)*	14 *(9%)*	38 *(10%)*
Sulfates	33 *(64%)*	93 *(58%)*	224 *(59%)*
Methylated	15 *(29%)*	53 *(33%)*	120 *(31%)*
Total (epi)catechin metabolites	**36 ± 9**[a]	**107 ± 27**[b]	**262 ± 26**[c]

Source: After Auger, C., Hara, Y., and Crozier, A., *J. Nutr.*, **138**, 1535S–1542S, 2008.

[a] Values for total (epi)gallocatechin and (epi)catechin metabolites with different superscripts are significantly different (P < 0.05).

107 ± 27 and 262 ± 26 μmol. Thus, even the highest dose (epi)catechins, unlike (epi gallocatechins, are still readily absorbed. The addition of a 5′-hydroxyl group to (epi)catechin, therefore, markedly reduces the extent to which the molecule can enter the circulatory system from the small intestine. It is also of note that at the three doses which were administered, the ratio of the urinary glucuronide, sulfate, and methylated (epi)catechin metabolites changed little (Table 2.7). This indicates that even at the highest intake the UGT, SULT, and COMT enzymes involved in the formation of the (epi)catechin metabolites do not become saturated and limit conversion.

Table 2.8 summarizes the varying plasma T_{max} values for (epi)catechin and (epi) gallocatechin metabolites obtained in studies with green tea flavan-3-ols in which very similar feeding procedures and methods of analysis were used. In the study by Stalmach et al. (2009) in which volunteers ingested 500 mL of Choladi green tea, a commercial bottled product containing vitamin C and 100 kcal, T_{max} ranged from 1.7 to 2.2 h. When an infusion prepared from green tea leaves was consumed by ileostomists, the T_{max} were of a shorter, ranging from 0.8 to 1.3 h. Initially, this was thought to reflect more rapid transport through the small intestine of the ileostomists because of the absence of an ileal brake (Stalmach et al. 2010). However, this

TABLE 2.8

Details of Feeding Studies and Times of Peak Plasma Concentrations (T_{max}) of (Epi)Catechin and (Epi)Gallocatechin Metabolites After the Consumption of Green Teas, Cocoa and a Polyphenol-Rich Juice[a]

	Green Tea (1)	Green Tea (2)	Polyphenol-Rich Juice (3)
Volunteers –with or without a colon	with	without	with
Volume consumed	500 mL	300 mL	350 mL
Calorie intake	144 kcal	0 kcal	51 kcal
Vitamin C content	100 mg	0 mg	168 mg
Flavan-3-ol intake	648 µmol	634 µmol	448 µmol
Plasma Metabolites			
(Epi)catechin-O-glucuronide	1.7 ± 0.2	0.8 ± 0.1	1.1 ± 0.2
(Epi)catechin-O-sulfate(s)	1.6 ± 0.2	1.3 ± 0.3	0.9 ± 0.1
O-Methyl-(epi)catechin-O-sulfate(s)	1.7 ± 0.2	1.3 ± 0.3	1.0 ± 0.0
(Epi)gallocatechin-O-glucuronide	2.2 ± 0.2	1.1 ± 0.5	0.6 ± 0.2

Source: Based on data of Stalmach, A. et al., *Mol. Nutr. Food Res.,* **53**, S44–S53, 2009; Stalmach, A. et al., *Mol. Nutr. Food Res.,* **54**, 323–334, 2010; and Borges, G. et al., *Mol. Nutr. Food Res.* **54**, S268–277, 2010.

[a] T_{max} data expressed in hours as mean values ± standard error.

is clearly not the only factor affecting absorption in the small intestine as short T_{max} times, 0.6–1.1 h, were also obtained when a juice drink containing a diversity of phenolic compounds, including green tea flavan-3-ols, was ingested by healthy subjects with a functioning colon (Borges et al. 2010) (Table 2.8). The rapid absorption of the flavan-3-ols in this drink is of interest because it occurred in the presence of substantial amounts of other phenolic components, including gallic acid, 5-O-caffeoylquinic acid, anthocyanins, flavanones, and dihydrochalcones. This suggests that there is not major competition for transport of these components across the gastrointestinal mucosa into the circulatory system. The varying (epi)catechin and (epi)gallocatechin metabolite T_{max} values observed in these studies indicate that other components in the drinks can impact on flavan-3-ol absorption. The sugar content of the drinks reflected in the different calorie intakes could play a role and the varying vitamin C contents of the drinks (Table 2.8) could also have an effect, but these are unlikely to be the only factors involved. Further experimentation is required as in general very little is known about the impact of other food constituents and dietary components on flavonoid bioavailability.

2.2.4.2 Studies with Ileostomists and Colonic Catabolism

As mentioned earlier, Stalmach et al. (2010) carried out an acute green tea feeding study with ileostomists. The volunteers drank tea containing 634 µmol of flavan-3-ols, a very similar intake to that used in their studies with healthy subjects

(Stalmach et al. 2009) that were discussed in Section 2.2.4. The plasma flavan-3-ol profiles were similar to those illustrated in Figure 2.7, which were obtained from healthy subjects with an intact functioning colon. Urinary excretion by the ileostomists was 8.0% of intake for (epi)gallocatechin metabolites and 27.4% of (epi)catechin metabolites, values that were similar to those observed with healthy subjects (see Table 2.6). This demonstrates unequivocally that the flavan-3-ol monomers are absorbed in the upper part of the gastrointestinal tract and suggests that absorption of (epi)catechin and (epi)gallocatehin after degallation of flavan-3-ol gallates excreted in bile does not contribute significantly to bioavailability of the intact flavan-3-ols in consumers with an intact colon.

Despite the substantial absorption of green tea flavan-3-ols in the upper gastro-intestinal tract, Stalmach et al. (2010) found that 69% of intake was recovered in 0–24 h ileal fluid as a mixture of native compounds and metabolites (Table 2.9). Thus, in volunteers with a colon, most of the ingested flavan-3-ols will pass from the small to the large intestine where their fate is a key part of the overall bioavailability equation.

To mimic events taking place in the large intestine, two sets of experiments were carried out by Roowi et al. (2010). First, 50 μmol of (–)-epicatechin, (–)-epigallocatechin and (–)-epigallocatechin-3-O-gallate were incubated under anaerobic conditions *in vitro* with fecal slurries and their degradation to phenolic acid by the microbiota was monitored. A limitation of *in vitro* fermentation models is that it may not fully depict the *in vivo* conditions. The use of fecal material may not fully represent the microbiota present in the colonic lumen and mucosa, where catabolism occurs *in vivo*. Obviously, the accumulation and retention of the degradation products in the fermentation vessel makes collection, identification, and quantification of the metabolites easier but is not necessarily representative of the events that occur *in vivo* where the actual concentration of a metabolite at any time interval is dependent on the combined rates of catabolism and absorption. However, the use of an *in vitro* model provides information on the types of breakdown products formed, helps elucidate the pathways involved, and allows for determining the rate of catabolism.

To complement the *in vitro* incubations, phenolic acids excreted in urine 0–24 h after (i) the ingestion of green tea and water by healthy subjects in a cross-over study and (ii) the consumption of green tea by ileostomists were also investigated (Roowi et al. 2010). The data obtained in these studies provided the basis for the proposed operation of the catabolic pathways illustrated in Figure 2.8. Some of these catabolites, such as 4′-hydroxyphenylacetic acid and hippuric acid, were detected in urine from subjects with an ileostomy, indicating that they are produced in the body by additional routes unrelated to colonic degradation of flavan-3-ols. It is, for instance, well known that there are pathways to hippuric acid (*N*-benzoyl-glycine) from compounds such as benzoic acid, quinic acid (Clifford et al. 2000), tryptophan, tyrosine, and phenylalanine (Self et al. 1960; Grumer 1961; Bridges et al. 1970). Nonetheless, the elevated urinary excretion of hippuric acid and 4′-hydroxyphenyl-acetic acid, occurring after green tea consumption, is likely to be partially derived from flavan-3-ol degradation. Earlier research showing statistically significant increases in urinary excretion of hippuric acid after consumption of both green and

TABLE 2.9
Quantities of Flavan-3-ols and Metabolites Recovered in Ileal Fluid 0–24 h After the Ingestion of 300 mL of Green Tea[a]

Flavan-3-ols and Metabolites (Number of Isomers)	Amount Ingested (μmol)	Recovered in Ileal Fluid	
		(μmol)	% of Amount Ingested
(+)-Catechin	18	1.2 ± 0.2	6.8 ± 1.2
(−)-Epicatechin	69	7.9 ± 1.7	11 ± 2.5
(+)-Gallocatechin	50	13 ± 2.1	27 ± 4.2
(−)-Epigallocatechin	190	35 ± 5.8	18 ± 3.1
(−)-Epigallocatechin-3-O-gallate	238	116 ± 5.1	49 ± 2.1
(+)-Gallocatechin-3-O-gallate	5.2	4.6 ± 0.3	89 ± 5.2
(−)-Epicatechin-O-gallate	64	29 ± 3.3	45 ± 5.2
Total parent flavan-3-ols	*634*	*206 ± 11*	*33 ± 1.8*
(Epi)catechin-O-glucuronide (1)		0.5 ± 0.2	
(Epi)catechin-O-sulfates (3)		72 ± 4.1	
O-Methyl-(epi)catechin-O-sulfates (5)		5.4 ± 0.5	
Total (epi)catechin metabolites		*78 ± 4.3*	*90 ± 5.0*
(Epi)gallocatechin-O-glucuronide (1)		1.5 ± 0.4	
(Epi)gallocatechin-O-sulfates (3)		108 ± 5.2	
O-Methyl-(epi)gallocatechin-O-sulfates (3)		40 ± 4.3	
O-Methyl-(epi)gallocatechin-O-glucuronide (1)		1.5 ± 0.5	
Total (epi)gallocatechin metabolites		*151 ± 8.9*	*63 ± 3.7*
(Epi)gallocatechin-3-O-gallate-O-sulfate (1)		1.8 ± 0.4	
O-Methyl-(epi)gallocatechin-3-O-gallate-O-sulfates (2)		0.9 ± 0.2	
Total (−)-epigallocatechin-3-O-gallate metabolites		*2.8 ± 0.5*	*1.1 ± 0.2*
(Epi)catechin-3-O-gallate-O-sulfate (1)		0.4 ± 0.1	
Methyl-(epi)catechin-3-O-gallate-O-sulfates (2)		*0.4 ± 0.1*	
Total (−)-epicatechin-3-O-gallate metabolites		*0.9 ± 0.2*	*1.4 ± 0.3*
Total metabolites		*232 ± 13*	*37 ± 2.1*
Total parent flavan-3-ols and metabolites		439 ± 13	69 ± 2.0

Source: Stalmach, A. et al., *Mol. Nutr. Food Res.,* **54**, 323–334, 2010.

[a] Data expressed as mean value ± standard error (*n* = 5).

black tea by human subjects (Clifford et al. 2000; Mulder et al. 2005) supports this hypothesis.

Quantitative estimates of the extent of ring fission of the flavan-3-ol skeleton are difficult to assess because, as discussed earlier, the production of some of the urinary phenolic acids was not exclusive to colonic degradation of flavan-3-ols. If these compounds are excluded along with pyrogallol and pyrocatechol, which are derived from cleavage of the gallate moiety from (−)-epigallocatechin-3-O-gallate rather than ring fission,

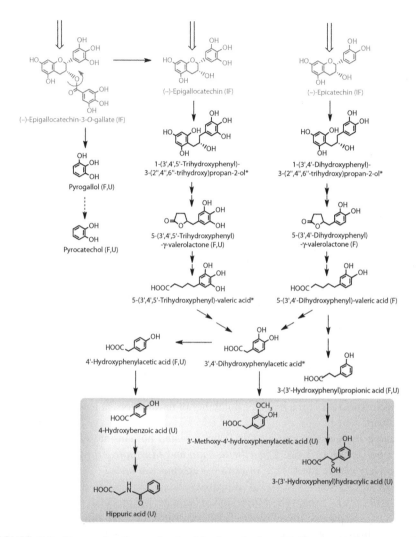

FIGURE 2.8 Proposed pathways involved in the colonic catabolism and urinary excretion of green tea flavan-3-ols. Following consumption of green tea more than 50% of the ingested flavan-3-ols (gray structures) pass into the large intestine. When incubated with fecal slurries these compounds are catabolized by the colonic microflora probably via the proposed pathways illustrated. Analysis of urine after green tea consumption indicates that some of the colonic catabolites enter the circulation and undergo further phase II metabolism before being excreted in urine. Structures in the gray box indicate such catabolites that are detected in urine but not produced by fecal fermentation of (−)-epicatechin, (−)-epigallocatechin, or (−)-epigallocatechin-3-*O*-gallate. The dotted arrow between pyrogallol and pyrocatechol indicate this is a minor conversion. Double arrows indicate conversions where the intermediate(s) did not accumulate and are unknown. Compounds detected in ileal fluid after green tea consumption (IF); catabolites detected in fecal slurries (F) and in urine (U); potential intermediates that did not accumulate in detectable quantities in fecal slurries (*). An alternative route to 3-(3′-hydroxyphenyl) hydracrylic acid is from cinnamic acid by the addition of water across the double bond in a hydratase catalyzed reaction. (After Roowi, S. et al., *J. Agric. Food Chem.*, **58**, 1296–1304, 2010.)

the excretion of the remaining urinary phenolic acids, namely 4-hydroxybenzoic acid, 3'-methoxy-4'-hydroxyphenylacetic acid, 3-(3'-hydroxyphenyl)hydracrylic acid, and 5-(3',4',5'-trihydroxyphenyl)-γ-valerolactone, after ingestion of green tea was 210 μmol compared with 38 μmol after drinking water (Roowi et al. 2010). The 172 μmol difference between these figures corresponds to 27% of the 634 μmol of flavan-3-ols present in the ingested green tea (Table 2.6) (Stalmach et al. 2010). Added to this is the ~8% excretion of glucuronide, sulfate, and methylated flavan-3-ols originating from absorption in the small intestine. This estimate of a 35% recovery is nonetheless a minimum value, because with the analytical methodology used, some urinary catabolites will have escaped detection (Roowi et al. 2010). This will include glucuronide and sulfate metabolites of 5-(3',4',5'-trihydroxyphenyl)-γ-valerolactone, 5-(3',4'-dihydroxyphenyl)-γ-valerolactone, and 5-(3',5'-dihydroxyphenyl)-γ-valerolactone, which were detected after green tea consumption with a cumulative 0–24 h excretion corresponding to 16% of flavan-3-ol intake (Li et al. 2000; Meng et al. 2002; Sang et al. 2008). More recently, in a similar study in which urine was collected for 24 h after green tea intake, valerolactone metabolites were excreted in quantities equivalent to 36% of intake (Del Rio et al. 2010). When added to the 35% recovery of Roowi et al. (2010), this gives a total of 71% of intake. While this figure is obviously an approximation because of factors such as different volunteers, flavan-3-ol intakes and analytical methodologies, it does demonstrate that, despite substantial modification as they pass through the body, there is a very high urinary recovery of flavan-3-ols, principally in the form of colon-derived catabolites. Moreover, the excretion of colonic valerolactones has been shown to continue far beyond 24 h after ingestion, as some of these metabolites have been detected in urine after 54 h (Del Rio et al. 2010). This observation is consistent with a possible underestimation of the real bioavailability of flavan-3-ols, at least when green tea is consumed.

Research by Kutschera et al. (2011) has isolated human intestinal bacteria that can metabolize flavan-3-ols. *Eggerthella lenta* and *Flavonifractor plantii* were able to reductively cleave the C-ring of both (−)-epicatechin ands (+)-catechin, producing the 3R and 3S isomers of 1-(3',4'-dihydroxyphenyl)-3-(2'',4'',6''-trihydroxyphenyl) propan-2-ol), respectively. The conversion of (+)-catechin proceeded five times faster than that of (−)-epicatechin; *F. plantii* further converted 1-(3',4'-dihydroxyphenyl)-3-(2'',4'',6''-trihydroxyphenyl)propan-2-ol to 5-(3',4'-dihydroxyphenyl)-γ-valerolactone and 5-(3',4'-dihydroxyphenyl)-γ-valeric acid (see Figure 2.8).

Establishing relationships between flavan-3-ol metabolism and the human gut microbiome is lagging behind the rapid developments in metabolomics and microbiomics (Turnbaugh and Gordon 2008). Microbiome–metabolome links may be investigated *in vivo*, as was demonstrated in a small human cohort study, where several microbial species were found to correlate with metabolites appearing in polyphenol bioconversion pathways (Li et al. 2008). Moreover, obtaining information about the genetic determinants of flavan-3-ol microbial metabolism rather than phylogeny of the gut microbial species may be a more appropriate approach for studying these links. Unlike polysaccharides, polyphenols are not an energy source for microbial growth, other than possibly via certain aliphatic catabolites, so from this perspective, they may have a modest impact on the composition of the colonic microflora, although there may be a change in metabolic activity that could be measured at the RNA and protein level. Shifts in gut microbial composition may occur because of other effects of polyphenols, such as antimicrobial activity and knock-on effects (van Duynhoven et al. 2011), and there is some evidence of prebiotic effects (Goto et al. 1998; Tzounis et al. 2008, 2011).

So far, only a few microbial polyphenol bioconversion products have been associated with systemic biological effects in the host or at the local level at the gut wall, but human trials are revealing an increasing number of metabolites that appear at high levels in the colon and systemic circulation (Crozier et al. 2010). The biological relevance for most of these metabolites is unknown, and systematic approaches are required to address this.

2.2.4.3   *In Vivo* and *In Vitro* Stability of Green Tea Flavan-3-ols

There are reports in the literature that green tea flavan-3-ols are poorly bioavailable because of instability under digestive conditions with >80% losses being observed with *in vitro* digestion models simulating gastric and small intestine conditions (Zhu et al. 1997; Yoshino et al. 1999; Record and Lane 2001; Green et al. 2007). It is clear that the data obtained in these investigations do not accurately reflect the *in vivo* fate of flavan-3-ols following ingestion as they are at variance with the high urinary excretion observed in green tea feeding studies (Manach et al. 2005; Stalmach et al. 2009, 2010) and the substantial recovery of flavan-3-ols in ileal fluid (Auger et al. 2008; Stalmach et al. 2010).

2.2.5   PROANTHOCYANIDINS

There have been numerous feeding studies with animals and humans indicating that the oligomeric and polymeric proanthocyanidins are not absorbed to any degree. A study in which ileostomists consumed apple juice indicates that most pass unaltered to the large intestine (Kahle et al. 2007) where they are catabolized by the colonic microflora yielding a diversity of phenolic acids (Manach et al. 2005; Espín et al. 2007), including 3-(3′-hydroxyphenyl)propionic acid and 4-O-methyl-gallic acid (Déprez et al. 2000; Gonthier et al. 2003; Ward et al. 2004) which are absorbed into the circulatory system and excreted in the urine. There is one report based on data obtained from an *in vitro* model of gastrointestinal conditions that procyanidins degrade, yielding more readily absorbable flavan-3-ol monomers (Spencer et al. 2000). However, subsequent *in vivo* studies have not supported this conclusion (Donovan et al. 2002; Rios et al. 2002; Tsang et al. 2005), suggesting not more than about 10% of procyanidin dimers are converted to monomers in this way (Appeldoorn et al. 2009; Stoupi et al. 2010b).

There is a report of minor quantities of procyanidin B_2 being detected in enzyme-treated human plasma collected after the consumption of cocoa (Holt et al. 2002). The T_{max} and pharmacokinetic profile of the B_2 dimer were similar to those of flavan-3-ol monomers, but the C_{max} was ~100-fold lower (Holt et al. 2002). Urpi-Sarda et al. (2009) also detected and quantified procyanidin B_2 in human and rat urine after cocoa intake.

Recent studies using procyanidin B_2 and [^{14}C]procyanidin B_2 have provided information on *in vitro* gut flora catabolism (Appeldoorn et al. 2009; Stoupi et al. 2010a,

2010b) and rodent pharmacokinetics (Stoupi et al. 2010c). Following oral dosing of rats with a [^{14}C]procyanidin B$_2$ dimer, approximately 60% of the radioactivity was excreted in urine after 96 h with the vast majority in a form(s) very different from the intact procyanidin dosed (Stoupi et al. 2010c). This observation is consistent with the *in vitro* studies that show extensive catabolism by the gut microflora. The scission of the interflavan bond represents a minor route, and the dominant products are a series of phenolic acids having one or two phenolic hydroxyls and between one and five aliphatic carbons in the side chain (Appeldoom et al. 2009; Stoupi et al. 2010a, 2010b). There were, in addition, some C_6–C_5 catabolites with a side chain hydroxyl, and associated lactones, and several diaryl-propan-2-ols, most of which are also produced from the flavan-3-ol monomers via the proposed routes illustrated in Figure 2.8, while 3′,4′-dihydroxyphenylacetic acid is derived from cleavage of the C-ring of the upper flavan-3-ol unit. However, the findings of Stoupi et al. (2010b) also indicate that a feature of flavan-3-ol catabolism is conversion to C_6-C_5 valerolactones and progressive β-oxidation to C_6-C_3 and C_6-C_1 products, broadly in keeping with the routes illustrated in Figure 2.9.

The *in vitro* fate of procyanidin B$_2$ in fecal slurries is also much more complex than the routes indicated in Figures 2.8 and 2.9 as the dimer also yields 24 "dimeric" catabolites, i.e., having a mass greater than the constituent monomer (–)-epicatechin (290 amu), and which early in the incubation collectively accounted for some 20% of the substrate (Stoupi et al. 2010b). Clearly, these catabolites retain the interflavan bond. One was identified tentatively as either 6- or 8-hydroxy-procyanidin B$_2$. Thirteen were characterized as having been microbially reduced in at least one of the epicatechin units. Five contained an apparently unmodified epicatechin unit, but in at least one case, this was shown to consist of the B-ring of the "upper" epicatechin unit and the A-ring of the "lower." It is not known whether these unique catabolites are produced *in vivo*, and if so, whether they are absorbed (Stoupi et al. 2010b).

The potential biological effects of procyanidins are generally attributed to their more readily absorbed colonic breakdown products, the phenolic acids, although there is a lack of detailed study in this area. There is, however, a dissenting view as trace levels of procyanidins, in contrast to (–)-catechin and (+)-epicatechin, inhibit platelet aggregation *in vitro* and suppress the synthesis of the vasoconstricting peptide endothelin-1 by cultured endothelial cells (Corder et al. 2008). Supporting this view is a study in which individual procyanidins were fed to rats after which dimers through to pentamers were detected in the plasma, which was extracted with 8 M urea, rather than the more traditional methanol/acetonitrile, which was proposed to have prevented or reversed the binding of procyanidins to plasma proteins (Shoji et al. 2006). The procyanidins were, however, administered by gavage at an extremely high dose, 1 g/kg body weight, and it remains to be determined if procyanidins can be similarly detected in urea-extracted plasma after the ingestion by humans of more dietary relevant quantities in cocoa or chocolate products.

2.2.6 Black Tea Theaflavins and Thearubigins

Although consumed far more extensively in Europe than green tea, the absorption of flavan-3-ols from black tea, their gut microbial catabolism and human metabolism

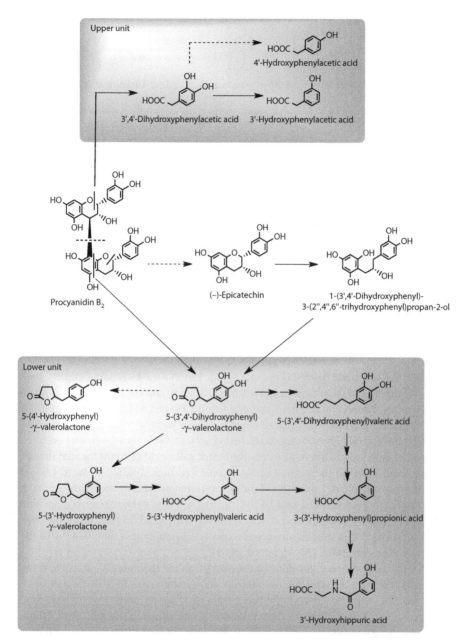

FIGURE 2.9 Proposed pathways for human microbial degradation of procyanidin B₂ dimer. Main routes are indicated with solid arrows, minor pathways with dotted arrows. Double arrows indicate conversions where the intermediate(s) are unknown. Metabolites derived from upper and lower units are grouped within the shaded rectangles. (After Appeldoorn, M.M. et al., *J. Agric. Food Chem.*, **57**, 1084–1092, 2009; Stoupi, S. et al., *Arch. Biochem. Biophys.*, 501, 73–78, 2010a and Stoupi, S. et al., *Mol. Nutr. Food Res.*, **54**, 747–759, 2010b.)

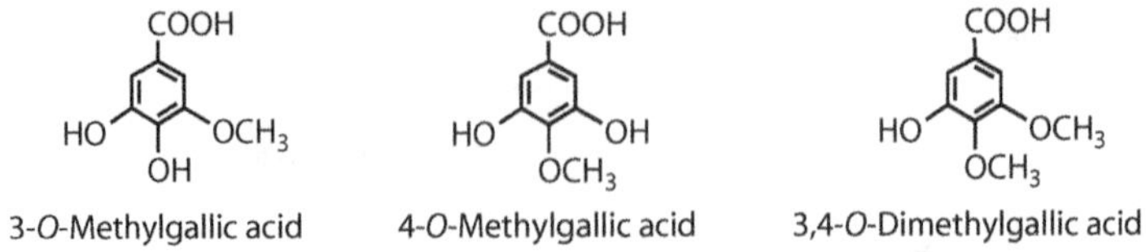

3-O-Methylgallic acid   4-O-Methylgallic acid   3,4-O-Dimethylgallic acid

FIGURE 2.10 Metabolites of gallic acid excreted in urine after the consumption of black tea.

have been less extensively and less thoroughly studied. In part, this can be explained by the lack of knowledge of the thearubigins and the lack of commercial standards for thearubigins or theaflavins. Such studies as have been performed focused on the absorption of flavan-3-ol monomers (Henning et al. 2004). The appearance in urine of the gallic acid metabolites 3-O-methylgallic acid, 4-O-methylgallic acid, and 3,4-di-O-methylgallic acid (Figure 2.10) has also been reported and used as an index of black tea consumption (Hodgson et al. 2000, 2004), although these metabolites are also to be expected following not only green and black tea consumption but also the ingestion of certain fruits, such as grapes, and the associated wines. The absorption and metabolism of flavan-3-ol monomers from black tea are not obviously different from that observed after green tea consumption, although pro rata the dose of flavan-3-ols is much reduced.

To date, only one study has investigated the absorption of mixed theaflavins (theaflavin 17.7%, theaflavin-3-O-gallate 31.8%, theaflavin-3'-O-gallate 16.7%, and theaflavin-3,3'-digallate 31.4%) (Figure 2.4) (Mulder et al. 2001). An extremely high dose (700 mg mixed theaflavins, equivalent to about 30 cups of black tea) was given to two healthy volunteers, one male and one female. Plasma and urine concentrations were analyzed by HPLC–MS^2 after enzymatic deconjugation with β-glucuronidase and sulfatase, followed by extraction into ethyl acetate. Only theaflavin was detected because the enzyme treatment also removed ester gallate. Maximum theaflavin concentrations detected in the plasma of the female and male volunteers were 1.0 and 0.5 µg/L, respectively (1.8 and 0.9 fM), and maximum urine concentrations were 0.6 and 4.2 µg/L, respectively (1.1 and 7.4 fM), all at 2 h. These values should be doubled to correct for the relatively poor recovery observed with standard theaflavin, but even so, the total amount of theaflavin excreted was considerably less than 0.001% of the very large dose consumed (Mulder et al. 2001).

Attempts to investigate gut flora catabolism of mixed thearubigins *in vitro*, using conditions that were suitable for flavan-3-ols and proanthocyanidin B$_2$ (Stoupi et al. 2010a, 2010b), failed to observe the production of phenolic acids, even when using a low-protein medium to minimize the effects of protein binding (Knight 2004; Stoupi 2010). In contrast, volunteer studies have clearly demonstrated that the consumption of black tea beverage results in a substantially increased excretion of hippuric acid relative to baseline, suggesting that a combination of gut flora catabolism and post-absorption metabolism results in a significant production of benzoic acid (Clifford et al. 2000; Daykin et al. 2005; Mulder et al. 2005). The yield of benzoic acid excreted as hippuric acid is such that it points to thearubigins and theaflavins serving as substrates *in vivo*, and being degraded to phenolic acids. The yield of hippuric acid was not significantly affected by the presence of caffeine in the black tea or by the addition of milk, but varied by approximately four-fold between individuals for a given intake of black tea (Clifford et al. 2000).

2.2.6.1 Milk and Matrix Effects

The influence of milk on black tea polyphenol absorption has been investigated by several groups. In one study, nine healthy participants drank 400 mL of black tea (the average daily UK consumption) prepared from 3 g of black tea (the average content of a UK tea bag). The beverage was diluted before consumption by the addition of 100 mL of either water (control) or semiskimmed cow's milk. Plasma was sampled for up to 3 h and samples were analyzed colorimetrically for flavan-3-ols (using dimethylaminocinnamaldehyde) and by HPLC with fluorescence detection after enzymatic deconjugation. The addition of milk produced no significant effects on the ~0.3–0.4 µM concentrations of total plasma flavan-3-ols (Kyle et al. 2007).

A second study in which 12 healthy volunteers consumed 3 g of freeze-dried tea solids reconstituted in either 500 mL of water or 500 mL water plus 100 mL semiskimmed milk produced essentially identical results for plasma total flavan-3-ols as measured with dimethylaminocinnamaldehyde (C_{max} 0.17–0.18 µM), but interestingly suggested that the rate of elimination of total flavan-3-ols after consumption of black tea with milk was somewhat slower than for black tea alone, and both were significantly slower than elimination after consumption of green tea ($t_{e\frac{1}{2}}$ = 8.6, 6.9, and 4.8 h, respectively) (van het Hof et al. 1998). In the absence of data for the elimination flavan-3-ols given intravenously, such differences in the elimination rate constants must be interpreted cautiously because they might reflect a situation where the rate of absorption is the limiting factor on the rate of excretion.

In contrast to the above, a third study using five healthy volunteers who consumed either 300 mL of black tea beverage or 300 mL of black tea beverage to which 100 mL of milk was added concluded that an increase in the plasma antioxidant potential was completely abolished by the addition of milk (Serafini et al. 1996). In this study, not only was the rate of milk addition appreciably higher but efficacy was judged by measuring the effects on human plasma antioxidant capacity (TRAP). When these data are taken in combination with the results from the other studies, they suggests that (i) polyphenol metabolites in human plasma have little effect on plasma antioxidant capacity, as discussed more fully elsewhere (Clifford 2004; Clifford and Brown 2006) and (ii) that the TRAP assay is not specific for polyphenols. A similar study in which volunteers consumed cocoa with and without added milk clearly demonstrated that added milk significantly reduced urinary excretion of (epi)catechin metabolites (Mullen et al. 2009) was discussed in Section 2.2.3.1. It is also of note that the consumption of cocoa with milk had a significant effect on the excretion of microbial degradation products by both rats and humans (Urpi-Sarda et al. 2010). Nine out of 15 phenolic catabolites analyzed in urine were affected by milk consumption with a decrease in seven and an increase in two catabolites.

2.3 SUMMARY

Monomeric and polymeric flavan-3-ols are major dietary components with monomers being found in especially high concentrations in green tea. Proanthocyanidin oligomers and polymers have a widespread occurrence. In contrast thearubigins, which recent research has shown to comprise ~5000 "polymeric" structures, are found exclusively in fermented black tea along with the dimer-like compounds, theaflavins. The body treats flavan-3-ols, and other dietary (poly)phenols, as xenobiotics and following ingestion the monomer (–)-epicatechin is absorbed in the proximal GI tract, appearing transitorily in the circulatory system as glucuronide, sulfate, and methylated metabolites before being excreted in amounts equivalent to 30% of intake. Its 5′-hydroxy analog, (–)-epigallocatechin, is absorbed much less readily, especially at high doses.

Only minute amounts of proanthocyanidins and theaflavins enter the bloodstream and, along with the fraction of monomers that are not absorbed in the small intestine and metabolites excreted in bile, they pass into the large intestine. In the colon, both the monomers and procyanidins are degraded by the action of the microflora yielding characteristic products such as γ-valerolactones and a range of simpler phenolic acid structures. Recent *in vitro* and *in vivo* studies have begun to provide information on the likely catabolic routes involved in the production of these compounds. The colonic bacteria involved include *Eggerthella lenta* and *Flavonifractor plantii*. Colonic catabolism of black tea theaflavins and thearubigins has not been investigated to the same degree, but it appears that they are degraded by as yet underdetermined pathways and some significant portion excreted as hippuric acid.

The biological effects of these microbial catabolites is only now beginning to be investigated, but it is clear that they appear in very high levels in the colon, and in the systemic circulation at higher levels than the intact flavan-3-ols (Crozier et al. 2010). The biological relevance for most of these metabolites is unknown, and systematic approaches are required to address this.

REFERENCES

Appeldoorn, M.M., Vincken, J.-P., Aura, A.-M. et al. (2009). Procyanidin dimers are metabolized by human microbiota with 2-(3,4-dihydroxyphenyl)acetic acid and 5-(3,4-dihydroxyphenyl)-γ-valerolactone as the major metabolites. *J. Agric. Food Chem.*, **57**, 1084–1092.

Aron, P.M. and Kennedy, J.A. (2008). Flavan-3-ols: Nature, occurrence and biological activity. *Mol. Nutr. Food Res.*, **52**, 79–104.

Auger, C., Al Awwadi, N., Bornet, A. et al. (2004). Catechins and procyanidins in Mediterranean diets. *Food Res. Int.*, **37**, 233–245.

Auger, C., Hara, Y., and Crozier, A. (2008). Bioavailability of Polyphenon E flavan-3-ols in humans with an ileostomy. *J. Nutr.*, **138**, 1535S–1542S.

Baba, S., Osakabe, N., Yasuda, A. et al. (2000). Bioavailability of (–)-epicatechin upon intake of chocolate and cocoa in human volunteers. *Free Rad. Res.*, **33**, 635–641.

Bartolomè, B., Monagas, M., Garrido, I. et al. (2010). Almond (*Prunus dulcis* [Mill.] D.A. Webb) polyphenols: From chemical characterization to targeted analysis of phenolic metabolites in humans. *Arch. Biochem. Biophys.*, **501**, 124–133.

Borges, G., Mullen, W., Mullan, A. et al. (2010). Bioavailability of multiple components following acute ingestion of a polyphenol-rich juice drink. *Mol. Nutr. Food Res.* **54**, S268–277.

Bridges, J.W., French, M.R., Smith, R.L. et al. (1970). The fate of benzoic acid in various species. *Biochem. J.*, **118**, 47–51.

Chow, H.H., Cai, Y., Alberts, D.S. et al. (2001). Phase I pharmacokinetic study of tea polyphenols following single-dose administration of epigallocatechin gallate and Polyphenon E. *Cancer Epidemiol. Biomarkers Prev.*, **10**, 53–58.

Clifford, M.N. (2004). Diet-derived phenols in plasma and tissues and their implications for health. *Planta Med.*, **12**, 1103–1114.

Clifford, M.N. and Brown, J.E. (2006). Dietary flavonoids and health—broadening the perspective. In O. Andersen and K.R. Markham (eds.), *Flavonoids: Chemistry, Biochemistry and Applications*. CRC Press, Boca Raton, FL, pp. 320–370.

Clifford, M.N. and Crozier, A. (2011). Phytochemicals in teas and tisanes and their bioavailability. In A. Crozier, H. Ashihara, and F. Tomás-Barberán (eds.), *Teas, Cocoa and Coffee*. Blackwell Publishing, Oxford, pp. 45–98.

Clifford, M.N., Copeland, E.L., Bloxsidge, J.P. et al. (2000). Hippuric acid as a major excretion product associated with black tea consumption. *Xenobiotica*, **30**, 317–326.

Corder, R. (2008). Red wine, chocolate and vascular health: Developing the evidence base. *Heart*, **94**, 821–823.

Crozier, A., Jaganath, I.B., and Clifford, M.N. (2006). Phenols, polyphenols and tannins: An overview. In A. Crozier, M.N. Clifford, H. Ashihara (eds.), *Plant Secondary Metabolites: Occurrence, Structure and Role in the Human Diet*. Blackwell Publishing, Oxford, pp. 1–24.

Crozier, A., Del Rio, D., and Clifford, M.N. (2010). Bioavailability of dietary flavonoids and phenolic compounds. *Mol. Aspects Med.*, **31**, 446–67.

Crozier, A., Jaganath, I.B., and Clifford, M.N. (2009). Dietary phenolics: Chemistry, bioavailability and effects on health. *Nat. Prod. Rep.*, **26**, 1001–1043.

Day, A.J., Mellon, F., Barron, D. et al. (2001). Human metabolites of dietary flavonoids: Identification of plasma metabolites of quercetin. *Free Rad. Res.* **35**, 941–952.

Day, A.J. and Morgan M.R.A. (2003). Methods of polyphenol extraction from biological fluids and tissues. In C. Santos-Buelga and G. Williamson (eds.), *Methods in Polyphenol Analysis*. Royal Society of Chemistry, Cambridge, pp. 17–47.

Daykin, C.A., Van Duynhoven, J.P., Groenewegen, A. et al. (2005). Nuclear magnetic resonance spectroscopic based studies of the metabolism of black tea polyphenols in humans. *J Agric. Food Chem.*, **53**, 1428–1434.

Del Rio, D., Stewart, A.J., Mullen, W. et al. (2004). HPLC-MS^n analysis of phenolic compounds and purine alkaloids in green and black tea. *J. Agric. Food Chem.* **52**, 2807–2815.

Del Rio, D., Calani, L., Cordero, C. et al. (2010). Bioavailability and catabolism of green tea flavan-3-ols in humans. *Nutrition*, **11/12**, 1110–1116.

Del Rio, D., Calani, L., Dall'Asta, M. et al. (2011). Polyphenolic composition of hazelnut skins. *J. Agric. Food Chem.*, **59**, 9935–9941.

Déprez, S., Brezillon, C., Rabot, S. et al. (2000). Polymeric proanthocyanidins are catabolized by human colonic microflora into low-molecular-weight phenolic acids. *J. Nutr.*, **130**, 2733–2738.

Donovan, J.L., Manach, C., Rios, L. et al. (2002). Procyanidins are not bioavailable in rats fed a single meal containing a grape seed extract or the procyanidin dimer B3. *Br. J. Nutr.*, **87**, 299–306.

Donovan, J.L., Crespy, V., Oliveira, M. et al. (2006a). (+)-Catechin is more bioavailable then (−)-catechin: Relevance to the bioavailability of catechin from cocoa. *Free Rad. Res.*, **40**, 1029–1034.

Donovan, J.L., Manach, C., Faulks, R.M. et al. (2006b). Absorption and metabolism of dietary secondary metabolites. In A. Crozier, M.N. Clifford, and H. Ashihara (eds.), *Plant Secondary Metabolites. Occurrence, Structure and Role in the Human Diet.* Blackwell Publishing, Oxford, pp. 303–351.

Drynan, J.W., Clifford, M.N., Obuchowicz, J. et al. (2010). The chemistry of low molecular weight black tea polyphenols. *Nat. Prod. Rep.* **27**, 417–462.

Espín, J.C., García-Conesa, M.T., and Tomás-Barberán, F.A. (2007). Nutraceuticals: Facts and fiction. *Phytochemistry*, **68**, 2986–3008.

Gonthier, M.P., Donovan, J.L., Texier, O. et al. (2003). Metabolism of dietary procyanidins in rats. *Free Rad. Biol. Med.*, **35**, 837–844.

Gosnay, S.L., Bishop, J.A., New, S.A. et al. (2002). Estimation of the mean intakes of fourteen classes of dietary phenolics in a population of young British women aged 20–30 years. *Proc. Nutr. Soc.*, **61**, 125A.

Gossai, D. and Lau-Cam, C.A. (2009). Assessment of the effect of type of dairy product and chocolate matrix on the oral absorption of momomeric chocolate flavanols in a small animal model. *Pharmazie*, **64**, 202–209.

Goto, K., Kanaya, S., Nishikawa, T. et al. (1998). The influence of tea catechins on fecal flora of elderly residents in long-term care facilities. *Ann. Long-Term Care*, **6**, 43–48.

Green, R.J., Murphy, A.S., Schulz, B. et al. (2007). Common tea formulations modulate *in vitro* digestive recovery of green tea catechins. *Mol. Nutr. Food Res.*, **51**, 1152–1162.

Grumer, H.D. (1961). Formation of hippuric acid from phenylalanine labelled with carbon-14 in phenylketonuric subjects. *Nature*, **189**, 63–64.

Gu, L., Kelm, M.A., Hammerstone, J.F. et al. (2004). Concentrations of proanthocyanidins in common foods and estimations of normal consumption. *J. Nutr.*, **134**, 613–617.

Gu, L., Laly, M., Chang, H.C. et al. (2005). Isoflavone conjugates are underestimated in tissues using enzymatic hydrolysis. *J. Agric. Food Chem.*, **53**, 6858–6863.

Hammerstone, J.F., Lazarus, S.A., Mitchell, A.E. et al. (1999). Identification of procyanidins in cocoa (*Theobroma cocao*) and chocolate using high-performance liquid chromatography/mass spectrometry. *J. Agric. Food Chem.*, **47**, 490–496.

Haslam, E. (1998). *Practical Polyphenolics. From Structure to Molecular Recognition and Physiological Action.* Cambridge University Press, Cambridge.

Henning, S.M., Niu, Y., Lee, N.H. et al. (2004). Bioavailability and antioxidant activity of tea flavanols after consumption of green tea, black tea, or a green tea extract supplement. *Am. J. Clin. Nutr.*, **80**, 1558–1564.

Ho, Y., Lee, Y.-L., and Hsu, K.-Y. (1995). Determination of (+)-catechin in plasma by high-performance liquid chromatography using fluorescence detection. *J. Chromatogr. B*, **665**, 383–389.

Hodgson, J.M., Morton, L.W., Puddey, I.B. et al. (2000). Gallic acid metabolites are markers of black tea intake in humans. *J. Agric. Food Chem.*, **48**, 2276–2280.

Hodgson, J.M., Chan, S.Y., Puddey, I.B. et al. (2004). Phenolic acid metabolites as biomarkers for tea- and coffee-derived polyphenol exposure in human subjects. *Br. J. Nutr.*, **91**, 301–306.

Holt, R.R., Lazarus, S.A., Sullards, M.C. et al. (2002). Procyanidin dimer B2 [epicatechin-(4β-8)-epicatechin] in human plasma after the consumption of a flavanol-rich cocoa. *Am. J. Clin. Nutr.*, **76**, 798–804.

Kahle, K., Huemmer, W., Kempf, M. et al. (2007). Polyphenols are extensively metabolized in the human gastrointestinal tract after apple juice consumption. *J. Agric Food Chem.*, **55**, 10695–10614.

Keogh, J.B., McInerney, J., and Clifton P.M. (2007). The effect of milk protein on the bioavailability of cocoa polyphenols. *J. Food Sci.*, **72**, S230–233.

Kida, T., Suzuki, Y., Matsumoto, N. et al. (2000). Identification of biliary metabolites of (–)-epigallocatechin gallate in rats. *J. Agric. Food Chem.*, **48**, 4151–4155.

Knight, S. (2004). *Metabolism of Dietary Polyphenols by Gut Flora*. PhD Dissertation, University of Surrey, UK.

Kohri, T., Nanjo, F., Suziki, M. et al. (2001). Synthesis of (−)-[4-³H]epigallocatechin gallate and its metabolic fate in rats after intravenous administration. *J. Agric. Food Chem.*, **49**, 1042–1048.

Kuhnert, N., Clifford, M.N., and Muller, A. (2010a). Oxidative cascade reactions yielding polyhydroxy-theaflavins and theacitrins in the formation of black tea thearubigins: Evidence by tandem LC̃MS. *Food Funct.*, **1**, 180–199.

Kuhnert, N., Drynan, J.W., Obuchowicz, J. et al. (2010b). Mass spectrometric characterization of black tea thearubigins leading to an oxidative cascade hypothesis for thearubigin formation. *Rapid Commun. Mass Spectrom.*, **24**, 3387–3404.

Kutschera, M., Engst, W., Blaut, M. et al. (2011). Isolation of catechin–converting human intestinal bacteria. *J. Appl. Microbiol.* **111**, 165–175.

Kyle, J.A., Morrice, P.C., McNeill, G. et al. (2007). Effects of infusion time and addition of milk on content and absorption of polyphenols from black tea. *J. Agric. Food Chem.*, **55**, 4889–4894.

Li, C., Lee, M.J., Sheng, S. et al. (2000). Structural identification of two metabolites of catechins and their kinetics in human urine and blood after tea ingestion. *Chem. Res. Toxicol.*, **3**, 177–184.

Li, M., Wang, B., Zhang, M. et al. (2008). Symbiotic gut microbes modulate human metabolic phenotypes. *Proc. Natl. Acad. Sci. USA*, **105**, 2117–2122.

Luck, G. Liao, H., Murray, N.J. et al. (1994). Polyphenols, astringency and proline-rich proteins. *Phytochem.* **37**, 357–371.

Manach, C., Williamson, G., Morand, C. et al. (2005). Bioavailability and bioefficacy of polyphenols in humans. I. Review of 97 bioavailability studies. *Am. J. Clin. Nutr.*, **81**, 230S–242S.

Meng, X., Sang, S., Zhu, N. et al. (2002). Identification and characterization of methylated and ring-fission metabolites of tea catechins formed in humans, mice, and rats. *Chem. Res. Toxicol.*, **15**, 1042–1050.

Miller, K,B., Hurst, W.J., Payne, M.J. et al. (2008). Impact of alkalization on the antioxidant and flavanol content of commercial cocoa powders. *J. Agric. Food Chem.*, **56**, 8527–8533.

Mulder, T.P., van Platerink, C.J., Wijnand Schuyl, P.J. et al. (2001). Analysis of theaflavins in biological fluids using liquid chromatography–electrospray mass spectrometry. *J. Chromatogr. B*, **760**, 271–279.

Mulder, T.P., Rietveld, A.G., and van Amelsvoort, J.M. (2005). Consumption of both black tea and green tea results in an increase in the excretion of hippuric acid into urine. *Am. J. Clin. Nutr.*, **81**, 256S–260S.

Mullen, W., Borges, G., Donovan, J. L. et al. (2009). Milk decreases urinary excretion but not plasma pharmacokinetics of cocoa flavan-3-ol metabolites in humans. *Am. J Clin. Nutr.*, **89**, 1784–1791.

Neilson, A.P., George, J.C., Janle, E.M. et al. (2009). Influence of chocolate matrix composition of cocoa flavan-3-ol bioaccessibility *in vitro* and bioavailability in humans. *J. Agric. Food. Chem.*, **57**, 9418–9426.

Neilson, A.P., Sapper, T.N., Janle, E.M. et al. (2010). Chocolate matrix factors modulate the pharmacokinetic behaviour of cocoa flavan-3-ol phase II metabolites following oral consumption by Sprague-Dawley rats. *J. Agric. Food Chem.*, **58**, 6685–6691.

Ottaviani, J.I,. Momma, T.Y., Heiss, C. et al. (2011). The sterochemical configuration of flavanols influences the level and metabolism of flavanols in humans and their biological activity *in vivo*. *Free Rad. Biol. Med.*, **50**, 237–244.

Record, I.R. and Lane, J.M. (2001). Simulated intestinal digestion of green and black teas. *Food Chem.*, **73**, 481–486.

Reeve, D.R. and Crozier, A. (1980). Quantitative analysis of plant hormones. In J. MacMillan, ed., *Hormonal Regulation of Development 1. Molecular Aspects of Plant Hormones.* Encyclopedia of Plant Physiology New Series, Vol. 9, Springer-Verlag, Heidelberg, pp. 203–280.

Rein, D., Lotito, S., Holt, R.R. et al. (2000). Epicatechin in plasma: *In vivo* determination and effect of chocolate consumption on plasma oxidative status. *J. Nut.*, **130**, 2109S–2114S.

Richelle, M., Tavazzi, I., Enslen, M. et al. (1999). Plasma kinetics in man of epicatechin from black chocolate. *Eur. J. Clin. Nut.* **53**, 22–26.

Rios, L.Y., Bennett, R.N., Lazarus, S.A. et al. (2002). Cocoa procyanidins are stable during gastric transit in humans. *Am. J. Clin. Nut.*, **76**, 1106–1110.

Rios, L.Y., Gonthier, M.P., Remesy, C. et al. (2003). Chocolate intake increases urinary excretion of polyphenol-derived phenolic acids in healthy human subjects. *Am. J. Clin. Nutr.,* **77**, 912–918.

Roowi, S., Stalmach, A., Mullen, W. et al. (2010). Green tea flavan-3-ols: Colonic degradation and urinary excretion of catabolites by humans. *J. Agric. Food Chem.*, **58**, 1296–1304.

Roura, E., Andrés-Lacueva, C., Jáuregui, O. et al. (2005). Rapid liquid chromatography tandem mass spectrometer assay to quantify plasma (–)-epicatechin metabolites after ingestion of a standard portion of cocoa beverage in humans. *J. Agric. Food Chem.*, **53**, 6190–6194.

Roura, E., Almajano, M.P., Mata-Bilbao, M.L. et al. (2007a). Human urine: Epicatechin metabolites and antioxidant activity after cocoa intake. *Free Rad. Res.*, **41**, 943–949.

Roura, E., Andés-Lacueva, C., Estruch, R. et al. (2007b). Milk does not affect the bioavailability of cocoa powder flavonoid in healthy human. *Ann. Nutr. Metab.,* **51**, 493–498.

Roura, E., Andrés-Lacueva, C., Estruch, R. et al. (2008). The effects of milk as a food matrix for polyphenols on the excretion profile of cocoa (–)-epicatechin metabolites in healthy human subjects. *Br. J. Nut.,* **100**, 846–851.

Sang, S., Lee, M.J., Yang, I. et al. (2008). Human urinary metabolite profile of tea polyphenols analyzed by liquid chromatography/electrospray ionization tandem mass spectrometry with data-dependent acquisition. *Rapid Comm. Mass Spectrom.*, **22**, 1567–1578.

Schroeter, H., Holt, R.R., Orozco, T.J. et al. (2003). Milk and absorption of dietary flavanols. *Nature*, **426**, 787–788.

Self, H.L., Brown, R.R., and Price, J.M. (1960). Quantitative studies on the metabolites of tryptophan in the urine of swine. *J. Nutr.*, **70**, 21–25.

Serafini, M., Ghiselli, A., and Ferro-Luzzi, A. (1996). *In vivo* antioxidant effect of green and black tea in man. *Eur. J. Clin. Nutr.,* **50**, 28–32.

Serafini, M., Bugianesi, R., Maiaini, G. et al. (2003). Plasma antioxidants from chocolate. *Nature* **424**, 1013.

Seto, R., Nakamura, H., Nanjo, F. et al. (1997). Preparation of epimers of tea catechins by heat treatment. *Biosci. Biotechnol. Biochem.*, **61**, 1434–1439.

Shahrzad, S. and Bitsch, I. (1998). Determination of gallic acid and its metabolites in human plasma and urine by high-performance liquid chromatography. *J. Chromatogr. B*, **705**, 87–95.

Shahrzad, S., Aoyagi, K., Winter, A. et al. (2001). Pharmacokinetics of gallic acid and its relative bioavailability from tea in healthy humans. *J. Nutr.* **131**, 1207–1210.

Shoji, T., Masumoto, S., Moriichi, N. et al. (2006). Apple procyanidin oligomers absorption in rats after oral administration: Analysis of procyanidins in plasma using the porter method and high-performance liquid chromatography/tandem mass spectrometry. *J. Agric. Food Chem.,* **54**, 884–892.

Spencer, J.P., Chaudry, F., Pannala, A.S. et al. (2000). Decomposition of cocoa procyanidins in the gastric milieu. *Biochem. Biophys. Res. Commun.,* **272**, 236–241.

Stalmach, A., Troufflard, S., Serafini M. et al. (2009). Absorption, metabolism and excretion of Choladi green tea flavan-3-ols by humans. *Mol. Nutr. Food Res.,* **53**, S44–S53.

Stalmach, A., Mullen, W., Steiling, H. et al. (2010). Absorption, metabolism, efflux and excretion of green tea flavan-3-ols in humans with an ileostomy. *Mol. Nutr. Food Res.,* **54**, 323–334.

Stoupi, S. (2010). *In Vitro and In Vivo Metabolic Studies of Dietary Flavan-3-ols.* PhD Dissertation, University of Surrey, UK.

Stoupi, S., Williamson, G., Drynan, J.W. et al. (2010a). Procyanidin B2 catabolism by human fecal microflora: Partial characterization of 'dimeric' intermediates. *Arch. Biochem. Biophys.,* **501**, 73–78.

Stoupi, S., Williamson, G., Drynan, J.W. et al. (2010b). A comparison of the *in vitro* biotransformation of (–)-epicatechin and procyanidin B2 by human faecal microbiota. *Mol. Nutr. Food Res.,* **54**, 747–759.

Stoupi, S., Williamson, G., Viton, F. et al. (2010c). *In vivo* bioavailability, absorption, excretion, and pharmacokinetics of [^{14}C]procyanidin B2 in male rats. *Drug Met. Disp.,* **38**, 287–291.

Tomás-Barberán, F.A., Cienfuegos-Jovellanos, E., Marín, A. et al. (2007). A new process to develop a cocoa powder with higher flavonoid monomer content and enhanced bioavailability in healthy humans. *J. Agric. Food Chem.,* **55**, 3926–3935.

Tsang, C., Auger, C., Mullen, W. et al. (2005). The absorption, metabolism and excretion of flavan-3-ols and procyanidins following the ingestion of a grape seed extract by rats. *Br. J. Nutr.,* **94**, 170–181.

Turnbaugh, P.J and Gordon, J.I. (2008). An invitation to the marriage of metagenomics and metabolomics. *Cell,* 134, 708–713.

Tzounis, X., Vulevic, J., Kuhnle, G.G. et al. (2008). Flavanol monomer-induced changes to the human faecal microflora. *Br. J. Nutr.,* **99**, 782–792.

Tzounis, X., Rodriguez-Mateos, A., Vulevic, J. et al. (2011). Prebiotic evaluation of cocoa-derived flavanols in healthy humans by using a randomized, controlled, double-blind, crossover intervention study. *Am. J. Clin. Nutr.,* **93**, 62–72.

Urpi-Sarda, M., Monagas, M., Khan, N. et al. (2009). Epicatechin, procyanidins and phenolic microbial metabolites after cocoa intake in humans and rats. *Anal. Bioanal. Chem.,* **394**, 1545–1556.

Urpi-Sarda, M., Llorach, R., Khan, N. et al. (2010). Effect of milk on the urinary excretion of microbial phenolic acid after cocoa powder consumption in humans. *J. Agric. Food Chem.,* **58**, 4706–4711.

Unno, T., Kondo, K., Itakura, H. et al. (1996). Analysis of (–)-epigallocatechin gallate in human serum obtained after ingesting green tea. *Biosci. Biotechnol. Biochem.,* **60**, 2066–2068.

van Duynhoven, J., Vaughan, E.E., Jacobs, D.M. et al. (2011). Metabolic fate of polyphenols in the human superorganism. *Proc. Natl. Acad. Sci. USA,* **108** (Suppl. 1), 4531–4538.

van het Hof, K.H., Kivits, G.A., Weststrate, J.A. et al. (1998). Bioavailability of catechins from tea: The effect of milk. *Eur. J. Clin. Nutr.,* **52**, 356–359.

Vega, C. and Grover, M.K. (2011). Physico-chemical properties of acidified skim milk gels containing cocoa flavanols. *J. Agric. Food Chem.,* **59**, 6740–6747.

Walle, T, Hsieh, F., De Legge, M.H. et al. (2004). High absorption but very low bioavailability of oral resveratrol in humans. *Drug Met. Disp.,* **32**, 1377–1382.

Wang, J.F., Schramm, D.D., Holt, R.R. et al. (2000). A dose-response effect from chocolate consumption on plasma epicatechin and oxidative damage. *J. Nutr.,* **130**, 2115S–2119S.

Ward, N.C, Croft, K.D., Puddey, I.B. et al. (2004). Supplementation with grape seed polyphenols results in increased urinary excretion of 3-hydroxyphenylpropionic acid, an important metabolite of proanthocyanidins in humans. *J. Agric. Food Chem.* **52**, 5545–5549.

Williamson G, Clifford MN. (2010). Colonic metabolites of berry polyphenols: The missing link to biological activity? *Br. J. Nutr.,* **104**, S48–S66.

Woods, E., Clifford M.N., Gibbs, M. et al. (2003). Estimation of mean intakes of 14 classes of dietary phenols in a population of male shift workers. *Proc. Nutr. Soc.* **62**, 60A.

Yoshino, K., Suzuki, M., Sasaki, K. et al. (1999). Formation of antioxidants from (–)-epigallocatechin gallate in mild alkaline fluids, such as authentic intestinal juice and mouse plasma. *J. Nutr. Biochem.,* **10**, 223–229.

Zhu, Q.A., Zhang, A. Tsang, D. et al. (1997). Stability of green tea catechins. *J. Agric. Food Chem.,* **45**, 4624–4628.

3 Anthocyanins: Understanding Their Absorption and Metabolism

Ronald L. Prior

CONTENTS

3.1 INTRODUCTION

Anthocyanins are water soluble plant pigments responsible for the blue, purple, and red colors of many plant tissues. They occur primarily as *O*-glycosides of their respective aglycone anthocyanidin-chromophores. Although there are about 17 anthocyanidins found in nature, only six of them—cyanidin, delphinidin, petunidin, peonidin, pelargonidin, and malvidin—are ubiquitously distributed. The differences in chemical structure of these six common anthocyanidins occur at the 3′ and 5′ positions of the B-ring (Figure 3.1). Except for the 3-deoxyanthocyanidins such as luteolinidin and apigeninidin in sorghum (Figure 3.2) (Wu and Prior 2005b), the aglycones are rarely found in fresh plant materials. The sugar moiety is attached mainly at the 3-position on the *C*-ring or in some cases at the 5-, or 7-position on the A-ring. Glucose, galactose, arabinose, rhamnose, and xylose are the most common

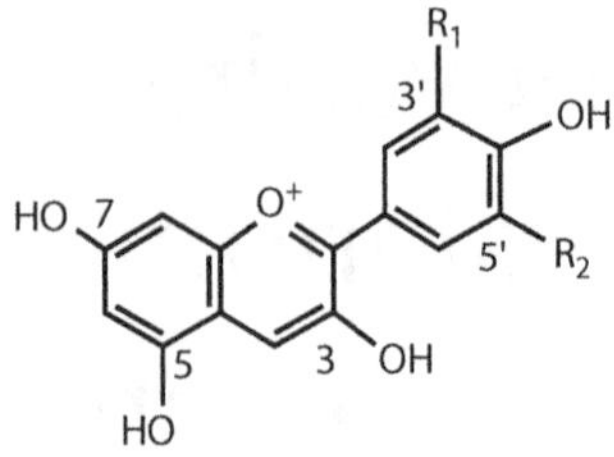

Anthocyanidin	R_1	R_2
Pelargonidin	H	H
Cyanidin	OH	H
Delphinidin	OH	OH
Peonidin	OCH_3	H
Petunidin	OCH_3	OH
Malvidin	OCH_3	OCH_3

FIGURE 3.1 Structures of the six major anthocyanidins. Sugar moieties are generally on position 3 of the C-ring.

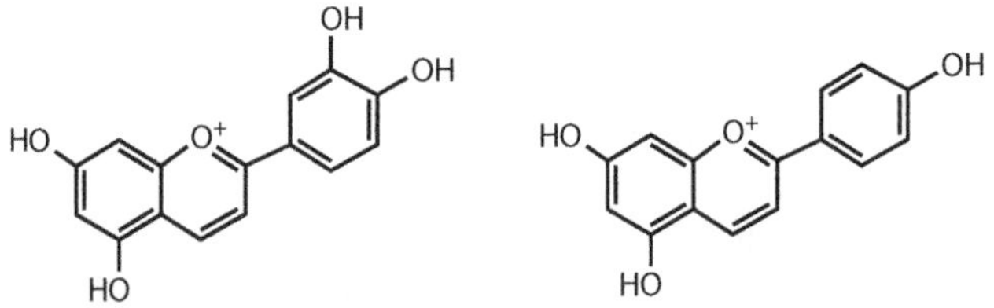

FIGURE 3.2 The 3-deoxyanthocyanidins luteolinidin and apigeninidin are unusual as they occur in sorghum as aglycones.

sugars that are bonded to anthocyanidins as mono-, di-, or tri-saccharide forms (Figure 3.3). The sugar moieties may also be acylated by a range of aromatic or aliphatic acids. Common acylating agents are cinnamic acids (Figure 3.3). More than 600 naturally occurring anthocyanins have been reported (Anderson 2002) and they are known to vary in: (i) the number and position of hydroxyl and methoxyl groups on the basic anthocyanidin skeleton; (ii) the identity, number, and positions at which sugars are attached; and, (iii) the extent of sugar acylation and the identity of the acylating agent (Prior 2004). An example of some of the complex structures that can occur in foods are in publications by Wu and Prior (2005a, 2005b).

Unlike other subgroups of flavonoids with a similar C_6-C_3-C_6 skeleton, anthocyanins have a positive charge in their structure at acidic pH. In solution, the anthocyanin actually occurs in equilibrium with essentially four molecular forms: the flavylium cation, the quinoidal base, the hemiacetal base, and chalcone (Cooke et al. 2005) (Figure 3.4). The relative amounts of these four forms vary with both

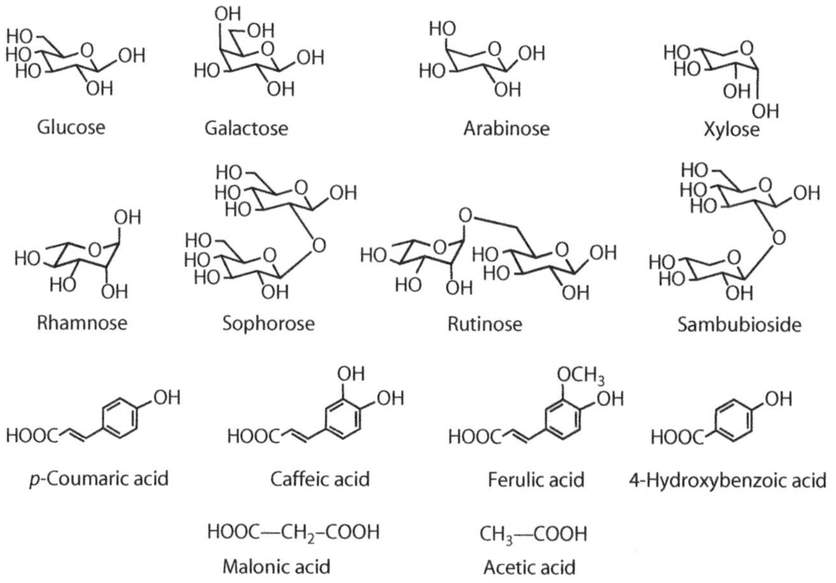

FIGURE 3.3 Selected sugars and aromatic or aliphatic acids that commonly occur in anthocyanin structures.

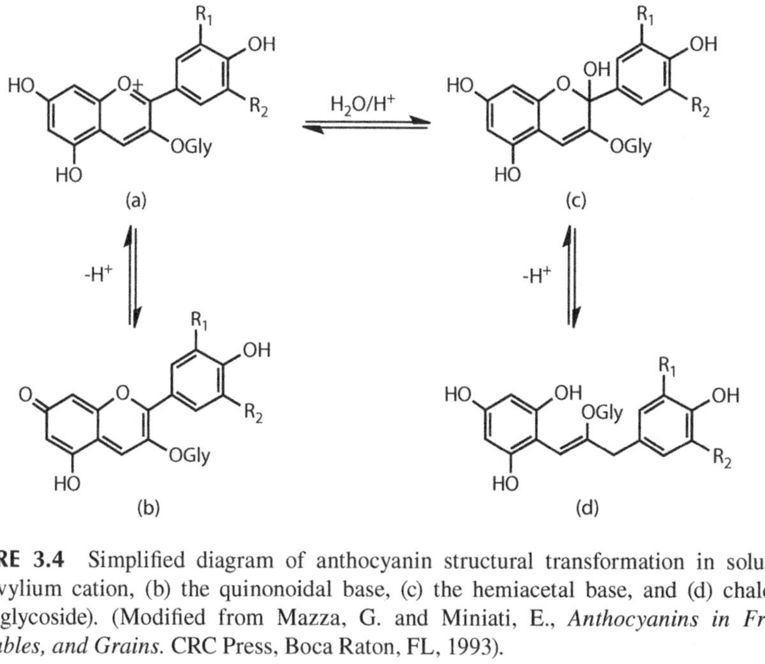

FIGURE 3.4 Simplified diagram of anthocyanin structural transformation in solution. (a) flavylium cation, (b) the quinonoidal base, (c) the hemiacetal base, and (d) chalcone (Gly: glycoside). (Modified from Mazza, G. and Miniati, E., *Anthocyanins in Fruits, Vegetables, and Grains.* CRC Press, Boca Raton, FL, 1993).

pH and structure of the anthocyanins (Brouillard and Delaporte 1977; Brouillard 1982; Mazza and Miniati 1993). Anthocyanins exist primarily as the stable flavylium cation only when the pH is less than two. This uniqueness in the chemical structure is one of the key factors that may affect the absorption, metabolism, bioavailability, and, consequently, the biological responses to anthocyanins.

A number of reviews of anthocyanin absorption and metabolism have been published previously (Scalbert and Williamson 2000; Crozier et al. 2004, 2009; Galvano et al. 2004, 2007; Manach et al. 2005; Prior and Wu 2006; McGhie and Walton 2007). This review will focus principally on more recent results published in the last 3–5 years.

3.2 INTAKE AND BIOAVAILABILITY ESTIMATES OF ANTHOCYANINS

3.2.1 INTAKE

Of the various classes of flavonoids, the potential daily dietary intake of anthocyanins at 100 + mg is perhaps the most substantial. However, in practice, the average consumption in the United States was calculated to be much lower (~12 mg/day) (Wu et al. 2006a), which is probably due to the low consumption fruits, especially berries. However, other estimates have been much higher (Wang and Stoner 2008). The content in fruits varies considerably between 0.25 and 700 mg/100 g fresh weight (Wu et al. 2006a). Not only does the concentration vary, but the specific anthocyanins present in foods are also quite different (Wu and Prior 2005a, 2005b).

3.2.2 SITES OF ANTHOCYANIN ABSORPTION

Anthocyanins were shown to be absorbed in the glycosidic form (Cao and Prior 1999; Cao et al. 2001). This is in contrast with other flavonoids that are deglycosylated prior to absorption. During the past decade, several studies have been completed aimed at understanding the site(s) of absorption of anthocyanins.

Studies with animals have shown that the stomach is a site of anthocyanin absorption. Talavera et al. (2003) found that after administration of a high concentration of blackberry anthocyanins, intact anthocyanins were observed in plasma from the gastric vein and aorta. Cyanidin-3-O-glucoside appeared in bile after as little as 20 min. This study seemed to demonstrate that anthocyanin glycosides were quickly and efficiently absorbed from the stomach and rapidly excreted into bile as both intact and metabolized forms. This rapid and apparent high absorption of anthocyanins from the stomach raises several questions as to the disposition of anthocyanins once they are absorbed. Grape anthocyanins were shown to reach the brain within minutes of their introduction into the stomach (Passamonti et al. 2005). In another study, mavidin-3-O-glucoside appeared in both portal and systemic plasma of rats after only 6 min (Passamonti et al. 2003). Anthocyanins are capable of permeating the gastric mucosa, possibly through a bilitranslocase-mediated mechanism (Passamonti et al. 2003).

In addition to the stomach, Talavera et al. (2004) found that anthocyanins were also absorbed efficiently from the small intestine. Using *in situ* perfusion of the jejunum and ileum in rats, they observed that the absorption of anthocyanins was influenced by the chemical structure of the anthocyanin and varied from 10.7% (malvidin-3-O-glucoside) to 22.4% (cyanidin-3-O-glucoside). Regardless of the anthocyanins perfused, only the glycoside forms were recovered in the intestinal lumen. In a more recent paper in which the intestinal absorption of anthocyanins was studied using an *in vitro* chamber model (Matuschek et al. 2006), the authors found that the highest absorption of anthocyanins occurred with jejunual tissue (55.3 ± 7.6%). Minor absorption occurred with duodenal tissue (10.4 ± 7.6%), with no absorption recorded in tissue from the ileum or colon. In the plasma of rats after oral administration of 100 mg delphinidin-3-O-glucose/kg body weight (Ichiyanagi et al. 2004a, 2004b), two peaks of delphinidin-3-O-glucoside were observed at 15 and 60 min. The second peak seemed to appear in relation to the time at which the anthocyanins reached the small intestine from the stomach, indicating that delphinidin-3-O-glucoside was likely absorbed from the jejunum as well as from the stomach.

González-Barrio et al. (2010) studied the fate of anthocyanins following the consumption of 300 g of raspberries by healthy human volunteers and subjects with an ileostomy. The three main raspberry anthocyanins were excreted in urine in both healthy and ileostomy volunteers 0–7 h after ingestion, in quantities corresponding to <0.1% of intake, indicating an apparent low level of absorption in the small intestine. With ileostomy volunteers 40% of anthocyanins were recovered in ileal fluid with the main excretion period being the first 4 h after raspberry consumption.

3.2.3 IN VIVO AND IN VITRO BIOAVAILABILITY ESTIMATES

The lack of radioactive tracers has limited the capability of getting accurate estimates of anthocyanin bioavailability. Hence, various *in vivo* and *in vitro* techniques including appearance rates in the blood and urine, disappearance from the gastrointestinal tract, etc. have been used in studying this question, but none really give true estimates of bioavailability.

Anthocyanins are absorbed intact without cleavage of the sugar to form the aglycones. Anthocyanins have been considered to have a very low bioavailability. Plasma levels of anthocyanins are in the range of 1–120 nM following a meal high in anthocyanins, but fasting plasma levels are generally non-detectable (Galvano et al. 2008; Kay et al. 2009). Much remains to be learned as to why plasma anthocyanin levels are so low after supplementation compared with those of other dietary flavonoids. Clinical studies carried out with humans consuming different types of fruits containing anthocyanins have shown that the concentration of anthocyanins in plasma ranged 10–50 nM (after a 50 mg of aglycone equivalent dose) and the T_{max} was 1.5 h. At intakes of 188–3570 mg total cyanidin glycosides, the C_{max} was in the range of 2.3–96 nM (Prior and Wu 2006; Galvano et al. 2007). The proportion of the dose that appears in the urine is usually quite small (<0.1% of dose) (Cao et al. 2001; Wu et al. 2004, 2005), but proportions of dose excreted have been reported up to 0.37% of the intake (Galvano et al. 2007). The time of appearance of anthocyanins in the plasma (typical T_{max} times are 0.2–1.5 h) (Prior and Wu 2006) is consistent with absorption at the stomach but also at the level of the small intestine (Talavera et al. 2003; Felgines et al. 2005, 2007; de Pascual-Teresa et al. 2010). Maximal urinary excretion is usually achieved in less than 4 h. As will be discussed later, the bioavailability estimates of anthocyanins are likely underestimated because of the numerous metabolites that are formed with some metabolites likely not identified to date.

Two recent studies in humans with anthocyanins from red wine and acai have confirmed these parameters. After a 12-h overnight fast, seven healthy volunteers received 12 g of an anthocyanin extract from red wine in a test meal. Anthocyanins and their metabolites appeared in plasma about 30 min after ingestion of the test meal and reached their maximum value around 1.6 h later for glucosides and 2.5 h for glucuronides. Total urinary excretion of red wine anthocyanins was $0.05 \pm 0.01\%$ of the administered dose within 24 h. About 94% of the excreted anthocyanins found in urine was within 6 h (Garcia-Alonso et al. 2009). The absorption of anthocyanins in acai pulp or acai juice has also been studied in human volunteers by Mertens-Talcott et al. (2008). Anthocyanin plasma C_{max} values of 5.2 and 2.5 nM, T_{max} times of 2.2 and 2.0 h, and area under the curve (AUC) values of 19.1 and 7.4 nmol/h/L were obtained with pulp and juice, respectively. Total anthocyanins were quantified as cyanidin-3-O-glucoside. Using non-linear mixed effect modeling, dose volume identified as a significant predictor of relative oral bioavailability in a negative nonlinear relationship for acai pulp and juice.

Bioavailability of cyanidin-3-O-glucoside was estimated in C57BL6J mice given cyanidin-3-O-glucoside by determining blood concentrations following either gavage (500 mg/kg) or tail vein injection (1 mg/kg) (Marczylo et al. 2009). After oral administration, peak concentrations of anthocyanins occurred within 30 min after administration. Systemic bioavailabilities for the parent cyanidin-3-O-glucoside was estimated to be 1.7% and for total anthocyanins, 3.3%. This compares to most studies reported in humans in which the urinary excretion of anthocyanins ranged from 0.018 to 0.37% of the dose (Galvano et al. 2007). After oral or intravenous administration, cyanidin-3-O-glucoside half-lives in the different body fluids (blood, urine, and bile) ranged from 0.7 to 1.8 h and in tissues (heart, lung, kidney, liver, prostate, brain, and gastrointestinal mucosa) from 0.3 to 0.7 h (Marczylo et al. 2009). In the gastrointestinal mucosa and liver, the predominant species after oral administration was cyanidin-3-O-glucoside, but after intravenous dosing, the majority of the anthocyanins were methylated and glucuronidated metabolites of cyanidin-3-O-glucoside (Marczylo et al. 2009).

However, the apparent absorption efficiency differs depending upon whether one measures disappearance from the gastrointestinal tract or appearance in the blood or excretion in the urine. Wu et al. (2006) observed that about 58% of a dose of anthocyanins disappeared from the gastrointestinal tract within 4 h after a meal as determined by recovery of the anthocyanins in the duodenum, ileum, colon, and cecum of pigs. However, the extent of disappearance varied considerable depending upon the anthocyanin with 98% of the cyanidin-3-O-glucose disappearing but only 22% of the cyanidin-3-O-sambubioside. In this case, a majority of the anthocyanins that disappeared were likely degraded in the gut and were not absorbed.

The bioavailability of sweet cherry anthocyanins was assessed *in vitro* using a digestion process involving pepsin-HCl digestion (to simulate gastric digestion) and pancreatin digestion with bile salts (to simulate small intestine conditions) and dialyzed to assess serum- and colon-available fractions (Fazzari et al. 2008). After pepsin digestion, all the anthocyanins were recovered. Following pancreatic digestion and dialysis, the anthocyanin content in the fraction available to the serum (IN fraction) was 15–21%, and in the fraction available to the colon (OUT fraction), 52–67%. Anthocyanins present in commercial blackcurrant juice were shown to remain stable during *in vitro* digestion in gastric fluid regardless whether pepsin was added into the medium or not. They also remained stable during *in vitro* digestion in simulated intestinal fluid without pancreatin, but the stability of anthocyanins in intestinal fluid containing pancreatin was reduced (Uzunovic and Vranic 2008).

3.3 MECHANISMS OF ABSORPTION

Understanding the mechanisms of absorption, tissue distribution, metabolism, and excretion of dietary anthocyanins will help in understanding the apparent paradox between their low concentrations in cells and their bioactivity.

The roles played by transport mechanisms at the intestinal and blood–brain barrier (BBB) are largely unknown. Dreiseitel et al. (2009) evaluated the effects of 16 anthocyanins and anthocyanidins on human efflux transporter multidrug resistance protein 1 (MDR1) and breast cancer resistance protein (BCRP), using dye efflux, ATPase, and, for BCRP, vesicular transport assays. All of the tested compounds interacted with the BCRP transporter *in vitro*. It is known that BCRP actively transports various anticancer drugs and restricts the uptake of the food carcinogen 2-amino-1-methyl-6-phenylimidazo[4,5-*b*]pyridine from the gut lumen. BCRP may be an important part of the intestinal barrier protecting the body from food-associated carcinogens. Of the anthocyanin compounds tested, seven emerged as potential BCRP substrates (malvidin, petunidin, malvidin-3-*O*-galactoside, malvidin-3,5-*O*-diglucoside, cyanidin-3-*O*-galactoside, peonidin-3-*O*-glucoside, and cyanidin-3-*O*-glucoside) and 12 as potential inhibitors of BCRP (cyanidin, peonidin, cyanidin-3,5-*O*-diglucoside, malvidin, pelargonidin, delphinidin, petunidin, delphinidin-3-*O*-glucoside, cyanidin-3-*O*-rutinoside, malvidin-3-*O*-glucoside, pelargonidin-3,5-*O*-diglucoside, and malvidin-3-*O*-galactoside). Malvidin, malvidin-3-*O*-galactoside, and petunidin exhibited bimodal activities serving as BCRP substrates at low concentrations and, at higher concentrations, as BCRP inhibitors. Although the anthocyanidins studied may alter pharmacokinetics of drugs that are BCRP substrates, they are less likely to interfere with activities of MDR1 substrates. The present data suggest that several anthocyanins may be actively transported out of intestinal tissues and endothelia, limiting their bioavailability in plasma and brain (Dreiseitel et al. 2009). Anthocyanidins may not be important *in vivo* as they are unstable and not found in plasma or other body tissues.

Absorption of anthocyanins in the intestine was tested using a Caco-2 cell model and an anthocyanin extract, rich in malvidin-3-*O*-glucoside obtained from red grape skins (Faria et al. 2009). Cells that were pretreated for 96 h with anthocyanins (200 µg/mL) showed an increase in their own transport by 50%. Expression of facilitative glucose transporters 2 was increased (60%) in Caco-2 cells pretreated with anthocyanins, in comparison with controls as assessed by RT-PCR.

Vanzo et al. (2008) studied the renal uptake of dietary anthocyanins and the underlying molecular mechanism using a solution containing anthocyanins extracted from red grapes of *Vitis vinifera,* which was introduced into the isolated stomach of anesthetized rats. After 10 and 30 min, plasma, liver, and kidney were analyzed for their anthocyanin content. While anthocyanins in the liver were at apparent equilibrium with plasma both after 10 and 30 min, kidney anthocyanins were 3- and 2.3-fold higher than in plasma, after 10 and 30 min, respectively. The transport activity of bilitranslocase in kidney basolateral membrane vesicles was competitively inhibited by malvidin-3-*O*-glucoside [K(i) = 4.8 ± 0.2 µM]. From this data, the authors concluded that anthocyanin uptake from blood into kidney tubular cells is likely to be mediated by the kidney isoform of this organic anion membrane transporter.

3.4 FACTORS AFFECTING ANTHOCYANIN ABSORPTION

3.4.1 ACYLATION EFFECTS ON ANTHOCYANIN ABSORPTION

In feeding studies using whole foods, nonacylated anthocyanins were more bioavailable than their acylated counterparts, but the extent to which the plant matrix determines relative bioavailability of anthocyanins has not been known. Using juice of purple carrots to circumvent matrix effects, acylated anthocyanins comprised 76% of total anthocyanins in the juice, yet their bioavailability in humans was found to be significantly less than that of non-acylated anthocyanins (Charron et al. 2009). Peak plasma concentrations of non-acylated anthocyanins were four-fold higher than that for acylated anthocyanins (Kurilich et al. 2005; Charron et al. 2009). Because the treatments were consumed as juice, it could be discerned that the difference in bioavailability of acylated versus non-acylated anthocyanins was not primarily caused by interactions with the plant matrix but by acylation status of the anthocyanin (Charron et al. 2009).

3.4.2 EFFECTS OF DEGRADATION ON ESTIMATES OF BIOAVAILABILITY

The proportion of ingested anthocyanins reaching the systemic circulation is a small percentage of their ingested dose, which may be due to physiochemical degradation *in vivo*. Woodward et al. (2009) investigated the effect of anthocyanin structure on their stability under simulated *in vitro* physiological conditions and their degradation and recovery following routine pre-analytical sample extraction and storage. The hydroxylation of the B-ring was shown to mediate the degradation of anthocyanins to their phenolic acid and aldehyde constituents. Their data indicate that significant portions of ingested anthocyanins are likely to degrade to phenolic acids and aldehydes *in vivo*.

Almost all of the *in vitro* mechanistic studies of anthocyanins on bioactivity outcomes have been done using anthocyanin aglycones and/or glycosides, even though it is not clear that these are the biological active components. Kay et al. (2009) conducted intestinal epithelial cell culture experiments using Caco-2 cells, which indicated that after a 4-h incubation of anthocyanins in cell-free culture media, 57% of the initial cyanidin-3-*O*-glucoside and 96% of cyanidin had degraded. The level of degradation was not statistically different from that of cultured cell incubations, suggesting that degradation was spontaneous. Degradation products were identified as 3,4-dihydroxybenzoic acid (protocatechuic acid) and 2,4,6-trihydroxybenzalde-hyde (phloroglucinaldehyde). Similar results were confirmed using two other buffer matrices (phosphate and Hank's buffers). In the cultured cell media system, the degradation products 3,4-dihydroxybenzoic acid and 2,4,6-trihydroxybenzaldehyde were further metabolized to glucuronide and sulfate conjugates. These data suggest a significant proportion of intestinal metabolites of anthocyanins are likely to be conjugates of their degradation products.

These observations are beginning to be confirmed *in vivo*. Vitaglione et al. (2007) demonstrated that 3,4-dihydroxybenzoic acid was the main metabolite present in the bloodstream and was excreted in the feces of human volunteers consuming one liter of blood orange juice providing mainly cyanidin-3-*O*-glucoside. Future efforts to establish the biological activities of anthocyanins should therefore include the investigation of phenolic acids and other products of degradation, along with their respective metabolites. Recent data obtained with rats has also shown that a signifi-cant portion of the phenolic acids are conjugated *in vivo* (Prior et al. 2010), but the extent of conjugation during absorption and excretion varies considerably depending upon the specific phenolic acid.

3.4.3 Effects of Different Anthocyanin-*O*-Glycosides on Absorption

In a study in which pigs were fed black raspberries, while cyanidin-3-*O*-glucoside disappearance from the small intestine after the meal was 98%, the total recovery of cyanidin-3-glucoside and its possible metabolites in the urine did not show a corresponding increase compared with that of other parent anthocyanins (Wu et al. 2006b). Higher recovery of anthocyanins in the gut was associated with increased recovery in the urine. This likely indicates that the disappearance of cyanidin-3-*O*-glucoside in the small intestine was caused by degradation rather than absorption and movement into the circulation. However, the ratio of the anthocyanin di- or triglyco-sides did not change within the different segments of the small intestine compared with that in the black raspberries. The significant increase of cyanidin-3-*O*-sambubi-oside in the cecum and colon that was observed may be the result of deglycosylation of cyanidin-3-sambubioside-5-rhamnoside because the rhamnose linkage has been found to be cleaved more easily than other sugars (Wu et al. 2006b).

In earlier studies by Wu et al. (2004, 2005), anthocyanins with more complex sugar conjugates were not cleared from the blood as rapidly as cyanidin-3-*O*-gluco-side. The ratios for the plasma anthocyanin AUC, adjusted for amount consumed, were 1.0:2.0:6.2 for cyanidin-3-*O*-glucoside:cyanidin-3-*O*-rutinoside:cyanidin-3-*O*-sambubioside. Ratios for the same anthocyanins excreted in urine were 1.0:1.7:2.9. The relative amounts in the gastrointestinal tract were reflected in plasma and urine. This suggests that degradation of anthocyanins within the gastrointestinal tract may be a limiting factor for their absorption and if the anthocyanins are more stable within the gastrointestinal tract, greater quantities can be absorbed. One might expect the converse with low recovery in the gastrointestinal tract associated with reduced absorption and decreased amounts found in the urine.

Blackcurrant anthocyanins were dissolved in water with or without the addition of oatmeal and orally administered to rats, providing approximately 250 mg total anthocyanins per kilogram body weight (Walton et al. 2009). The relative concentration of rutinosides in the digesta increased during their passage through the gastrointestinal tract, while the glucosides decreased. This suggests that the rutinosides were not absorbed as well as glucosides or that the rutinosides were more stable than glucosides in the gastrointestinal tract.

3.4.4 GLUCOSE AND OTHER FACTORS

Neither 24-h urinary anthocyanin excretion nor plasma anthocyanin concentration was significantly affected by simultaneous ingestion of glucose in rats (Felgines et al. 2008). However, *in vitro* in Caco-2 cells, an anthocyanin extract rich in malvidin-3-*O*-glucoside caused a 60% reduction in glucose absorption (Faria et al. 2009). Thus, the anthocyanidin-glycoside seems to interact or bind with the glucose transporter. The aglycone, malvidin, did not have an effect on glucose absorption. In the presence of ethanol, anthocyanin absorption was stimulated after 60 min of incubation (Faria et al. 2009). Anthocyanins from either orange or purple carrots did not alter the absorption of β-carotene in human subjects (Arscott et al. 2010).

In rats, a viscous food matrix (oatmeal) was shown to delay urinary anthocyanin excretion implying that absorption was decreased (Walton et al. 2009). The plasma T_{max} occurred faster when anthocyanins were consumed with water (0.25 h) than after an anthocyanin meal with oatmeal (1.0 h). Also, the maximum anthocyanin excretion in urine occurred later after consumption of anthocyanins with oatmeal than after consumption with water (3 h compared with 2 h). Whether these effects are on total anthocyanin absorption or a reflection of effects on intestinal transit time is not clear.

The effect of consumption of 200 g strawberries with and without cream on the absorption and excretion of pelargonidin-3-glucoside and its metabolites was studied in humans (Mullen et al. 2008). The quantities excreted over the 0–24 h collection period were not influenced significantly by consumption of cream. However, during the 0–2 h period of excretion, the excretion of anthocyanin metabolites was significantly lower when the strawberries were eaten with cream, whereas the reverse occurred during with the 5–8 h excretion period. This effect seemed to be the result of delayed gastric emptying and extended mouth to cecum transit time.

In a study with rats fed blueberry anthocyanins, the duration of time over which anthocyanins were consumed also seemed to affect anthocyanin absorption (Del Bo et al. 2010). In this study, the anthocyanin profile in plasma, urine, feces, brain, and liver was evaluated by LC-MS/MS. The quantities of anthocyanins excreted in the urine and not in the feces increased after 8 weeks on the blueberry diet compared with that after 4 weeks of consumption. Urinary excretion of hippuric acid increased significantly after 4 and 8 weeks of blueberry anthocyanin consumption, suggesting that the extent of anthocyanin metabolism by gut microflora and excretion can be altered by duration of dietary blueberry consumption.

3.5 METABOLISM OF ANTHOCYANINS

Cao et al. (2001) were the first to demonstrate the excretion of unmetabolised, intact anthocyanins from elderberry excreted in the urine. Wu et al. (2002) also identified four urinary anthocyanin metabolites from elderberry, namely (1) peonidin-3-O-glucoside; (2) peonidin-3-O-sambubioside; (3) peonidin-O-monoglucuronide; and (4) cyanidin-3-O-glucoside-O-monoglucuronide. Peonidin-O-glycosides are not present in elderberry, so the presence of these compounds in the urine is a result of methylation of cyanidin-O-glycosides during the absorption/excretion process. Miyazawa et al. (1999) were not able to detect conjugated or methylated anthocyanins in plasma of humans but did observe the presence of peonidin-3-glucoside in the liver of rats following the consumption of red fruit anthocyanins (cyanidin-3-O-glucoside; cyanidin-3-O-diglucoside). The formation of the peonidin metabolites likely takes place in the liver through the catechol-O-methyltransferase reaction. Delphinidin would be the only other anthocyanidin that might undergo this methylation reaction as malvidin and petunidin already are methylated in the 3' position (see Figure 3.1).

Felgines et al. (2003) found pelargonidin-3-O-glucoside from strawberries excreted in the urine along with five metabolites: three monoglucuronides of pelargonidin, one sulfoconjugate of pelargonidin and pelargonidin itself. Total urinary excretion of strawberry anthocyanin metabolites was much higher than that observed with other anthocyanins (1.80 ± 0.29% of the ingested glucoside). More than 80% of the anthocyanin excreted was as the monoglucuronide form. This is the first time that sulfoconjugates of any anthocyanins have been reported in humans.

Wu et al. (2004) identified 11 metabolites from neonatal pigs following marionberry feeding. They were glucuronidated and/or methylated forms of the original anthocyanins (cyanidin-3-O-glucoside, cyanidin-3-O-rutinoside, and pelargonidin-3-O-glucoside). The extent of metabolism of cyanidin-3-O-rutinoside was lowest of the three parent anthocyanins as most of the rutinoside was excreted in the original unmetabolized form, whereas majority of the cyanidin-3-O-glucoside and pelargonidin-3-O-glucoside were excreted in the form of metabolites. Cyanidin-3-O-glucoside and cyanidin-3-O-rutinoside had similar apparent total excretion rates relative to dose, whereas pelargonidin-3-O-glucoside had a much higher total urinary excretion that the cyanidin-based anthocyanins. Thus, different aglycones and/or sugar moieties may influence the absorption and metabolism of anthocyanins.

3.6 SUMMARY

Only a small portion of the anthocyanins appear to be absorbed (<0.1% of the dose). Anthocyanins appear in the plasma within 30 min of consumption with maximal concentrations at 0.2–1.5 h after a meal. Time of absorption can be altered by transient time of the meal through the gut and by the food matrix. A large part of the disappearance of anthocyanins from the gut following a meal is due to degradation of the anthocyanins. 3,4-Dihydroxybenzoic acid (protocatechuic acid) and other phenolic acids have been identified as probable metabolites. Acylation of the anthocyanins appears to decrease absorption. Anthocyanin-O-glycosides seem to bind to the glucose transporter, but the presence or absence of glucose in the meal does not seem to alter the amount of anthocyanins absorbed. Although much has been learned during the past few years about the absorption/metabolism of anthocyanins, much more is to be learned about the possible active forms of metabolites that are responsible for some of the health benefits associated with consumption of anthocyanins.

REFERENCES

Anderson, O.M. (2002). Anthocyanin occurrences and analysis. *Proceedings of the International Workshop on Anthocyanins: Research and Development of Anthocyanins.* Adelaide, South Australia.

Arscott, S.A., Simon, P.W., and Tanumihardjo, S.A. (2010). Anthocyanins in purple-orange carrots (*Daucus carota* L.) do not influence the bioavailability of β-carotene in young women. *J. Agric. Food Chem.*, **58**, 2877–2881.

Brouillard, R. (1982). Chemical structure of anthocyanins. In P. Markakis (ed.), *Anthocyanins as Food Colors*. Academic Press, New York, pp. 1–40.

Brouillard, R. and Delaporte, B. (1977). Chemistry of anthocyanin pigments. 2. Kinetic and thermodynamic study of proton transfer, hydration, and tautometric reactions of malvidin-3-glucoside. *J. Am. Chem. Soc.*, **99**, 8461–8468.

Cao, G. and Prior, R.L. (1999). Anthocyanins are detected in human plasma after oral administration of an elderberry extract. *Clin. Chem.*, **45**, 574–576.

Cao, G., Muccitelli, H.U., Sanchez-Moreno, C. et al. (2001). Anthocyanins are absorbed in glycated forms in elderly women: a pharmacokinetic study. *Am. J. Clin. Nutr.*, **73**, 920–926.

Charron, C.S., Kurilich, A.C., Clevidence, B.A. et al. (2009). Bioavailability of anthocyanins from purple carrot juice: effects of acylation and plant matrix. *J. Agric. Food Chem.*, **57**, 1226–1230.

Cooke, D., Steward, W.P., Gescher, A.J. et al. (2005). Anthocyanins from fruits and vegetables—does bright colour signal cancer chemopreventive activity? *Eur. J. Cancer*, **41**, 1931–1940.

Crozier, A., Borges, G., and Stewart, A.J. (2004). Absorption, metabolism and potential bioavailability of dietary flavonols and anthocyanins. *Agro Food Ind. Hi-Tech*, **15**, 16–19.

Crozier, A., Jaganath, I.B., and Clifford, M.N. (2009). Dietary phenolics: chemistry, bioavailability and effects on health. *Nat. Prod. Rep.*, **26**, 965–1096.

de Pascual-Teresa, S., Moreno, D.A., and Garcia-Viguera, G. (2010). Flavanols and anthocyanins in cardiovascular health: a review of current evidence. *Int. J. Mol. Sci.*, **11**, 1679–1703.

Del Bo, C., Ciappellano, S., Klimis-Zacas, D. et al. (2010). Anthocyanin absorption, metabolism, and distribution from a wild blueberry-enriched diet (*Vaccinium angustifolium*) is affected by diet duration in the Sprague-Dawley rat. *J. Agric. Food Chem.* **58**, 2491–2497.

Dreiseitel, A., Oosterhuis, B., Vukman, K.V. et al. (2009). Berry anthocyanins and anthocyanidins exhibit distinct affinities for the efflux transporters BCRP and MDR1. *Br. J. Pharmacol.*, **158**, 1942–1950.

Faria, A., Pestana, D., Azevedo, J. et al. (2009). Absorption of anthocyanins through intestinal epithelial cells—Putative involvement of GLUT2. *Mol. Nutr. Food Res.*, **53**, 1430–1437.

Fazzari, M., Fukumoto, L., Mazza, G. et al. (2008). *In vitro* bioavailability of phenolic compounds from five cultivars of frozen sweet cherries (*Prunus avium* L.). *J. Agric. Food Chem.*, **56**, 3561–3568.

Felgines, C., Talavera, S., Gonthier, M.P. et al. (2003). Strawberry anthocyanins are recovered in urine as glucuro- and sulfoconjugates in humans. *J. Nutr.*, **133**, 1296–1301.

Felgines, C., Talavera, S., Texier, O. et al. (2005). Blackberry anthocyanins are mainly recovered from urine as methylated and glucuronidated conjugates in humans. *J. Agric. Food Chem.* **53**, 7721–7727.

Felgines, C., Texier, O., Besson, C. et al. (2007). Strawberry pelargonidin glycosides are excreted in urine as intact glycosides and glucuronidated pelargonidin derivatives in rats. *Br. J. Nutr.*, **98**, 1126–1131.

Felgines, C., Texier, O., Besson, C. et al. (2008). Influence of glucose on cyanidin 3-glucoside absorption in rats. *Mol. Nutr. Food Res.*, **52**, 959–964.

Galvano, F., La Fauci, L., Lazzarino, G. et al. (2004). Cyanidins: metabolism and biological properties. *J. Nutr. Biochem.*, **15**, 2–11.

Galvano, F., La Fauci, L., Vitaglione, P. et al. (2007). Bioavailability, antioxidant and biological properties of the natural free-radical scavengers cyanidin and related glycosides. *Ann. Ist. Super Sanita*, **43**, 382–393.

Galvano, F., Vitaglione, P., Li Volti, G. et al. (2008). Protocatechuic acid: the missing human cyanidins' metabolite. *Mol. Nutr. Food Res.*, **52**, 386–387.

Garcia-Alonso, M., Minihane, A.M., Rimbach, G. et al. (2009). Red wine anthocyanins are rapidly absorbed in humans and affect monocyte chemoattractant protein 1 levels and antioxidant capacity of plasma. *J. Nutr. Biochem.*, **20**, 521–529.

González-Barrio, R., Borges, G., Mullen, W. et al. (2010). Bioavailability of anthocyanins and ellagitannins following consumption of raspberries by healthy humans and subjects with an ileostomy. *J. Agric. Food Chem.*, **58**, 3933–3939.

Ichiyanagi, T., Rahman, M.M., Kashiwada, Y. et al. (2004a). Absorption and metabolism of delphinidin 3-O-β-D-glucoside in rats. *Biofactors* **21**, 411–413.

Ichiyanagi, T., Rahman, M.M., Kashiwada, Y. et al. (2004b). Absorption and metabolism of delphinidin-3-O-β-glucopyranoside in rats. *Free Rad. Biol. Med.*, **36**, 930–937.

Kay, C.D., Kroon, P.A., and Cassidy A. (2009). The bioactivity of dietary anthocyanins is likely to be mediated by their degradation products. *Mol. Nutr. Food Res.*, **53**, S92–S101.

Kurilich, A.C., Clevidence, B. A., Britz, S.J. et al. (2005). Plasma and urine responses are lower for acylated vs nonacylated anthocyanins from raw and cooked purple carrots. *J. Agric. Food Chem.*, **53**, 6537–6542.

Manach, C., Williamson, G., Morand, C. et al. (2005). Bioavailability and bioefficacy of polyphenols in humans. I. Review of 97 bioavailability studies. *Am. J. Clin. Nutr.*, **81**, 230S–242S.

Marczylo, T.H., Cooke, D., Brown, K. et al. (2009). Pharmacokinetics and metabolism of the putative cancer chemopreventive agent cyanidin-3-glucoside in mice. *Cancer Chemother. Pharmacol.*, **64**, 1261–1268.

Matuschek, M.C., Hendriks, W.H., McGhie, T.K. et al. (2006). The jejunum is the main site of absorption for anthocyanins in mice. *J. Nutr. Biochem.*, **17**, 31–36.

Mazza, G. and Miniati. E. (1993). *Anthocyanins in Fruits, Vegetables, and Grains.* CRC Press, Boca Raton, FL.

McGhie, T.K. and Walton, M.C. (2007). The bioavailability and absorption of anthocyanins: towards a better understanding. *Mol. Nutr. Food Res.*, **51**, 702–713.

Mertens-Talcott, S.U., Rios, J., Jilma-Stohlawetz, P. et al. (2008). Pharmacokinetics of anthocyanins and antioxidant effects after the consumption of anthocyanin-rich acai juice and pulp (*Euterpe oleracea* Mart.) in human healthy volunteers. *J. Agric. Food Chem.*, **56**, 7796–7802.

Miyazawa, T., Nakagawa, K., Kudo, M. et al. (1999). Direct intestinal absorption of red fruit anthocyanins, cyanidin-3-glucoside and cyanidin-3,5-diglucoside, into rats and humans. *J. Agric. Food Chem.*, **47**, 1083–1091.

Mullen, W., Edwards, C.A., Serafini, M. et al. (2008). Bioavailability of pelargonidin-3-O-glucoside and its metabolites in humans following the ingestion of strawberries with and without cream. *J. Agric. Food Chem.*, **56**, 713–719.

Passamonti, S., Vrhovsek, U. Vanzo, A. et al. (2003). The stomach as a site for anthocyanins absorption from food. *FEBS Lett.*, **544**, 210–213.

Passamonti, S., Vrhovsek, U. Vanzo, A. et al. (2005). Fast access of some grape pigments to the brain. *J. Agric. Food Chem.*, **53**, 7029–7034.

Prior, R.L. (2004). Absorption and metabolism of anthocyanins: Potential health effects. In M. Meskin, W.R. Bidlack, A.J. Davies, D.S. Lewis, and R.K. Randolph, eds., *Phytochemicals: Mechanisms of Action.* CRC Press, Boca Raton, FL, pp. 1–19.

Prior, R.L. and Wu. X. (2006). Anthocyanins: structural characteristics that result in unique metabolic patterns and biological activities. *Free Rad. Res.* **40**, 1014–1028.

Prior, R.L., Rogers, T.R., Khanal, R.C. et al. (2010). Urinary excretion of phenolic acids in rats fed cranberry. *J. Agric. Food Chem.*, **58**, 3940–3949.

Scalbert, A. and Williamson, G. (2000). Dietary intake and bioavailability of polyphenols. *J. Nutr.*, **130**, 2073S–2085S.

Talavera, S., Felgines, C., Texier, O. et al. (2003). Anthocyanins are efficiently absorbed from the stomach in anesthetized rats. *J. Nutr.*, **133**, 4178–4182.

Talavera, S., Felgines, C., Texier, O. et al. (2004). Anthocyanins are efficiently absorbed from the small intestine in rats. *J. Nutr.*, **134**, 2275–2279.

Uzunovic, A. and Vranic, E. (2008). Stability of anthocyanins from commercial blackcurrant juice under simulated gastrointestinal digestion. *Bosn. J. Basic Med. Sci.*, **8**, 254–258.

Vanzo, A., Terdoslavich, M., Brandoni, A. et al. (2008). Uptake of grape anthocyanins into the rat kidney and the involvement of bilitranslocase. *Mol. Nutr. Food Res.*, **52**, 1106–1116.

Vitaglione, P., Donnarumma, G., Napolitano, A. et al. (2007). Protocatechuic acid is the major human metabolite of cyanidin-glucosides. *J. Nutr.*, **137**, 2043–2048.

Walton, M.C., Hendriks, W. H., Broomfield, A.M. et al. (2009). Viscous food matrix influences absorption and excretion but not metabolism of blackcurrant anthocyanins in rats. *J. Food Sci.*, **74**, H22–H29.

Wang, L.S. and Stoner, G.D. (2008). Anthocyanins and their role in cancer prevention. *Cancer Lett.*, **269**, 281–290.

Woodward, G., Kroon, P., Cassidy, A. et al. (2009). Anthocyanin stability and recovery: implications for the analysis of clinical and experimental samples. *J. Agric. Food Chem.*, **57**, 5271–5278.

Wu, X., Beecher, G.R., Holden, J.M. et al. (2006a). Concentrations of anthocyanins in common foods in the United States and estimation of normal consumption. *J. Agric. Food Chem.*, **54**, 4069–4075.

Wu, X. and Prior, R.L. (2005a). Identification and characterization of anthocyanins by high-performance liquid chromatography-electrospray ionization-tandem mass spectrometry in foods in the United States: vegetables, nuts, and grains. *J. Agric. Food Chem.*, **53**, 3101–3113.

Wu, X. and Prior, R.L. (2005b). Systematic identification and characterization of anthocyanins by HPLC-ESI-MS/MS in common foods in the United States: Fruits and berries. *J. Agric. Food Chem.*, **53**, 2589–2599.

Wu, X., Cao, G., and Prior, R.L. (2002). Absorption and metabolism of anthocyanins in human subjects following consumption of elderberry or blueberry. *J. Nutr.*, **132**, 1865–1871.

Wu, X., Pittman, H.E., McKay, S. et al. (2005). Aglycones and sugar moieties alter anthocyanin absorption and metabolism after berry consumption in weanling pigs. *J. Nutr.*, 135, 2417–2424.

Wu, X., Pittman, H.E., and Prior, R.L. (2004). Pelargonidin is absorbed and metabolized differently than cyanidin after marionberry consumption in pigs. *J. Nutr.*, **134**, 2603–2610.

Wu, X., Pittman, H.E., and Prior, R.L. (2006b). Fate of anthocyanins and antioxidant capacity in contents of the gastrointestinal tract of weanling pigs following black raspberry consumption. *J. Agric. Food Chem.*, **54**, 583–589.

4 Bioavailability of Flavonols and Flavones

Mariusz Konrad Piskula, Kaeko Murota, and Junji Terao

CONTENTS

4.1 INTRODUCTION

There are numerous reports stating that consumption of foods of plant origin may be beneficial to human health. In addition to traditional vitamins, selenium, and fiber, the effects have been linked with the intake of flavonoids and other phytochemicals, which, evidence suggests, can reduce the incidence of certain cancers as well as several diseases associated with malfunctions in the cardiovascular system (see Crozier et al. 2009; Del Rio et al. 2010a). The question on how such food components act at the systemic level in the organism involves recognition of several processes preceding this potential activity. These processes are associated with the digestion, absorption, and metabolism of flavonoids and related compounds. Flavonoids are xenobiotics forming a non-nutritive part of the diet, so they are rapidly metabolized and eliminated from the body. Moreover, it is metabolites, rather than the native compounds, that are present in the systemic circulation and have the potential to interact with organ and cellular functions (Del Rio et al. 2010b).

Xenobiotics absorbed from the intestinal lumen into enterocytes, are subject to a combination of phase II sulfation, glucuronidation, and/or methylation reactions, before entering the portal vein. In plasma, the metabolites are bound to albumin and distributed via the circulatory system to sites of action within the body. The liver is a site for further metabolism of xenobiotics with many of the oxidative reactions being catalyzed by cytochrome P450 systems in the smooth endoplasmic reticulum of hepatocytes. A series of phase I reactions (oxidation, reduction, or hydrolysis) can introduce or deprotect hydroxyl and amino groups, making them available for additional phase-II metabolism, including possible sulfation and conjugation with glucuronic acid, glutathione, or glycine. The resulting water-soluble metabolites are excreted via the renal route in urine. Alternatively, if the molecular weight of metabolites is > 500–600 Da, they may be subjected to enterohepatic circulation, being removed from the circulatory system in the liver and transported in bile back to the duodenum. Here they would again undergo regular digestive processes (Hackett 1986) being hydrolyzed and at least a portion of the liberated derivatives being reabsorbed and remetabolized (Boyer et al. 1999). As a consequence, flavonoids like most xenobiotics have low bioavailability.

There are several factors that can influence the bioavailability of flavonoids: the food matrix, accompanying liquids, stomach contents, intestinal peristalsis, pH of the alimentary tract, characteristics of a carrier, and blood and lymph flow. The other factors are dependent upon the basic flavonoid structure: presence or absence of glycosylation, type of conjugating sugar and bond linking the aglycone and sugar moiety, site of glycosylation and the number of sugar moieties. Bioavailability is described by three parameters: area under the curve (AUC) describing the relationship between the metabolite concentration in blood versus time; T_{max} (time at which concentration of the compound in blood has reached a maximum); and C_{max} (the value of the maximum plasma concentration). The other important parameter is the biological half-life ($t_{1/2}$), which is the time at which the compound concentration falls to 50% of the C_{max} in the course of its elimination from the circulatory system.

Bioavailability is also dependent upon the manner of administration. It can be via the following routes: oral, intraperitoneal, subcutaneous, intradermal, intraintestinal, or intravenous. Intravenous administration guarantees 100% absorption or bioavailability equal to 100%. The absolute bioavailability of a given compound is the relationship between the bioavailability measured after a particular type of administration to its bioavailability after intravenous administration. Bioavailability studies on human subjects dealing with food provide information on the relative bioavailability with reference to the administered dose.

4.2 CHEMISTRY AND OCCURRENCE OF FLAVONES AND FLAVONOLS

Flavones and flavonols are flavonoids, the basic structure of which is a 2-phenyl-benzo-γ-pyrone skeleton formed by two phenyl rings (A and B) linked with a heterocyclic pyrone ring (C). Flavones have a double bond at the 2–3 position in the C-ring, whereas flavonols possess an additional hydroxyl group at the 3-position of the C-ring (Figure 4.1). The flavonoid skeleton is derived from condensation of three molecules of malonyl-CoA with 4-coumaroyl-CoA. The A ring is made from the three malonyl residues formed during the glucose metabolic pathway. The B

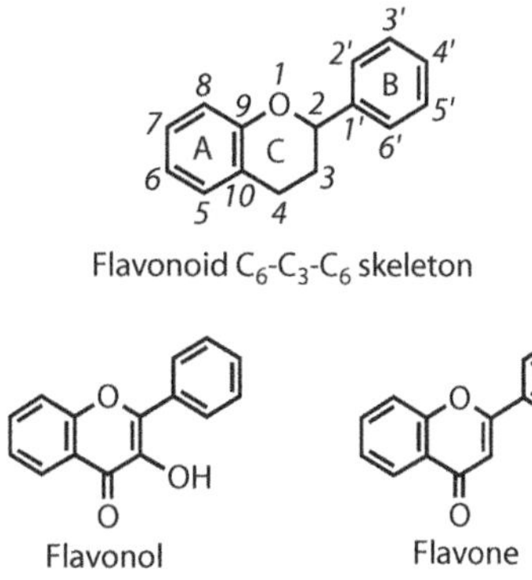

Flavonoid C₆-C₃-C₆ skeleton

Flavonol   Flavone

FIGURE 4.1 Structures of flavones and flavonols.

ring is formed from 4-coumaroyl-CoA produced via the shikimic acid pathway from phenylalanine (Crozier et al. 2009).

In plants, flavones and flavonols are usually linked to sugar moieties to form glycoside conjugates with the sugar residue being bound through either an oxygen or a carbon atom. Most plant flavones and flavonols are O-glycosides (Karakaya and Nehir 1999), although C-glycosides are not uncommon. Typically, glucose is the sugar substituent, but xylose, arabinose, galactose rhamnose, and glucuronic acid conjugates also occur. Usually, glycosylation takes place at C-3 and less frequently at the C-5, C-7, C-3′, and C-4′ positions (Herrmann 1988). The most abundant flavone C-glucosides are apigenin-C-8-glucoside (vitexin), apigenin-C-6-glucoside (isovitexin), (March et al. 2006), luteolin-C-8-glucoside (orientin), luteolin-C-6-glucoside (isoorientin) (Waridel et al. 2001), chrysoeriol-8-C-gluoside (scoparin), and chrysoeriol-6,8-C-digluoside (stellarin-2) (Gattuso et al. 2006) (Figure 4.2).

Flavones and flavonols are usually colorless, but they can contribute to some extent to the yellow color of flowers. They occur principally in white- or cream-colored flowers, where they serve as attractants for bees and other pollinating insects (Harborne 1982). In flower lobules, they are linked to anthocyanins, providing copigmentation and color stabilization. Quercetin has a widespread occurrence in food products and is therefore the most commonly consumed flavonol (Herrmann 1976) and one of the most studied flavonoids. Quercetin is present largely in the glycosidic form (Herrmann 1976, 1988) as its 3-O-glucoside (apples, pears, plums, cherries, grapes, apricots, lettuce, strawberries, tomatoes, leek, and onions), 3-O-rutinoside (rutin) (buckwheat, apples, cherries, tomatoes, and beans), 3-O-rhamnoside (apples, plums, rhubarb, and sweet cherries), and 3-O-galactoside (apples, quince, and strawberries) (Herrmann 1976, 1988; Macheix et al. 1990) (Figure 4.3). High intakes have been reported for other flavonols such as kaempferol (strawberries, cherries, blackcurrants, beans, leeks, raspberries, grapes, broccoli, chives, and onions) (Herrmann 1976, 1988; Macheix et al. 1990; Hertog et al. 1992). Lower amounts of myricetin are present in dark grapes, broad bean, red

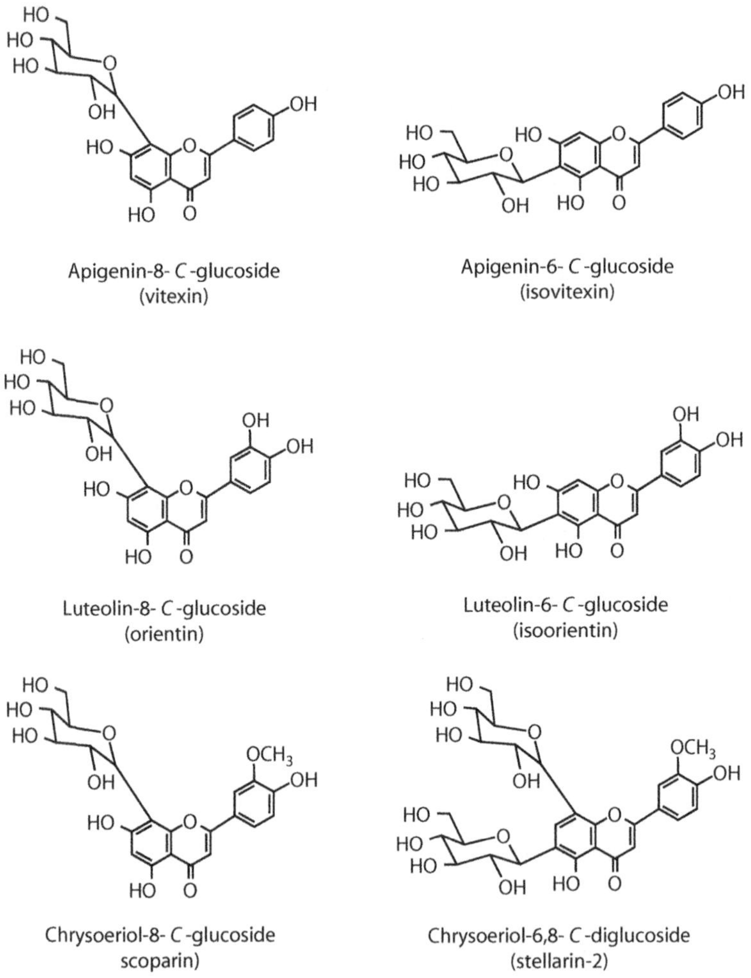

FIGURE 4.2 Flavone *C*-glucosides.

wine, and tea infusions (Hertog et al. 1992, 1993; Trichopoulou et al. 2000; Wang and Halliwell 2001; Tsanova-Savova and Ribarova 2002). Flavones are also an important group of flavonoids represented by apigenin (celery, parsley, carrots, and chicory), luteolin (celery, carrots, chicory, and lettuce), or orientin and vitexin (citrus fruits) (Hertog et al. 1992; Crozier et al. 1997; Trichopoulou et al. 2000). When focusing on the daily consumption of flavonols and flavones, a study conducted in the Netherlands limited to three flavonols (quercetin, myricetin, and kaempferol) and two flavones (luteolin and apigenin) provided an estimated intake of 23–25 mg per capita, with quercetin (16 mg/day)—the most consumed compound (Hertog et al. 1993). These intakes are low compared to the amounts of flavan-3-ols and anthocyanins that can be ingested by regular consumers of cocoa products and red wines.

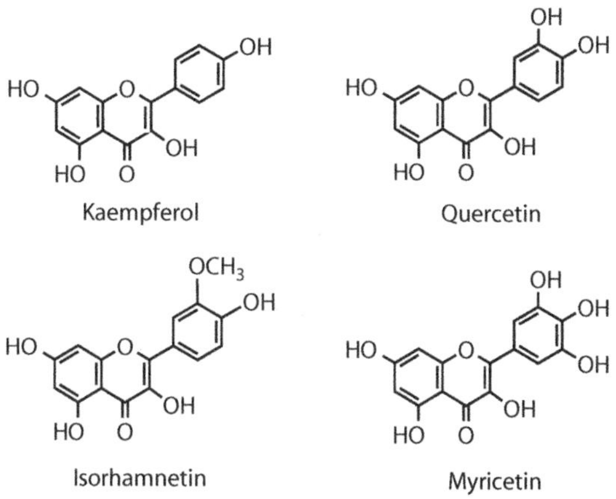

FIGURE 4.3 Structures of common flavonols.

4.3 BIOAVAILABILITY OF FLAVONES AND FLAVONOLS

The bioavailability of almost all the main subclasses of flavonoids has been studied, and Manach et al. (2005) compared the data obtained with the different groups. They used results from several human studies with the restriction that they were carried out using a single dose and that the sources were appropriately characterized. Different flavonoid doses had been ingested in the published studies, so the data were converted to a dose of 50 mg, assuming that the bioavailability increases relative to intake in a linear manner. This approach led to the conclusion that the best-absorbed flavonoids were isoflavones, with flavan-3-ols, flavanones, and flavonols being much less bioavailable when expressed as AUC (μmol h/L). The respective bioavailability parameters of quercetin-O-glucosides and quercetin-O-rutinoside were estimated to be 1.46 ± 0.45 μM and 0.20 ± 0.06 μM for the peak plasma concentration (C_{max}) and $2.5 \pm 1.2\%$ and $0.7 \pm 0.3\%$ for urinary excretion. Nevertheless, two points must be stressed. First, because of the wide range of molecular weights of individual flavonoids, 50 mg can correspond to intakes of somewhat different molar doses. Second, only the metabolites with a parental flavonoid structure were considered, leaving unaccounted other metabolites, especially the many low-molecular-weight derivatives that are the products of extensive flavonoid catabolism by intestinal microorganisms (see Chapter 10).

4.4 DEGLYCOSYLATION IN THE DIGESTIVE TRACT

After ingestion, the first step in the absorption of dietary flavonoid glucosides is deglycosylation, which occurs in the lumen of the small intestine. Hydrolysis of flavonoid O-glucosides is catalyzed by cytosolic β-glycosidase (CBG) or lactase-phlorizin hydrolase (LPH) (Day et al. 2003; Nemeth et al. 2003). Partial deglycosylation of flavonol and flavone glycosides from parsley (quercetin, kaempferol, isorhamnetin, apigenin, luteolin, and chrysoeriol) has been reported to take place in the stomach of rats after administration of an aqueous parsley extract (Pforte et al. 1999). Gastrointestinal (GI) tract epithelial cells are the only cells of the organism in direct contact with flavonoid glucosides (Depeint et al. 2002). Moreover, the type of sugar, its linkage to the aglycone, and the food matrix in which the conjugate is ingested, all impact upon flavonoid bioavailability (Graefe et al. 2001).

Cleavage of flavonoid aglycones from the sugar moiety is necessary if absorption is to take place. When this does not occur, conjugates not absorbed in the small intestine pass to the lower digestive tract where they are subject to the action of the resident enterobacteria (Boyer et al. 2005). Release of quercetin from quercetin-3-O-rutinoside prior to absorption in rats is due to the action of cecal microflora (Manach et al. 1997) and a similar situation occurs in humans where, following consumption of tomato juice, 85% of the ingested quercetin-3-O-rutinoside passed to the colon (Jaganath et al. 2006). There it is converted to quercetin by bacterial α-rhamnosidase, β-glucosidase, or endo-β-glucosidase (Kim et al. 1998). Only limited absorption of the released quercetin occurs in the large intestine, most of the aglycone undergoes bacterial-mediated C-ring fission and is converted to a range of phenolic acids (Jaganath et al. 2006, 2009).

Even when flavonoid aglycones are released in the GI tract, other components of the diet can affect their bioaccessibility. In particular, proteins can lower absorption because of strong binding to flavonoids. However, no differences in quercetin and kaempferol absorption were noted when black tea was consumed with or without milk (Hollman 2001). Less-studied C-glucosides such as vitexin and homoorientin are not liberated to their respective aglycones with the action of β-glycosidase (Ravise and Chopin 1981). Moreover, they were shown to be inhibitors of β-glycosidases, thereby affecting the absorption of other compounds that require the action of this enzyme before uptake in to the circulatory system. Zhang et al. (2007a) orally administered bamboo leaf extracts containing orientin, homoorientin, vitexin, and isovitexin (Figure 4.2) to rats, but the flavones did not accumulate in detectable quantities in either plasma or urine. It was concluded that this was because the flavone aglycones were not liberated from the C-glucosides, unlike O-glucosides, as a consequence of not being substrates for β-glycosidase or LPH enzymes. The data also suggest that flavone C-glucosides are unlikely to pass through enterocytes by means of active transport.

Further evidence for the very limited bioavailability of flavone C-glucosides was obtained in a human feeding study with rooibos tea, which contains dihydrochalcone C-glucosides and lower concentrations of the flavone C-glucosides—orientin, isoorientin, vitexin, and isovitexin. No flavones or dihydrochalcones were detected in 0–24 h plasma and while trace quantities of sulfated, methylated, and glucuronidated dihydrochalcones, which in some instances had retained the C-glucoside moiety, were excreted in urine, equivalent to <0.3% of intake, no urinary flavone metabolites were detected (Stalmach et al. 2009).

4.5   ABSORPTION AND METABOLIC CONVERSIONS

Absorption of the quercetin aglycone is dependent upon its solubility in the digestive environment. This was demonstrated in experiments with combinations of water and polypropylene glycol (Piskula and Terao 1998). Azuma et al. (2002) also studied different vehicles based on lipids and emulsifiers (10% lecithin; 20% soybean oil and 3% sucrose fatty acid ester; 3% polyglycerol fatty acid ester; or 3% sodium taurocholate) to improve quercetin absorption. It was concluded that quercetin dispersion in micelles can be a crucial factor in its absorption from the alimentary tract, and that the mixture of soybean oil and sucrose fatty acid ester was the most effective at solubilizing quercetin.

Hollman et al. (1995) reported that the bioavailability of quercetin-O-glucosides from onion flesh was higher than that of a crystalline standard of quercetin which had limited solubility. In contrast, a more recent study by Wiczkowski et al. (2008) showed that the bioavailability of the quercetin aglycone from the dry skin of shallots was much higher than that of quercetin-O-glucosides from the fleshy shallot tissue (Figure 4.4). Wiczkowski et al. (2008) suggested that *in planta* flavonols are dispersed in the tissues, so low solubility is not a limiting factor. However, the food matrix in which dietary flavones and flavonols are consumed may be a key factor in their bioavailability.

Two mechanisms of flavonoid transport through intestinal enterocytes have been proposed. First, free aglycones already present in the food, or those released from glucosides with the involvement of intestinal enzymes during digestion in the intestinal lumen, can pass into enterocytes via passive transport. Flavones and flavonols possess a hydrophobic coplanar structure, resulting in a high affinity with cellular membranes; therefore, simple diffusion is most probably the main mechanism for their uptake into intestinal epithelial cells.

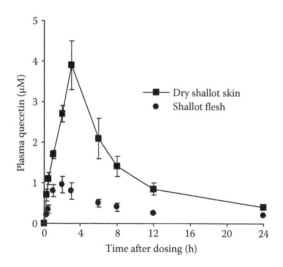

FIGURE 4.4 Plasma quercetin concentration in subjects who consumed 4.63 µmol quercetin per kg/body weight in i) dry shallot skin and ii) shallot flesh. Data expressed as mean values ± standard error ($n = 9$). (After Wiczkowski, W. et al., *J. Nutr.*, **138**, 885–888, 2008. With permission.).

In the second mechanism, flavonoid glucosides are transported directly to entero-cytes through a sodium-dependent glucose transporter SGLT1 (Day et al. 2000, 2003). However, an investigation, in which SGLT1 was expressed in *Xenopus laevis* oocytes, indicated that SLGT1 does not transport flavonoids and that glycosylated flavonoids, and some aglycones, have the capability to inhibit the glucose transporter (Kottra and Daniel 2007). SGLT1 is therefore unlikely to play a major part in flavo-noid transport to intestinal epithelial cells.

In intestinal epithelial cells, free flavones and flavonols are subjected to the action of phase-II enzymes forming glucuronide, sulfate, and/or methylated metabolites (Day et al. 2002). A portion of the metabolite pool enters the circu-latory systems and is transported via the portal vein to the liver where further phase II metabolism can take place, before urinary excretion. Flavonoid metab-olites produced in enterocytes also undergo efflux and are returned to the GI tract through various transporters present on the basolateral membrane. Several studies have demonstrated that multidrug resistance-associated protein-2 (MRP-2) participates in the efflux of flavonoids from intestinal epithelial cells (Williamson et al. 2007).

Flavonoid metabolites are also removed from the circulatory system in the liver, transported to bile, and returned into the GI tract via enterohepatic circulation. Conjugated metabolites that escape from these efflux systems are conveyed to the body via the circulation and are finally excreted into the urine. Several studies identified various conjugated metabolites present in the plasma and urine of humans after intake of flavonol-rich foods (Day et al. 2002; Hong and Mitchell 2006). Figure 4.5 shows an example of the numerous quercetin metabolites that appear in human plasma and urine after feeding fried onions, a rich source of quercetin-*O*-glucosides (Mullen et al. 2004). Most conjugated quercetin metabolites in plasma were concentrated not into lipoproteins but into plasma subfractions containing serum albumin (Murota et al. 2007). Studies on the binding properties of quercetin with human plasma proteins suggest strongly that conjugated quercetin metabolites circulate in the bloodstream using serum albumin as a carrier protein.

4.6 MOLECULES RESPONSIBLE FOR CELLULAR UPTAKE AND TRANSPORT OF FLAVONOIDS

Recent studies revealed that various protein molecules including MRPs and breast cancer resistance protein (BCRP) participate in the cellular uptake and transport of conjugated flavonoid metabolites. MRP-2, which is mentioned earlier, is mostly expressed on the brush-border membrane of the small intestine and is responsible for the efflux of flavonoid metabolites from intestinal epithelial cells. MRP-2 also acts as a transporter protein for the efflux from hepatocytes. BCRP is also located on the brush-border membrane and serves as a transporter of flavonoid metabolites for the efflux to the GI tract, similar to MRP-2 (Brand et al. 2008). Table 4.1 summarizes the membrane proteins responsible for the cellular transport of flavonoids.

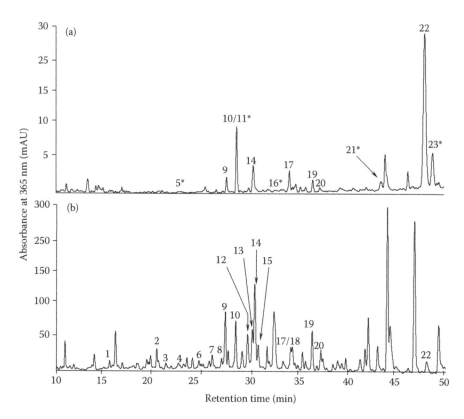

FIGURE 4.5 HPLC of quercetin metabolites in (a) a plasma extract and (b) urine obtained from a human volunteer after consumption of fried onions. Peaks were identified as follows: [1, 6, and 9] quercetin-*O*-diglucuronides, [2 and 4] methylquercetin-*O*-diglucuronides, [3, 7, and 8] quercetin-*O*-glucoside-*O*-glucuronides, [5] quercetin-3,4′-*O*-diglucoside, [10] quercetin-3-*O*-glucuronide, [11] quercetin-3-*O*-glucoside, [12 and 15] quercetin-*O*-glucoside-*O*-sulfate, [13 and 14] quercetin-*O*-glucuronide-*O*-sulfate, [16] isorhamnetin-3-*O*-glucoside, [17] isorhamnetin-3-*O*-glucuronide, [18] quercetin-4′-*O*-glucuronide, [19] quercetin-3′-*O*-glucuronide, [20] isorhamnetin-4′-*O*-glucuronide, [20] isorhamnetin-4′-*O*-glucuronide, [21] quercetin, [22] quercetin-3′-*O*-sulfate, and [23] quercetin-*O*-sulfate. * Indicates peaks detected in only one of six volunteers. (After Mullen, W. et al., *J. Chromatogr. A*, **1058**, 163–168, 2004. With permission.)

4.7 TISSUE DISTRIBUTION OF CONJUGATED METABOLITES AND THE POSSIBILITY OF THEIR DECONJUGATION

[2-^{14}C]Quercetin-4′-glucoside fed by gavage to rats was converted to methylated and glucuronide metabolite in the proximal GI tract before being degraded to phenolic acid such as 3-hydroxyphenylacetic acid in the colon. As xenobiotics, the metabolites were rapidly removed from the bloodstream without accumulating to any extent despite more that 50% of the ingested radioactivity being excreted in urine within 6 h of ingestion (see Figure 4.6). In this acute study, there was minimal accumulation

TABLE 4.1
Membrane Proteins Participating in The Cellular Uptake and Transport of Flavonoids

Transporter	Localization	Flavonoid	References
Sodium-glucose cotransporter 1 (SGLT-1)	Small intestine	Quercetin-4′-O-glucoside	Walgren et al. (2000) Day et al. (2003)
Bilitranslocase	Liver, stomach, small intestine, kidney, endothelial cells	Anthocyanin Quercetin aglycone	Passamonti et al. (2002) Passamonti et al. (2009)
Monocarboxylate transporter 1 (MCT1)	Small intestine	(−)-Epigallocatechin-3-O-gallate	Vaidyanathan and Walle (2003)
Organic anion transporter (OAT)	Small intestine	Flavonoid conjugates	Hu et al. (2003) Chen et al. (2005)
P-glycoprotein (MDR)	Blood–brain barrier	Flavonoid conjugates	Youdim et al. (2004)
Multidrug-resistance protein 1 (MRP-1)	Small intestine	(−)-Epigallocatechin-3-O-gallate	Vaidyanathan and Walle (2003) Hong et al. (2003)
Multidrug-resistance protein 2 (MRP2)	Small intestine, liver	Flavonoid conjugates	Hu et al. (2003) Chen et al. (2005) O'Leary et al. (2003)
Multidrug-resistance protein 3 (MRP3)	Small intestine, liver	Flavonoid conjugates	Zhang et al. (2007)
Breast cancer-resistance protein (BCRP)	Small intestine	Flavonoid conjugates	Brand et al. (2008)

of radioactivity in body tissues out with the GI tract, including the brain (Mullen et al. 2008).

The bioavailability of flavonols has also been evaluated in rats and pigs fed a diet containing high doses of quercetin (de Boer et al. 2005; Biegler et al. 2008). In these studies, quercetin and its conjugated metabolites were detected in tissues with the highest concentrations being found in the lungs and the lowest in the brain, white fat, and spleen. In addition to conjugated metabolites, quercetin aglycone as also found in rat tissues in this investigation. Most of the quercetin, however, could be derived from artificial deconjugation during the extraction procedure. A recent study with rats has also demonstrated low level accumulation of quercetin metabolites in the brain although the mechanism of how these compounds cross the blood-brain barrier is not fully understood (Ishisaka et al. 2011). In contrast, human neutrophils release luteolin from its glucuronide conjugate through β-glucuronidase activity (Shimoi et al. 2001). Kawai et al. (2008) demonstrated that lipopolysaccaride-stimulated macrophages enhanced β-glucuronidase activity that catalyzes the conversion of quercetin-3-O-glucuronide to quercetin. These findings indicate that conversion of flavonoid conjugates to their respective aglycones, due to enhanced β-glucuronidase

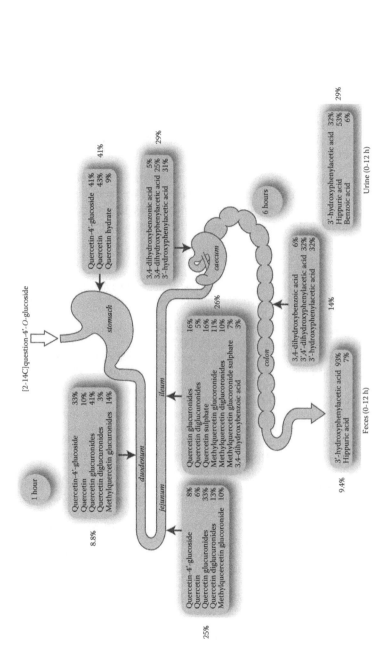

FIGURE 4.6 Schematic of the metabolism and catabolism of [2-14C]quercetin-4'-O-glucoside in rats following its ingestion, movement down the GI tract and subsequent appearance in feces and excretion in urine. Figures for quercetin-4'-O-glucoside and its metabolites inside the boxes represent their percentage of total radioactivity in the individual sections of the GI tract and/or feces and urine. Figures outside the boxes indicate radioactivity as a percentage of the amount ingested. (After Mullen, W. et al., *J. Agric. Food Chem.*, **56**, 12137–12137, 2008).

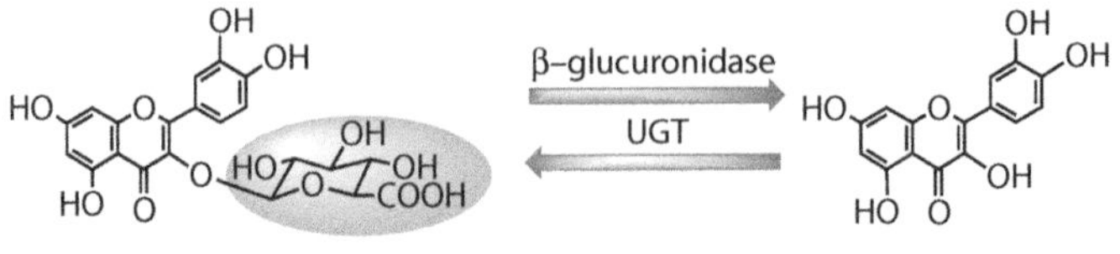

FIGURE 4.7 Conversion between quercetin-3-O-glucuronide and quercetin. Quercetin serves as a substrate of uridine-5'-diphosphate glucuronosyltransferase (UGT) activity in phase-II enzyme-catalyzed detoxification. Conjugated metabolites such as quercetin-3-O-glucuronide can be converted to quercetin by β-glucuronidase activity in inflamed tissues.

activity, may occur when tissues are exposed to abnormal conditions such as inflammation. Subsequently, the liberated aglycone may be reconverted to its glucuronide via cellular uridine-5'-diphosphate glucuronosyltransferase activity prior to being excreted from the cells (Terao et al. 2011) (Figure 4.7).

REFERENCES

Azuma, K., Ippoushi, K., Ito, H. et al. (2002). Combination of lipids and emulsifiers enhances the absorption of orally administered quercetin in rats. *J. Agric. Food Chem.,* **50,** 1706–1712.

Bieger, J., Cermak, R., Blank, R. et al. (2008). Tissue distribution of quercetin in pigs after long-term dietary supplementation. *J. Nutr.,* **138,** 1417–1420.

Boyer, J., Brown, D., Manach, C. et al. (1999). Part of quercetin absorbed in the small intestine is conjugated and further secreted in the intestinal lumen. *Am. J. Physiol.* **277,** G120–G126.

Boyer, J., Brown, D., and Liu, R.H. (2005). *In vitro* digestion and lactase treatment influence uptake of quercetin and quercetin glucoside by the Caco-2-cell monolayer. *Nutr. J.,* **4,** 1.

Brand, W., van der Wel, P.A., Rein, M.J. et al. (2008). Metabolism and transport of the citrus flavonoid hesperetin in Caco-2 cell monolayers. *Drug Metab. Dispos.,* **36,** 1794–1782.

Chen, J., Wang, S., Jia, X. et al. (2005). Disposition of flavonoids via recycling comparison of intestinal versus hepatic disposition. *Drug Metab. Dispos.,* **33,** 1777–1784.

Crozier, A., Lean, M.E.J., McDonald, M.S. et al. (1997). Quantitative analysis of the flavonoid content of commercial tomatoes, onions, lettuce and celery. *J. Agric. Food Chem.,* **45,** 590–595.

Crozier, A., Jaganath, I.B., and Clifford, M.N. (2009). Dietary phenolics: Chemistry, bioavailability and effects on health. *Nat. Prod. Rep.,* **26,** 1001–1043.

Day, A.J., Canada, F.J., Diaz, et al. (2000). Dietary flavonoid and isoflavone glycosides are hydrolysed by the lactase site of lactase phlorizin hydrolase. *FEBS Lett.,* **468,** 166–170.

Day, A.J., Mellon F., Barron, D. et al. (2002). Human metabolism of dietary flavonoids: Identification of plasma metabolites of quercetin. *Free Radic. Res.,* **35,** 941–952.

Day, A.J., Gee, J.M., Dupont, M.S. et al. (2003). Absorption of quercetin-3-glucoside and quercetin-4'-glucoside in the rat small intestine: The role of lactase phlorizin hydrolase and the sodium-dependent glucose transporter. *Biochem. Pharmacol.,* **65,** 1199–1206.

Del Rio, D., Borges, G., and Crozier, A. (2010a). Berry flavonoids and phenolics: Bioavailability and evidence of protective effects. *Br. J. Nutr.,* **104,** S67–S90.

Del Rio, D., Costa, L.G., Lean, M.E.J. et al. (2010b). Polyphenols and health: What compounds are involved. *Nutr. Met. Cardiovasc. Disease* **20,** 1–6.

Depeint, F., Gee, J.M., Williamson, G. et al. (2002). Evidence for consistent patterns between flavonoid structures and cellular activities. *Proc. Nutr. Soc.,* **61**, 97–103.

de Boer, V.C., Dihal, A.A., van den Wounde, H. et al. (2005). Tissue distribution of quercetin in rats and pigs. *J. Nutr.,* **135**, 1718–1725.

Gattuso, G., Caristi, C., Gargiulli, C. et al. (2006). Flavonoid glycosides in Bergamot juice (*Citrus bergamia Risso*). *J. Agric. Food Chem.,* **54**, 3929–3935.

Graefe, E.U., Wittig, J., Mueller, S. et al. (2001) Pharmacokinetics and bioavailability of quercetin glycosides in humans. *J. Clin. Pharmacol.,* **41**, 492–499.

Harborne, J.B. (1982). *Introduction to Ecological Biochemistry.* Academic Press, London.

Hackett, A.M. (1986). The metabolism of flavonoid compounds in mammals. In V. Cody, E. Middleton, and J.B. Harborne (eds). *Plant Flavonoids in Biology and Medicine. Biochemical Pharmacological and Structure Activity Relationship.* Alan Riss Inc., New York, pp. 177–194.

Herrmann, K. (1976). Flavonols and flavones in food plants: A review. *J. Food Technol.,* **11**, 433–448.

Herrmann, K. (1988). On the occurrence of flavonol and flavone glycosides in vegetables. *Z. Lebensm. Unters. Forsch.,* **186**, 1–5.

Hertog, M.G.L., Hollman, P.C.H., and Katan, M.B. (1992). Content of potentially anticarcinogenic flavonoids of 28 vegetables and 9 fruits commonly consumed in the Netherlands. *J. Agric. Food Chem.,* **40**, 2379–2383.

Hertog, M.G.L., Hollman, P.C.H., and van de Putte, B. (1993a). Content of anticarcinogenic flavonoids of tea infusions, wines and fruit juices. *J. Agric. Food Chem.,* **41**, 1242–1246.

Hertog, M.G., Hollman, P.C., Katan, M.B. et al. (1993b). Intake of potentially anticarcinogenic flavonoids and their determinant in adults in the Netherlands. *Nutr. Cancer,* **20**, 21–29.

Hollman, P.C.H., de Vries, J.H.M., van Leeuwen, S.D. et al. (1995). Absorption of dietary quercetin glycosides and quercetin in healthy ileostomy volunteers. *Am. J. Clin. Nutr.,* **62**, 1276–1282.

Hollman, P.C.H. (2001). Evidence for health benefits of plant phenols: Local or systemic effects? *J. Sci. Food Agric.,* **81**, 842–852.

Hong J, Lambert JD, Lee SH, et al. (2003). Involvement of multidrug resistance-associated proteins in regulating cellular levels of (–)-epigallocatechin-3-gallate and its methyl metabolites. *Biochem. Biophys. Res. Commun.,* **310**, 222–227.

Hong, Y.-J. and Mitchell, A.E. (2006). Identification of glutathione-related quercetin metabolites in humans. *Chem. Res. Toxicol.,* **19**, 1525–1532.

Hu, M., Chen, J., and Lin, H. (2003). Metabolism of flavonoids via enteric recycling: Mechanistic studies of disposition of apigenin in the Caco-2 cell culture model. *J. Pharmacol. Exp. Ther.,* **307**, 314–321.

Ishisaka, A., Ichikawa, S., Sakakibara, H. et al. (2011). Accumulation of orally administrated quercetin to brain tissue and its antioxidative effects in rats. *Free Radic. Biol. Med.,* **51**, 1329–1336.

Jaganath, I.B., Mullen, W., Edwards, C.A. et al. (2006). The relative contribution of the small and large intestine to the absorption and metabolism of rutin in man. *Free Radic. Res.,* **40**, 1035–1046.

Jaganath, I.B., Mullen, W., Lean, M.E.J. et al. (2009). *In vitro* catabolism of rutin by human fecal bacteria and the antioxidant capacity of its catabolites. *Free Radic. Biol. Med.,* **47**, 1180–1189.

Karakaya, S. and Nehir, E.L.S. (1999). Quercetin, luteolin, apigenin, and kaempferol contents of some foods. *Food Chem.,* **66**, 289–292.

Kawai, Y., Nishikawa, T., Shiba, Y. et al. (2008). Macrophage as a target of quercetin glucuronides in human atherosclerotic arteries. *J. Biol. Chem.,* **283**, 9424–9434.

Kim, D.H., Jung, E.A., Sohng, I.S. et al. (1998). Intestinal bacterial metabolism of flavonoids and its relation to some biological activities. *Arch. Pharm. Res.*, **21**, 17–23.

Kottra, G. and Daniel, H. (2007). Flavonoid glycosides are not transported by the human Na+/glucose transporter when expressed in *Xenopus laevis* oocytes, but effectively inhibit electrogenic glucose uptake. *J. Pharmacol. Exp. Ther.*, **322**, 829–835.

Macheix, J., Fleuriet, A., and Billot, J. (1990). *Fruit phenolics*. CRC Press, Boca Raton, FL.

March, R.E., Lewars, E.G., Stadey, C.J. et al. (2006). A comparison of flavonoid glycosides by electrospray tandem mass spectrometry. *Int. J. Mass Spectrom.* **248**, 61–85.

Manach, C., Morand, C., Demigne, C. et al. (1997). Bioavailability of rutin and quercetin in rats. *FEBS Lett.*, **409**, 12–16.

Manach, C., Williamson, G., Morand, C. et al. (2005). Bioavailability and bioefficacy of polyphenols in humans. I. Review of 97 bioavailability studies. *Am. J. Clin. Nutr.*, **81**, 230S–242S.

Mullen, W., Boitier, A., Stewart, A.J. et al. (2004). Flavonoid metabolism in human plasma and urine after the consumption of red onion: Analysis by liquid chromatography with photodiode array and scan tandem mass spectrometric detection. *J. Chromatogr. A*, **1058**, 163–168.

Mullen, W., Rouanet, J.-M., Auger, C. et al. (2008). The bioavailability of [2-^{14}C]quercetin-4′-glucoside in rats. *J. Agric. Food Chem.*, **56**, 12137–12137.

Murota, K., Hotta, A., Ido, H. et al. (2007). Antioxidant capacity of albumin-bound quercetin metabolites after onion consumption in humans. *J. Med. Invest.*, **54**, 370–374.

Nemeth, K., Plumb, G.W., Berrin, J.G. et al. (2003). Deglycosylation by small intestinal epithelial cell β-glucosidases is a critical step in the absorption and metabolism of dietary flavonoid glycosides in humans. *Eur. J. Nutr.*, **42**, 29–42.

O'Leary, K.A., Day, A.J., Needs, P.W. et al. (2003). Metabolism of quercetin-7- and quercetin-3-glucuronide by an *in vitro* hepatic model: The role of human β-glucuronidase sulfotransferase, catechol-*O*-methyltransferase and multi-resistant protein 2 (MRP-2) in flavonoid metabolism. *Biochem. Pharmacol.*, **65**, 479–491.

Passamonti, S., Vrhovsek, U., and Mattivi, F. (2002). The interaction of anthocyanins with bilitranslocase. *Biochem. Biophys. Res. Commun.*, **296**, 631–636.

Passamonti, S., Terdoslavich, M., Franca, R. et al. (2009). Bioavailability of flavonoids: A review of their membrane transport and the function of bilitranslocase in animal and plant organisms. *Curr. Drug Metab.*, **10**, 369–394.

Pforte, H., Hempel, J., and Jacobasch, G. (1999). Distribution pattern of a flavonoid extract in the gastrointestinal lumen and wall of rats. *Nahrung*, **43**, 205–208.

Piskula, M.K. and Terao, J. (1998). Quercetin's solubility affects its accumulation in rat plasma after oral administration. *J. Agric. Food Chem.*, **46**, 4313–4317.

Ravisé, A. and Chopin, J. (1981). Influence of the structure of phenolic compounds on the inhibition of the growth of *Phytophthora parasifica* and the activity of parasitogenic enzyme V. Flavone *O*- and *C*-glycosides. *Phytopath. Z.*, **100**, 257–269.

Shimoi, K., Saka, N., Nozawa, R. et al. (2001). Deconjugation of a flavonoid, luteolin monoglucuronide, during inflammation. *Drug Metab. Dispos.*, **29**, 1521–1524.

Stalmach, A., Mullen, W., Pecorari, M. et al. (2009). Bioavailability of *C*-linked dihydrochalcone and flavone glucosides in humans following ingestion of unfermented and fermented Rooibos teas. *J. Agric. Food Chem.*, **57**, 7104–7111.

Terao, J., Murota, K., and Kawai, Y. (2011). Conjugated quercetin glucuronides as bioactive metabolites and precursors of aglycone *in vivo*. *Food Funct.*, **2**, 11–17.

Trichopoulou, A., Vasilopoloulou, E., Hollman, P.C.H. et al. (2000). Nutritional composition and flavonoid content of edible wild greens and green pies: A potential rich source of antioxidant nutrients in the Mediterranean diet. *Food Chem.*, **70**, 319–323.

Tsanova-Savova, S. and Ribarova, F. (2002). Free and conjugated myricetin, quercetin, and kaempferol in Bulgarian red wines. *J. Food Compos. Anal.*, **15**, 639–645.

Vaidyanathan, J.B. and Walle, T. (2003). Cellular uptake and efflux of the tea flavonoid (–)-epicatechin-3-gallate in the human intestinal cell line Caco-2. *J. Pharmacol. Exp. Ther.,* **307**, 745–752.

Wang, H. and Helliwell, K. (2001). Determination of flavonols in green and black tea leaves and green tea infusions by high-performance liquid chromatography. *Food Res. Int.,* **34**, 223–227.

Walgren, R.A., Lin, J.T., Kinne, R.K. et al. (2000). Cellular uptake of dietary flavonoids quercetin-4'-β-glucoside by sodium-dependent glucose transporter SGLT1. *J. Pharmacol. Exp. Ther.,* **294**, 837–843.

Waridel, P., Wolfender, J. L., Ndjoko, K. et al. (2001). Evaluation of quadrupole time-of flight tandem mass spectrometry and ion-trap multiple-stage mass spectrometry for the differentiation of C-glycosidic flavonoid isomers. *J. Chromatogr. A,* **926**, 29–41.

Wiczkowski, W., Romaszko, J., Bucinski, A. et al. (2008). Quercetin from shallots (*Allium cepa* L. var. aggregatum) is more bioavailable than its glucosides. *J. Nutr.,* **138**, 885–888.

Williamson, G., Aeberli, I., Miguet, L. et al. (2007). Interaction of positional isomers of quercetin glucuronides with the transporter ABCC2 (cMOAT, MRP-2). *Drug Metab. Dispos.,* **35**, 1262–1291.

Youdim, K.A., Qaiser, M.Z., Begley, D.J. et al. (2004). Flavonoid permeability across an *in situ* model of the blood-brain barrier. *Free Radic. Biol. Med.,* **36**, 592–604.

Zhang, Y., Tie, X., Bao, B. et al. (2007a). Metabolism of flavone C-glucosides and p-coumaric acid from antioxidant of bamboo leaves (AOB) in rats. *Br. J. Nutr.,* **97**, 484–494.

Zhang, I., Lin, G., Kovacs, B. et al. (2007b). Mechanistic study on the intestinal absorption and disposition of baicalein. *Eur. J. Pharm. Sci.,* **31**, 221–231.

5 Bioavailability of Isoflavones in Humans

Aedín Cassidy, José Peñalvo, and Peter Hollman

CONTENTS

5.1 INTRODUCTION

Bioavailability or pharmacokinetics of isoflavones is based on data from absorption, metabolism, distribution, and excretion (ADME) studies conducted both in humans and animals. Following the consumption of pure compounds, isoflavone-rich extracts, or foods/beverages containing high levels of isoflavones, the parent compounds and their metabolites can be detected in plasma and urine of human volunteers. Numerous studies attest to the fact that following ingestion, soy isoflavones attain maximal plasma concentrations (C_{max}) within 4–8 h (T_{max}) and are then eliminated from the body through the bile and kidneys with a terminal elimination half-life ($t_{1/2}$) that is on average ~8 h (Setchell et al. 2001, 2003a,b; Cassidy 2006). Available data suggest that they are more efficiently absorbed than other subclasses of flavonoids, with C_{max} levels of ~2 μM and mean relative urinary excretions of 42% for daidzein and 16% for genistein, after a 50 mg isoflavone intake (Manach et al. 2005).

After ingestion, the parent isoflavone glycosides are hydrolyzed by intestinal glucosidases, which partially release the aglycones, daidzein, genistein, and glycitein (Figures 5.1 and 5.2). These may be absorbed or converted to a number of metabolites, including equol and *p*-ethyl phenol (Cassidy 2006). There is evidence from

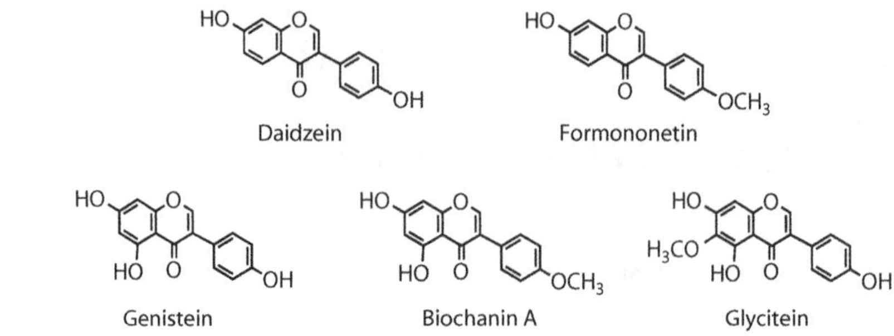

FIGURE 5.1 Structures of the main isoflavone aglycones.

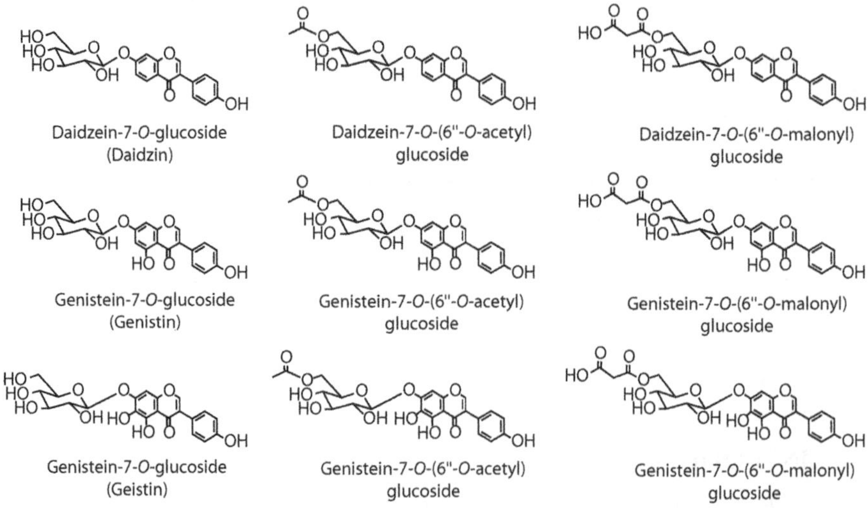

FIGURE 5.2 Some of complex array of isoflavone glycosides in foods and supplements.

several studies that high concentrations of isoflavones can be found in tissues; breast tissue of premenopausal women and in the prostate gland of men (Morton et al. 1997; Maubach et al. 2004; Bolca et al. 2010).

Most of the previously published studies of the serum pharmacokinetics of soy isoflavones have focused on a single food type, purified isoflavone aglycones or glucosides, or stable-isotopically labeled tracers (Setchell et al. 2001, 2003a,b; Cassidy 2006). As with pharmacological compounds, demonstrating efficacy and understanding potential risks of soy and its isoflavones requires knowledge of their bioavailability.

5.2   INITIAL ISOFLAVONE ABSORPTION

There are a complex array of isoflavones in soy foods and supplements, although they are primarily present as β-glucosides, which also occur as acetyl- and malonyl-glucoside (Figures 5.1 and 5.2). Levels of glucosides, which can be acetylated or

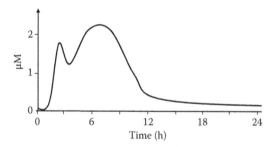

FIGURE 5.3 A typical human isoflavone plasma pharmacokinetic profile 0–24 h after acute ingestion of soy.

malonylated, vary in different products, while industrial processing and cooking procedures result in hydrolysis of the glycosides, especially the acetyl- and malonyl-forms. Glycosides are all poorly absorbed in the small intestine because of their hydrophilic nature and high molecular weight (Xu et al. 1995; Liu and Hu 2002). Fermented soy products, or supplements in which the soy extract has been hydrolyzed, contain predominantly the aglycone forms of isoflavones; however, following ingestion, the plasma profile of isoflavone metabolites attained is the same independently of the form ingested (Setchell et al. 2003a) (Figure 5.3). Initial hydrolysis is, therefore, necessary to release the free aglycones, which are then rapidly absorbed via passive diffusion across the intestinal brush border (Scalbert and Williamson 2000) with high efficiency (Liu and Hu 2002). Glycoside hydrolysis occurs in the small intestine as well as in the colon, which is reflected by a typical biphasic appearance of isoflavone metabolites in plasma following soy ingestion (Setchell et al. 2002a; Rowland et al. 2003). The first phase corresponds to isoflavone absorption in the proximal small intestine, which occurs within the first hour following consumption, and the membrane bound enzyme lactase phlorizin hydrolase on the luminal side of the brush border in the small intestine has been shown to play a key role in the hydrolysis (Day et al. 1998). High interindividual variability may exist for this hydrolysis capacity and subjects who poorly absorbed lactose have been reported to have impaired absorption of isoflavone glucosides within the small intestine (Day et al. 2000; Nemeth et al. 2003; Tamura et al. 2008). The gut bacterium also extensively hydrolyses glycosides into aglycones and can compensate for impaired hydrolysis in the small intestine (Tamura et al. 2008).

5.3 ISOFLAVONE METABOLISM AND SECONDARY ABSORPTION

After initial absorption, isoflavones undergo extensive first pass metabolism, which accounts for their relatively low bioavailability (Chen et al. 2003). During phase II metabolism, they undergo glucuronidation and sulfation by glucuronosyl-transferases and sulfotransferases in the intestine and/or liver (Setchell et al. 2001; Larkin et al. 2008). The resulting glucuronide and sulfate conjugates can be transported via the systemic circulation to tissues from where they can be excreted via the kidneys or secreted into bile and returned to the intestine. After deconjugation by intestinal bacteria, isoflavone aglycones can be reabsorbed, return to liver via the portal vein for reconjugation, and either undergo further enterohepatic circulation or renal excretion. This intestinal conjugation of isoflavones, their secretion back into the intestinal lumen, and further reabsorption and reconjugation constitutes enteric recycling (Liu and Hu 2002; Chen et al. 2003), which in combination with enterohepatic recycling significantly prolongs systemic exposure to isoflavones (Turner et al. 2003). Enterohepatic recycling also contributes to the biphasic appearance of isoflavones in plasma. Aglycones that are not reabsorbed reach the large intestine, along with any conjugates from liver or intestinal biotransformation that are not deconjugated (Liu and Hu 2002), together with the fraction of isoflavone that has not been hydrolyzed or absorbed in the small intestine (Decroos et al. 2005). In the large intestine, the microbiota further degrade isoflavones (Hollman 2001); both daidzein and genistein can be further metabolized to secondary metabolites with significant interest in the potential health effects of equol, which is a metabolite of daidzein (Figure 5.4).

In the circulation, aglycones represent only a fraction of the total plasma isoflavones. Aglycones have been recovered in small proportions, generally <5% of the total metabolites (Doerge et al. 2000; Setchell et al. 2001). The main metabolites are 7-O-glucuronides and 4'-O-glucuronides of daidzein and genistein, with small proportions of sulfate ethers. Additional metabolites that have been identified in human plasma or urine include glucuronides and sulfates of the aglycones dihydrodaidzein, dihydrogenistein, 4-dihydroequol, O-desmethylangolensin, and 6-hydroxy-O-desmethylangolensin (Figure 5.5) (Kelly et al. 1993, Rowland et al. 2003).

The plasma T_{max} of daidzein and genistein metabolites is typically reached 6–8 h after isoflavone intake (Setchell et al. 2001, 2003a,b; Rowland et al. 2003; Cassidy 2006, Larkin et al. 2008). Although the two isoflavones have similar half-lives, plasma genistein metabolites levels are consistently higher than those of daidzein

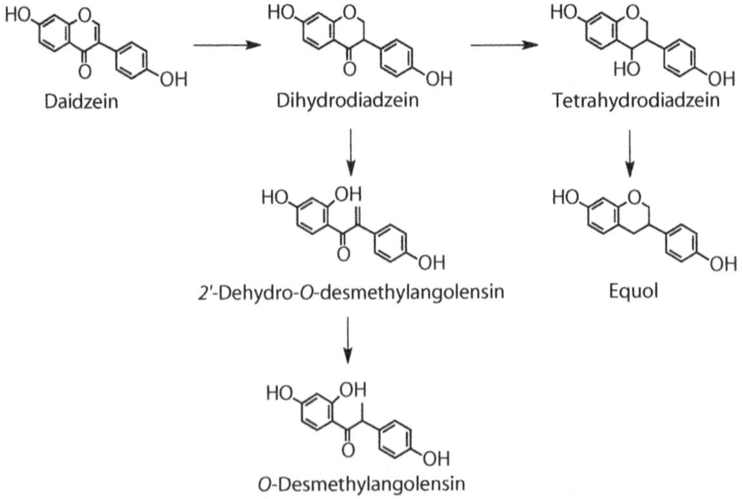

FIGURE 5.4   Proposed pathways for the metabolism of daidzein.

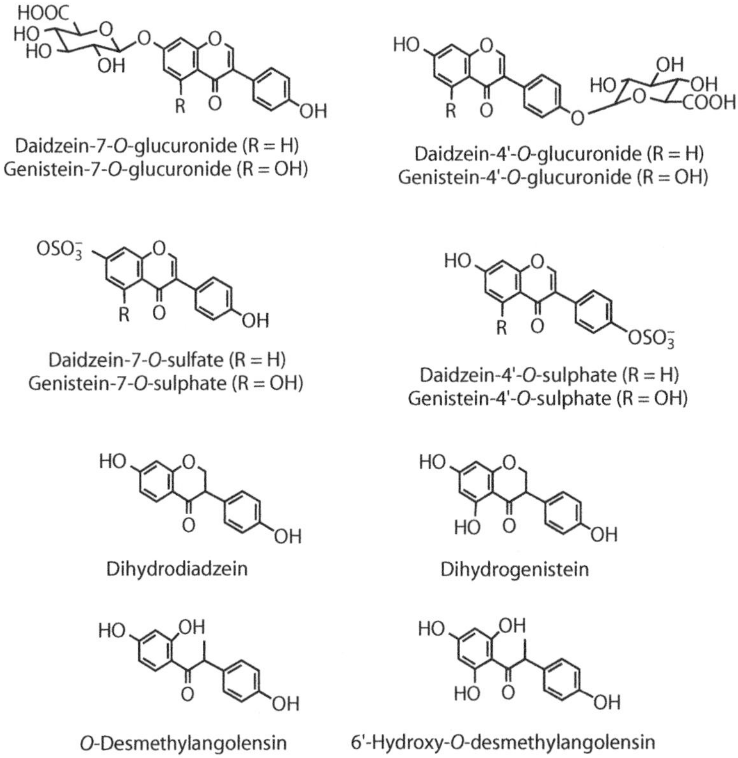

FIGURE 5.5 Structures of some of the metabolites of daidzein and genistein that have been detected in plasma and urine.

derivatives with faster clearance rates, and higher volume of distribution explaining the lower plasma daidzein levels (Setchell et al. 2003a,b; Larkin et al. 2008).

5.4 DISTRIBUTION, ELIMINATION, AND RECOVERY OF ISOFLAVONES

To date isoflavones have been quantified in a number of biological fluids and tissues, including plasma, urine, bile, faeces, saliva, breast aspirates, and prostatic fluid. In urine, isoflavones are excreted almost exclusively as acidic conjugates, predominately glucuronides, with smaller amounts of sulfate conjugates (Adlercreutz et al. 1993). Following ingestion of isoflavones, most recovery of daidzein and genistein metabolites occurs in the following 24 h, with higher urinary levels of the former (Larkin et al. 2008), perhaps as a consequence of the greater water solubility of daidzein.

The proportional recovery of urinary isoflavones relative to the amount ingested is generally 10–50%, potentially indicating significant bacterial degradation or the presence of other unidentified metabolites (Shelnutt et al. 2002; Rowland et al. 2003; Faughnan et al. 2004; Larkin et al. 2008).

There is currently limited data on the concentrations and profiles of isoflavone metabolites in human tissues. They can accumulate in breast tissue and milk (Franke et al. 1998; Maubach et al. 2004; Bolca et al. 2010), cross the blood–brain barrier and placenta (Adlercreutz et al. 1999; Setchell and Cassidy 1999), and concentrations in prostatic fluid were two fold higher than levels present in plasma (Morton et al. 1997; Hedlund et al. 2006).

5.5   ROLE OF THE GUT MICROBIOTA

The importance of the microflora in the metabolism of isoflavones was first observed in the 1980s when studies on germ-free animals showed that synthesis of equol was dependent on the gut microflora and antibiotic administration obliterated its formation (Axelson et al. 1982, 1984; Setchell et al. 1984). It has been suggested that the inability of some individuals to metabolize daidzein to equol is due to the absence of certain bacterial species in the intestine, and ongoing research has focused on the identification of strains of bacteria that can convert daidzein to equol, and to date, several have been identified (Atkinson et al. 2004). Given the complexity of the metabolic pathways for daidzein, the involvement of more than one species of bacteria in the metabolism of isoflavones is conceivable (Atkinson et al. 2004; Lampe 2009). A number of strains involved in isoflavone metabolism have been identified. For example, *Escherichia coli* HGH21 and the gram-positive strain HGH6 convert daidzein-7-*O*-glucoside and genistein-7-*O*-glucoside to their respective aglycones (Hur et al. 2000) and further reduce them to dihydrodaidzein and dihydrogenistein (Figure 5.4), respectively (Hur et al. 2000; Wang et al. 2005a,b). Recent reports have shown a microbial consortium, consisting of four associated bacteria, which were identified as *Enterococcus faecium* strain EPI1, *Lactobacillus mucosae* strain EPI2, *Fingoldia magna* strain EPI3, and a species related to *Veillonella* spp., are capable of producing equol from daidzein (Decroos et al. 2005). Other bacteria reported being capable of equol production are *Streptococcus intermedius* spp., *Ruminococcus productus* spp., and *Bacteroides ovatus* (Setchell et al. 2005).

5.5.1   THE EQUOL-PRODUCER PHENOTYPE

Equol does not naturally occur in plants but is a specific bacterial metabolite, which is found in high concentrations in urine and plasma following consumption of isoflavone-rich foods (Axelson et al. 1982, 1984). Using stable isotope-labeled compounds, it was conclusively shown that it is a metabolite specifically formed following daidzein consumption (Setchell et al. 2003b).

The consistent observation that not all adults synthesize equol in response to challenges from soy foods or isoflavones has led to the realization that there are two distinct subgroups of the population, defined as "equol producers" and "nonequol producers." The factors governing equol production remain poorly understood (Setchell et al. 2002b; Lampe 2009), but emerging data from clinical studies suggest that the ability to produce equol following ingestion of soy isoflavones may be a significant factor in the clinical effectiveness to soy diets. It is well established that only 30–50% of any given population group can produce equol following ingestion

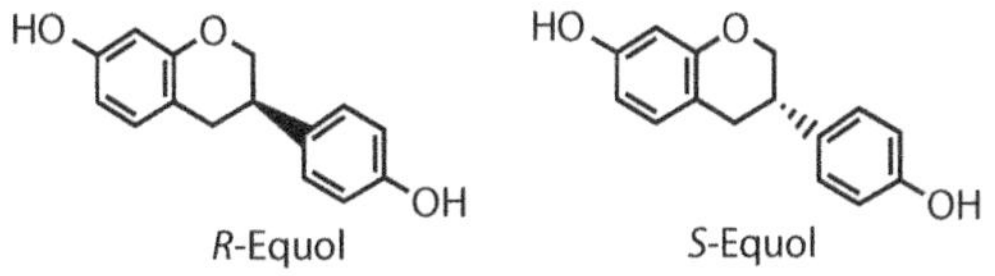

FIGURE 5.6 Structures of *R*- and *S*-equol.

of soy foods (Cassidy et al. 1994; Lampe et al. 2001; Lampe 2009; Frankenfeld et al. 2006) and the capacity to produce equol has been associated with circulating reproductive hormones (Cassidy et al. 1994) and positively associated with bone mineral density in response to isoflavone intervention (Setchell et al. 2002b; Fujioka et al. 2004; Frankenfeld et al. 2006; Wu et al. 2007; Lampe 2009). Equol production may enhance the action of isoflavones as it has a lower affinity for serum proteins, greater affinity for estrogen receptors compared to its precursors, daidzein and dihydrodaidzein and exerts superior antioxidant activity (Shutt and Cox, 1972; Hodgson et al. 1996; Arora et al. 1998). The definition of what constitutes an equol producer has recently been refined, and recent data suggest that the log10 transformed urinary equol:daidzein ratio provides a clearer distinction of equol producer status than equol concentrations *per se* (Setchell and Cole 2006); a demarcation of −1.75 in the urinary ratio of equol:daidzein to define equol producer status, and this methodology has successfully been used for determining daidzein-metabolizing phenotypes in previous studies (Frankenfeld et al. 2004).

Equol exists in two enantiomeric forms, *S*- and *R*-equol (Figure 5.6), and data indicate that only the *S* isomer is present in humans (Setchell et al. 2002b), leading to the suggestion that the equol-forming bacterial process is enantioselective (Setchell et al. 2005). Recent data on the pharmacokinetics of stable-labeled tracers of pure *S*- and *R*-equol suggest they are more bioavailable than their parent compound, daidzein; both are rapidly absorbed, attain a high circulating C_{max} level (1.6–1.7 µM after a 20 mg intake) and have a $T_{1/2}$ of 7 h (Setchell et al. 2009). Using healthy adults, it was shown that both enantiomers are highly bioavailable (65–83%) (Setchell et al. 2005).

5.6 FACTORS AFFECTING ISOFLAVONE BIOAVAILABILITY

A number of factors can influence the absorption of isoflavones including age, dose, food matrix and chemical composition, background diet, and genetic characteristics. Inherent and genetic characteristics, which determine pathways of absorption and metabolism may contribute to variability in isoflavone bioavailability, but to date, there is limited data examining their relative influences. Both genetic predisposition and shared environment have been associated with overall intestinal microflora profiles, and in a population-based study, the equol producer phenotype was weakly correlated with familial relationship (Frankenfeld et al. 2004). Some studies suggest a relationship between background diet including total fat, meat, fiber, carbohydrate, and equol production, but overall, the evidence is equivocal (Rowland et al. 2003; Lampe 2009). The question whether there are specific components of the diet that influence bacterial conversion of daidzein to equol therefore remains to be established.

5.6.1 Chemical Composition

In Asian countries where soy is consumed as a staple, it remains to be determined if chemical composition of the soy food alters absorption and metabolism and, thus, the potential biological efficacy of soy isoflavones. Asian populations have traditionally consumed mainly fermented soy products, and since these foods contain a higher proportion of aglycone isoflavones, it has been suggested that they may be more bioavailable since these aglycone isoflavones do not require hydrolysis in the intestine prior to absorption. A comparison of the pharmacokinetics of two soy foods, namely Tempeh, a fermented food that contains mostly aglycones of daidzein and genistein and textured vegetable protein (TVP), which contains mainly the glucoside conjugates, provides further evidence for differences in pharmacokinetics relative to soy food type. Circulating isoflavone metabolite levels following tempeh ingestion were more rapidly achieved than after intake of soy TVP (Cassidy 2006).

Although in their purified form, daidzein and genistein aglycones are more rapidly absorbed into the systemic circulation (Izumi et al. 2000; Setchell et al. 2001), other data suggest that the overall systemic bioavailability of the pure aglycone compounds is lower than that of their 7-O-glucosides (Setchell et al. 2001; Faughnan et al. 2004). Data from studies comparing supplements versus food sources are equivocal. One study suggested supplements are more bioavailable than soy foods, other study suggested that plasma C_{max} levels were higher following soy foods than supplements, and a third one showed no difference between soy flour and a soy extract (Anupongsanugool et al. 2005; Gardner et al. 2008; Vergne et al. 2008). These differences in pharmacokinetics would presumably affect tissue exposure and overall bioavailability; however, the nature of these effects has not yet been established. What is very clear is that the glucoside conjugates are biologically inactive *per se* and are not absorbed intact, with the initial hydrolysis of the sugar moiety being a rate limiting step (Setchell et al. 2002a). Conjugation may also determine the extent to which the isoflavones are further metabolized with suggestions that the longer transit times for consumed glycosides may increase residence time in the large intestine, providing an extended opportunity for bacterial metabolism (Zubik and Meydani 2003). The capacity for metabolism is high as levels of isoflavone aglycones in humans are low even after repeated intake of soy. The percentage of aglcones in the circulation is similar following consumption of different isoflavone forms including soy nuts, soy milk, or tempeh (1.4–2.3%) (Setchell personal communication).

5.6.2 Dose Response

Several studies have shown a nonlinear relationship between the ingested dose of isoflavones, plasma C_{max} values and the bioavailability measured as area under the curve (AUC) and percentage of dose excreted through urine (Setchell et al. 2001, 2003a; Faughnan et al. 2004), which suggests that the uptake is a rate-limiting step and is saturable (Faughnan et al. 2004). A nonlinear increase in AUC and a decrease in the percentage of dose excreted in urine was also observed following intakes of

15, 30, and 60 mg total isoflavones, as glycosides, in one study and intakes of 0.45–1.8 mg of isoflavone aglycones per kg/BW fed as soy milk in another (Setchell et al. 2001, 2003a,b; Faughnan et al. 2004).

All these data suggest there may be limited benefit in ingesting high levels of isoflavones, although whether or not high levels of intake, achievable from supplements, results in dose-response increases requires further investigation.

5.6.3 FOOD MATRIX

Humans absorb isoflavones from a range of different soy-rich foods in a similar manner, but data suggest that the food matrix has an effect on altering the pharmacokinetic profiles. Numerous studies have used a liquid matrix, like soy milk or specific soy protein drinks (Coward et al. 1993; Xu et al. 1994; King et al. 1998; Watanabe et al. 1998; Setchell et al. 2001; Richelle et al. 2002; Shelnutt et al. 2002; Cassidy 2006) and have determined that the plasma T_{max} is around 6 h for daidzein and genistein. These studies suggest that the bioavailability and pharmacokinetics of isoflavones are influenced mostly by the type of food matrix or form in which they are ingested. A liquid matrix, such as soy milk, yields a faster absorption rate and higher C_{max} than a solid matrix. Solubility of a substance in the intestine influences the rate of absorption, and since the isoflavones in soymilk are mainly hydrophilic β-glucoside conjugates, and therefore in solution, this probably accounts for the faster absorption rates and shorter duration T_{max} times following soy milk consumption than attained with solid soy food matrices such as TVP. Stomach emptying occurs later following the ingestion of solid foods compared to liquid food matrices (Cadwallader 1983).

5.6.4 AGE

During the first few months of life, infants are unable to convert the isoflavone, daidzein to equol, which has been attributed to the absence of a mature gut microflora and inactivity of metabolic enzymes (Setchell et al. 1997). In children, one study suggested greater absorption in children compared to adults, following consumption of soy nuts (0.62 mg/kg/bw total isoflavones) (Halm et al. 2007). The effect of age on isoflavone metabolism at later stages of life is not known, although in small studies no differences in isoflavone pharmacokinetics were observed between pre- and postmenopausal women (Setchell et al. 2003a; Faughnan et al. 2004). However, there is no data in older age groups, and poor bioavailability of micronutrients is common in the elderly, in part due to subtle alterations in intestinal morphology and reduced intestinal absorption (Drozdowski and Thompson 2006). For older women, there is evidence to suggest that they metabolize some drugs less effectively than premenopausal women, an age effect not observed in men. Data from a large study of familial correlations in relation to soy isoflavone metabolizing phenotypes showed that older subjects were less likely to produce equol than younger individuals (Frankenfeld et al. 2004). This has important implications since the predominant interest in isoflavone consumption relates to ageing, specifically menopausal health.

118 Flavonoids and Related Compounds: Bioavailability and Function

5.7 SUMMARY

From the available evidence, it is clear that in healthy adults, isoflavones are absorbed rapidly and efficiently. More studies are required to characterize the profile of conjugated and unconjugated isoflavones present in human plasma following different levels and forms of isoflavone intake. Further knowledge of factors, including genetic determinants, affecting the bioavailability of isoflavone is an important area for research.

REFERENCES

Adlercreutz, H., Fotsis, T., Lampe, J. et al. (1993). Quantitative determination of lignans and isoflavonoids in plasma of omnivorous and vegetarian women by isotope dilution gas chromatography-mass spectrometry. *Scand. J. Clin. Lab. Invest.*, **215** (Suppl.), 5–18.

Adlercreutz, H.,Yamada, T., Wahala, K. et al. (1999). Maternal and neonatal phytoestrogens in Japanese women during birth. *Am. J. Obstet. Gynecol.*, **180**, 737–473.

Anupongsanugool, E., Teekachunhatean, S., Rojanasthien, N. et al. (2005). Pharmacokinetics of isoflavones, daidzein and genistein, after ingestion of soy beverage compared with soy extract capsules in postmenopausal Thai women. *BMC Clin. Pharmacol.*, **5**, 2.

Arora, A., Nair M.G., Strasburg, G.M. et al. (1998). Antioxidant activities of isoflavones and their biological metabolites in a liposomal system. *Arch. Biochem. Biophys.*, **356**, 133–141.

Atkinson, C., Berman, S., Humbert, O. et al. (2004). *In vitro* incubation of human feces with daidzein and antibiotics suggests interindividual differences in the bacteria responsible for equol production. *J. Nutr.*, **134**, 596–599.

Axelson, M., Kirk, D.N., Farrant, R.D. et al. (1982). The identification of the weak oestrogen equol [7-hydroxy-3-(4'-hydroxyphenyl)chroman] in human urine. *Biochem. J.*, **201**, 353–735.

Axelson, M., Sjovall, J., Gustafsson, B.E. et al. (1984). Soya—a dietary source of the non-steroidal oestrogen equol in man and animals. *J. Endocrinol.*, **102**, 49–56.

Bolca, S., Urpi-Sarda, M., Blondeel, P. et al. (2010). Disposition of soy isoflavones in normal human breast tissue. *Am. J. Clin. Nutr.*, **91**, 976–984.

Cadwallader, D.E. (1983). *Biopharmaceutics and Drug Interactions.* 3rd ed. Raven Press, New York.

Cassidy, A. (2006). Factors affecting the bioavailability of soy isoflavones in humans. *J. AOAC Int.*, **89**, 1182–1188.

Cassidy, A., Bingham, S., Setchell, K.D. et al. (1994). Biological effects of a diet of soy protein rich in isoflavones on the menstrual cycle of premenopausal women. *Am. J. Clin. Nutr.*, **60**, 333–340.

Chen, J., Lin, H., Hu, M. et al. (2003). Metabolism of flavonoids via enteric recycling: Role of intestinal disposition. *J. Pharmacol. Exp. Ther.*, **304**, 1228–1235.

Coward, L., Barnes, N., Setchell, K.D.R. et al. (1993). Genistein, daidzein, and their β-glycoside conjugates: Antitumor isoflavones in soybean foods from American and Asian diets. *J. Agric. Food Chem.*, **41**, 1961–1967.

Day, A.J., DuPont, M.S., Ridley, S. et al. (1998). Deglycosylation of flavonoid and isoflavonoid glycosides by human small intestine and liver β-glucosidase activity. *FEBS Lett.*, **436**, 71–75.

Day, A.J., Canada, F.J., Diaz, J.C. et al. (2000). Dietary flavonoid and isoflavone glycosides are hydrolysed by the lactase site of lactase phlorizin hydrolase. *FEBS Lett.*, **468**, 166–170.

Decroos, K., Vanhemmens, S., Cattoir, S. et al. (2005). Isolation and characterisation of an equol-producing mixed microbial culture from a human faecal sample and its activity under gastrointestinal conditions. *Arch. Microbiol.,* **183**, 45–55.

Doerge, D.R., Chang, H.C., Churchwell, M.I. et al. (2000). Analysis of soy isoflavone conjugation *in vitro* and in human blood using liquid chromatography-mass spectrometry. *Drug Metab. Dispos.,* **28**, 298–307.

Drozdowski, L. and Thomson, A.B. (2006). Aging and the intestine. *World J. Gastroenterol.,* **12**, 7578–8754.

Faughnan, M.S., Hawdon, A., Ah-Singh, E. et al. (2004). Urinary isoflavone kinetics: The effect of age, gender, food matrix and chemical composition. *Br. J. Nutr.,* **91**, 567–574.

Franke, A.A., Custer, L.J., Wang, W. et al. (1998). HPLC analysis of isoflavonoids and other phenolic agents from foods and from human fluids. *Proc. Soc. Exp. Biol. Med.,* **217**, 263–273.

Frankenfeld, C.L., Atkinson, C., Thomas, W.K. et al. (2004). Familial correlations, segregation analysis, and nongenetic correlates of soy isoflavone-metabolizing phenotypes. *Exp. Biol. Med.,* **229**, 902–913.

Frankenfeld, C.L., McTiernan, A., Thomas, W.K. et al. (2006). Postmenopausal bone mineral density in relation to soy isoflavone-metabolizing phenotypes. *Maturitas,* **53**, 315–324.

Fujioka, M., Uehara, M., Wu, J. et al. (2004). Equol, a metabolite of daidzein, inhibits bone loss in ovariectomized mice. *J. Nutr.,* **134**, 2623–2627.

Gardner, C.D., Chatterjee, L.M., and Franke, A.A. (2009). Effects of isoflavone supplements vs. soy foods on blood concentrations of genistein and daidzein in adults. *J. Nutr. Biochem.,* **20**, 227–234.

Halm, B.M., Ashburn, L.A, and Franke, A.A. (2007). Isoflavones from soya foods are more bioavailable in children than adults. *Br. J. Nutr.,* **98**, 998–1005.

Hedlund, T.E., van Bokhoven, A., Johannes, W.U. et al. (2006). Prostatic fluid concentrations of isoflavonoids in soy consumers are sufficient to inhibit growth of benign and malignant prostatic epithelial cells in vitro. *Prostate,* **66**, 557–566.

Hodgson, J.M., Croft, K.D., Puddey, I.B. et al. (1996). Soybean isoflavonoids and their metabolic products inhibit *in vitro* lipoprotein oxidation in serum. *J. Nutr. Biochem.,* **7**, 664–669.

Hollman, P. (2001). Evidence for health benefits of plant phenols: Local or systemic effects? *J. Sci. Food Agric.,* **81**, 842–852.

Hur, H.G., Lay, J.O., Beger, K.D. et al. (2000). Isolation of human intestinal bacteria metabolizing the natural isoflavone glycosides daidzin and genistin. *Arch. Microbiol.,* **174**, 422–428.

Izumi, T., Piskula, M.K., Osawa, S. et al. (2000). Soy isoflavone aglycones are absorbed faster and in higher amounts than their glucosides in humans. *J. Nutr.,* **130**, 1695–1699.

Kelly, G.E., Nelson, C., Waring, M.A. et al. (1993). Metabolites of dietary (soya) isoflavones in human urine. *Clin. Chim. Acta,* **223**, 9–22.

King, R.A. and Bursill, D.B. (1998). Plasma and urinary kinetics of the isoflavones daidzein and genistein after a single soy meal in humans. *Am. J. Clin. Nutr.,* **67**, 867–872.

Lampe, J.W. (2009). Is equol the key to the efficacy of soy foods? *Am. J. Clin. Nutr.,* **89**, S1664–S1667.

Lampe, J.W., Skor, H.E., Li, S. et al. (2001). Wheat bran and soy protein feeding do not alter urinary excretion of the isoflavan equol in premenopausal women. *J. Nutr.,* **131**, 740–744.

Larkin, T., Price, W.E., and Astheimer, L. (2008). The key importance of soy isoflavone bioavailability to understanding health benefits. *Crit. Rev. Food Sci. Nutr.,* **48**, 538–552.

Liu, Y. and Hu, M. (2002). Absorption and metabolism of flavonoids in the caco-2 cell culture model and a perused rat intestinal model. *Drug Metab. Dispos.,* **30**, 370–377.

Manach, C., Williamson, G., Morand, C. et al. (2005). Bioavailability and bioefficacy of polyphenols in humans. I. Review of 97 bioavailability studies. *Am. J. Clin. Nutr.,* **81**, 230S–242S.

Maubach, J., Depypere, H.T., Goeman, J. et al. (2004). Distribution of soy-derived phytoestrogens in human breast tissue and biological fluids. *Obstet. Gynecol.,* **103**, 892–898.

Messina, M., Watanabe, S., and Setchell, K.D.R. (2009). Report on the 8th international symposium on the role of soy in health promotion and chronic disease prevention and treatment. *J. Nutr.,* **139**, 796S–802S.

Morton, M.S., Chan, P.S., Cheng, C. et al. (1997). Lignans and isoflavonoids in plasma and prostatic fluid in men: Samples from Portugal, Hong Kong, and the United Kingdom. *Prostate* **32**, 122–128.

Németh, K., Plumb, G.W., Berrin, J.G. et al. (2003). Deglycosylation by small intestinal epithelial cell β-glucosidases is a critical step in the absorption and metabolism of dietary flavonoid glycosides in humans. *Eur. J. Nutr.,* **42**, 29–42.

Richelle, M., Pridmore-Merten, S., Bodenstab, S. et al. (2002). Hydrolysis of isoflavone glycosides to aglycones by β-glycosidase does not alter plasma and urine isoflavone pharmacokinetics in postmenopausal women. *J. Nutr.,* **132**, 2587–2592.

Rowland, I., Faughnan, M., Hoey, L. et al. (2003). Bioavailability of phyto-oestrogens. *Br. J. Nutr.,* **89**, S45–S58.

Scalbert, A. and Williamson, G. (2000). Dietary intake and bioavailability of polyphenols. *J. Nutr.,* **130**, 2073S–2085S.

Setchell, K.D. and Cassidy, A. (1999). Dietary isoflavones: Biological effects and relevance to human health. *J. Nutr.,* **129**, 758S–767S.

Setchell, K.D. and Cole, S.J. (2006). Method of defining equol-producer status and its frequency among vegetarians. *J. Nutr.,* **136**, 2188–2193.

Setchell, K.D., Borriello, S.P., Hulme, P. et al. (1984). Nonsteroidal estrogens of dietary origin: Possible roles in hormone-dependent disease. *Am. J. Clin. Nutr.,* **40**, 569–578.

Setchell, K.D., Zimmer-Nechemias, L., Cai, J. et al. (1997). Exposure of infants to phytooestrogens from soy-based infant formula. *Lancet,* **350**, 23–27.

Setchell, K.D., Brown, N.M., Desai, P. et al. (2001). Bioavailability of pure isoflavones in healthy humans and analysis of commercial soy isoflavone supplements. *J. Nutr.,* **131**, 1362S–1375S.

Setchell, K.D., Brown, N.M., Zimmer-Nechemias, L. et al. (2002a). Evidence for lack of absorption of soy isoflavone glycosides in humans, supporting the crucial role of intestinal metabolism for bioavailability. *Am. J. Clin. Nutr.,* **76**, 447–453.

Setchell, K.D., Brown, N.M., and Lydeking-Olsen, E. (2002b). The clinical importance of the metabolite equol—a clue to the effectiveness of soy and its isoflavones. *J. Nutr.,* **132**, 3577–3584.

Setchell, K.D., Brown, N.M., Desai, P. et al. (2003a). Bioavailability, disposition, and dose-response effects of soy isoflavones when consumed by healthy women at physiologically typical dietary intakes. *J. Nutr.,* **133**, 1027–1035.

Setchell, K.D., Faughnan, M.S., Avades, T. et al. (2003b). Comparing the pharmacokinetics of daidzein and genistein with the use of ^{13}C-labeled tracers in premenopausal women. *Am. J. Clin. Nutr.,* **77**, 411–419.

Setchell, K.D.R., Clerici, C., Lephart, E. et al. (2005). S-equol, a potent ligand for estrogen receptor beta, is the exclusive enantiomeric form of the soy isoflavone metabolite produced by human intestinal bacterial flora. *Am. J. Clin. Nutr.,* **81**, 1072–1079.

Setchell, K.D., Zhao, X., Shoaf, S.E. et al. (2009). The pharmacokinetic behavior of the soy isoflavone metabolite S-(–)equol and its diastereoisomer R-(+)equol in healthy adults determined by using stable-isotope-labeled tracers. *Am. J. Clin. Nutr.,* **90**, 1029–1037.

Shelnutt, S.R., Cimino, C.O., Wiggins, P.A. et al. (2002). Pharmacokinetics of the glucuronide and sulfate conjugates of genistein and daidzein in men and women after consumption of a soy beverage. *Am. J. Clin. Nutr.,* **76**, 588–594.

Shutt, D.A. and Cox, R.I. (1972). Steroid and phytoestrogen binding to sheep uterine receptors *in-vitro. J. Endocrinol.,* **52**, 299–307.

Tamura, A., Shiomi, T., Hachiya, S. et al. (2008). Low activities of intestinal lactase suppress the early phase absorption of soy isoflavones in Japanese adults. *Clin. Nutr.,* **27**, 248–253.

Turner, N.J., Thomson, B.M., and Shaw, I.C. (2003). Bioactive isoflavones in functional foods: The importance of gut microflora on bioavailability. *Nutr. Rev.,* **61**, 204–213.

Vergne, S., Bennetau-Pelissero, C., Lamothe, V. et al. (2008). Higher bioavailability of isoflavones after a single ingestion of a soya-based supplement than a soya-based food in young healthy males. *Br. J. Nutr.,* **99**, 333–344.

Wang, X.L., Hur, H.G., Lee, J.H. et al. (2005a). Enantioselective synthesis of *S*-equol from dihydrodaidzein by a newly isolated anaerobic human intestinal bacterium. *App. Environ. Microbiol.,* **71**, 214–219.

Wang, X.L., Shin, K.H., Hur, G.H. et al. (2005b). Enhanced biosynthesis of dihydrodaidzein and dihydrogenistein by a newly isolated bovine rumen anaerobic bacterium. *J. Biotechnol.,* **115**, 261–269.

Watanabe, S., Yamaguchi, M., Sobue, T. et al. (1998). Pharmacokinetics of soybean isoflavones in plasma, urine and feces of men after ingestion of 60 g baked soybean powder (kinako). *J. Nutr.,* **128**, 1710–1715.

Wu, J., Oka, J., Ezaji, J. et al. (2007). Possible role of equol status in the effects of isoflavone on bone and fat mass in postmenopausal Japanese women: A double-blind, randomized, controlled trial. *Menopause,* **14**, 866–874.

Xu, X., Wang, H.J., Murphy, P.A. et al. (1994). Daidzein is a more bioavailable soymilk isoflavone than is genistein in adult women. *J. Nutr.,* **124**, 825–832.

Xu, X., Harris, K.S., Wang, H.J. et al. (1995). Bioavailability of soybean isoflavones depends upon gut microflora in women. *J. Nutr.,* **125**, 2307–2315.

Zubik, L. and Meydani, M. (2003). Bioavailability of soybean isoflavones from aglycone and glucoside forms in American women. *Am. J. Clin. Nutr.,* **77**, 1459–1465.

6 Dietary Hydroxycinnamates and Their Bioavailability

*Angelique Stalmach, Gary Williamson,
and Michael N. Clifford*

CONTENTS

6.1   INTRODUCTION

6.1.1   Hydroxycinnamic Acids and Derivatives

Hydroxycinnamic acids are a class of (poly)phenolic compounds with a C6–C3 skeleton structure of *trans*-phenyl-3-propenoic acid, with one or more hydroxyl groups attached to the phenyl moiety, some of which may be methylated. The main hydroxycinnamic acids in the diet are *p*-coumaric, caffeic, ferulic, and sinapic acids, usually found as glycosides or esters of quinic acid, shikimic acid, or tartaric acid (Clifford 2000; Manach et al. 2004), as well as ether- and ester-linked to polysaccharides forming the structure of plant cell walls (Sun et al. 2002). Structures of major hydroxycinnamic acids in the diet are shown in Figure 6.1. Quinic esters of mono- and dihydroxycinnamic acids (chlorogenic acids) are found in abundance in coffee beverage, artichoke, cherries, blueberries, aubergine, and apples (Herrmann 1989; Clifford 2000; Manach et al. 2004); mono- and dicaffeoyl and *p*-coumaroyl tartaric acids can be found in wine; endives, and spinach (Singleton et al. 1978; Herrmann 1989); glycosylated hydroxycinnamic acids are found in kale leaf, spinach, and tomatoes (Herrmann 1989); malic acid esters are found in radishes and chicory (Herrmann 1989); and rosmarinic acid, which is 3-(3′,4′-dihydroxyphenyl)lactic acid ester of caffeic acid, is commonly found in rosemary, oregano, thyme, and sage (Herrmann 1989).

Positional isomers occur depending on the position of the hydroxycinnamic acid to the ester. For example, caffeic acid attached to a quinic acid moiety can result in 1-, 3-, 4-, or 5-*O*-caffeoylquinic acid isomers, found in various proportions depending on the food source (Herrmann 1989). Hydroxycinnamic acids in the diet occur naturally in the *trans* form although geometrical isomerization to the *cis* form occurs easily through UV irradiation (Clifford et al. 2008). Plant tissues that are exposed to intense UV irradiation contain *cis*-isomers of at least some chlorogenic acids, for example, *cis*-5-*O*-*p*-coumaroylquinic acid in herbal Aster (Clifford et al. 2006) and a wide range of *cis*-chlorogenic acid isomers in maté (Jaiswal et al. 2010).

6.1.2   Effect of Processing and Storage

Food processing and storage result in the transformation of hydroxycinnamic acids and their conjugates. Heat treatment through microwaving, boiling, and oven-baking resulted in 46, 60, and 100% loss of chlorogenic acids (quinic esters of hydroxycinnamic acids) in potatoes (Dao and Friedman 1992), whereas the roasting process of coffee beans results in a degradation of chlorogenic acids, with the loss of caffeoylquinic, feruloylquinic, and dicaffeoylquinic acids ranging from 60% to almost 100% with increasing levels of roasting (Trugo and Macrae 1984). The roasting

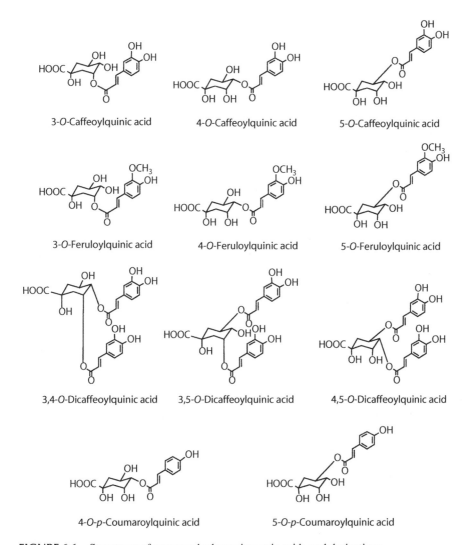

FIGURE 6.1 Structures of common hydroxycinnamic acids and derivatives.

process yields the production of volatile compounds such as furans, aldehydes, indoles, thiols, pyrazines, and pyridines providing coffee with its characteristic aroma and taste (Rocha et al. 2003; Charles-Bernard et al. 2005a, 2005b; Lopez-Galilea et al. 2006). Non-volatile melanoidins produced during roasting constitute up to 25–30% of the dry matter of the beverage (Borrelli et al. 2002). These compounds result from the non-enzymatic browning Maillard reaction and are thought to include the degradation products of chlorogenic acids that have been incorporated into melanoidins upon roasting (Dorfner et al. 2003; Bekedam et al. 2008a, 2008b; Gniechwitz et al. 2008). Enzymatic reaction during brewing process of beer-making also leads to the degradation of hydroxycinnamic acids into volatile phenols following fermentation with specific strains of yeasts (Vanbeneden et al. 2006). Similarly, the reaction product 2-*S*-glutathionyl caftaric acid occurs as a result of the oxidation of caftaric acid in presence of grape polyphenol oxidase, oxygen, and glutathione during wine-making (Singleton et al. 1985; Cheynier et al. 1986). Other undesirable compounds, polycyclic aromatic hydrocarbons (Houessou et al. 2005; Houessou et al. 2007) and heterocyclic aromatic amines (Casal et al. 2004) are produced during the roasting process. It has also been reported that the presence of cinnamoylquinic acid lactones formed during the roasting of coffee beans through elimination of a water molecule of the quinic moiety and subsequent formation of a lactone ring (Farah et al. 2005) contribute to the bitter taste of the beverage (Frank et al. 2006).

6.2   DIETARY LEVELS INGESTED

6.2.1   Dietary Intake

Daily dietary intake of hydroxycinnamic acids has been estimated in various populations. In a German population, the total daily amount of phenolic acids (hydroxycinnamic and hydroxybenzoic acids combined) reported was 222 mg, with 206 mg being caffeic acid and derivatives (Radtke et al. 1998). In Finnish adults, the daily amount estimate has been reported as 641 mg (Ovaskainen et al. 2008), whereas in a UK population, the daily amount of hydroxycinnamic acids ingested was estimated to be 39–56% of the total amount of simple phenols, polyphenols and tannins (451–598 mg/d) (Clifford 2004). In all cases, coffee was reported as the main contributor to the daily intake of hydroxycinnamic acids among regular coffee drinkers (Radtke et al. 1998; Clifford 2004; Ovaskainen et al. 2008). This is due to the high levels of chlorogenic acids in coffee beverage, with a single serving providing between 20 and 675 mg depending on the type of roast and the volume consumed (Clifford 1999).

6.2.2   Major Food Sources

Tomatoes contain *O*-glycosylated caffeic, ferulic, and *p*-coumaric acids in the range of 11–37 mg/kg fresh weight, whereas spinach contains cinnamoylglucose in the range of 16–64 mg/kg, and lettuce and endive leaves contain tartaric and malic acid conjugates in the range of 2–163 mg/kg fresh weight (Herrmann 1989). In fruits and vegetables, sources of 5-*O*-caffeoylquinic acid are apples, pears, peaches, aubergines, artichokes, potatoes, and carrots, with ranges comprised between 15 and 2200 mg/kg fresh weights (Herrmann 1989; Manach et al. 2004). In other sources such as cherries, plums, kale, cabbages, and Brussels sprouts, the main caffeoylquinic acid is the 3-isomer. Extensive reviews are available (Clifford 1999, 2000) and critically evaluated data are available online in the Phenolexplorer database (http://www.phenol-explorer.eu/).

6.3   BIOAVAILABILITY OF FREE HYDROXYCINNAMIC ACIDS

6.3.1   Absorption and Transport Mechanisms

Several studies have investigated the metabolic fate of free and conjugated hydroxycinnamic acids in cultured cell systems, animal models and humans. The absorption of hydroxycinnamic acids has been shown to take place throughout the gastrointestinal tract.

6.3.1.1 Absorption and Transport through Stomach

Following the incubation of 2.25 µmol of ferulic acid in a rat stomach, 74% of the dose administered disappeared after a 25-min incubation period, and free as well as sulfate, glucuronide, and sulfoglucuronide metabolites of ferulic acid were detected in the circulation (portal vein and celiac artery) and excreted in bile and urine (Zhao et al. 2004). In 2006, the same group carried out a similar experiment with various hydroxycinnamic acids and concluded that the absorption efficiency in the stomach increased in the order: 5-O-caffeoylquinic acid, caffeic acid, ferulic acid, and p-coumaric acid (Konishi et al. 2006). They reported that absorption efficiency agreed with the affinity order for the active and saturable monocarboxylic acid transporter (MCT) observed in Caco-2 cells. This group reported that the isoform MCT1 did not appear to be involved in the transport of phenolic acids across the transepithelal intestinal wall (Watanabe et al. 2006), but Haughton et al. (2007) have shown convincingly that benzoic acid and the three isomeric monohydroxybenzoic acids are transported by cells expressing this isoform.

6.3.1.2 Absorption and Transport through Small Intestine and Colon

6.3.1.2.1 Culture Cell Models

The mechanisms of transport of hydroxycinnamic acids through the intestinal epithelium have been identified using Caco-2 cells, a cell monolayer absorption model of a line of heterogeneous human epithelial colorectal adenocarcinoma cells, which has been widely used in bioavailability research as a model for assessing intestinal transport of a variety of (poly)phenolic compounds (Walgren et al. 1998; Walle et al. 1999; Vaidyanathan and Walle 2001; Konishi et al. 2003a; Kobayashi and Konishi 2008; Poquet et al. 2008b). Following a 21-day seeding period, the cells are able to form monolayers and differentiate and have demonstrated efflux and phase II metabolic processes (Vaidyanathan and Walle 2001; Kern et al. 2003b; Poquet et al. 2008a).

Konishi and Shimizu (2003) suggested that the transport of ferulic acid across the epithelium occurs via MCT. The same group also reported that caffeic acid was primarily transported via passive paracellular diffusion and to a lesser extent via MCT, whereas 5-O-caffeoylqinic acid transport exclusively occurs via paracellular diffusion and is restricted by the tight junction (Konishi and Kobayashi 2004). Other studies claimed the existence of a Na^+-dependent active transport system for cinnamic acids across the intestinal brush border in rats (Wolffram et al. 1995; Ader et al. 1996), which was not supported by Konishi and Shimizu (2003), arguing that the permeation of ferulic acid across Caco-2 cell monolayers was constant in the presence and absence of a proton gradient. They also highlighted the effect of hydroxylation on a decreasing affinity for the MCT-carrier. Recently, Poquet et al. (2008b) showed that ferulic acid was mainly transported by passive transcellular diffusion through the intestinal epithelium, and to a smaller extent, via MCT-carrier. The discrepancies in the outcomes observed between Konishi et al. and Poquet et al.

may have resulted from the type of cell culture used, which consisted in a mono-culture of Caco-2 absorptive-type cells in the work of Konishi et al., as opposed to a co-culture with HT29-MTX goblet-type cells in the work of Poquet et al. Previous investigators have reported that intestinal permeability was generally higher in co-culture when compared with mono-culture for compounds undergoing passive transport (Hilgendorf et al. 2000). Besides the use of different cell cultures, the con-centration of ferulic acid loaded on the apical side of the inserts also varied widely between the two experiments, which in the case of Konishi and Kobayashi's work corresponded to 170 times that of Poquet's. Determination of the partition coefficient (log P values) for various polyphenolic compounds has also provided information regarding their lipophilicity and mode of transport across the intestinal barrier. It is determined as the ratio of concentrations of a compound in two immiscible solvents, usually water and octanol. Although log P values do not necessarily correlate to their *in vivo* bioavailability, epithelial transport of polyphenolic compounds determined using this technique, across Caco-2 cells and *in situ* perfusion of rat jejuno-ileal segments were consistent with a passive diffusion mechanism (Rothwell et al. 2005; Silberberg et al. 2006b; Tian et al. 2009). Transport is also likely to be transcellular rather than paracellular and determined by the law of nonionic diffusion, being dependant on the pKa of a compound and pH of the medium (Garcia-Conesa et al. 1999; Poquet et al. 2008b).

6.3.1.2.2   Animal Models

In situ and *ex vivo* perfusion of jejunum and ileum sections of rat intestine with caf-feic, ferulic, cinnamic and *p*-coumaric acids have shown that free hydroxycinnamic acids are largely absorbed through the jejunum (Wolffram et al. 1995; Spencer et al. 1999; Garrait et al. 2006) and to a smaller extent, through the ileum (Spencer et al. 1999; Garrait et al. 2006). The efficiency of absorption through the jejunum appeared in the order: caffeic acid < *p*-coumaric acid < ferulic acid (Spencer et al. 1999), with a net absorption of the perfused doses on the serosal side accounting for 19.5% for caffeic acid (Lafay et al. 2006b) and 56.1% for ferulic acid (Adam et al. 2002). This is in agreement with results obtained from Caco-2 cell models (Konishi et al. 2003b, 2006), suggesting an increased absorption of O-methylated compounds across the epithelium. *Ex vivo* studies on rat also showed the ability of hydroxycinnamic acids to cross the colonic epithelium for absorption and are transferred across mainly in an unconjugated form (Garrait et al. 2006; Poquet et al. 2008b).

6.3.2   SITES OF METABOLISM AND METABOLIC PROCESSES

The metabolism of hydroxycinnamic acids upon uptake has been investigated using various cell models, which provided further information regarding the sites of metabolism. Table 6.1 provides a summary of studies investigating the bioavailabil-ity of hydroxycinnamic acids in humans and animal models.

6.3.2.1   Gastrointestinal Tract

The human gastric and intestinal epithelia have been suggested as sites for phase II enzyme reactions, leading to sulfation, glucuronidation, and/or

TABLE 6.1

Bioavailability Studies of Hydroxycinnamic Acids in Animal Models and Humans

Species	Compound Tested and Matrix	Dose (μmol/kg bw)[a]	Delivery Method	Metabolites	C_{max} Plasma (μmol/L)	T_{max} Plasma (h)	% Excreted (24 h)	Reference
Rats	CA	1852	Oral (gavage)	3-HPPA, FA, VA	n.a.	n.a.	13, 14, 3.6	Booth et al. (1957)
	FA	1145		3-HPPA, feruloylglycine, DHFA, VA, vanilloylglycine				
	m-CoA	1355		3-HHA, m-CoA-GlcA				
	DHCA	1221		3-HPPA, feruloylglycine, DHFA, VA, vanilloylglycine				
	DHFA	1134		3-HPPA, feruloylglycine, DHFA, VA, vanilloylglycine				
Rats	Free FA	258	Oral	FA + FA-GlcA	n.a.	n.a.	10.5	Choudhury et al. (1999)
			i.v.	FA + FA-GlcA			11.5	
	5-CQA	141	Oral	nd	n.a.	n.a.	n.d.	
			i.v.	5-CQA + 5-CQA-GlcA			9.2	
Rats	CA	700	Oral (gastric intubation)	CA-GlcA	26.1	2 h	n.a.	Azuma et al. (2000)
				CA-S-GlcA	13			
				FA-S	8.5			
				FA-S-GlcA	6.5			
				FA-GlcA	4			
				FA	1.6			
				CA	1.2			

(Continued)

TABLE 6.1 (CONTINUED)

Bioavailability Studies of Hydroxycinnamic Acids in Animal Models and Humans

Species	Compound Tested and Matrix	Dose (µmol/kg bw)[a]	Delivery Method	Metabolites	C_{max} Plasma (µmol/L)	T_{max} Plasma (h)	% Excreted (24 h)	Reference
Rats	CA mixed in diet	1625/day (8 days)	Oral	CA, FA, iFA m-CoA HPPA HBA, HHA Hippuric acid	41.3, 7.3, 4.5 n.d. 1.4 not reported 54.2	n.a.	12.8 1.1 4 0.1 22.9	Gonthier et al. (2003)
Rats	FA mixed in diet	26.5	Oral	FA FA-GlcA FA-S + FA-GlcA-S	1.68 0.87 2.73	0.5 h	43.4 (4.5 h)	Rondini et al. (2002)
Rats	FA	70	Oral	FA-S-GlcA > FA > FA-GlcA > FA-S (plasma at 15 min) FA-S-GlcA > FA-GlcA > FA > FA-S (urine at 16 h)	109.5	0.25 h	72 (16 h)	Zhao et al. (2003)
Rats	p-CoA	100	Oral (gastric intubation)	p-CoA/p-CoA-S, p-CoA-GlcA, p-CoA-S-GlcA	165.7/72.6 (portal vein)	10 min	n.a.	Konishi et al. (2004)
Rats	p-CoA CIN	233	Oral	n.a.	n.a.	n.a.	24 0.3	Garrait et al. (2006)
Rats	DHCA	100	Oral	DHCA-GlcA, DHCA-S, DHFA, DHFA-S, FA, FA-S, iFA, iFA-S	not reported	n.a.	<1	Poquet et al. (2008)

Species	Substrate	Dose	Route	Metabolites				Reference
Humans	CA	79	Oral	Vanilloylglycine	12	n.a.	n.a.	Booth et al. (1957)
				3-HBA	3.9–4.6			
				FA	3.7			
				Feruloylglycine	2.5			
				DHFA	1.4–1.8			
				VA	1.1–1.6			
Humans	CA in apple juice	80	Oral	VA, CA, FA, iFA	11.4 (48 h)	n.a.	n.a.	Jacobson et al. (1983)
Humans	FA in tomatoes	2.2	Oral	FA + FA-GlcA	11–25	n.a.	n.a.	Bourne et al. (1998)
Humans	CA	40	Oral	Not reported	11	n.a.	n.a.	Olthof et al. (2001)

[a] If not stated, body weights (bw) of 7-week-old rats, adult rats, and humans taken as 200 g, 450 g, and 70 kg. C_{max}: peak plasma concentration; T_{max}: time to reach C_{max}; 5-CQA: 5-caffeoylquinic acid; CA: caffeic acid; CIN: cinnamic acid; m-CoA: m-coumaric acid; p-CoA: p-coumaric acid; DHCA: dihydrocaffeic acid; DHFA: dihydroferulic acid; FA: ferulic acid; iFA: isoferulic acid; HBA: hydroxybenzoic acid; HHA: hydroxyhippuric acid; HPPA, hydroxyphenylpropionic acid; VA: vanillic acid; GlcA: glucuronide; S: sulfate; n.a., not analyzed; n.d.: not detected; i.v.: intravenous.

O-methylation (Lin et al. 1999; Chen et al. 2003a), with various isoforms of the enzymes detected along the gastrointestinal tract (Grams et al. 2000; Nishimura and Naito 2006; Teubner et al. 2007). Hydroxycinnamic acids absorbed in the stomach are mainly transported across the gastric wall as unconjugated species (Zhao et al. 2004; Konishi et al. 2006). Perfusion of a rat jejuno-ileal segment with caffeic acid (50 µM) led to the recovery of free and conjugated ferulic acid in the effluent fluxes accounting for 0.5% of the perfused flux, which demonstrates the ability of the small intestine to carry out phase II metabolic processes (Lafay et al. 2006b). Similarly, perfusion of isolated rat jejunum and ileum segments with caffeic, ferulic, and *p*-coumaric acids (ca. 75 µM) resulted in the recovery of conjugated hydroxycinnamic acids in the serosal fluids (Spencer et al. 1999). The levels of caffeic acid glucuronides identified and quantified were twofold higher than caffeic acid, with little ferulic acid and *p*-coumaric acid conjugates detected on the serosal side of the rat jejunum (Spencer et al. 1999). Similarly, following perfusion of rat stomach and small intestine, ferulic acid and *p*-coumaric acid were recovered mainly as the free form in the portal/mesenteric vein plasma (Konishi et al. 2004, 2006; Zhao et al. 2004; Silberberg et al. 2006b), indicating that the liver is the main site of metabolism for these hydroxycinnamic acids. Incubation of Caco-2 cells with methyl esters of various hydroxycinnamic acids and diferulic acid resulted in the formation of methylated, sulfated, and glucuronidated hydroxycinnamic acids, indicating the presence of intra- and extracellular esterases as well as intracellular sulfotransferases (SULT), glucuronosyltransferases (UGT), and catechol-*O*-methyltransferases (COMT) (Kern et al. 2003b). Uptake of the free hydroxycinnamic acids by Caco-2 cells was in the order: *p*-coumaric acid (ca. 10%) < caffeic acid (ca. 30%) < sinapic acid (42%) < ferulic acid (ca. 50%), and excreted metabolites in the medium (present as sulfates) were in the order: *p*-coumaric acid (not detected) < caffeic acid (ca. 4%) < ferulic acid (ca. 15%) < sinapic acid (ca. 20%) (Kern et al. 2003b). Perfusion of rat jejuno-ileal segments with ferulic acid and caffeic acid resulted in the presence of mainly conjugated ferulic acid detected in plasma of the mesenteric vein (Adam et al. 2002), with conjugates of caffeic acid also found present in the portal vein (Konishi et al. 2005), suggesting the enterocytes as a site for hydroxycinnamic acid metabolism upon intestinal absorption. Incubation of ferulic and dihydrocaffeic acids by rat colon everted sacs resulted in the recovery of mainly the free compounds on the serosal side, followed by glucuronides, methyl conjugates (in the case of dihydrocaffeic acid), and to a smaller extent sulfates (Poquet et al. 2008a, 2008b). HT29-MTX goblet-like cells appeared to have a major role in the metabolism of ferulic acid, with an increased capacity to conjugate ferulic acid compared to Caco-2 cells not capable of glucuronidation of these compounds (Kern et al. 2003b; Poquet et al. 2008b). Sulfation of caffeic and ferulic acids by human intestinal S9 homogenates appeared to be the preferential pathway of hydroxycinnamic acid metabolism over glucuronidation, accounting for more than 95% of the conjugates produced (Wong et al. 2010).

Besides phase II conjugation, the small intestine and colon are also capable of carrying out hydrogenation of the aliphatic carbon double bond of caffeic and ferulic acid to produce dihydrocaffeic and dihydroferulic acids (Peppercorn and Goldman 1972; Ohmiya et al. 1986; Rechner et al. 2004; Poquet et al. 2008b).

6.3.2.2 Liver

After absorption of free hydroxycinnamic acids in the stomach and small intestine, the liver appears to be the main site of metabolism, with sulfated and/or glucuronidated conjugates detected (Zhao et al. 2004; Konishi et al. 2006).

As observed in small intestinal cells, caffeic acid appears to be extensively methylated by the liver (Moridani et al. 2002; Mateos et al. 2006), yielding ferulic acid, which suggests that COMT preferentially methylates this $3',4'$-dihydroxy structure at the $3'$-position of the phenol ring. The liver has also demonstrated the ability to conjugate various hydroxycinnamic acids, suggesting the presence of sulfotransferases and glucuronosyltransferases in hepatocytes (Mateos et al. 2006; Poquet et al. 2008a; Wong et al. 2010). Sulfation activity appeared more extensive than glucuronidation in human liver S9 homogenates, with efficiency of conjugation in the order: caffeic acid > ferulic acid > dihydrocaffeic acid > isoferulic acid > dihydroferulic acid (Wong et al. 2010). In this study, the only glucuronide produced by the liver homogenates was isoferulic acid-$3'$-O-glucuronide. Similarly, when incubated with HepG2 cells, caffeic acid was the most extensively metabolized compared to ferulic acid and its quinic ester, 5-O-caffeoylquinic acid, with metabolites produced being free and glucuronidated ferulic acid, as well as sulfated and glucuronidated caffeic acid (Mateos et al. 2006). By contrast, incubation of ferulic acid and 5-O-caffeoylquinic acid resulted in the production of ferulic acid glucuronide and caffeoylquinic acid isomers, respectively. Although caffeic acid appeared more extensively metabolized, ferulic acid was increasingly taken up by hepatic cells (Mateos et al. 2006), as was observed in the gastrointestinal tract (Spencer et al. 1999; Kern et al. 2003b; Konishi et al. 2006).

Besides phase II conjugation, the liver has demonstrated deconjugation as well as hydrogenation/dehydrogenation processes, with the incubation of caffeic acid, yielding dihydrocaffeic and dihydroferulic acids following hydrogenation of the aliphatic chain (Moridani et al. 2002). The reverse reaction has also been observed with incubation of dihydrocaffeic acid resulting in the production of caffeic acid, as well as ferulic acid and dihydroferulic acid (Moridani et al. 2002; Poquet et al. 2008a). Possibly, ferulic acid is also subject to demethylation, yielding caffeic acid after incubation with rat hepatocytes (Moridani et al. 2002).

6.3.2.3 Biliary Excretion and Efflux Mechanisms

Excretion of hydroxycinnamic acids in bile was reported following absorption and uptake by hepatocytes, although it does not appear to be a major route of excretion. Following perfusion/incubation of caffeic and ferulic acids in the rat upper gastrointestinal tract, metabolites, mainly as conjugates, were recovered in bile in the range of 0.4–7% of the dose perfused/incubated (Adam et al. 2002; Zhao et al. 2004; Lafay et al. 2006b; Silberberg et al. 2006a). This contributes to an enterohepatic circulation of metabolites, which upon efflux into the lumen can be reabsorbed by the epithelium and further metabolized (Williamson et al. 2000). Efflux mechanisms have been identified in the intestine and liver. Multidrug resistance protein (MRP) and P-glycoprotein (P-gp) efflux pumps are members of the ATP binding cassette (ABC) transporter family, the primary role of which is to transport molecules at the expense of ATP hydrolysis in the cells. These transporters are involved in the efflux of anticancer drugs, preventing active compounds from entering the cells, thus causing the drugs to be less effective than intended (Sharom 1997; Leslie et al. 2001). These efflux transporters have been identified in various tissues such as intestine, liver, kidneys, and brain and have a high polymorphism across populations (Zhang et al. 2006). *In vitro* studies have shown the role of such efflux transporters in the intestinal disposition of flavonoids (Walgren et al. 2000; Chen et al. 2002; Walle and Walle 2003; Zhang et al. 2004). Isoforms of MRP have been identified, with MRP2, located at the apical side of the hepatocytes, being responsible for the biliary excretion of polyphenols such as quercetin and resveratrol (O'Leary et al. 2003; Leslie et al. 2005; Lancon et al. 2007). Perfusion of rat intestinal segments and incubation of Caco-2/HT29-MTX cocultures with caffeic and ferulic acids resulted in the production of conjugated metabolites, which were secreted on the apical side of the intestinal epithelium (accounting for 0.5–6% of the dose perfused), implying absorption, metabolism, and efflux occurring *in situ* (Lafay et al. 2006b; Silberberg et al. 2006a, 2006b; Poquet et al. 2008b).

Altogether, the results suggest that uptake and metabolism of hydroxycinnamic acids depend on a range of factors such as the chemical structure of the compound, whether it is transported as free or as an ester conjugate, the model used to investigate transport and metabolism (cell culture or animal), pKa value, which will determine the preferential region of absorption and metabolism along the gastrointestinal tract, and the presence of specific isoforms of conjugating enzymes.

6.3.3 BIOAVAILABILITY

Azuma et al. (2000) reported that after oral ingestion of either caffeic acid or 5-*O*-caffeoylquinic acid by rats, metabolites in plasma were found in the form of glucuronide and/or sulfate conjugates of caffeic acid and ferulic acid. Studies in rats fed ferulic acid reported the presence of sulfated, glucuronidated and apparently sulfoglucuronidated ferulic acids as the main metabolites excreted, although the exact structure of the sulfoglucuronidated metabolite has yet to be elucidated because of the presence of a single hydroxyl group on the ferulic acid. Urinary recoveries of these metabolites were reported as 10.5% (Choudhury et al. 1999), 43% (Rondini et al. 2002), and 72% (Zhao et al. 2003b) of the amounts ingested (64 μmol, and ~9 and 7 μmol, respectively). The different recoveries may have resulted from the use of different enzymatic treatments, which consisted of glucuronidase only (Choudhury et al. 1999) as opposed to a mixture of sulfatase and glucuronidase (Rondini et al. 2002; Zhao et al. 2003b). Bourne and Rice-Evans (1998) found a urinary recovery of 11–25% of ferulic acid in the glucuronidated and free form in the urine of human volunteers, following the ingestion of ca. 155 μmol of ferulic acid as a single intake of tomatoes. The maximum of excretion was reached 7 h after supplementation. Similarly, Jacobson et al. (1983) reported a recovery of ca. 10% of metabolites as vanillic, caffeic, ferulic, and isoferulic acids following the ingestion of 1 g of caffeic acid, and 11% of 2.8 mmol of caffeic acid was excreted in urine of ileostomy volunteers (Olthof et al. 2001). However, these studies were limited by the lack of availability of sulfated standards at the time. Following intake of hydroxycinnamic acids as the free form or as part of a diet resulted in an extensive metabolism of these compounds upon absorption, with sulfation, glucuronidation, O-methylation, and glycination being reported in plasma and urine of rats and humans (Table 6.1). Hydroxycinnamic acids appeared rapidly absorbed with maximum plasma concentrations of metabolites detected from 10 min (Konishi et al. 2004, 2005) to 2 h (Azuma et al. 2000) in circulation (T_{max} values). Maximum plasma concentrations ranged from 1 μM (Azuma et al. 2000) to 166 μM (Konishi et al. 2004), and urinary excretion accounting for 11–30.5% of caffeic acid intake (Booth et al. 1957; Olthof et al. 2001), 10.5–72% of ferulic acid (Choudhury et al. 1999; Zhao et al. 2003b), and 24% of p-coumaric acid (Garrait et al. 2006). These studies suggest that hydroxycinnamic acids are highly absorbed and also subject to extensive metabolism.

6.3.3.1 Effect of Bound Hydroxycinnamic Acids and Esterification on Their Bioavailability

The bioavailability of hydroxycinnamic acids depends on the matrix in which they are ingested, as well as whether ingested as the free or esterified form. For example, hydroxycinnamic acids are rarely found as free acids in plant products, but are naturally bound to structural cell components in cereals and grains, present as carbohydrate esters in rice, as organic acid esters (such as quinic esters in coffee, tartaric esters in white wine, or conjugated with phenolic acids) and occurring as dimers (Clifford 1999, 2000; Robbins 2003; Manach et al. 2004; Crozier et al. 2009). The effect of esterification on the bioavailability of hydroxycinnamic acids is summarized in Table 6.2.

6.3.3.2 Effect on Absorption

Lower urinary recoveries were observed following supplementation of ferulic acid contained in a high-bran cereal in volunteers, with ca. 3% of ferulic acid excreted in the free and metabolized forms (Kern et al. 2003a). Following intake of corn bran by rats, only 0.4–0.5% of ferulic acid was excreted in urine (Zhao et al. 2005), and following ingestion of flour and milled fraction diets, 3–8% was excreted in rats compared to an excretion accounting for 39–52% of the dose ingested when diets were supplemented with the pure compound (Adam et al. 2002). Similarly, when equimolar amounts of ferulic acid were administered to rats in its free form, as an oligosaccharide ester, and as part of a hemicellulosic component of cell walls, urinary excretion of ferulic acid metabolites accounted for 72, 54, and 20%, respectively (Zhao et al. 2003b), accompanied by increased recovery of the non-absorbed compounds excreted in faeces (Zhao et al. 2003a). This highlights the reduced capacity of bound hydroxycinnamic acids to be absorbed through the enterocytes in the gastrointestinal wall. Similarly, caffeic acid was ca. 2–3 times better absorbed in the upper gastrointestinal tract compared with 5-O-caffeoylquinic acid (Olthof et al. 2001; Lafay et al. 2006b), with the permeation rate of 5-O-caffeoylquinic acid across Caco-2 cell monolayers being lower than that of caffeic acid (Konishi and Kobayashi 2004). The rate of uptake of 5-O-caffeoylquinic acid in the stomach, small intestine and liver was reported to be less than that of caffeic acid (Spencer et al. 1999; Konishi et al. 2006; Mateos et al. 2006), highlighting the effect of esterification on a reduced absorption capacity. This probably reflects the increased hydrophilic characteristics

TABLE 6.2

Effect of Esterification on The Bioavailability of Hydroxycinnamic Acids[a]

Species	HC + Ester	Dose (µmol/kg bw)	Administration	Absorption Rate (%)	Plasma C_{max} (µmol/L)	Plasma T_{max}	Urine Excretion (%)	Recovery in Faeces (%)	Reference
Rats	Free FA	100–2176	21-day diet	n.a.	n.d.–7.6	n.a.	38.6–51.8	0.2–3.5	Adam et al. (2002)
	FA in whole wheat flour and milled fractions	33–541			n.d.–0.3	n.a.	3.2–8.4	6.0–38.4	
Rats	FA	7.72 mmol/kg diet	Prepared from refined corn bran	99% (SI)	11	n.a.	n.a.	n.d.	Zhao et al. (2003)
	FAA			40% (SI)	12			3	
	FAXn			57% (CE) 44% (CE) 23% (CO)	ca. 1			33	
Rats	FA	70	Single dose orally in water	n.a.	109.5	0.25 h	72	1%	Zhao et al. (2003)
	FAA				23	0.5 h	54	1%	
	FAXn				1.3	12 h	20	20%	
Rats	FA	116	Single meal with 5% refined corn bran	n.a.	n.a.	n.a.	2.3	81 (72 h)	Zhao et al. (2005)
	p-CoA	13					0.4	64 (72 h)	
Rats	CA	100	Single dose (gastric incubation)	n.a.	11.24 (free)/31.23 (conjugates) (portal vein)	10 min	n.a.	n.a.	Konishi et al. (2005)
	RA				1.36 (free)/0.86 (conjugates) (portal vein)				

Rats	CA 5-CQA	11.25	Jejunoileal perfusion	19.5 8	0.71 (FA), 0.14 (iFA) 0.12 (iFA)	n.a.	n.a.	80.9 (CE) 92 (CE)	Lafay et al. (2006)
Humans	CA 5-CQA	40	Single dose in water	95 33	n.a.	n.a.	10.7 (CA) <1 (CA), 0.3 (5-CQA)	n.a.	Olthof et al. (2001)
Humans	FA (80%) + SA (6%)	19	Single serving of breakfast cereals	na	0.15–0.21 (FA + conjugates) 0.01–0.04 (SA + conjugates)	1–3 h	3.13 (FA)/2.76 (SA)	n.a.	Kern et al. (2003)

[a] If not stated, body weights (bw) of 7-week-old rats, adult rats, and humans taken as 200 g, 450 g, and 70 kg. C_{max}: peak plasma concentration; T_{max}: time to reach C_{max}; 5-CQA: 5-caffeoylquinic acid; CA: caffeic acid; CE: cecum; CO: colon; p-CoA, p-coumaric acid; FA: ferulic acid; FAA: 5-O-feruloyl-l-arabinofuranose; FAXn: feruloyl-arabinoxylan; HC: hydroxycinnamic acid; iFA, isoferulic acid; RA, rosmarinic acid; SA: sinapic acid; SI: small intestine; n.d.: not detected; n.a.: not analyzed.

of the quinic ester, responsible for altering its permeability across the epithelium (Poquet et al. 2008b). Chemical structure, number, and position of hydroxyl groups, conjugation and efflux mechanisms are also thought to play a role in the epithelial transport (Liu and Hu 2002; Rothwell et al. 2005; Passamonti et al. 2009; Tian et al. 2009). Similarly, ferulic acid covalently bound to complex carbohydrates is less well absorbed in the upper gastrointestinal tract compared with the free form, with recoveries in the feces of up to 81% of the dose ingested (Adam et al. 2002; Zhao et al. 2003a, 2003b, 2005). This, however, did not affect the metabolic profile of excretion, with the majority of bound and esterified ferulic acid excreted as ferulic acid sulfate or sulfoglucuronide (74–91%) (Zhao et al. 2003b; Rondini et al. 2004).

6.3.3.3   Effect on Metabolism

In bioavailability studies where esters of hydroxycinnamic acids were fed orally in a single dose, no free or conjugated form of quinic esters (Choudhury et al. 1999; Azuma et al. 2000; Takenaka et al. 2000; Olthof et al. 2001; Rechner et al. 2001b; Nardini et al. 2002) or tartaric esters (Nardini et al. 2009) were detected in plasma or urine. A range of methylated, sulfated, and glucuronidated hydroxycinnamic acids were reported, implying hydrolysis of the ester moiety prior to metabolism. Previous investigators have reported the high stability of chlorogenic acids when incubated with various gastrointestinal digestive fluids (Takenaka et al. 2000; Olthof et al. 2001; Rechner et al. 2001b; Farah et al. 2006). Therefore, hydrolysis of the quinic moiety must occur upon absorption, by action of intra or extracellular esterases or ultimately by microflora in the colon. In contrast, the ester of caffeic acid to the side chain hydroxyl of 3-(3′,4′-dihydroxyphenyl)lactic acid does not require hydrolysis of the ester moiety prior to metabolism, because the sulfates and/or glucuronides and its methylated derivative have been reported in plasma and urine of humans and rats following the administration of a single dose of rosmarinic acid (Baba et al. 2004) or perilla (Baba et al. 2005).

Hydroxycinnamic acids are rapidly absorbed along the gastrointestinal tract when ingested as the free form (Bourne et al. 2000; Zhao et al. 2004; Konishi et al. 2006; Poquet et al. 2008b), as the presence of free hydroxycinnamic acids in a complex food matrix increases their bioaccessibility (Mateo Anson et al. 2009b). When present in the diet as esters or bound to cell walls, hydroxycinnamic acids are hydrolyzed by esterases present throughout the digestive tract (Buchanan et al. 1996; Kroon et al. 1997; Chesson et al. 1999; Plumb et al. 1999; Andreasen et al. 2001a, 2001b; Couteau et al. 2001; Kern et al. 2003b; Gonthier et al. 2006; Lafay et al. 2006b), prior to absorption and release of hydroxycinnamic acids into the circulatory system. Alternatively, the bioprocessing of wheat bran also increases the bioaccessibility of hydroxycinnamic acids initially bound to the food matrix, resulting in an increase in their bioavailability (Mateo Anson et al. 2009a). The presence of an ester moiety reduces the absorption of hydroxycinnamic acids, as indicated earlier. Regarding their metabolism, chlorogenic acids appear to be conjugated by catechol-O-methyltransferase, sulfotransferase, and glucuronosyltransferase, occurring in the small intestine and / or liver (Choudhury et al. 1999; Yang et al. 2005; Stalmach et al. 2010). Metabolism occurring in the liver showed the production of 5-O-caffeoylquinic isomers (Mateos et al. 2006) as well as methylated derivatives

(Moridani et al. 2002) in *in vitro* models. However, once the hydroxycinnamic acid is released from the ester moiety, the metabolic pathway appears similar to that of the free compounds, with the presence of conjugated hydroxycinnamic acids reported in circulation and excreted in urine following ingestion of esters (Table 6.1).

6.4 BIOAVAILABILITY OF CHLOROGENIC ACIDS

6.4.1 ABSORPTION THROUGHOUT THE GASTROINTESTINAL TRACT

Intact chlorogenic acids are not efficiently absorbed and transported across the small intestinal epithelium (Spencer et al. 1999; Azuma et al. 2000; Konishi and Kobayashi 2004; Konishi et al. 2006; Lafay et al. 2006b), but are mainly metabolized in the colon with hydrolysis of the quinic moiety (Plumb et al. 1999; Couteau et al. 2001; Gonthier et al. 2006) prior to the release of the hydroxycinnamic acids in circulation. 5-*O*-Caffeoylquinic acid can however be absorbed to a small extent in the upper gastrointestinal tract, with less than 0.1 μM plasma concentration detected in the portal vein and abdominal artery of rats following gastric infusion of 2.25 μmol of 5-*O*-caffeoylquinic acid (Konishi et al. 2006). Similarly, 16.3% of the dose infused in rats was absorbed from the gastric lumen, and recovered intact at concentrations of 3.3 and 1.6 μM in the gastric vein and aorta respectively (Lafay et al. 2006a).

A number of studies have used the ileostomy model to measure the intestinal absorption of dietary polyphenols. Ileostomy subjects are deprived of a functional colon, surgically removed for medical reasons, and upon complete surgical recovery, ileostomists are as healthy as individuals with an intact colon, with similar dietary habits (Kennedy et al. 1982). Upon ingestion of 385 μmol of chlorogenic acids contained in instant coffee, 446 μmol contained in apple juice, and 2.8 mmol of 5-*O*-caffeoylquinic acid by ileostomy volunteers, the recoveries obtained in ileal effluents ranged from 3.6% for the caffeoylquinic acid lactones to 11–46% for the *p*-coumaroylquinic acids, 10–67% for the caffeoylquinic acids, 46% for the dicaffeoylquinic acids, and 77% for the feruloylquinic acids (Olthof et al. 2001; Kahle et al. 2005; Stalmach et al. 2010). A wide variation in the levels recovered in ileal effluent is observed between the different studies, but taken as total chlorogenic acids, about one third was absorbed in the stomach and/or small intestine (Olthof et al. 2001; Stalmach et al. 2010), with chlorogenic acids being stable in the gastrointestinal tract (Takenaka et al. 2000; Olthof et al. 2001; Rechner et al. 2001b; Farah et al. 2006). The remaining two thirds of chlorogenic acids pass through the colon where the colonic bacteria carry out further metabolic processes (Gonthier et al. 2003, 2006; Olthof et al. 2003).

The presence of low amounts of intact 5-*O*-caffeoylquinic acid in plasma and urine following oral ingestion of coffee (Ito et al. 2005; Stalmach et al. 2009), artichoke (Azzini et al. 2007), and pure compounds in humans and rats (Olthof et al. 2001; Gonthier et al. 2003) suggests that although the bioavailability of quinic esters of hydroxycinnamic acids is low, these compounds are able to cross the gastrointestinal epithelium in their intact form to a certain extent and are subject to phase II metabolism.

6.4.2 Identification of Metabolites in Plasma and Urine

Following the intake of chlorogenic acids contained in various foods and matrices, a wide variety of metabolites were identified in plasma and urine and were mainly sulfated, methylated, and glucuronidated hydroxycinnamic acids as well as glycine conjugates, hydrogenated hydroxycinnamic acids, and lower molecular weight phenolic acids, but also intact chlorogenic acids (Table 6.3). The wide range of metabolites identified reflects the extensive metabolism of chlorogenic acids upon absorption. Analysis of the various metabolites has been carried out based on the use of enzymes (β-glucuronidase and sulfatase) for the identification of conjugates, as well as high-performance liquid chromatography with mass spectrometry, which provides a greater degree of information and accuracy compared with the enzymatic treatment, but can be less quantitative without appropriate standards.

Reported peak plasma concentrations of free and conjugated hydroxycinnamic acids were generally under 1 μM, reaching the circulation rapidly under an hour following oral administration. Urinary excretion accounted for up to 29% of intake (Stalmach et al. 2009), indicating that coffee chlorogenic acids and their products of metabolism are bioavailable to a much greater extent than many other dietary flavonoids and phenolic compounds when based on urinary data (Manach et al. 2005; Donovan et al. 2006).

6.4.3 Bioavailability of Intact Chlorogenic Acids

The pharmacokinetic profile of intact chlorogenic acids in plasma ranged from as little as 2 nM of 5-O-caffeoylquinic acid following a single serving of coffee (Stalmach et al. 2009) to 5.9 μM of 5-O-caffeoylquinic acid following a single serving of green coffee extract (Farah et al. 2008). Peak plasma concentrations were reached under 4 h following intake indicating an absorption in the upper gastrointestinal tract, although a large interindividual variation was observed following intake of the green coffee extract (Farah et al. 2008). Urinary excretion of intact chlorogenic acids ranged from 0.29% for 5-O-caffeoylquinic acid to 4.9% for feruloylquinic acids (Olthof et al. 2001, 2003; Ito et al. 2005; Stalmach et al. 2009).

From Table 6.3, it appears that chlorogenic acids are differentially bioavailable depending on the nature of the compound and preparation ingested. The intake of 5-O-caffeoylquinic acid contained in coffee beverage, green coffee extract, or artichoke heads resulted in a wide variation in the pharmacokinetic profile, with C_{max} values ranging from 2.2 nM to 5.9 μM, area under the curve (AUC) values of 4.1 nmol h/L to 17.9 μmol h/L, and T_{max} values of 0.7–3.3 h (Azzini et al. 2007; Monteiro et al. 2007; Farah et al. 2008; Stalmach et al. 2009). Apart from the potential matrix effect arising from the form in which 5-O-caffeoylquinic acid was ingested as well as the levels ingested (from 119 to 1068 μmol), it appeared that the group of Farah and Monteiro reported a much higher bioavailability of 5-O-caffeoylquinic acid in circulation compared with Stalmach et al. and Azzini et al. When expressed as a percentage of intake, C_{max} values of intact circulating 5-O-caffeoylquinic acid accounted for ca. 0.01% of the dose following instant coffee and artichoke consumption, compared with much greater recoveries of 0.9–14.8% after brewed coffee or green coffee extract observed

TABLE 6.3
Metabolic and Pharmacokinetic Profiles of Chlorogenic Acids in Humans

Matrix	Dose (µmol/kg bw)[a]	Metabolites	C_{max} Plasma (µmol/L)	AUC Plasma (µmol·h/L)	T_{max} Plasma (h)	% Excreted	Reference
Horsetail extract for 3 days	0.7 (CA eq)	FA, feruloylglycine, DHFA, DHCA	n.a.	n.a.	n.a.	ca. 10–11 (total)	Graefe and Veit (1999)
6 cups of coffee at 4 h intervals	35.0 (CA eq)	FA, iFA, DHFA, VA	n.a.	n.a.	n.a.	5.9 (total)	Rechner et al. (2001)
Coffee beverage (single intake)	3.9 (CQA eq)	CA	0.5	n.a.	1	n.a.	Nardini et al. (2002)
Artichoke leaf extracts (single intake)	8.5 (CA eq) /12.2 (CA eq) in cross-over	CA, DHCA, FA, iFA, DHFA	36.10^{-3} (CA) $/203.10^{-3}$ (DHFA)	109.10^{-3} (CA) $/918.10^{-3}$ (DHFA)	0.77 (FA) – 6.34 (DHFA)	Total 4.7/4.0	Wittemer et al. (2005)
Coffee beverage with breakfast (single intake)	10.0 (CQA eq)	5-CQA, CA, m-CoA	n.a.	n.a.	n.a.	2.3, 0.3, 0.4, 1.4 (total)	Ito et al. (2005)
Cooked artichoke heads with olive oil (single intake)	18.8 (CGA eq)	CGA, DHCA, CA, DHFA, FA	18.10^{-3} (total CGA) – 111.10^{-3} (total DHFA)	28.10^{-3} (CGA) – 645.10^{-3} (CA)	0.7 (CGA) – 8.0 (DHCA)	n.a.	Azzini et al. (2007)
Coffee beverage (single 190 mL-serving)	48.5	CQAs, diCQAs, CA, DHCA, FA, iFA, p-CoA, GA, HBA, VA, SA	0.92 (3,4-diCQA) – 3.14 (5-CQA)	1.65 (3-CQA) – 8.10 (5-CQA) 17.11 (total CGA)	1.4 (CA) – 2.3 (5-CQA, 3,5- & 4,5-diCQA)	Not reported	Monteiro et al. (2007)

(Continued)

TABLE 6.3 (CONTINUED)
Metabolic and Pharmacokinetic Profiles of Chlorogenic Acids in Humans

Matrix	Dose (μmol/kg bw)[a]	Metabolites	C_{max} Plasma (μmol/L)	AUC Plasma (μmol.h/L)	T_{max} Plasma (h)	% Excreted	Reference
Green coffee extract (single intake)	6.4 (CGA eq)	CQAs, diCQAs, CA, FA, iFA, p-CoA, SA, HBA, GA, VA, DHCA	0.4 (p-CoA) – 5.9 (5-CQA)	3.0 (3-CQA) – 17.9 (5-CQA) – 45.6 (total CGA)	2.5 (p-CoA) – 4.0 (3-CQA)	5.5 (total)	Farah et al. (2008)
Coffee beverage (single 200 mL-serving)	5.8 (5-CQA eq)	CA-S, FA-S, iFA-S, iFA-GlcA, DHFA, DHCA, DHFA-S, DHCA-S, FQAs, CQA-S, CQAL-S, DHCA-GlcA, DHFA-GlcA, DHiFA-GlcA, feruloylglycine, 5-CQA	2.10^{-3} (5-CQA) – 385.10^{-3} (DHFA)	4.10^{-3} (5-CQA) – 2648.10^{-3} (DHCA-S)	3.6 (CQAL-S) – 5.2 (DHCA)	29.1 (Total)	Stalmach et al. (2009)
Coffee beverage (single 400 mL-serving)	12.9 (CQA eq)	CA, FA, iFA, DHCA, DHFA	81.10^{-3} (5-CQA) – 550.10^{-3} (DHFA)	n.a.	1.7 (CA) – 10 (DHFA)	n.a.	Renouf et al. (2010)
5-CQA supplement (single intake)	40 (CQA eq)	5-CQA, traces of 3- and 4-CQA	n.a.	n.a.	n.a.	0.29 (5-CQA)	Olthof et al. (2001)
Coffee beverage (single 200 mL-serving)	5.4 (5-CQA eq)	CA-S, FA-S, iFA-S, iFA-GlcA, DHFA-S, DHCA-S, FQAs, CQAL-S, feruloylglycine	n.a.	n.a.	n.a.	8.0 (total)	Stalmach et al. (2010)

[a] If not stated, body weights (bw) taken as 70 kg.

AUC: area under the plasma concentration curve; C_{max}: peak plasma concentration; T_{max}: time to reach C_{max}; 5-CQA: 5-O-caffeoylquinic acid; CA: caffeic acid; CGA: chlorogenic acid; m-CoA: m-coumaric acid; p-CoA: p-coumaric acid; p-CoQA: p-coumaroylquinic acid; CQAL: caffeoylquinic acid lactone; diCQA: dicaffeoylquinic acid; DHCA: dihydrocaffeic acid; DHFA: dihydroferulic acid; DHiFA: dihydro(iso)ferulic acid; FA: ferulic acid; FQA: feruloylquinic acid; GA: gallic acid; HBA: hydroxybenzoic acid; SA: sinapic acid; VA: vanillic acid; eq: equivalent; n.a.: not analyzed; n.d.: not detected; GlcA: glucuronide; S, sulfate.

by Monteiro et al. and Farah et al. These authors also reported a high C_{max} value for the dicaffeoylquinic acids of 6.6 μM following ingestion of ~43 μmol, corresponding to ~46% of intake. Neither of the other two investigators reported the presence of diC-QAs in plasma, although Azzini et al. did not mention this in their analysis. The other main difference in the pharmacokinetic profile of intact chlorogenic acids observed between the groups is that of the feruloylquinic acids, which were detected in the plasma of volunteers following the consumption of instant coffee (Stalmach et al. 2009), but not following the intake of green coffee extract (Farah et al. 2008), despite similar levels ingested. The differences observed between the groups may arise from the matrix in which chlorogenic acids were ingested (beverages or food vs extract) as well as the methods of analysis used (enzymatic treatment versus use of HP LC-MS in selective ion monitoring mode for the detection and quantification of metabolites). Although the bioavailability of intact chlorogenic acids requires further investigation, current and past studies reporting the absorption and metabolism of these compounds in humans do not generally report a high bioavailability (Table 6.3), and Farah and Monteiro is the first group reporting C_{max} values in the μM range associated with intact chlorogenic acids.

6.4.4 BIPHASIC PHARMACOKINETIC PROFILE OF ABSORPTION

During their passage through the body, there is extensive metabolism of chlorogenic acids following their ingestion. The 0–24 h excretion of the parent compounds and their metabolites in humans corresponded to 0.29% of intact 5-O-caffeoylquinic acid (Olthof et al. 2001) and to up to 29% of total chlorogenic acids ingested (Stalmach et al. 2009) (Table 6.3). Within one hour of intake, as a result of absorption in the small intestine, low plasma C_{max} values were obtained with unmetabolized chlorogenic acids as well as conjugated hydroxycinnamic acids (Nardini et al. 2002; Wittemer et al. 2005; Azzini et al. 2007; Stalmach et al. 2009; Renouf et al. 2010) (Table 6.3). More substantial absorption occurred, presumably from the large intestine, with T_{max} values of 5–10 h being attained for free and conjugated dihydroferulic and dihydro-caffeic acids (Wittemer et al. 2005; Azzini et al. 2007; Stalmach et al. 2009; Renouf et al. 2010). This differential in the metabolism has been reported in studies examining the bioavailability of chlorogenic acids from artichoke heads and extracts (Rechner et al. 2001a; Wittemer et al. 2005; Azzini et al. 2007), as well as in coffee beverage (Stalmach et al. 2009; Renouf et al. 2010). Metabolites found in plasma and urine of volunteers following coffee and artichoke extracts are similar (Table 6.3). The same biphasic profile of chlorogenic acid metabolites could also be observed. Peak plasma concentration for the unmetabolized chlorogenic acids as well as conjugated hydroxycinnamic acids was reached under an hour following intake and ranged from of 2.2 nM for 5-O-caffeoylquinic acid (Stalmach et al. 2009) to 0.5 μM for caffeic acid and its conjugates (Nardini et al. 2002). Free and conjugated dihydrocaffeic and dihydroferulic acids reached their maximum concentration within 5–8 h, ranging from 41 to 325 nM for free and conjugated dihydrocaffeic acid and from 111 nM to 550 nM of dihydroferulic acid (Wittemer et al. 2005; Azzini et al. 2007; Stalmach et al. 2009; Renouf et al. 2010). This biphasic profile of absorption results, in part, from a metabolism taking place in different sites of the gastrointestinal tract.

6.4.5 SITES OF METABOLISM

Although some studies have reported a lack of hydrolysis of chlorogenic acids taking place in the upper gastrointestinal tract and/or liver (Plumb et al. 1999; Azuma et al. 2000; Olthof et al. 2001), the rapid circulation of metabolites as determined by the short T_{max} values (Table 6.3) suggests the presence of rate limiting intracellular esterases capable of releasing the hydroxycinnamic acids prior to absorption and conjugation. Kern et al. (2003b) and Andreasen et al. (2001a) reported the presence of intracellular cinnamoyl esterases located in the small intestine and capable of hydrolysing methyl esters of hydroxycinnamic acids. The main site of hydrolysis of chlorogenic acids and other esters of free and bound hydroxycinnamic acids is the colon where these compounds are subject to the action of a microbial metabolism (Buchanan et al. 1996; Kroon et al. 1997; Andreasen et al. 2001a; Couteau et al. 2001). Upon their release from the quinic moiety, the absorption and metabolism of the hydroxycinnamic acids are carried out as described in Section 6.3.2.

Regarding the metabolism of intact chlorogenic acids, rat liver and small intestine were reported to exert various metabolic processes such as methylation and glucuronidation, catalyzed by COMT and UGT, respectively (Yang et al. 2005). In the plasma of rats, 1,5-O-dicaffeoylquinic acid was found methylated and glucuronidated and glucuronides of 5-O-caffeoylquinic acid were detected in the urine following intravenous injection (Choudhury et al. 1999; Yang et al. 2005, 2006). Following coffee intake, traces of sulfated caffeoylquinic acids and their lactone derivatives were excreted in the urine of humans (Stalmach et al. 2009, 2010).

Following uptake in the upper gastrointestinal tract, metabolites of chlorogenic acids have been detected in ileal effluent of ileostomists fed a single cup of coffee, predominantly as sulfates and glucuronides of caffeoylquinic, feruloylquinic and lactones of caffeoylquinic acids (Stalmach et al. 2010). The majority of conjugated metabolites recovered in ileal fluid comprised sulfates (77%), with only 7% as glucuronides. Those results are in keeping with the literature, suggesting the human small intestine epithelium as a site for phase II enzyme reactions (Lin et al. 1999; Chen et al. 2003a), as well as enteric recycling of metabolites (Chen et al. 2003b; Poquet et al. 2008b).

6.4.6 EXTENSIVE METHYLATION, SULFATION, AND ISOMERIZATION OF CHLOROGENIC ACID METABOLITES

Following their ingestion, chlorogenic acids are subject to extensive metabolism leading to the release, absorption, and phase II conjugation of various hydroxycinnamic acids (Table 6.3). The presence of ferulic and isoferulic acids was reported in plasma and urine samples of rats and humans following the intake of caffeoylquinic acids (Graefe and Veit 1999; Azuma et al. 2000; Rechner et al. 2001a; Wittemer et al. 2005; Lafay et al. 2006a, 2006b; Azzini et al. 2007), suggesting an extensive methylation of the caffeic moiety, taking place either in the small intestine (Kern et al. 2003b) or liver (Moridani et al. 2002; Mateos et al. 2006). This metabolic process appears to occur preferentially in the meta position of the phenyl ring (Lafay et al. 2006b; Poquet et al. 2008a). Similarly, methylation of intact chlorogenic acids was reported in rats, with methylated dicaffeoylquinic acids identified in plasma, urine, and bile after oral administration (Yang et al. 2005). This extensive methylation pathway could explain the recovery of feruloylquinic acids but no or little caffeoylquinic acids in circulation and in urine of humans after coffee consumption (Stalmach et al. 2009, 2010). Besides methylation, sulfation of hydroxycinnamic acids also emerges as a preferential pathway of metabolism in humans, with a regioselectivity preference for sulfation of the 3'-hydroxyl of caffeic and dihydrocaffeic acids (Poquet et al. 2008a; Stalmach et al. 2009; Wong et al. 2010).

Yang et al. (2005, 2006) investigated the extensive metabolism of 1,5-O-dicaffeoylquinic acid upon oral and intravenous administration to rats and reported the production of methylated and methyl-glucuronidated isomers in plasma and urine. Similarly, Kahle et al. (2007) reported an isomerization occurring in the human gastrointestinal tract after the consumption of 4- and 5-O-caffeoylquinic acids in apple juice, with the detection in the ileostomy fluid of 1- and 3-O-caffeoylquinic acids. Such isomerization has been observed following incubation of 5-O-caffeoylquinic acid with ileal effluent at pH > 6 (Farah et al. 2006). The liver has also been identified as a site of isomerization following incubation of 5-O-caffeoylquinic acid with HepG2 cells (Mateos et al. 2006). Isomerization from the *trans* to the *cis* configuration of hydroxycinnamic acids *ex vivo* has also been reported, with the putative identification of *cis*-ferulic acid-β-D-glucuronide following incubation of everted sacs of rat colon with ferulic acid (Poquet et al. 2008b).

6.4.7 EXCRETION

As previously mentioned, intact chlorogenic acids are not readily bioavailable to the body, but are extensively metabolized, with total urinary excretion of the metabolites accounting for up to 29% of the amount ingested (Stalmach et al. 2009). This is without taking into account the phenolic acids deriving from a colonic metabolism, with a urinary excretion of these compounds which can exceed 50% of the amount of chlorogenic acids ingested (Gonthier et al. 2003, 2006; Olthof et al. 2003; Farah et al. 2008). Following intake of chlorogenic acids contained in various matrices, their urinary excretion did not exceed 5% of the amounts consumed (Cremin et al. 2001; Olthof et al. 2001; Gonthier et al. 2003; Ito et al. 2005; Monteiro et al. 2007; Farah et al. 2008; Stalmach et al. 2009). In two studies investigating the bioavailability of coffee chlorogenic acids, only traces of caffeoylquinic acids were detected in the urine of humans, despite millimolar concentrations of the caffeoylquinic and dicaffeoylquinic acids seemingly quantified in plasma (Monteiro et al. 2007; Farah et al. 2008). Previous studies have reported the presence of an esterase activity in animals' kidneys (Searle 1986; Tanaka and Suzuki 1994), suggesting that 5-O-caffeoylquinic acid could undergo hydrolysis of the quinic moiety prior to being excreted in urine. This could partially explain the presence of caffeoylquinic acid in the plasma of volunteers, which could not be detected in 24-h urine, although the low oral bioavailability of intact chlorogenic acids is mainly due to their limited absorption in the upper gastrointestinal tract and extensive metabolism, as previously mentioned.

Besides urinary excretion, the recovery of free and conjugated chlorogenic acids has been reported in the bile of rats following oral administration of 1,5-O-dicaffeoylquinic acid (Yang et al. 2005). Although the authors did not include any quantitative information regarding the presence of dicaffeoylquinic acid metabolites in bile, they concluded that urinary and biliary excretion were two important elimination routes for 1,5-O-dicaffeoylquinic acid in rats. This was not confirmed by another study investigating the intestinal uptake of 5-O-caffeoylquinic acid in rats, with only caffeic acid detected in low concentration in the bile after 5-O-caffeoylquinic acid perfusion, accounting for less than 0.4% of the dose perfused (Lafay et al. 2006b).

6.4.8 COLONIC METABOLITES

From the observations of studies using ileostomists, up to two thirds of the chlorogenic acids ingested are not absorbed in the upper gastrointestinal tract but reach the colon (Olthof et al. 2001; Kahle et al. 2005; Stalmach et al. 2010) where the microflora are readily capable of hydrolysing the ester moiety (Chesson et al. 1999; Couteau et al. 2001; Gonthier et al. 2003). Following release of the hydroxycinnamic acid from the quinic moiety, the colonic microflora is capable of producing 3-(3',4'-di-hydroxyphenyl)propionic acid, 3-(3'-methoxy-4'-hydroxyphenyl)propionic acid, 3-(3'- and 4'-hydroxyphenyl)propionic acid, 3'- and 4'-hydroxyphenylacetic acid, 3',4'-dihydroxyphenylacetic acid, 3'-methoxy-4'-hydroxyphenylacetic acid, 3- and 4-hydroxybenzoic acid, 3,4-dihydroxybenzoic acid, 3-methoxy-4-hydroxybenzoic acid, 3'-hydroxycinnamic acid (m-coumaric acid), hippuric acid, and 3'-hydroxy-hippuric acid by means of dehydroxylation, demethylation, dehydrogenation, glycination, and hydrogenation (Choudhury et al. 1999; Gonthier et al. 2003, 2006; Olthof et al. 2003).

6.5 ANALYTICAL METHODS USED IN IDENTIFYING AND QUANTIFYING THE PRODUCTS OF METABOLISM IN PLASMA AND URINE

Most studies in Tables 6.1 to 6.3 investigating the metabolic fate of hydroxycinnamates in humans have used a combination of β-glucuronidase and sulfatase treatments prior to the analysis of metabolites in plasma and urine samples, which only allowed their quantification as aglycones (Graefe and Veit 1999; Cremin et al. 2001; Nardini et al. 2002; Kern et al. 2003a). Although rarely undertaken, the development of methods involving enzymatic hydrolysis and their efficiencies is of importance. Studies investigating the efficiencies of enzymes to deconjugate metabolites present in various tissues and body fluids have reported the impact of the methodology and emphasized the need of monitoring the release and quantification of aglycones to avoid an underestimation of metabolites (Cooper et al. 2001; Gu et al. 2005). The use of certain type of enzymes such as *Helix pomatia* exhibiting sulfatase activity have been reported to hydrolyse the quinic acid moiety of chlorogenic acids (Cremin et al. 2001; Ito et al. 2005; Manach et al. 2005; Monteiro et al. 2007), which may have contributed to the lack of detection of intact chlorogenic acids in previous bioavailability studies. Experiments investigating the efficiency of enzymatic deconjugation of synthesized glucuronidated and sulfated hydroxycinnamic acids highlighted a relatively high efficiency of a β-glucuronidase from *Escherichia coli* to deconjugate glucuronides (75–117%), wherease sulfatase from abalone entrails resulted in the recovery of the sulfate conjugates ranging from 24 to 80% (Guy et al. 2009). Similarly, the use of a chlorogenate esterase from *Aspergillus japonicus* resulted in recovery values for the ester compounds ranging from 40 to 123%, with the least efficient hydrolysis associated with 4-*O*-caffeoylquinic acid (40%), followed by 3-*O*-caffeoylquinic acid (70%) and 5-*O*-caffeoylquinic acid being totally hydrolyzed (Guy et al. 2009).

The recent progress obtained upon investigating the bioavailability of polyphenols lies in the use of mass spectrometry, a technique which allows a better identification of circulating and excreted metabolites following ingestion of certain dietary polyphenols (Kussmann et al. 2007), and in particular, there are validated procedures available to discriminate between and identify the individual chlorogenic acid regioisomers that are not commercially available. These procedures are also applicable to mammalian conjugates of the original chlorogenic acid if an ion trap mass spectrometer is used. (Clifford et al. 2003; 2005; Kuhnert et al. 2010).

These techniques, compared to the use of enzymatic treatments, provide a greater quality of information with an increased sensitivity of detection and quantification of metabolites in biological matrices, although the lack of commercially available authentic standards remains a limiting factor in the quantification of metabolites (Renouf et al. 2010). When standards are available, these lead to an accurate identification and quantification of metabolites and result in establishing metabolic pathways of ingested compounds in great detail. One example of the use of conjugated hydroxycinnamic acids (sulfates and glucuronides) for the identification and quantification of coffee chlorogenic acids detected in plasma, urine, and ileal effluent of volunteers is illustrated in Figure 6.2. These metabolic pathways provide to date, one of the most complete pictures of the chlorogenic acid metabolism occurring upon absorption (Stalmach et al. 2009, 2010).

6.6 SUMMARY

Hydroxycinnamic acids are major components of the diet and are mainly found conjugated with quinic acid in coffee beverage, one of the main dietary sources of chlorogenic acids. The main groups of chlorogenic acids in the beverage are the caffeoylquinic acids, dicaffeoylquinic acids, feruloylquinic acids, and *p*-coumaroylquinic acid, which are quinic esters of caffeic acid, ferulic acid, and *p*-coumaric acids, respectively.

Free hydroxycinnamic acids are absorbed throughout the gastrointestinal tract, although the jejunum appears to be the main site of absorption. Hydroxycinnamic acids permeate passively across the intestinal epithelium, as the free form mainly by a transcellular mechanism. Upon absorption, hydroxycinnamic acids are extensively metabolized mainly in the liver but also in the small intestine and colon, with sulfation, *O*-methylation/de-*O*-methylation, glucuronidation, hydrogenation/dehydrogenation, isomerization, dehydroxylation, and glycine conjugation taking place in the various sites. Caffeic acid appears more extensively metabolized compared to its methylated derivative, ferulic acid, but less absorbed and transported across the epithelium. *O*-Methylation appears to be the preferential pathway of metabolism for

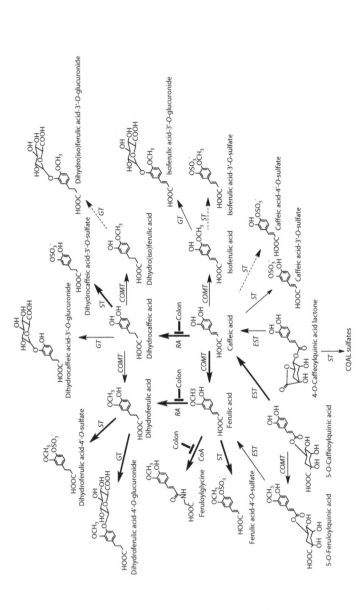

FIGURE 6.2 Metabolism of chlorogenic acids following the ingestion of coffee by volunteers. 5-*O*-caffeoylquinic acid and 5-*O*-feruloylquinic acid are illustrated structures but their respective 3- and 4-isomers would be metabolized in a similar manner and likewise with 4-*O*-caffeoylquinic acid lactone and 3-*O*-caffeoylquinic acid lactone. COMT: catechol-*O*-methyltransferase; ET: esterase; RA: reductase; GT: UDP-glucuronyltransferase; ST: sulfuryl-*O*-transferase; Co-A: coenzyme A. Bold arrows indicate major routes, dotted arrows minor pathways. Steps blocked in subjects with an ileostomy and hence occurring in the colon are indicated. (After Stalmach, A. et al., *Arch. Biochem. Biophys.,* 501, 98–105, 2010. With permission.)

caffeic acid in both the gastrointestinal and liver cells, occurring in the *meta* position of the phenyl ring. Similarly, sulfation on the 3′-position is also favored in compounds presenting a 3′,4′-dihydroxy structure, irrespective of the site of metabolism (liver or gastrointestinal tract).

The presence of a quinic moiety reduces the bioavailability of hydroxycinnamic acids, which are consequently not absorbed to the same extent by the body. The extensive metabolism following absorption of quinic esters of hydroxycinnamic acids also explains the low bioavailability of intact chlorogenic acids. Up to one third of chlorogenic acids are absorbed in the upper gastrointestinal tract, with the remaining two thirds reaching the colon where the colonic microflora are capable of hydrolyzing the quinic moiety to release the hydroxycinnamic acids and also of carrying out further metabolic processes. Upon absorption, chlorogenic acids are hydrolyzed, and sulfates and glucuronides of hydroxycinnamic acids are found circulating in plasma, with peak plasma concentration reaching up to 0.5 μM under an hour following ingestion. In rats, methyl and methyl-glucuronide derivatives of 1,5-*O*-caffeoylquinic acid were also identified in plasma. The pharmacokinetic profile of 5-*O*-caffeoylquinic acid following intakes of coffee, artichoke heads, and green coffee extract resulted in a ~3000-fold variation in the concentration reported. Besides hydroxycinnamic acids and intact chlorogenic acids, free and conjugated dihydrocaffeic and dihydroferulic acids were detected in plasma with peak concentrations reported up to 550 nM, reached 5–8 h following intake. This biphasic pharmacokinetic profile reflects the differential sites of absorption and metabolism, taking place both in the upper gastrointestinal tract and colon. Urinary excretion of metabolites accounted for up to 29% of chlorogenic acids intake, and intact chlorogenic acids were excreted in amounts ranging from 0.29 to 4.9% of intake. Lower molecular weight phenolic acids are also present in urine and derive from a colonic metabolism.

The use of powerful analytical tools such as mass spectrometry has recently allowed unraveling of the metabolic pathway of chlorogenic acids, allowing detailed understanding of the complex metabolic pathways involved in mammalian metabolism of hydroxycinnamic acids.

REFERENCES

Adam, A., Crespy, V., Levrat-Verny, M.-A. et al. (2002). The bioavailability of ferulic acid is governed primarily by the food matrix rather than its metabolism in intestine and liver in rats. *J. Nutr.*, **132**, 1962–1968.

Ader, P., Grenacher, B., Langguth P. et al. (1996). Cinnamate uptake by rat small intestine: Transport kinetics and transepithelial transfer. *Exp. Physiol.*, **81**, 943–955.

Andreasen, M.F., Kroon, P.A., Williamson, G. et al. (2001a). Esterase activity able to hydrolyze dietary antioxidant hydroxycinnamates is distributed along the intestine of mammals. *J. Agric. Food Chem.*, **49**, 5679–5684.

Andreasen, M.F., Kroon, P.A., Williamson, G. et al. (2001b). Intestinal release and uptake of phenolic antioxidant diferulic acids. *Free Radic. Biol. Med.*, **31**, 304–314.

Azuma, K., Ippoushi, K., Nakayama, M. et al. (2000). Absorption of chlorogenic acid and caffeic acid in rats after oral administration. *J. Agric. Food Chem.*, **48**, 5496–5500.

Azzini, E., Bugianesi, R., Romano F. et al. (2007). Absorption and metabolism of bioactive molecules after oral consumption of cooked edible heads of *Cynara scolymus L.* (cultivar *Violetto di Provenza*) in human subjects: A pilot study. *Br. J. Nutr.*, **97**, 963–969.

Baba, S., Osakabe, N., Natsume, M. et al. (2004). Orally administered rosmarinic acid is present as the conjugated and/or methylated forms in plasma, and is degraded and metabolized to conjugated forms of caffeic acid, ferulic acid and *m*-coumaric acid. *Life Sci.*, **75**, 165–178.

Baba, S., Osakabe, N., Natsume, M. et al. (2005). Absorption, metabolism, degradation and urinary excretion of rosmarinic acid after intake of *Perilla frutescens* extract in humans. *Eur. J. Nutr.*, **44**, 1–9.

Bekedam, E.K., Loots, M.J., Schols, H.A. et al. (2008a). Roasting effects on formation mechanisms of coffee brew melanoidins. *J. Agric. Food Chem.*, **56**, 7138–7145.

Bekedam, E.K., Schols, H.A., Van Boekel, M.A.J.S. et al. (2008b). Incorporation of chlorogenic acids in coffee brew melanoidins. *J. Agric. Food Chem.*, **56**, 2055–2063.

Booth, A.N., Emerson, O.H., Jones, F.T. et al. (1957). Urinary metabolites of caffeic and chlorogenic acids. *J. Biol. Chem.*, **229**, 51–59.

Borrelli, R.C., Visconti, A., Mennella, C. et al. (2002). Chemical characterization and antioxidant properties of coffee melanoidins. *J. Agric. Food Chem.*, **50**, 6527–6533.

Bourne, L., Paganga, G., Baxter, D. et al. (2000). Absorption of ferulic acid from low-alcohol beer. *Free Radic. Res.*, **32**, 273–280.

Buchanan, C.J., Wallace, G. and Fry, S.C. (1996). *In vivo* release of ^{14}C-labelled phenolic groups from intact spinach cell walls during passage through the rat intestine. *J. Sci. Food Agric.*, **71**, 459–469.

Carpenter, D.O., Arcaro, K., and Spink, D.C. (2002). Understanding the human health effects of chemical mixtures. *Environ. Health Persp. Suppl.*, **110**, 25.

Casal, S., Mendes, E., Fernandes, J.O. et al. (2004). Analysis of heterocyclic aromatic amines in foods by gas chromatography-mass spectrometry as their tert-butyldimethylsilyl derivatives. *J Chromatogr.*, **1040**, 105–114.

Charles-Bernard, M., Kraehenbuehl, K., Rytz, A. et al. (2005a). Interactions between volatile and nonvolatile coffee components. 1. Screening of nonvolatile components. *J. Agric. Food Chem.*, **53**, 4417–4425.

Charles-Bernard, M., Roberts, D.D., and Kraehenbuehl, K. (2005b). Interactions between volatile and nonvolatile coffee components. 2. Mechanistic study focused on volatile thiols. *J. Agric. Food Chem.*, **53**, 4426–4433.

Chen, G., Zhang, D., Jing, N. et al. (2003a). Human gastrointestinal sulfotransferases: Identification and distribution. *Toxicol. Appl. Pharmacol.*, **187**, 186–197.

Chen, J., Lin, H., and Hu, M. (2002). Metabolism of flavonoids via enteric recycling: Role of intestinal disposition. *J. Pharmacol. Exp. Ther.*, **304**, 1228–1235.

Chen, J., Lin, H.M., and Hu, M. (2003b). Metabolism of flavonoids via enteric recycling: Role of intestinal disposition. *J. Pharmacol. Exp. Ther.*, **304**, 1228–1235.

Chesson, A., Provan, G.J., Russell, W.R. et al. (1999). Hydroxycinnamic acids in the digestive tract of livestock and humans. *J. Sci. Food Agric.*, **79**, 373–378.

Cheynier V.F., Trousdale E.K., Singleton V.L. et al. (1986). Characterization of 2-S-glutathionyl caftaric acid and its hydrolysis in relation to grape wines. *J. Agric. Food Chem.*, **34**, 217–221.

Choudhury, R., Srai, S.K., Debnam, E. et al. (1999). Urinary excretion of hydroxycinnamates and flavonoids after oral and intravenous administration. *Free Radic. Biol. Med.*, **27**, 278–286.

Clifford, M.N. (1999). Chlorogenic acids and other cinnamates: Nature, occurrence and dietary burden. *J. Sci. Food Agric.*, **79**, 362–372.

Clifford, M.N. (2000). Chlorogenic acids and other cinnamates: Nature, occurrence, dietary burden, absorption and metabolism. *J. Sci. Food Agric.*, **80**, 1033–1043.

Clifford, M.N. (2004). Diet-derived phenols in plasma and tissues and their implications for health. *Planta Med.*, **70**, 1103–1114.

Clifford, M.N., Johnston, K.L., Knight, S. et al. (2003). A hierarchical scheme for LC-MS[n] analysis of the *cis* isomers of chlorogenic acids. *J. Agric. Food Chem.*, **51**, 2900–2911.

Clifford, M.N., Kirkpatrick, J., Kuhnert, N. et al. (2008). LC-MS[n] analysis of the *cis* isomers of chlorogenic acids. *Food Chem*, **106**, 379–385.

Clifford, M.N., Knight, S., and Kuhnert, N. (2005). Discriminating between the six isomers of dicaffeoylquinic acids by LC-MS[n]. *J. Agric. Food Chem.*, **53**, 3821–3832.

Clifford, M.N., Zheng, W., and Kuhnert, N. (2006). Profiling the chlorogenic acids of aster by HPLC-MS[n]. *Phytochem. Anal.*, **17**, 384–393.

Cooper, J., Currie, W., and Elliott, C.T. (2001). Comparison of the efficiences of enzymatic and chemical hydrolysis of (nortestosterone and diethylstilboestrol) glucuronides in bovine urine. *J. Chromatogr. B*, **757**, 221–227.

Couteau, D., McCartney, A.L., Gibson, G.R. et al. (2001). Isolation and characterization of human colonic bacteria able to hydrolyse chlorogenic acid. *J. Appl. Microbiol.*, **90**, 873–881.

Cremin, P., Kasim-Karakas, S., and Waterhouse, A.L. (2001). LC/ES-MS detection of hydroxycinnamates in human plasma and urine. *J. Agric. Food Chem.*, **49**, 1747–1750.

Crozier, A., Jaganath, I.B., and Clifford, M.N. (2009). Dietary phenolics: Chemistry, bioavailability and effects on health. *Nat. Prod. Rep.*, **26**, 1001–1043.

Dao, L., and Friedman, M. (1992). Chlorogenic acid content of fresh and processed potatoes determined by ultraviolet spectrophotometry. *J. Agric. Food Chem.*, **40**, 2152–2156.

Donovan, J.L., Manach, C., Faulks, R.M. et al. (2006). Absorption and metabolism of dietary plant secondary metabolites. In A. Crozier, H. Ashihara, and M.N. Clifford (eds.), *Plant Secondary Metabolites: Occurrence, Structure and Role in the Human Diet.* Blackwell Publishing, Oxford, pp. 303–351.

Dorfner, R., Ferge, T., Kettrup, A. et al. (2003). Real-time monitoring of 4-vinylguaiacol, guaiacol, and phenol during coffee roasting by resonant laser ionization time-of-flight mass spectrometry. *J. Agric. Food Chem.*, **51**, 5768–5773.

Farah, A., de Paulis, T., Trugo, L.C. et al. (2005). Effect of roasting on the formation of chlorogenic acid lactones in coffee. *J. Agric. Food Chem.*, **53**, 1505–1513.

Farah, A., Guigon, F., and Trugo, L.C. (2006). The effect of human digestive fluids on chlorogenic acids isomers from coffee. *Proceedings of the 21st International Conference on Coffee Science*. ASIC, Paris, pp. 93–96.

Farah, A., Monteiro, M., Donangelo, C.M. et al. (2008). Chlorogenic acids from green coffee extract are highly bioavailable in humans. *J. Nutr.*, **138**, 2309–2315.

Frank, O., Zehentbauer, G., and Hofmann, T. (2006). Bioresponse-guided decomposition of roast coffee beverage and identification of key bitter taste compounds. *Eur. Food Res. Technol.*, **222**, 492–508.

Garcia-Conesa, M.T., Wilson, P.D., Plumb, G.W. et al. (1999). Antioxidant properties of 4,4′-dihydroxy-3,3′-dimethoxy-β,β′-bicinnamic acid (8-8-diferulic acid, non-cyclic form). *J. Sci. Food Agric.*, **79**, 379–384.

Garrait, G., Jarrige, J.F., Blanquet, S. et al. (2006). Gastrointestinal absorption and urinary excretion of *trans*-cinnamic and *p*-coumaric acids in rats. *J. Agric. Food Chem.*, **54**, 2944–2950.

Gniechwitz, D., Reichardt, N., Ralph, J. et al. (2008). Isolation and characterisation of a coffee melanoidin fraction. *J. Sci. Food Agric.*, **88**, 2153–2160.

Gonthier, M.P., Remesy, C., Scalbert, A. et al. (2006). Microbial metabolism of caffeic acid and its esters chlorogenic and caftaric acids by human faecal microbiota *in vitro. Biomed. Pharmacother.*, **60**, 536–540.

Gonthier, M.P., Verny, M.A., Besson, C. et al. (2003). Chlorogenic acid bioavailability largely depends on its metabolism by the gut microflora in rats. *J. Nutr.*, **133**, 1853–1859.

Graefe, E.U. and Veit, M. (1999). Urinary metabolites of flavonoids and hydroxycinnamic acids in humans after application of a crude extract from *Equisetum arvense*. *Phytomedicine*, **6**, 239–246.

Grams, B., Harms, A., Braun, S. et al. (2000). Distribution and inducibility by 3-methylcholanthrene of family 1 UDP-glucuronosyltransferases in the rat gastrointestinal tract. *Arch. Biochem. Biophys.*, **377**, 255–265.

Gu, L.W., Laly, M., Chang, H.C. et al. (2005). Isoflavone conjugates are underestimated in tissues using enzymatic hydrolysis. *J. Agric. Food Chem.*, **53**, 6858–6863.

Guy, P.A., Renouf, M., Barron, D. et al. (2009). Quantitative analysis of plasma caffeic and ferulic acid equivalents by liquid chromatography tandem mass spectrometry. *J. Chromatogr. B*, **877**, 3965–3974.

Haughton, E., Clifford, M.N., and Sharp, P. (2007). Monocarboxylate transporter expression is associated with the absorption of benzoic acid in human intestinal epithelial cells. *J. Sci. Food Agric.*, **87**, 239–244.

Herrmann, K. (1989). Occurrence and content of hydroxycinnamic and hydroxybenzoic acid compounds in foods. *Crit. Rev. Food Sci. Nutr.*, **28**, 315-347.

Hilgendorf, C., Spahn-Langguth, H., Regardh, C.G. et al. (2000). Caco-2 versus Caco-2/HT29-MTX co-cultured cell lines: Permeabilities via diffusion, inside- and outside-directed carrier-mediated transport. *J. Pharm. Sci.*, **89**, 63–75.

Houessou, J.K., Benac, C., Delteil, C. et al. (2005). Determination of polycyclic aromatic hydrocarbons in coffee brew using solid-phase extraction. *J. Agric. Food Chem.*, **53**, 871–879.

Houessou, J.K., Maloug, S., Leveque, A.S. et al. (2007). Effect of roasting conditions on the polycyclic aromatic hydrocarbon content in ground Arabica coffee and coffee brew. *J. Agric. Food Chem.*, **55**, 9719–9726.

Ito, H., Gonthier, M.P., Manach, C. et al. (2005). Polyphenol levels in human urine after intake of six different polyphenol-rich beverages. *Br. J. Nutr.*, **94**, 500–509.

Jacobson, E.A., Newmark, H., Baptista, J. et al. (1983). A preliminary investigation of the metabolism of dietary phenolics in humans. *Nutr. Rep. Int.*, **28**, 1409–1417.

Jaiswal, R., Sovdat, T., Vivan, F. et al. (2010). Profiling and characterization by LC-MS^n of the chlorogenic acids and hydroxycinnamoylshikimate esters in Mate (*Ilex paraguariensis*). *J. Agric. Food Chem.*, **58**, 5471–5484.

Kahle, K., Huemmer, W., Kempf, M. et al. (2007). Polyphenols are intensively metabolized in the human gastrointestinal tract after apple juice consumption. *J. Agric. Food Chem.*, **55**, 10605–10614.

Kahle, K., Kraus, M., Scheppach, W. et al. (2005). Colonic availability of apple polyphenols: A study in ileostomy volunteers. *Mol. Nutr. Food Res.*, **49**, 1143–1150.

Kennedy, H.J., Lee, E.C.G., Claridge, G. et al. (1982). The health of subjects living with a permanent ileostomy. *Quart. J. Med.*, **51**, 341–357.

Kern, S.M., Bennett, R.N., Mellon, F.A. et al. (2003a). Absorption of hydroxycinnamates in humans after high-bran cereal consumption. *J. Agric. Food Chem.*, **51**, 6050–6055.

Kern, S.M., Bennett, R.N., Needs, P.W. et al. (2003b). Characterization of metabolites of hydroxycinnamates in the *in vitro* model of human small intestinal epithelium Caco-2 cells. *J. Agric. Food Chem.*, **51**, 7884–7891.

Kobayashi, S. and Konishi, Y. (2008). Transepithelial transport of flavanone in intestinal Caco-2 cell monolayers. *Biochem. Biophys. Res. Commun.* **368**, 23–29.

Konishi, Y., Hitomi, Y., Yoshida, M. et al. (2005). Pharmacokinetic study of caffeic and rosmarinic acids in rats after oral administration. *J. Agric. Food Chem.*, **53**, 4740–4746.

Konishi, Y., Hitomi, Y., and Yoshioka, E. (2004). Intestinal absorption of *p*-coumaric and gallic acids in rats after oral administration. *J. Agric. Food Chem.*, **52**, 2527–2532.

Konishi, Y. and Kobayashi, S. (2004). Transepithelial transport of chlorogenic acid, caffeic acid, and their colonic metabolites in intestinal Caco-2 cell monolayers. *J. Agric. Food Chem.*, **52**, 2518–2526.

Konishi, Y., Kobayashi, S. and Shimizu, M. (2003a). Transepithelial transport of *p*-coumaric acid and gallic acid in caco-2 cell monolayers. *Biosci. Biotechnol. Biochem.*, **67**, 2317–2324.

Konishi, Y., Kubo, K., and Shimizu, M. (2003b). Structural effects of phenolic acids on the transepithelial transport of fluorescein in Caco-2 cell monolayers. *Biosci. Biotechnol. Biochem.*, **67**, 2014–2017.

Konishi, Y. and Shimizu, M. (2003). Transepithelial transport of ferulic acid by monocarboxylic acid transporter in Caco-2 cell monolayers. *Biosci. Biotechnol. Biochem.*, **67**, 856–862.

Konishi, Y., Zhao, Z.H., and Shimizu, M. (2006). Phenolic acids are absorbed from the rat stomach with different absorption rates. *J. Agric. Food Chem.*, **54**, 7539–7543.

Kroon, P.A., Faulds, C.B., Ryden, P. et al. (1997). Release of covalently bound ferulic acid from fiber in the human colon. *J. Agric. Food Chem.*, **45**, 66–667.

Kuhnert, N., Jaiswal, R., Matei, M.F. et al. (2010). How to distinguish between feruloyl quinic acids and isoferuloyl quinic acids by liquid chromatography/tandem mass spectrometry. *Rapid Commun. Mass Spectrom.*, **24**, 1575–1582.

Kussmann, M., Affolter, M., Nagy, K. et al. (2007). Mass spectrometry in nutrition: Understanding dietary health effects at the molecular level. *Mass Spectrom. Rev.*, **26**, 727–750.

Lafay, S., Gil-Izquierdo, A., Manach C. et al. (2006a). Chlorogenic acid is absorbed in its intact form in the stomach of rats. *J. Nutr.*, **136**, 1192–1197.

Lafay, S., Morand, C., Manach, C. et al. (2006b). Absorption and metabolism of caffeic acid and chlorogenic acid in the small intestine of rats. *Br. J. Nutr.* **96**, 39–46.

Lancon, A., Hanet, N., Jannin, B. et al. (2007). Resveratrol in human hepatoma HepG2 cells: Metabolism and inducibility of detoxifying enzymes. *Drug Metab. Disp.*, **35**, 699–703.

Leslie, E.M., Deeley, R.G. and Cole, S.P.C. (2001). Toxicological relevance of the multidrug resistance protein 1, MRP1 (ABCC1) and related transporters. *Toxicology*, **167**, 3–23.

Leslie, E.M., Deeley, R.G. and Cole, S.P.C. (2005). Multidrug resistance proteins: Role of P-glycoprotein, MRP1, MRP2, and BCRP (ABCG2) in tissue defense. *Toxicol. Appl. Pharmacol.*, **204**, 216–237.

Lin, J.H., Chiba, M. and Baillie, T.A. (1999). Is the role of the small intestine in first-pass metabolism over-emphasised? *Pharmacol. Rev.*, **51**, 135–157.

Liu, Y. and Hu, M. (2002) Absorption and metabolism of flavonoids in the Caco-2 cell culture model and a perfused rat intestinal model. *Drug Metab. Disp.*, **30**, 370–377.

Lopez-Galilea, I., Fournier, N., Cid, C. et al. (2006). Changes in headspace volatile concentrations of coffee brews caused by the roasting process and the brewing procedure. *J. Agric. Food Chem.*, **54**, 8560–8566.

Manach, C., Scalbert, A., Morand, C. et al. (2004). Polyphenols: Food sources and bioavailability. *Am. J. Clin. Nutr.*, **79**, 727–747.

Manach, C., Williamson, G., Morand, C. et al. (2005). Bioavailability and bioefficacy of polyphenols in humans. I. Review of 97 bioavailability studies. *Am. J. Clin. Nutr.*, **81**, 230S–242S.

Mateo Anson, N., Selinheimo, E., Havenaar, R. et al. (2009a). Bioprocessing of wheat bran improves *in vitro* bioaccessibility and colonic metabolism of phenolic compounds. *J. Agric. Food Chem.*, **57**, 6148–6155.

Mateo Anson, N., van den Berg, R., Havenaar, R. et al. (2009b). Bioavailability of ferulic acid is determined by its bioaccessibility. *J. Cereal Sci.*, **49**, 296–300.

Mateos, R., Goya, L., and Bravo, L. (2006). Uptake and metabolism of hydroxycinnamic acids (chlorogenic, caffeic, and ferulic acids) by HepG2 cells as a model of the human liver. *J. Agric. Food Chem.*, **54**, 8724–8732.

Monteiro, M., Farah, A., Perrone D. et al. (2007). Chlorogenic acid compounds from coffee are differentially absorbed and metabolized in humans. *J. Nutr.*, **137**, 2196–2201.

Moridani, M.Y., Scobie, H., and O'Brien, P.J. (2002). Metabolism of caffeic acid by isolated rat hepatocytes and subcellular fractions. *Toxicol. Lett.*, **133**, 141–151.

Nardini, M., Cirillo, E., Natella, F. et al. (2002). Absorption of phenolic acids in humans after coffee consumption. *J. Agric. Food Chem.*, **50**, 5735–5741.

Nardini, M., Forte, M., Vrhovsek, U. et al. (2009). White wine phenolics are absorbed and extensively metabolized in humans. *J. Agric. Food Chem.*, **57**, 2711–2718.

Nishimura, M. and Naito, S. (2006). Tissue-specific mRNA expression profiles of human phase I metabolizing enzymes except for cytochrome P450 and phase II metabolizing enzymes. *Drug Met. Pharmacokin.*, **21**, 357–374.

O'Leary, K.A., Day, A.J., Needs, P.W. et al. (2003). Metabolism of quercetin-7-and quercetin-3-glucuronides by an *in vitro* hepatic model: The role of human β-glucuronidase, sulfotransferase, catechol-*O*-methyltransferase and multi-resistant protein 2 (MRP2) in flavonoid metabolism. *Biochem. Pharmacol.*, **65**, 479–491.

Ohmiya, K., Takeuchi, M., Chen, W. et al. (1986). Anaerobic reduction of ferulic acid to dihydroferulic acid by *Wolinella succinoyenes* from cow rumen. *Appl. Microbiol. Biotechnol.*, **23**, 274–279.

Olthof, M.R., Hollman, P.C. and Katan, M.B. (2001). Chlorogenic acid and caffeic acid are absorbed in humans. *J. Nutr.*, **131**, 66–71.

Olthof, M.R., Hollman, P.C.H., Buijsman, M. et al. (2003). Chlorogenic acid, quercetin-3-rutinoside and black tea phenols are extensively metabolized in humans. *J. Nutr.*, **133**, 1806–1814.

Ovaskainen, M.-L., Torronen, R., Koponen, J.M. et al. (2008). Dietary intake and major food sources of polyphenols in Finnish adults. *J. Nutr.*, **138**, 562–566.

Passamonti, S., Terdoslavich, M., Franca, R. et al. (2009). Bioavailability of flavonoids: A review of their membrane transport and the function of bilitranslocase in animal and plant organisms. *Curr. Drug Metab.*, **10**, 369–394.

Peppercorn, M.A. and Goldman, P. (1972). Caffeic acid metabolism by gnotobiotic rats and their intestinal bacteria. *Proc. Natl. Acad. Sci. U.S.A.*, **69**, 1413–1415.

Plumb, G.W., Garcia-Conesa, M.T., Kroon, P.A. et al. 1999). Metabolism of chlorogenic acid by human plasma, liver, intestine and gut microflora. *J. Sci. Food Agric.*, **79**, 390–392.

Poquet, L., Clifford, M.N., and Williamson, G. (2008a). Investigation of the metabolic fate of dihydrocaffeic acid. *Biochem. Pharmacol.*, **75**, 1218–1229.

Poquet, L., Clifford, M.N. and Williamson, G. (2008b). Transport and metabolism of ferulic acid through the colonic epithelium. *Drug Metab. Disp.*, **36**, 190–197.

Radtke, J., Linseisen, J., and Wolfram, G. (1998). Phenolic acid intake of adults in a Bavarian subgroup of the national food consumption survey. *Z. Ernahrungswiss.* **37**, 190–197.

Rechner, A.R., Pannala, A.S., and Rice-Evans, C.A. (2001a). Caffeic acid derivatives in artichoke extract are metabolised to phenolic acids *in vivo*. *Free Radic. Res.*, **35**, 195–202.

Rechner, A.R., Smith, M.A., Kuhnle, G. et al. (2004). Colonic metabolism of dietary polyphenols: Influence of structure on microbial fermentation products. *Free Radic. Biol. Med.*, **36**, 212–225.

Rechner, A.R., Spencer, J.P.E., Kuhnle, G. et al. (2001b). Novel biomarkers of the metabolism of caffeic acid derivatives *in vivo*. *Free Radic. Biol. Med.*, **30**, 1213–1222.

Renouf, M., Guy, P.A., Marmet, C. et al. (2010). Measurement of caffeic and ferulic acid equivalents in plasma after coffee consumption: Small intestine and colon are key sites for coffee metabolism. *Mol. Nutr. Food Res.*, **54**, 760–766.

Robbins, R.J. (2003). Phenolic acids in foods: An overview of analytical methodology. *J. Agric. Food Chem.*, **51**, 2866–2887.

Rocha, S., Maeztu, L., Barros, A. et al. (2003). Screening and disctinction of coffee brews based on headspace solid phase microextraction/gas chromatography/principal component analysis. *J. Sci. Food Agric.*, **84**, 43–51.

Rondini, L., Peyrat-Maillard, M.-N., Marsset-Baglieri, A. et al. (2002). Sulfated ferulic acid is the main *in vivo* metabolite found after short-term ingestion of free ferulic acid in rats. *J. Agric. Food Chem.,* **50**, 3037–3041.

Rondini, L., Peyrat-Maillard, M.N., Marsset-Baglieri, A. et al. (2004). Bound ferulic acid from bran is more bioavailable than the free compound in rat. *J. Agric. Food Chem.,* **52**, 4338–4343.

Rothwell, J.A., Day, A.J. and Morgan, M.R.A. (2005). Experimental determination of octanol-water partition coefficients of quercetin and related flavonoids. *J. Agric. Food Chem.,* **53**, 4355–4360.

Searle, J.B. (1986). A trimeric esterase in the common shrew. *J. Hered.,* **77**, 121–122.

Sharom, F.J. (1997). The P-glycoprotein efflux pump: How does it transport drugs? *J. Membr. Biol.,* **160**, 161–175.

Silberberg, M., Besson, C., Manach, C. et al. (2006a). Influence of dietary antioxidants on polyphenol intestinal absorption and metabolism in rats. *J. Agric. Food Chem.,* **54**, 3541–3546.

Silberberg, M., Morand, C., Mathevon, T. et al. (2006b). The bioavailability of polyphenols is highly governed by the capacity of the intestine and of the liver to secrete conjugated metabolites. *Eur. J. Nutr.,* **45**, 88–96.

Singleton, V.L., Salgues, M., Zaya, J. et al. (1985). Caftaric acid disappearance and conversion to products of enzymic oxidation in grape must and wine. *Am. J. Enol. Viticult.,* **36**, 50–56.

Singleton, V.L., Timberlake, C.F., and Lea, A.G.H. (1978). The phenolic cinnamates of white grapes and wine. *J. Sci. Food Agric.,* **29**, 403–410.

Spencer, J.P.E., Chowrimootoo, G., Choudhury, R. et al. (1999). The small intestine can both absorb and glucuronidate luminal flavonoids. *FEBS Lett.,* **458**, 224–230.

Stalmach, A., Mullen, W., Barron, D. et al. (2009). Metabolite profiling of hydroxycinnamate derivatives in plasma and urine after the ingestion of coffee by humans: Identification of biomarkers of coffee consumption. *Drug Metab. Disp.,* **37**, 1749–1758.

Stalmach, A., Steiling, H., Williamson, G. et al. (2010). Bioavailability of chlorogenic acids following acute ingestion of coffee by humans with an ileostomy. *Arch. Biochem. Biophys.,* **501**, 98–105.

Sun, R., Sun, X.F., Wang, S.Q. et al. (2002). Ester and ether linkages between hydroxycinnamic acids and lignins from wheat, rice, rye, and barley straws, maize stems, and fast-growing poplar wood. *Indust. Crops Prod.,* **15**, 179–188.

Takenaka, M., Nagata, T., and Yoshida, M. (2000). Stability and bioavailability of antioxidants in garland (*Chrysanthemum coronarium* L.). *Biosci. Biotechnol. Biochem.,* **64**, 2689–2691.

Tanaka, Y. and Suzuki, A. (1994). Enzymatic hydrolysis of zenarestat-1-*O*-acylglucuronide. *J. Pharm. Pharmacol.,* **46**, 235–239.

Teubner, W., Meinl, W., Florian, S. et al. (2007). Identification and localization of soluble sulfotransferases in the human gastrointestinal tract. *Biochem. J.,* **404**, 207–215.

Tian, X.J., Yang, X.W., Yang, X.D. et al. (2009). Studies of intestinal permeability of 36 flavonoids using Caco-2 cell monolayer model. *Int. J. Pharm.,* **367**, 58–64.

Trugo, L.C. and Macrae, R. (1984). A study of the effect of roasting on the chlorogenic acid composition of coffee using HPLC. *Food Chem.,* **15**, 21–-227.

Vaidyanathan, J.B. and Walle, T. (2001). Transport and metabolism of the tea flavonoid (–)-epicatechin by the human intestinal cell line Caco-2. *Pharm. Res.,* **18**, 1420–1425.

Vanbeneden, N., Delvaux, F., and Delvaux, F.R. (2006). Determination of hydroxycinnamic acids and volatile phenols in wort and beer by isocratic high-performance liquid chromatography using electrochemical detection. *J. Chromatogr.,* **1136**, 237–242.

Walgren, R.A., Karnaky, K.J., Lindenmayer, G.E. et al. (2000). Efflux of dietary flavonoid quercetin 4'-*O*-β-glucoside across human intestinal Caco-2 cell monolayers by apical multidrug resistance-associated protein-2. *J. Pharmacol. Exp. Ther.,* **294**, 830–836.

Walgren, R.A., Walle, U.K., and Walle, T. (1998). Transport of quercetin and its glucosides across human intestinal epithelial Caco-2 cells. *Biochem. Pharmacol.,* **55,** 1721–1727.

Walle, T. and Walle, U.K. (2003). The β-D-glucoside and sodium-dependent glucose transporter 1 (SGLT1)-inhibitor phloridzin is transported by both SGLT1 and multidrug resistance-associated proteins 1/2. *Drug Metab. Disp.,* **31,** 1288–1291.

Walle, U.K., French, K.L., Walgren, R.A. et al. (1999). Transport of genistein-7-glucoside by human intestinal Caco-2 cells: Potential role for MRP2. *Res. Commun. Mol. Pathol. Pharmacol.,* **103,** 45–56.

Watanabe, H., Yashiro, T., Tohjo, Y. et al. (2006). Non-involvement of the human monocarboxylic acid transporter 1 (MCT1) in the transport of phenolic acid. *Biosci. Biotechnol. Biochem.,* **70,** 1928–1933.

Williamson, G., Day, A.J., Plumb, G.W. et al. (2000). Human metabolic pathways of dietary flavonoids and cinnamates. *Biochem. Soc. Trans.,* **28,** 16–22.

Wittemer, S.M., Ploch, M., Windeck, T. et al. (2005). Bioavailability and pharmacokinetics of caffeoylquinic acids and flavonoids after oral administration of artichoke leaf extracts in humans. *Phytomedicine,* **12,** 28–38.

Wolffram, S., Weber, T., Grenacher, B. et al. (1995). A Na$^+$-dependent mechanism is involved in mucosal uptake of cinnamic acid across the jejunal brush border in rats. *J. Nutr.,* **125,** 1300–1308.

Wong, C., Meinl, W., Glatt, H. et al. (2010). *In vitro* and *in vivo* conjugation of dietary hydroxycinnamic acids by UDP-glucuronosyltransferases and sulfotransferases in humans. *J. Nutr. Biochem.,* **21,** 1060–1068.

Yang, B., Meng, Z.Y., Dong, J.X. et al. (2005). Metabolic profile of 1,5-dicaffeoylquinic acid in rats, an *in vivo* and *in vitro* study. *Drug Metab. Disp.,* **33,** 930–936.

Yang, B., Meng, Z.Y., Yan, L.P. et al. (2006). Pharmacokinetics and metabolism of 1,5-dicaffeoylquinic acid in rats following a single intravenous administration. *J. Pharm. Biomed. Anal.,* **40,** 417–422.

Zhang, L., Strong, J.M., Qiu, W. et al. (2006). Scientific perspectives on drug transporters and their role in drug interactions. *Mol. Pharm.,* **3,** 62–69.

Zhang, L., Zheng, Y., Chow, M.S.S. et al. (2004). Investigation of intestinal absorption and disposition of green tea catechins by Caco-2 monolayer model. *Int. J. Pharm.,* **287,** 1–12.

Zhao, Z., Egashira, Y., and Sanada, H. (2003a). Digestion and absorption of ferulic acid sugar esters in rat gastrointestinal tract. *J. Agric. Food Chem.,* **51,** 5534–5539.

Zhao, Z., Egashira, Y., and Sanada, H. (2003b). Ferulic acid sugar esters are recovered in rat plasma and urine mainly as the sulfoglucuronide of ferulic acid. *J. Nutr.,* **133,** 1355–1361.

Zhao, Z., Egashira, Y., and Sanada, H. (2004). Ferulic acid is quickly absorbed from rat stomach as the free form and then conjugated mainly in liver. *J. Nutr.,* **134,** 3083–3088.

Zhao, Z., Egashira, Y., and Sanada, H. (2005). Phenolic antioxidants richly contained in corn bran are slightly bioavailable in rats. *J. Agric. Food Chem.,* **53,** 5030–5035.

7 Bioavailability of Dihydrochalcones

Elke Richling

CONTENTS

7.1 INTRODUCTION

Dihydrochalcones are C-ring-open flavonoids with a 15-carbon skeleton (Figure 7.1) and are secondary metabolites found in several plants and plant products that are important components of the human diet. These compounds include phloretin and its glycoside conjugates phloretin-2′-*O*-glucoside (phloridzin) and phloretin-2′-*O*-(2″-*O*-xylosyl)glucoside from apples (*Malus domestica* Borkh.). In addition, aspalathin (2′,3,4,4′,6′-pentahydroxydihydrochalcone-3′-*C*-glucoside) and nothofagin (2′,4,4′,6′-tetrahydroxydihydrochalcone-3′-*C*-glucoside) are dihydrochalcone *C*-glycosides found in rooibos (*Aspalathus linearis*) tea. Naringin dihydrochalcone [2′,4,6′-trihydroxydihydrochalcone-4′-*O*-(2″-*O*-rhamnosyl)glucoside] and neohesperidin dihydrochalcone [2′,3,6′-trihydroxy-4-methoxydihydrochalcone-4′-*O*-(2″-*O*-rhamnosyl)glucoside] are artificial sweeteners derived from citrus fruits (Figure 7.1).

In order to have a physiological effect *in vivo*, ingested dihydrochalcones must reach the intestinal epithelium unchanged, be absorbed from the gastrointestinal tract, and accumulate in sufficiently high plasma concentrations in the systemic circulation.

7.2 BIOAVAILABILITY OF PHLORETIN AND DERIVATIVES

The dihydrochalcones consumed in the greatest quantities are the exclusively apple-derived compounds phloretin-2′-*O*-glucoside (phlorizin) and phloretin-2′-*O*-(2″-*O*-xylosyl)glucoside, which are reportedly present in concentrations of about 20–155 mg/kg in apples and 10–171 mg/L in apple juices (Vrhovsek et al. 2004; Kahle et al. 2005; Gerhauser 2008). Free phloretin has not been found in fresh fruits but is present in apple juices (Oleszek et al. 1988). Additionally, apple products such as

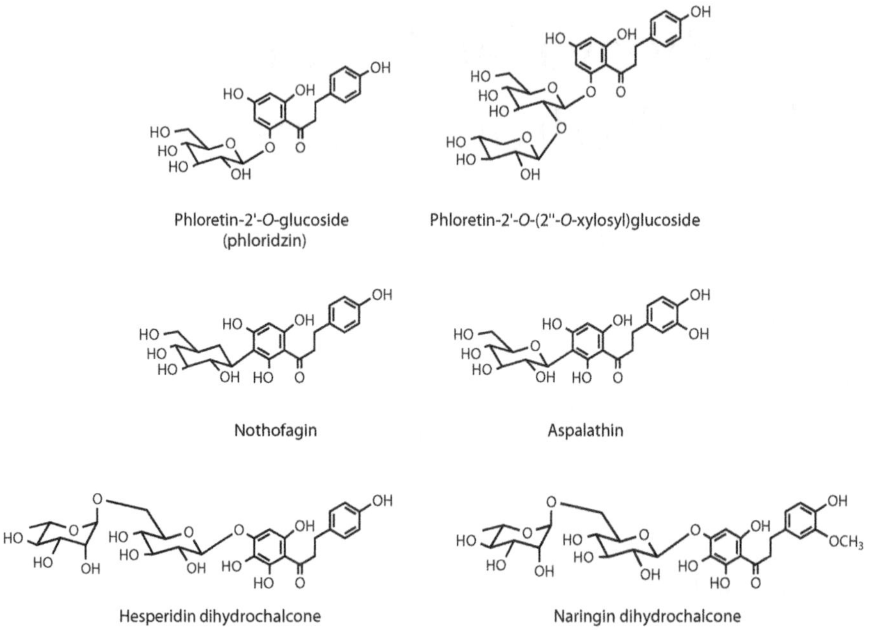

Phloretin-2'-*O*-glucoside
(phloridzin)

Phloretin-2'-*O*-(2''-*O*-xylosyl)glucoside

Nothofagin

Aspalathin

Hesperidin dihydrochalcone

Naringin dihydrochalcone

FIGURE 7.1   Structures of the most common dihydrochalcones contained in food.

apple juices and apple extracts contain low amounts of 3-hydroxy-phloretin-2'-*O*-glucoside, an oxidation product of phloretin-2'-*O*-glucoside (Vrhovsek et al. 2004). Apple-derived dihydrochalcones have received increased research interest because of their potential impact on health (Boyer and Liu 2004). Among many beneficial properties, they have been reported to possess antioxidant and anti-inflammatory activities and affect glucose absorption. These properties have been demonstrated in a diversity of *in vitro* assays, experiments with animals, and epidemiological studies (Johnston et al. 2002; Boyer and Liu 2004; Ehrenkranz et al. 2005; Lata et al. 2005; Jung et al. 2009).

Early investigations focused on the metabolism of phloretin in rats (Monge et al. 1984). Approximately 50% of the initial administered dose of phloretin was detected in urine, with the principle compounds being characterized as phloretin, 3-(4'-hydroxyphenyl)propionic acid and 1,3,5-trihydroxybenzene (phloroglucinol). The latter two compounds are colonic degradation products of phloretin.

It has been demonstrated with studies in which apple juice was consumed, that apple-derived phloretin-2'-*O*-glucoside can be cleaved by lactase phlorizin hydrolase of the brush-border membrane, by intestinal microbiota and/or cytosolic β-glucosidase in epithelial cells (Scalbert and Willamson 2000). In the small intestine, phloretin-2'-*O*-glucoside is transported by the sodium-dependent glucose transporter (SGLT-1) as an intact molecule (Scalbert and Williamson 2000). Rapid deglycosylation of phloretin-2'-*O*-glucoside and glucuronidation of phloretin by UDP-glucuronosyl transferase may occur in the intestinal epithelium or liver following absorption, with the metabolites formed effluxing back into the lumen of the gastrointestinal tract (Mizuma et al. 1998; Kahle et al. 2007). The *in vitro* cleavage rates of dihydrochalcone glycosides by small intestinal microbiota are shown in Figure 7.2, and it is clear that phloretin is released with phloretin-2′-*O*-glucoside being deglycosylated much more readily than phloretin-2′-*O*-(2″-*O*-xylosyl)glucoside (Kahle et al. 2011).

In the colon, phloretin is metabolized by the microbiota to 3-(4′-hydroxyphenyl) propionic acid and phloroglucinol, as shown by studies using a pig cecum model (Labib et al. 2006) and *in vitro* incubations with colonic bacteria such as *Eubacterium ramulus* (Schneider et al. 1999).

In another study, using an *in situ* intestinal perfusion model, Crespy et al. (2001) showed that phloretin and phloretin-2′-*O*-glucoside are absorbed in the small intestine. After perfusion, 80% of phloretin-2′-*O*-glucoside was hydrolyzed, with phloretin and its sulfated and glucuronide metabolites appearing in the lumen. In a further animal study, Crespy et al. (2002) recovered small quantities of phloretin from plasma; however, no phloretin-2′-*O*-glucoside was detected, indicating that the glucoside was hydrolyzed before absorption and metabolism. Skjevrak et al. (1986) investigated the metabolism of synthetic dihydrochalcones analogs, related to phloretin, in rats and concluded that independent of the hydroxylation pattern in the A-ring system, a ring scission reaction of dihydrochalcones occurs in the colon

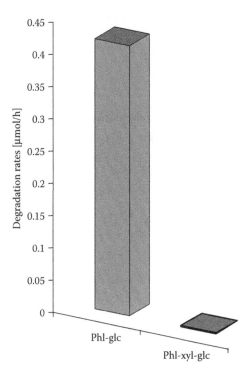

FIGURE 7.2 Degradation rates of phloretin-2′-*O*-β-D-glucopyranoside (Phl-glc) and Phloretin-2′-*O*-(2″-*O*-xylosyl)glucoside (Phl-xyl-glc) measured resulting in the release of phloretin, *in vitro* in ileostomy fluid. Data are mean values of triplicate determinations.

prior to absorption. These results imply that hydrolysis of phloretin-2'-O-glucoside to phloretin is not a rate determining step for its absorption.

Several studies have measured urinary excretion after the consumption of phloretin-2'-O-glucoside and phloretin-2'-O-(2''-O-xylosyl)glucoside in apple juice or apples. In a study with six volunteers, Kahle et al. (2011) found 0.63% of the ingested phloretin-2'-O-(2''-xylo)glucoside in urine, but phloretin-2'-O-glucoside was not detected. In another human study, DuPont et al. (2002) reported that after consumption of 1.1 liters of apple cider, containing 0.18 mg phloretin, 3.1 mg phloretin-2´-O-glucoside, and 1.5 mg phloretin-2'-O-(2''-O-xylosyl)glucoside, 21.5% of the administered dose of phloretin was recovered in urine. This figure is not comparable to that of Kahle et al. (2011) as samples were analyzed after being subjected to enzymatic hydrolysis, which would have cleaved conjugates and released phloretin.

To date, there have been few reports of the bioavailability of phloretin and phloretin glycosides in humans (DuPont et al. 2002; Kahle et al. 2007; Marks et al. 2009; Hagl et al. 2011). To investigate the colonic availability of dihydrochalcones, apple products were fed to subjects with an ileostomy after which ileal fluid collected in ileostomy bags was analyzed. In these studies, volunteers consumed apple cider, cloudy apple juice, or an apple smoothie containing phloretin-2'-O-glucoside and phloretin-2'-O-(2''-O-xylosyl)glucoside, but not the free aglycone, phloretin. In all three studies, the aglycone, phloretin-2'-O-(2''-xylo)glucoside, and up to three phloretin-O-glucuronides were detected in ileal effluent (Table 7.1). Thus, these compounds would presumably reach the colon of healthy subjects following consumption of the respective apple product. Surprisingly, the total amount of dihydrochalcones in the

TABLE 7.1
Data for the Recoveries of Dihydrochalcones and Metabolites in Ileal Fluid After Apple Smoothie, Cloudy Apple Juice, and Cider Consumption[a]

	Apple Smoothie (1)	Cloudy Apple Juice (2)	Apple Cider (3)
Total dihydrochalcones consumed	170.9	81.2	46.0
Dihydrochalcones in ileal fluid			
Phloretin	5.8	1.8	2.4
Phloretin-2'-O-glucoside	n.d.	n.d.	n.d.
Phloretin-2'-O-(2''-O-xylosyl)glucoside	31.5	12.7	3.2
Phloretin-2'-O-glucuronide	26.4	5.6	7.0
Phloretin-O-glucuronide-2	1.8	n.d.	0.3
Phloretin-O-glucuronide-3	15.3	n.d.	1.3
Other phloretin conjugates	n.d.	n.d.	3.5
Total in ileal fluid	**80.8** *(47.3%)*	**20.1** *(24.8%)*	**17.7** *(38.5%)*

n.d.: not detectable; (1) Hagl, S. et al., *Mol. Nutr. Food Res.*, **55**, 368–377, 2011; (2) Kahle, K. et al., *J. Agric. Food Chem.*, **55**, 10605–10614, 2007; (3) Marks, S.C. et al., *J. Agric. Food Chem.*, **57**, 2009–2015, 2009.

[a] Data expressed as mean values in μmol and italicized figures in parentheses as a percentage of the quantity ingested.

ileostomy bags after apple smoothie consumption reported by Hagl et al. (2011) was equivalent to 47.3% of the 170.9 µmol intake—exceeded the overall recovery reported by Kahle et al. (2007) (20.1 µmol, 24.8%) and Marks et al. (2009) (17.7 µmol, 38.5%) after the respective ingestion of cloudy apple juice (81.2 µmol) or apple cider (46 µmol) (Table 7.1).

Phloretin-2′-O-(2″-O-xylosyl)glucoside was found to be partially stable in the small intestine and was identified in all ileostomy bags studied. Hagl et al. (2011) detected the disaccharide in ileal fluid with a recovery of 56.3 ± 18.9%, whereas in the other studies, only about 20% of the ingested phloretin-2′-O-(2″-O-xylosyl) glucoside was recovered (Kahle et al. 2007, Marks et al. 2009). In studies by Hagl et al. (2011), phloretin-2′-O-glucoside present in the ingested apple smoothie was not detected in the ileal fluid, whereas phloretin and three phloretin-O-glucuronides that were found in ileal fluid were not present in the ingested apple smoothie. These results are consistent with the findings of Kahle et al. (2007) and Marks et al. (2009). In the studies by Kahle et al. (2006, 2007), the total amount of phloretin conjugates recovered in ileal samples was 20.1 µmol (24.8%) following consumption of one liter of cloudy apple juice containing 81.2 µmol of phloretin conjugates. Kahle et al. (2007) were the first to provide evidence for the position of phloretin glucuronidation using NMR.

In a study by Marks et al. (2009), the bioavailability of dihydrochalcones following the consumption of 500 mL of apple cider (46 µmol) was investigated in ileostomists and healthy subjects. The results showed that the phloretin glycosides, phloretin-2′-O-(2″-O-xylosyl)glucoside and phloretin-2′-O-glucoside, were metabolized to phloretin-2′-O-glucuronide and two partially characterized phloretin-O-glucuronide isomers in volunteers both with and without a colon. Glucuronides were detected in ileal fluid, urine, and plasma. In addition, minor amounts of phloretin conjugated with glucuronic acid and sulfate (phloretin-O-glucuronide-O-sulfate or phloretin-O-sulfates) and unconjugated phloretin were found in the ileal samples. In the Marks et al. study, the metabolite phloretin-2′-O-glucuronide was excreted in the urine of ileostomists 0–2 h after ingestion of cider, whereas in healthy subjects, the excretion rates remained comparably high for up to 8 h after intake. Interestingly, when both groups (healthy and ileostomists) were compared after consuming the same amount of dihydrochalcones (46 µmol), no significant differences were observed in their plasma and urine levels indicating that the dihydrochalcones were absorbed in the small intestine.

In healthy volunteers, the peak plasma concentration (C_{max}) of phloretin-O-glucuronide metabolites was 73 nM after the consumption of 46 µmol of dihydrochalcones in 500 mL cider (Marks et al. 2009). In contrast, Stracke et al. (2010) found that after the consumption of 1 kg of apples (50 µmol dihydrochalcones) the C_{max} was only 12.5 nM (see Table 7.2), thereby indicating the probable impact of the food matrix. Stracke et al. (2010) also found no significant difference in bioavailability of dihydrochalcones irrespective of whether the ingested apples were organic or conventionally grown. Volunteers consumed either 1 kg of apples as a single serving (an acute intervention) or 500 g of apples (cultivar Golden Delicious) daily for four weeks (chronic intervention). Plasma samples, collected either for a 24-h period or 24 h after the last apple consumption, respectively, were treated with glucuronidase/

TABLE 7.2

Plasma Pharmacokinetic Data for Phloretin and Phloretin Glucuronides After Apple and Apple Juice Consumption in Healthy Subjects and Ileostomists

Study	Total Consumption (µmol)	C_{max} (nM)	T_{max} (h)	AUC (nmol/h/L)
Stracke et al. (2010)[a] ($n = 6$)	50	12.5	3.0	125
Marks et al. (2009)				
Healthy ($n = 9$)	46	73	0.6	129
Ileostomists ($n = 3$)	46	93	0.5	140
Borges et al. (2010) ($n = 10$)	83	204	0.6 and 4.0	425

[a] Measured as phloretin aglycone after enzymatic hydrolysis.

sulfatase prior to analysis. In the chronic intervention study, the T_{max} for the enzymically released phloretin was between 2.8 and 3.2 h.

Recently, Borges et al. (2010) investigated the plasma and urine concentrations of dihydrochalcones after the consumption of a polyphenol-rich drink (350 mL) containing 68 µmol phloretin-2'-O-glucoside and 15 µmol phloretin-2'-O-(2"-O-xylosyl)glucoside, along with an array of other polyphenolic constituents. As in the previous studies of Marks et al. (2009), phloretin-2'-O-glucuronide was identified in the plasma samples with a T_{max} of 0.6 h and a second smaller peak at 4 h. Metabolites excreted in urine samples were phloretin-2'-O-glucuronide, a second phloretin-O-glucuronide, and a phloretin-O-glucuronide-O-sulfate with 4.0 µmol (4.9%) of the initial intake of phloretin recovered.

The metabolism of phloretin-2'-O-glucoside and phloretin-2'-O-(2"-O-xylosyl) glucoside following ingestion as they pass along the gastrointestinal tract are summarized in Figure 7.3.

7.3 BIOAVAILABILITY OF ASPALATHIN AND NOTHOFAGIN

The metabolism and bioavailability of both aspalathin and nothofagin after the consumption of rooibos tea infusions have been investigated by Courts et al. (2009), Stalmach et al. (2009), Laue et al. (2010), and van der Merwe et al. (2010). In one study, the dihydrochalcone-C-glucoside, aspalathin, was found in its unmetabolized form in plasma samples of healthy volunteers, whereas methylated derivatives, glucuronides and sulphates were detected in urine (Laue et al. 2010). In a separate study, Stalmach et al. (2009) investigated the bioavailability of aspalathin and nothofagin after the ingestion of unfermented and fermented rooibus teas by 10 healthy subjects. A 500 mL serving of the unfermented tea contained ~10-fold higher levels of aspalathin (90 µmol) and nothofagin (16 µmol) than the fermented tea

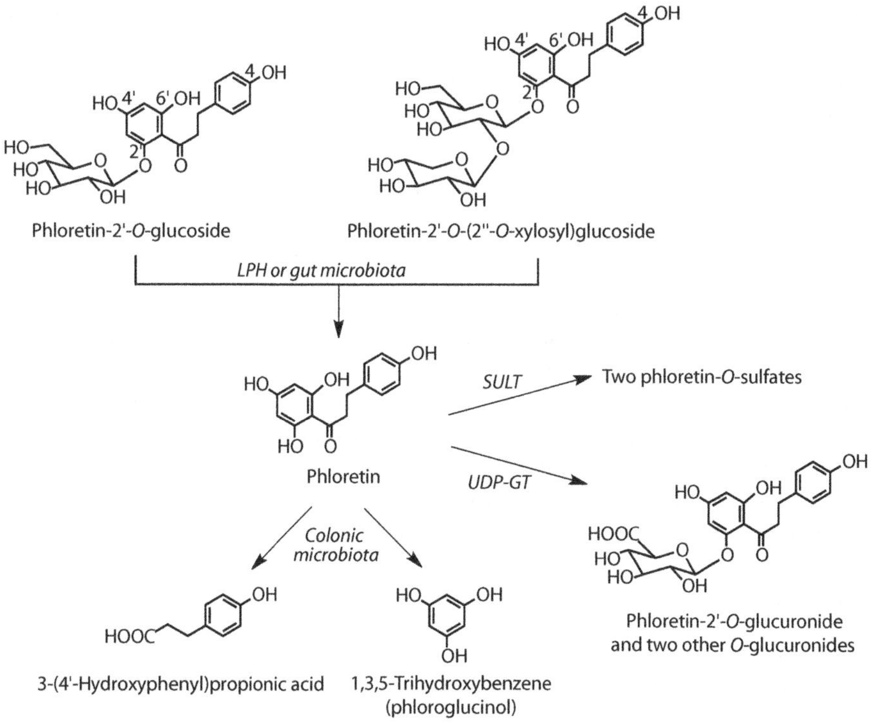

FIGURE 7.3 The metabolism of phloretin-2'-*O*-glucoside and phloretin-2'-*O*-(2"-*O*-xylosyl) glucoside following ingestion as they pass along the gastrointestinal tract. LPH: lactase phloridzin hydrolase; UDG-GT: UDP-glucuronosyltransferase; SULT: sulfotransferase.

(8.0 μmol aspalathin, 1.9 μmol nothofagin). Following consumption of the teas, no dihydrochalcones or metabolites were detected in plasma samples using HPLC-MS. Similarly, neither nothofagin nor its derivatives were detected in urine; however, a number of aspalathin metabolites were detected, namely aspalathin-*O*-glucuronide, *O*-methyl-aspalathin-*O*-glucuronide, aspalathin-sulfate, and *O*-methyl-aspalathin-sulfate. The total quantities of these metabolites were 317 nmol and 14.3 nmol for the unfermented and fermented teas, respectively, accounting for 0.35% and 0.02% of the ingested dose. Aspalathin-*O*-glucuronide was excreted in urine after consumption of unfermented rooibus tea but was not detected following consumption of fermented tea.

The findings of Stalmach et al. (2009) are in good agreement with those of Kreuz et al. (2008), who investigated the metabolism of aspalathin in pigs after the consumption of an unfermented rooibos tea extract containing 16.3% aspalathin. Courts and Williamson (2009) investigated the metabolism of aspalathin in six volunteers by measuring urinary metabolites after the consumption of rooibos tea. 3-*O*-Methylaspalathin (162 μg) and 3-*O*-methylaspalathin-*O*-glucuronide (87 μg) were identified in the urine within the first 2 h after ingestion, the amounts being equivalent to 0.74% of the ingested dose. These studies indicate that intact *C*-glycosides are poorly absorbed in the small intestine of mammals and therefore have low bioavailability. In an additional study, van der Merwe et al. (2010) have shown that aspalathin as well as nothofagin undergo *in vitro* hepatic conjugation to monoglucuronides and monosulfates, respectively.

7.4 BIOAVAILABILITY OF NEOHESPERIDIN AND NARINGIN DIHYDROCHALCONES

To date, there are no reports in the literature on the bioavailability of neohesperidin dihydrochalcone and naringin dihydrochalcone. However, the intestinal degradation of neohesperidin dihydrochalcone has been investigated using human colonic microbiota *in vitro* (Braune et al. 2005). The fecal bacteria converted the dihydrochalcone into 3-(3′-hydroxy-4′-methoxyphenyl)propionic acid and 3-(3′,4′-dihydroxyphenyl) propionic acid. Intermediates such as hesperetin dihydrochalcone-4′-O-glucoside and the aglycone hesperetin dihydrochalcone were also observed, indicating that neohesperidin dihydrochalcone is first cleaved by intestinal rhamnosidase then hydrolyzed to the corresponding propionic acids.

REFERENCES

Borges, G., Mullen, W., Mullen, A. et al. (2010). Bioavailability of multiple components following acute ingestion of a polyphenol-rich juice drink. *Mol. Nutr. Food Res.*, **54**, S268–S277.

Boyer J. and Liu R.H. (2004). Apple phytochemicals and their health benefits. *Nutr. J.*, **3**, 5–20.

Braune, A., Engst, W., and Blaut, M. (2005). Degradation of neohesperidin dihydrochalcone by human intestinal bacteria. *J. Agric. Food Chem.*, **53**, 1782–1790.

Courts, F. and Williamson, G. (2009). The *C*-glycosyl flavonoid, aspalathin, is absorbed, methylated and glucuronidated intact in humans. *Mol. Nutr. Food Res.*, **53**, 1104–1111.

Crespy V., Morand C., Besson C. et al. (2001). Comparison of the intestinal absorption of quercetin, phloretin and their glucosides in rats. *J. Nutr.*, **131**, 2109–2114.

Crespy, V., Aprikian, O., Morand, C. et al. (2002). Bioavailability of phloretin and phloridzin in rats. *J. Nutr.*, **132**, 3227–3220.

DuPont M.S., Bennett R.N., Mellon F.A. et al. (2002). Polyphenols from alcoholic apple cider are absorbed, metabolized and excreted by humans. *J. Nutr.*, **132**, 172–175.

Ehrenkranz, J.R.L., Lewis, N.G., Kahn, C.R. et al. (2005). Phlorizin: A review. *Diabetes Metabol. Res. Rev.*, **21**, 31–38.

Gerhauser C. (2008). Cancer chemopreventive potential of apples, apple juice, and apple components. *Planta Med.*, **74**, 1608–1624.

Hagl, S., Bergmann, H., Janzowski, C. et al. (2011). Colonic availability of polyphenols and D-(−)-quinic acid after apple smoothie consumption. *Mol. Nutr. Food Res.*, **55**, 368–377.

Johnston, K.L., Clifford, M.N., and Morgan, L.M. (2002). Possible role for apple juice phenolic, compounds in the acute modification of glucose tolerance and gastrointestinal hormone secretion in humans. *J. Sci. Food Agric.*, **82**, 1800–1805.

Jung, M., Triebel, S., Anke, T., et al. (2009). Influence of apple polyphenols on inflammatory gene expression. *Mol. Nutr. Food Res.*, **53**, 1263–1280.

Kahle K., Kraus M., and Richling E. (2005). Polyphenol profiles of apple juices. *Mol. Nutr. Food Res.*, **49**, 797–806.

Kahle, K., Kraus, M., Scheppach, W. et al. (2006). Studies on apple and blueberry fruit constituents: Do the polyphenols reach the colon? *Mol. Nutr. Food Res.*, **50**, 418–423.

Kahle, K., Huemmer, W., Kempf, M. et al. (2007). Polyphenols are intensively metabolized in the human gastrointestinal tract after apple juice consumption. *J. Agric. Food Chem.*, **55**, 10605–10614.

Kahle, K., Kempf, M., Schreier, P. et al. (2011). Intestinal transit and systemic metabolism of apple polyphenols. *Eur. J. Nutr.*, **50**, 507–522.

Knaup, B., Kahle, K., Erk, T. et al. (2007). Human intestinal hydrolysis of phenol glycosides—a study with quercetin and p-nitrophenol glycosides using ileostomy fluid. *Mol. Nutr. Food Res.*, **51**, 1423–1429.

Kreuz, A., Joubert, E., Waldmann, K.H. et al. (2008). Aspalathin, a flavonoid in *Aspalathus linearis* (rooibos), is absorbed by pig intestine as a C-glycoside. *Nutr. Res.*, **28**, 690–701.

Labib, S. (2006). Ex-vivo-*Studien zum intestinalen Metabolismus von Flavonoiden*. PhD thesis, University of Wuerzburg, Germany.

Lata, B., Przeradzka, M., and Binkowska, M. (2005). Great differences in antioxidant properties exist between 56 apple cultivars and vegetation seasons. *J. Agric. Food Chem.*, **53**, 8970–8978.

Laue, E., Gröll, S., Breiter, T. et al. (2010). Bioverfügbarkeit von Flavonoiden aus grünem Rooibos. *Lebensmittelchemie*, **64**, 121.

Marks, S.C., Mullen, W., Borges, G. et al. (2009). Absorption, metabolism, and excretion of cider dihydrochalcones in healthy humans and subjects with an ileostomy. *J. Agric. Food Chem.*, **57**, 2009–2015.

Mizuma, T. and Awazu, S. (1998). Inhibitory effect of phloridzin and phloretin on glucuronidation of *p*-nitrophenol, acetaminophen and 1-naphthol: Kinetic demonstration of the influence of glucuronidation metabolism on intestinal absorption in rats. *Biochim. Biophys. Acta,* **1425**, 398–404.

Monge, P., Solheim, E., and Scheline, R.R. (1984). Dihydrochalcone metabolism in the rat: Phloretin. *Xenobiotica,* **14**, 917–924.

Oleszek, W., Lee, Y.C., Jaworski, A.W. et al. (1988). Identification of some phenolic compounds in apples. *J. Agric. Food Chem.*, **36**, 430–432.

Scalbert, A. and Williamson, G. (2000). Dietary intake and bioavailability of polyphenols. *J. Nutr.*, **130**, 2073S–2085S.

Schneider, H., Schwiertz, A., Collins, M.D. et al. (1999). Anaerobic transformation of quercetin-3-glucoside by bacteria from the human intestinal tract. *Arch. Microbiol.*, **171**, 81–91.

Skjevrak, I., Solheim, E., and Scheline, R.R. (1986). Dihydrochalcone metabolism in the rat: Trihydroxylated derivatives related to phloretin. *Xenobiotica,* **16**, 35–45.

Stalmach, A., Mullen, W., Pecorari, M. et al. (2009). Bioavailability of *C*-linked dihydrochalcone and flavanone glucosides in humans following ingestion of unfermented and fermented rooibos teas. *J. Agric. Food Chem.*, **57**, 7104–7111.

Stracke, B.A., Rufer, C.E., Bub, A. et al. (2010). No effect of framing system (organic/conventional) on the bioavailability of apple (*Malus domestica* Borkh., cultivar Golden Delicious) polyphenols in healthy men: A comparative study. *Eur. J. Nutr.*, **49**, 301–310.

van der Merwe, J.D., Joubert, E., Manley, M. et al. (2010). *In vitro* hepatic biotransformation of aspalathin and nothofagin, dihydrochalcones of rooibos (*Aspalathus linearis*), and assessment of metabolite antioxidant activity. *J. Agric. Food Chem.*, **58**, 2214–2220.

Vrhovsek, U., Rigo, A., Tonon, D. et al. (2004). Quantitation of polyphenols in different apple varieties. *J. Agric. Food Chem.*, **52**, 6532–6538.

8 Occurrence, Bioavailability, and Metabolism of Resveratrol

Paola Vitaglione, Stefano Sforza, and Daniele Del Rio

CONTENTS

8.1 SOURCES OF DIETARY RESVERATROL

The stilbene resveratrol (3,5,4′-trihydroxy-stilbene) is a secondary metabolite with a C_6–C_2–C_6 structure that *in planta* is a phytoalexin produced in response to disease, injury, and stress (Langcake and Price 1977; Chong et al. 2009). It occurs as both the *cis* and *trans* isomers, and the main dietary sources are grapes red wines, peanuts (*Arachis hypogaea*), and to a lesser extent, berries, red cabbage (*Brassica oleacea*), spinach (*Spinaceae oleacea*), pistachio nuts (*Pistacia vera* L.), and certain herbs. Stilbenes occur in a variety of forms with *cis*- and *trans*-resveratrol-3-*O*-glucosides, commonly known as *cis*- and *trans*-piceid, being the major stilbenes in grapes and wines (Stervbo et al. 2007). Brazilian red wines have also been shown to contain *trans*-piceatannol (3,3′,4,5′-tetrahydroxystilbene) and its 3-*O*-glucoside *trans*-astringin (Vitrac et al. 2005). *Trans*-resveratrol is transformed by *Botrytis cinerea*, a fungal grapevine pathogen, to pallidol and resveratrol *trans*-dehydrodimer,

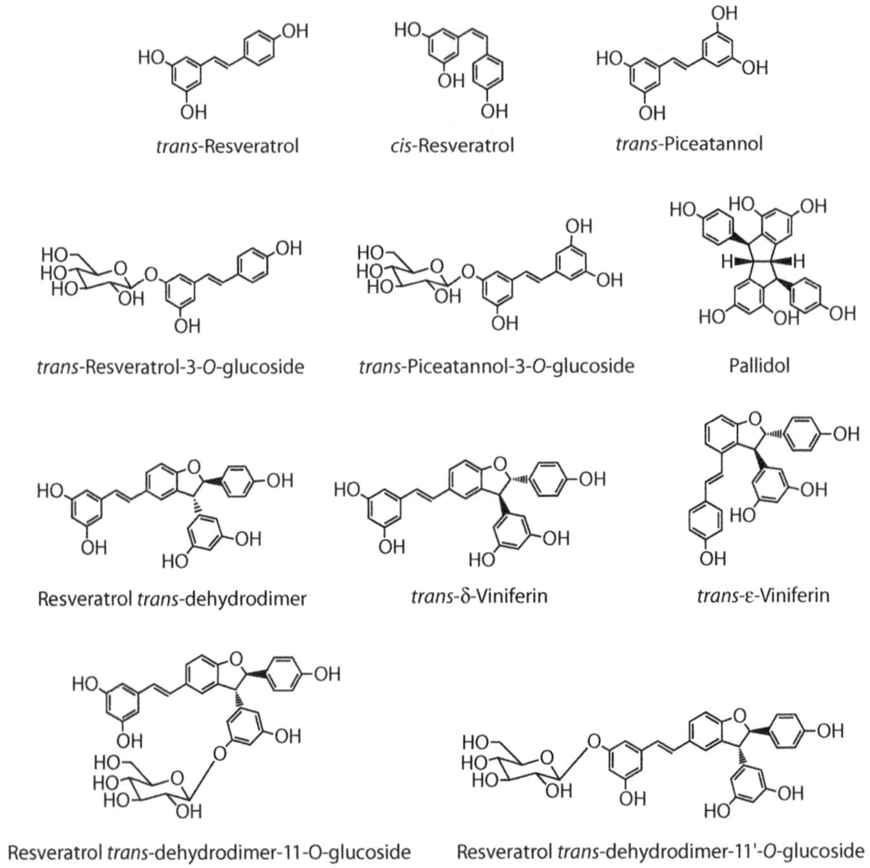

FIGURE 8.1 Structures of resveratrol and other stilbenes found in plants.

and both these compounds have been detected in grape cell cultures along with the 11-*O*- and 11′-*O*-glucosides of resveratrol *trans*-dehydrodimer (Waffo-Teguo et al. 2001). Viniferins are another family of oxidized resveratrol dimers (Langcake and Pryce 1977), and δ-viniferin and smaller amounts of its isomer ε-viniferin have been detected in *Vitis vinifera* leaves infected with *Plasmopara viticola* (downy mildew) (Pezet et al. 2003). Structures of these stilbene are illustrated in Figure 8.1. For further information on the structural diversity of naturally occurring stilbenes, interested readers should consult Shen et al. (2009).

Quantitative information of the stilbenes content of dietary products is limited to *cis*- and *trans*-resveratrol and *cis*- and *trans*-piceid, which have also been the focus of investigations on stilbene bioavailability.

8.1.1 GRAPES AND WINES

Grapes produced by *Vitis vinifera* are probably the most common dietary source of resveratrol, which occurs mainly in the skin where its concentration peaks before the fruit reach maturity (Stervbo et al. 2007). As resveratrol acts as a phytoalexins in vines, the levels in grapes vary greatly as its production is induced in response to stress factors, such as skin damage, ultraviolet (UV) irradiation, and pathogens (Stervbo et al. 2007). Organic grapes, which are more susceptible to pathogen infection, are often characterized by a higher resveratrol content than conventionally grown grapes (Tinttunen et al. 2001). Cultivation methods, processing, the region of cultivation, and the cultivar (Adrian et al. 2000) also contribute to different resveratrol concentrations in grapes. The *trans*-resveratrol content of two different grape cultivars was found to vary from 0.1 mg/kg to a few milligrams per kilogram (Creasy et al. 1988), whereas *trans*-piceid was present in higher amounts ranging from ~1 to 10 mg/kg (Burns et al. 2002). As a general rule, for both grapes and wines, the concentration of the four derivatives follows the following order: *trans*-piceid > *cis*-piceid ≥ *trans*-resveratrol > *cis*-resveratrol (Sato et al. 1997), and red grapes usually have a much higher content than white grapes (Romero-Perez et al. 1999). The total resveratrol in grapes is, therefore, extremely variable, ranging from < 1 to 10–20 mg/kg.

Red wines contain the highest levels of resveratrol (Pervaiz 2003) because the wine-making method is based on prolonged contact of the must with the grape skins during fermentation. Because of the high variability in the resveratrol content of grapes coupled with the relative efficacy of different wine-making procedures in extracting resveratrol, concentrations are also extremely variable in wines (Stervbo et al. 2007). Thus, attempts to assess the effects of grape variety and/or different regional methods of wine production on the resveratrol content in red wines, unsurprisingly, has yielded ambiguous results. However, there are indications that wines produced from Pinot Noir grapes may contain slightly higher levels of *trans*-resveratrol with concentrations ranging from 0 to 15 mg/L (Stervbo et al. 2007). The same variability also affects *cis*-resveratrol and *trans*- and *cis*-piceid levels. The *cis*-resveratrol content of red wines ranges from undetectable to 5 mg/L, *trans*-piceid varies from undetectable to ~30 mg/L, and *cis*-piceid from undetectable to ~15 mg/L (Stervbo et al. 2007).

Although data on white wines are scarce, free *trans*-resveratrol contents range from zero to ~0.2 mg/L, *cis*-resveratrol is usually not present in detectable amounts while *cis*- and *trans*-piceid occur in concentrations of up to 0.1 and 0.3 mg/L, respectively (Romero-Perez et al. 1996). The total resveratrol content can, therefore, range from 0 to 60 mg/L for red wines and zero to < 1 mg/L for white wines. These values make the concept of an average resveratrol content of wines meaningless, which in turn hampers epidemiological studies aimed at determining the resveratrol intake of a given population.

8.1.2 Peanuts

Peanuts are another source of dietary resveratrol, with a content comparable to grapes. Boiled peanuts have *trans*-resveratrol concentrations of 2–8 mg/kg, whereas roasted peanuts have a reduced content of <1 mg/kg (Sobolev and Cole 1999; Sanders et al. 2000). The *trans*-resveratrol content of peanut butter is higher than that of roasted peanuts but is still typically <1 mg/kg (Romero-Perez et al. 1996). In contrast to grapes, peanuts contain much lower amounts of *trans*-piceid with respect to *trans*-resveratrol

(Ibern-Gomez et al. 2000). This finding might be related to the different matrix of the two food commodities: more hydrophilic in grapes, favouring the formation of more polar derivatives such as piceid, and more hydrophobic in peanuts, favouring the formation of the less polar-free resveratrol. As with grapes, resveratrol levels in peanuts are affected by cultivation and farming practices and environmental stresses (Rudolf and Resurreccion 2005; Potrebko and Resurreccion 2009).

8.1.3   BERRIES

Some berries also contain low concentrations of resveratrol. Fresh cranberries have been reported to have a resveratrol content of 0.5–0.8 mg/kg (Borowska et al. 2009), whereas cranberry juices contain slightly more than 1 mg/kg (Wang et al. 2002). In blueberries, high regional variations and generally very low levels have been measured, with the highest being 0.03 mg/kg (Lyons et al. 2003). *Trans*-resveratrol has not been detected in some blueberry varieties. However, more consistent amounts of resveratrol, up to 16 mg/L, were reported in blueberry juices, as well as commercial blueberry powders, which contained more than 200 mg/kg (Lyons et al. 2003). After heating, up to 50% of the original resveratrol disappears (Wang et al. 2002).

8.1.4   TOMATOES

Resveratrol has also been found in tomato skin but at highly variable concentrations depending on the tomato variety (Ragab et al. 2006). The total resveratrol in tomato skin ranged from undetectable to almost 2780 mg/kg in dried skin, which corresponds to ~7 mg/kg in fresh tomato fruit. This value is the highest reported, with the lowest being a total absence of resveratrol. Clearly, the presence of resveratrol in tomatoes deserves more investigation because of the worldwide consumption of this food commodity. *Trans*-resveratrol was found to be the dominant form in tomatoes, ranging between 83% and 100% of the total resveratrol content (Ragab et al. 2006). The prevalence of the free form compared to the glycosylated form might be explained by the hydrophobicity of the tissue where resveratrol accumulates, as tomato skins have low water content.

8.1.5   COCOA-BASED PRODUCTS

Resveratrol is also found in dark chocolate (~0.4 mg/kg *trans*-resveratrol and ~1 mg/kg *trans*-piceid) and cocoa liquor (~0.5 mg/kg *trans*-resveratrol and 1.2 mg/kg *trans*-piceid) (Counet et al. 2006). Concentrations of *trans*-resveratrol and its glucoside were recently determined in 19 top-selling commercial cocoa-containing products on the American market. The highest average *trans*-resveratrol content was found in cocoa powders (1.85 mg/kg), with the lowest in dark chocolates (0.35 mg/kg), milk chocolates (0.10 mg/kg), and chocolate syrups (0.09 mg/kg). Concentrations of *trans*-piceid were 3–5 times higher than the aglycone in all of these products. The amount of total resveratrol was found to be closely related to the amount of non-fat cocoa solids. On a per serving basis, cocoa-containing and chocolate products were estimated to rank third after red wines and grape juice, and before peanuts and peanut-derived products, as potential dietary sources of resveratrol (Hurst et al. 2008).

8.1.6 HOPS AND BEER

Hops (*Humulus lupulus*) have also been found to contain resveratrol, with the distribution of the various derivatives resembling that occurring in grapes, for example, 0.5 mg/kg of *trans*-resveratrol, 2 mg/kg of *trans*-piceid, up to 0.9 mg/kg *cis*-piceid, and an absence of *cis*-resveratrol, in Tomahawk hop pellets (Callemien et al. 2005). Once again, marked variations in *trans*-resveratrol and *trans*-piceid levels were observed among different hop varieties, with the concentration of *trans*-resveratrol ranging 0.03–2.3 mg/kg and *trans*-piceid from 0.7 to 11 mg/kg (Jerkovic and Collin 2007). Interestingly, hop pelletization was reported to induce substantial resveratrol degradation in some cultivars (Jerkovic and Collin 2008). The presence of resveratrol in hops raises a question about its eventual presence in beer. Although the literature on the subject is scarce, the resveratrol content in beer seems to be low, probably due to the boiling process used during its production which degrades the stilbene. Approximately 5 µg/L resveratrol was detected in four commercial beers (Jerkovic et al. 2008), making them of little consequence as a dietary source of the stilbene.

8.1.7 JAPANESE KNOTWEED

The Itadori plant (*Polygonum cuspidatum*), known in Western countries as Japanese knotweed, is undoubtedly the richest source of resveratrol. This plant is a noxious weed in Europe and America, whereas in Asia, aqueous infusions of the dried root are used to make a herbal tea traditionally reputed to be a remedy for many diseases—the term "Itadori" means "well being" in Japanese (Burns et al. 2002). The dried root contains massive amounts of piceid (2.3–5.3 g/kg) and resveratrol (3.0–3.8 g/kg) as well as related compounds, such as resveratrol-4′-*O*-glucoside (2.8–5.3 g/kg) and piceatannol-4′-*O*-glucoside (1.2–2.8 g/kg) (Vastano et al. 2000). In Itadori tea, The presence of > 9 mg/L of *trans*-piceid and slightly less than 1 mg/L of *trans*-resveratrol have been reported in Itadori tea (Burns et al. 2002); total resveratrol is likely to be even higher, given the presence of resveratrol-4′-*O*-glucoside and other stilbenes. Considering the easy cultivation of Japanese knotweed and its high concentrations of resveratrol derivatives, this plant is suitable material for isolating large amounts of these potentially bioactive natural compounds (Burns et al. 2002; Zhang et al. 2009).

8.1.8 DIETARY INTAKE OF RESVERATROL

In planta resveratrol is phytoalexin that provides plants with a natural protection against, harmful microbes, and attacking insects. It does, however, have a limited distribution being produced by only ~70 species most of which are not used as food sources and, therefore, are limited interest in a dietary context. Although resveratrol and related stilbenes do occur in common food commodities such as grapes, wines, cocoa, and peanuts, the amounts ingested are typically very low and extremely variable. This makes it virtually impossible to accurately estimate the actual intake of different human groups in epidemiology studies. Because of the very low intake of resveratrol from dietary products, it is improbable that the stilbene has any impact on health despite the beneficial effects reported in model animal systems (Jang et al. 1997; Bradamante et al. 2004; Baur and Sinclair 2006; Baur et al. 2006). The protective effects of red wine consumption are regularly attributed to resveratrol in both the press and scientific literature (Kaeberlein and Rabinovitch 2006). However, this is highly unlikely as the concentration of resveratrol in red wines are low and for humans to ingest the quantity of resveratrol that affords protective effects in animals they would have to drink in excess of 100 L of red wine per day (Corder et al. 2003).

8.2 BIOAVAILABILITY AND METABOLISM OF RESVERATROL

8.2.1 IN VITRO STUDIES AND ANIMAL MODELS

In general, in vitro cellular models are a poor means of unravelling the bioavailability of resveratrol and other (poly)phenolic compounds. Such models make use of cell lines that mimic intestinal epithelia, but they fail to take into account the large range of chemical reactions that (poly)phenolics undergo in the gastrointestinal tract before being absorbed.

This is true for the widely used human intestinal Caco-2 cell line, which while useful does not fully represent the physiological events involved in polyphenol absorption. The uptake of trans-piceid and its aglycone, trans-resveratrol, across the apical membrane of the Caco-2 cells have been used to evaluate rates of transport (Henry et al. 2005). The cellular uptake of trans-resveratrol was four-times greater than that of its glucoside, and only 48% of the incorporated trans-resveratrol and 32% of the incorporated trans-piceid were still present in the cells after 15 s of incubation, indicating a very rapid efflux of the two compounds. In addition, trans-resveratrol was found to cross the apical membrane of the cells via passive diffusion, whereas trans-piceid more likely crosses the membrane via active transport. However, when considering complete transepithelial transport from lumen to interstitial fluid, trans-piceid was able to cross the epithelial cell at a higher rate, and it underwent deglycosylation to trans-resveratrol (Henry-Vitrac et al. 2006). The glucuronidation and sulfation of resveratrol has been observed in transport studies with Caco-2 cells (Kaldas et al. 2003). These biotransformations occur at the epithelial level and strongly influence the entrance and efflux rates into and out of Caco-2 cells (Maier-Salamon et al. 2006), demonstrating the inadequacy of this in vitro model for studying and fully understanding the bioavailability of resveratrol.

The absorption of resveratrol across the jejunum and ileum in a rat intestinal model was investigated as a function of time in a study that also examined conjugation and metabolism of the stilbenes (Kuhnle et al. 2000). A small amount of resveratrol was detected on the serosal side of the enterocytes (0.03 nmol/cm jejunum), whereas a resveratrol-O-glucuronide, the major product (99%), was transferred across the intestinal epithelium. In isolated preparations of luminally and vascularly perfused rat small intestine the vascular uptake of luminally administered resveratrol was 20.5% (Andlauer et al. 2000). The major form of the absorbed resveratrol was the glucuronide (16.8%), which was also the main luminal metabolite (11.2%). Lower amounts of resveratrol-O-sulfate were found on the luminal and vascular side, 3.0% and 0.3%, respectively, whereas only minute amounts of resveratrol and resveratrol glycosides (1.9%) were found in the intestinal tissue.

Key studies on the bioavailability of resveratrol in animal models began with the research of Bertelli et al. (1996) who administered red wine to rats in both acute and chronic 15 day feeds. Resveratrol concentrations were measured in the plasma, heart, liver, and kidneys. Tissue concentrations showed significant cardiac bioavailability and strong affinity for the liver and kidneys. However, more recent experiments took into account the presence of resveratrol conjugates and, occasionally, the appearance of microbial metabolites. Wenzel et al. (2005) reported the formation of resveratrol-3-*O*-sulfate, resveratrol-4′-*O*-sulfate, resveratrol-3,5-*O*-disulfate, resveratrol-3,4′-*O*-disulfate, resveratrol-3,4′,5-*O*-trisulfate, resveratrol-3-*O*-glucuronide, after the ingestion of *trans*-resveratrol by Wistar rats in two feeding experiments with a duration of 8 weeks. Total resveratrol recovery from the urine and feces of rats fed 50 mg resveratrol per kilogram per day was 15% and 13%, respectively. For rats fed a higher dosage of 300 mg resveratrol per kilogram per day, recovery was 54% and 17%, respectively.

HPLC-MS^2 methodology has been used to elucidate the structures of the major metabolites of resveratrol in the urine of Sprague–Dawley rats after oral administration of 20 mg/kg resveratrol. The main metabolites of were a resveratrol-*O*-glucuronide, resveratrol-*O*-sulfate, dihydroresveratrol, and dihydroresveratrol-*O*-sulfate. Dihydroresveratrol is believed to be a product of degradation of resveratrol by the colonic microflora (Wang et al. 2005; Juan et al. 2010).

Meng et al. (2004) described the urinary and plasma levels of resveratrol and its metabolites after ingestion of grape juice and *trans*-resveratrol. Administration of grape juice for 4 days to female CF-1 mice resulted in urinary excretion of resveratrol during the study period equal to ~1–2% of the ingested dose. After ingestion of either 2 or 5 mg/kg *trans*-resveratrol by female Wistar rats, resveratrol was present in plasma mainly as conjugates at 0.5, 1.5, and 4 h with the aglycone representing ~10% of total resveratrol. Recent feeding studies with rats detected dihydroresveratrol, which, on one occasion, was the most abundant resveratrol metabolite in the colon (Alfaras et al. 2010; Juan et al. 2010).

Resveratrol absorption and catabolism studies have also been carried out with pigs whose anatomy and physiology of the gastrointestinal system is similar to that of humans (Guillotteau et al. 2010). Metabolites identified in plasma include a resveratrol-*O*-diglucuronide, two isomers of resveratrol-*O*-sulfoglucuronide, two isomers of resveratrol-*O*-glucuronide, and a resveratrol-*O*-sulfate (Azorín-Ortuño et al. 2011).

8.2.2 Human Studies

Several human studies have investigated *trans*-resveratrol bioavailability under varying experimental conditions. Differences include the source of resveratrol (pure compound or dietary), dosage (physiological or pharmacological), duration of supplementation (single or multiple ingestions), type of subjects, and whether supplementation occurred in fasting subjects or following a meal. Reviews on the topic have also been published by Wenzel and Somoza (2005), Baur and Sinclair (2006), and Cottart et al. (2010).

Table 8.1 summarizes some of the most recent studies on *trans*-resveratrol bioavailability in humans. All intervention studies in humans have shown high absorption and a rapid metabolism and excretion of resveratrol in urine and faeces. Gastric

TABLE 8.1

Summary of Studies on Resveratrol Bioavailability in Humans[a]

Subjects (n)	Resveratrol Source, Dose, and Experiment Conditions	t-Res and Metabolites (Mean Concentrations, C_{max}, % Dose Ingested)			Major Findings	Reference
		Plasma	Urine	Feces/Tissue		
25 healthy	Red wine (res dose unknown) Fasting Standard meal Fat/Lean meal	C_{max}: **t-Res**: 0–0.03 µM **t-Res-4'-glcUA**: 0–3.9 µM **t-Res-3-glcUA**: 0–2.7 µM	—	—	**t-Res** in red wines is absorbed with no influence of meal or its fat content	Vitaglione et al. (2005)
40 healthy	Res in capsules (2.2 to 4.4 mmol)	C_{max}: **Res**: 0.04–0.23 µM **Res-glcUA**: 0.56–2.90 µM **Res-S**: 0.75–4.78 µM	*% of dose ingested:* **Res**: < 0.04% **Res-glcUA**: 11.0–3.8% **Res-S**: 11.4–4.9%	Feces **Res**: 0–0.10 µmol/g dry weight **Res met**: < 1% **Res**	C_{max}: **Res-glcUA/Res-S** 3–8 fold than **Res** AUC: **Res-metabolites** 23–fold higher than **Res** Urinary excretion within 4 h post ingestion	Bocoock et al. (2007a)
24 healthy	Pure **t-Res** – 1.8 mmol 1) fasting 2) high fat meal	**t-Res** (C_{max} and t_{max}) 1) 0.21 µM and 0.5 h 2) 0.19 µM and 2.0 h	—	—	Absorption of *t*-Res delayed by the presence of food	Vaz-da-Silva et al. (2008)
40 healthy	Pure **t-Res** in caplets (6 per d for 2 days + 1) 0.1–0.7 mmol/day	**t-Res** (C_{max}) 0.02–0.28 µM	—	—	**t-Res** half life: 1–3 h (single dose) and 2–5 h (repeated dose) Bioavailability greater after morning administration (circadian variation)	Almeida et al. (2009)

Subjects	Treatment	Parameters		Tissue	Effects	Reference
8 healthy	Pure t-Res in caplets 8.8 mmol + SB or HFB 8.8 mmol + SB + 0.5 g Q 8.8 mmol + SB + 0.5 g Q + 5 mL ethanol	t-Res C_{max}: 3.0–5.6 μmol/L t_{max}: 3–5 h AUC (0-12 h): 8.6–17.7 μmol/h/L	—	—	HFB decreased AUC and C_{max} by 45–46% compared to SB. Quercetin did not influence bioavailability of Res	La Porte et al. (2010)
20 Colorectal cancer patients	Pure t-Res in capsules (1 per d x 8 days) 2.2–4.4 mmol/day	C_{max}: Res: not detected Res-GlcUA: 0.04–0.5 Res-S: 0.4–1.2 Res-S-GlcUA: 13–22	—	Colon tissue (nmol/g) Res: Normal: 2.7–338 Tumor: 4.5–48 Res-glcUA: Normal: 0.2–23 Tumor: 0.2–0.4 Res-S: Normal: 0.9–18 Tumor: 1.0–3.4 Res-S-GlcUA: Normal: 0.0–23 Tumor: 1.0–25	Higher concentrations of all metabolites are found in normal tissue proximal to tumor. Tumor cell proliferation is reduced 5% by intake of resveratrol	Patel et al. (2010)

[a] t-Res: *trans*-resveratrol; Res-glcUA: resveratrol-*O*-glucuronide; Res-S: resveratrol-*O*-sulfate, resveratrol-*O*-sulfate; Res-S-glcUA: resveratrol-*O*-sulfate-*O*-glucuronide.

absorption explains the peak of free resveratrol in the plasma 30 min after ingestion (Vitaglione et al. 2005; Bocoock et al. 2007a; Vaz-da-Silva et al. 2008; Almeida et al. 2009). However, the peak plasma concentration (C_{max}) of free resveratrol is typically extremely low: < 37 nmol/L after oral ingestion at physiological doses ranging from 0.25 to 25 mg (0.0011–0.11 mmol) (Goldberg et al. 2003; Walle et al. 2004; Vitaglione et al. 2005). In acute feeds with doses of 0.1–5.0 g (0.44–22 mmol), an increasing *trans*-resveratrol dosage increased the C_{max} up to ~280 nmol/L (Bocoock et al. 2007a; Vaz-da-Silva et al. 2008; Almeida et al. 2009). The general consensus is that resveratrol-*O*-glucuronides and sulfates are the major plasma and urine metabolites, with the sulfates being predominant (Boocock et al. 2007a, 2007b; Urpi-Sarda et al. 2007; Burkon and Somoza 2008).

Resveratrol-3-*O*-sulfate and resveratrol-3-*O*-sulfate, resveratrol-4′-*O*-sulfate a resveratrol-*O*-disulfate, the 3-*O* or 4′-*O*-glucuronides, and one *O*-glucuronide-*O*-sulfate were found in the plasma of both healthy humans and colorectal cancer patients ingesting 0.5 and/or 1 g (2.2 and 4.4 mmol) of *trans*-resveratrol (Boocock et al. 2007b; Patel et al. 2010). In addition to these metabolites, two novel *trans*-resveratrol-*C/O*-conjugated diglucuronides were detected in plasma and urine of subjects who ingested averagely 84 mg of *trans*-piceid (corresponding to 50 mg of *trans*-resveratrol) (Burkon and Somoza 2008). Interestingly, Burkon and Somoza also showed for the first time that up to 50% of the plasma *trans*-resveratrol-3-*O*-sulfate, *trans*-resveratrol-*O*-disulfates, and the novel diglucuronides were bound to lipoproteins by noncovalent hydrogen bonding.

8.2.3 FACTORS INFLUENCING TRANS-RESVERATROL METABOLISM *IN VIVO*

The dose of resveratrol ingested influences the ratio of sulfate and glucuronide metabolites in urine. At resveratrol intakes of 50–5000 mg, sulfate derivatives are the quantitatively dominating metabolites in human plasma and urine (Bocoock et al. 2007a; Urpi-Sarda et al. 2007; Burkon and Somoza 2008; Patel et al. 2010). Increasing the dosage up to 300 mg/kg b.w. in feed with rats resulted in a major shift from sulfation to glucuronidation (Wenzel et al. 2005). This observation may be explained by a major affinity of resveratrol for sulfotransferases compared to UDP-glucuronosyltransferase (UGT), which may change in the presence of a high amount of substrate. This hypothesis has not yet been evaluated in humans.

Bocoock et al. (2007a) reported that plasma resveratrol sulfate and glucuronide levels were below the limit of quantification in healthy subjects following the administration of 0.5–5 g *trans*-resveratrol, but the same molecule was the major plasma metabolite in colorectal cancer patients (Patel et al. 2010). Physiopathological modifications influencing the metabolism of *trans*-resveratrol cannot be ruled out.

The possibility that the coingestion of resveratrol with quercetin might influence catabolism of the stilbenes was highlighted in an *in vitro* study in which quercetin inhibited resveratrol sulfation in both the liver and duodenum (De Santi et al. 2000). There is, however, no evidence to suggest that this occurs *in vivo* as a twice daily intake of 500 mg of quercetin did not influence *trans*-resveratrol pharmacokinetics in eight healthy subjects (La Porte et al. 2010). It must be pointed out that the applied analytical method involved enzyme hydrolysis of samples and so was not capable of distinguishing between different resveratrol conjugates. Thus, it was not possible to rule out a change in the ratio of different resveratrol metabolites when the stilbene was provided with or without quercetin. In the same study, coadministration of quercetin and resveratrol with a small amount of alcohol (5%) did not impact on absorption of the stilbene.

In a two-way crossover study, 24 healthy subjects were administered with a single, 400 mg dose of *trans*-resveratrol in conjunction with either a standard high fat-meal or an 8-h fast (Vaz-da-Silva et al. 2008). Despite large interindividual variability in the *trans*-resveratrol pharmacokinetic parameters, the rate of absorption of *trans*-resveratrol was significantly delayed by the presence of food. However, the extent of absorption, as reflected by the area under the plasma concentration–time curve (AUC), was unaffected. In the previously mentioned study by La Porte et al. (2010), 2 g of *trans*-resveratrol was administered twice daily along with either a standard or a high-fat breakfast. The high-fat breakfast significantly decreased both the plasma AUC and C_{max} values by ~45%, when compared with the standard breakfast.

These observations contrast with the findings of a study in which *trans*-resveratrol bioavailability was investigated using volunteers who consumed a moderate amount of red wine in association with an absence of food, a standard meal, and meals with either a high- or low-lipid content. In this case, the bioavailability of *trans*-resveratrol in red wine was not influenced by either the meal or by the kind and/or the quantity of lipids in the meal (Vitaglione et al. 2005). However, this may reflect a dose effect as the resveratrol intake was 0.48 mg, compared to the respective 400 mg and 2 g ingested by volunteers in the Vaz-da-Silva et al. (2008) and La Porte et al. (2010) studies.

8.2.4 SUMMARY OF RESVERATROL ABSORPTION AND METABOLISM

Figure 8.2 summarizes the metabolic fate of *trans*-resveratrol after oral ingestion in humans based on a combination of the available literature data on resveratrol bioavailability and *in vivo* metabolism. Resveratrol can be partially absorbed in the stomach, thereby rapidly reaching the bloodstream and entering the liver via the portal vein. At the enterocyte and hepatocyte level, *trans*-resveratrol is rapidly conjugated with glucuronic acid and sulfate due to the actions of UGT and sulfotransferases, respectively (Henry-Vitrac et al. 2006). After being produced in the liver, resveratrol-*O*-sulfates and resveratrol-*O*-glucuronides can be recirculated to the small intestine through the bile (enterohepatic circulation), where they can be further sulfated and/or glucuronidated, resulting in the presence of disulfates, diglucuronides, and sulfate-glucuronide metabolites in the bloodstream 12 h post-ingestion. In addition, resveratrol-*O*-sulfates and resveratrol-*O*-glucuronides can reach the large intestine, where hydrolysis by the microbiota results in the release of *trans*-resveratrol and the formation of dihydroresveratrol (Jung et al. 2009). Along with *trans*-resveratrol, dihydroresveratrol can be absorbed into the circulatory system, returned to the liver where it is glucuronidated and sulfated (Walle et al. 2004). Recent animal studies have supported this hypothesis with dihydroresveratrol and its metabolites being found in the colon, plasma, and urine of rats after the ingestion of either dihydroresveratrol or resveratrol (Alfaras et al. 2010; Juan et al. 2010). Because the colonic microflora is able to hydrolyze

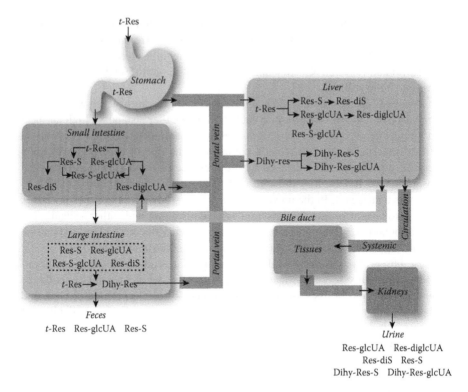

FIGURE 8.2 Human metabolism of resveratrol. *t*-Res: *trans*-resveratrol; Res-S: resveratrol-*O*-sulfate; Res-glcUA: resveratrol-*O*-glucuronide; Res-S-glcUA: resveratrol-*O*-sulfate-*O*-glucuronide; Res-diglcUA: resveratrol-*O*-diglucuronide; Res-diS: resveratrol-*O*-disulfate; Dihy-res: dihydroresveratrol; Dihy-res-S: dihydroresveratrol-*O*-sulfate; Dihy-res-glcUA: dihydroresveratrol-*O*-glucuronide.

resveratrol conjugates, free resveratrol and, to a lesser extent, resveratrol metabolites are excreted in feces with high interindividual variability (Boocock et al. 2007a).

Regarding tissue accumulation, resveratrol and its metabolites have been shown in animal studies to accumulate in the liver (Vitrac et al. 2003; Sale et al. 2004; Vitaglione et al. 2009). The only human study evaluating stilbene sequestration was performed with colorectal cancer patients (Patel et al. 2010). In this study *trans*-resveratrol (0.5 or 1 g) was administered to 20 patients with histologically confirmed resectable colorectal cancer for 8 days before surgical intervention. Blood and normal and malignant biopsy tissue samples were obtained pre-dosing at diagnostic endoscopy and post-dosing at resection surgery. Cell proliferation, as reflected by Ki-67 staining, was compared in pre- and post-intervention tissue samples. Resveratrol and its metabolite resveratrol-3-*O*-glucuronide were recovered from tissues at maximal mean concentrations of 674 and 86.0 nmol/g, respectively. Their levels were consistently higher in tissues originating in the right side of the colon compared with the left. As noted by Patel et al. (2010), this could be explained by fecal transit involving the cecum and then the right side across the transverse to the sigmoid colon and rectum on the left. Resveratrol tissue concentrations were credibly related to those in the feces, with higher concentrations in right-sided tissues as they come into contact with fecal resveratrol earlier than those on the left side. Finally, consumption of resveratrol reduced tumour cell proliferation by 5% ($p = 0.05$), and although the biological importance of such a minor decrease is doubtful, this observation implies that resveratrol can exert pharmacological effects in the human large intestine.

In conclusion:

- Resveratrol is readily absorbed and rapidly metabolized and excreted in urine and feces.
- The major plasma and urine metabolites of resveratrol are glucuronidated and sulfated derivatives, with the latter being most prevalent, and both are found bound primarily to LDL proteins by noncovalent hydrogen interactions in the plasma.
- Quercetin (up to 500 mg) and alcohol do not influence resveratrol absorption and pharmacokinetics up to an oral intake of 2 g.
- A high-fat meal does not influence the absorption of resveratrol from a moderate intake of red wine (up to 0.48 mg *trans*-resveratrol) but does delay absorption when the dose is 400 mg and modifies plasma pharmacokinetics, reducing resveratrol AUC by 45%, at very high (2 g) doses.
- High systemic levels of resveratrol conjugates and the recovery of high concentrations of resveratrol in colorectal cancer tissue concomitant with reduced cell proliferation suggest a benefit of oral administration of pharmacological doses of resveratrol.

REFERENCES

Adrian, M., Jeandet, P., Douillet-Breuil, A.C. et al. (2000). Stilbene content of mature *Vitis vinifera* berries in response to UV-C elicitation. *J. Agric. Food Chem.*, **48**, 6103–6105.

Alfaras, I., Juan, M.E., and Planas, J.M. (2010). *Trans*-resveratrol reduces precancerous colonic lesions in dimethylhydrazine-treated rats. *J. Agric. Food Chem.*, **58**, 8104–8110.

Almeida, L., Vaz-da-Silva, M., Falcão, A. et al. (2009). Pharmacokinetic and safety profile of *trans*-resveratrol in a rising multiple-dose study in healthy volunteers. *Mol. Nutr. Food Res.*, **53** (Suppl 1), S7–S15.

Andlauer, W., Kolb, J., Siebert, K. et al. (2000). Assessment of resveratrol bioavailability in the perfused small intestine of the rat. *Drugs Exp. Clin. Res.*, **26**, 47–55.

Azorín-Ortuño, M., Yañéz-Gascón, M.J., Pallarés, F.J. et al. (2011). Pharmacokinetic study of trans-resveratrol in adult pigs. *J. Agric. Food Chem.*, **58**, 11165–11171.

Baur, J.A., Pearson, K.J., Price, N.L. et al. (2006). Resveratrol improves health and survival of mice on a high-calorie diet. *Nature*, **444**, 337–342.

Baur, J.A. and Sinclair, D.A. (2006). Therapeutic potential of resveratrol: the *in vivo* evidence. *Nature Rev. Drug Discov.*, **5**, 493–506.

Bertelli, A.A., Giovannini, L., Stradi, R. et al. (1996). Kinetics of *trans*- and *cis*-resveratrol (3,4′,5-trihydroxystilbene) after red wine oral administration in rats. *Int. J. Clin. Pharmacol. Res.*, **16**, 77–81.

Boocock, D.J., Faust, G.E., Patel, K.R. et al. (2007a). Phase I dose escalation pharmacokinetic study in healthy volunteers of resveratrol, a potential cancer chemopreventive agent. *Cancer Epidemiol. Biomark. Prev.,* **16**, 1246–1252.

Boocock, D.J., Patel, K.R., Faust, G.E. et al. (2007b). Quantitation of *trans*-resveratrol and detection of its metabolites in human plasma and urine by high performance liquid chromatography. *J. Chromatogr. B,* **848**, 182–187.

Borowska, E.J., Mazur, B., Kopciuch, R.G. et al. (2009). Polyphenol, anthocyanin and resveratrol mass fractions and antioxidant properties of cranberry cultivars. *Food Technol. Biotechnol.,* **47**, 56–61.

Bradamante, S., Barenghi, L., and Villa, A. (2004). Cardiovascular protective effects of resveratrol. *Cardiovasc. Drug. Rev.,* **22**, 169–188.

Burkon, A. and Somoza, V. (2008). Quantification of free and proteinbound trans-resveratrol metabolites and identification of *trans*-resveratrol-*C/O*-conjugated diglucuronides— two novel resveratrol metabolites in human plasma. *Mol. Nutr. Food Res.,* **52**, 549–557.

Burns, J., Yokota, T., Ashihara, H. et al. (2002). Plant foods and herbal sources of resveratrol. *J. Agric. Food Chem.,* **50**, 3337–3340.

Callemien, D., Jerkovic, V., Rozenberg, R. et al. (2005). Hop as an interesting source of resveratrol for brewers: Optimization of the extraction and quantitative study by liquid chromatography/atmospheric pressure chemical ionization tandem mass spectrometry. *J. Agric. Food Chem.,* **53**, 424–429.

Chong, J., Poutaraud, A., and Hugueney, P. (2009). Metabolism and role of stilbenes in plants. *Plant Sci.,* **17**,143–155.

Corder, R., Crozier, A., and Kroon, P.A. (2003). Drinking your health? It's too early to say. *Nature,* **426**, 119.

Cottart, C.H., Nivet-Antoine, V., Laguillier-Morizot, C. et al. (2010). Resveratrol bioavailability and toxicity in humans. *Mol. Nutr. Food Res.,* **54**, 7–16.

Counet, C., Callemien, D., and Collin, S. (2006). Chocolate and cocoa: New sources of *trans*-resveratrol and *trans*-piceid. *Food Chemistry* 98: 649–657.

Creasy, L.L. and Cofee, M. (1988). Phytoalexin production potential of grape berries. *J. Am. Soc. Hort. Sci.,* **113**, 230–234.

De Santi, C., Pietrabissa, A., Spisni, R. et al. (2000). Sulfation of resveratrol, a natural product present in grapes and wine, in the human liver and duodenum. *Xenobiotica,* **30**, 609–617.

Goldberg, D.M., Yan, J., and Soleas, G.J. (2003). Absorption of three wine-related polyphenols in three different matrices by healthy subjects. *Clin. Biochem.,* **36**, 79–87.

Guilloteau, P., Zabielski, R., Hammon, H.M. et al. (2010). Nutritional programming of gastrointestinal tract development. Is the pig a good model for human? *Nutr. Res. Rev.,* **23**, 4–22.

Henry, C., Vitrac, X., Decendit, A. et al. (2005). Cellular uptake and efflux of *trans*-piceid and its aglycone *trans*-resveratrol on the apical membrane of human intestinal Caco-2 cells. *J. Agric. Food Chem.,* **53**, 798–803.

Henry-Vitrac, C., Desmoulière, A., Girard, D. et al. (2006). Transport, deglycosylation, and metabolism of *trans*-piceid by small intestinal epithelial cells. *Eur. J. Nutr.,* **45**, 376–382.

Hurst, W.J., Glinski, J.A., Miller, K.B. et al. (2008). Survey of the *trans*-resveratrol and *trans*-piceid content of cocoa-containing and chocolate products. *J. Agric. Food Chem.,* **56**, 8374–8378.

Ibern-Gomez, M., Roig-Perez, S., Lamuela-Raventos, R.M. et al. (2000). Resveratrol and piceid levels in natural and blended peanut butters. *J. Agric. Food Chem.,* **48**, 6352–6354.

Jang, M., Cai, L., Udeani, G.O. et al. (1997). Cancer chemopreventive activity of resveratrol: a natural product derived from grapes. *Science,* **275**, 218–220.

Jerkovic, V. and Collin, S. (2007). Occurrence of resveratrol and piceid in American and European hop cones. *J. Agric. Food Chem.,* **55**, 8754–8758.

Jerkovic, V. and Collin, S. (2008). Fate of resveratrol and piceid through different hop processings and storage times. *J. Agric. Food Chem.*, **56**, 584–590.

Jerkovic, V., Nguyen, F., Timmermans, A. et al. (2008). Comparison of procedures for resveratrol analysis in beer: Assessment of stilbenoids stability through wort fermentation and beer aging. *J. Inst. Brew.*, **114**, 143–149.

Juan, M.E., Alfaras, I., and Planas, J.M. (2010). Determination of dihydroresveratrol in rat plasma by HPLC. *J. Agric. Food Chem.*, **58**, 7472–7475.

Jung, C.M., Heinze, T.M., Schnackenberg, L.K. et al. (2009). Interaction of dietary resveratrol with animal–associated bacteria. *FEMS Microbiol. Lett.*, **297**, 266–273.

Kaeberlein, M. and Rabinovitch, P.S. (2006). Grapes versus gluttony. Nature, **444**, 280–281.

Kaldas, M.I., Walle, U.K., and Walle, T. (2003). Resveratrol transport and metabolism by human intestinal Caco-2 cells. *J. Pharm. Pharmacol.*, **55**, 307–312.

Kuhnle, G., Spencer, J.P., Chowrimootoo, G. et al. (2000). Resveratrol is absorbed in the small intestine as resveratrol glucuronide. *Biochem. Biophys. Res. Commun.*, **272**, 212–217.

La Porte, C., Voduc, N., Zhang, G. et al. (2010). Steady-state pharmacokinetics and tolerability of *trans*-resveratrol 2000 mg twice daily with food, quercetin and alcohol (ethanol) in healthy human subjects. *Clin. Pharm.*, **49**, 449–454.

Langcake, P. and Pryce, R.J. (1977). The production of resveratrol and viniferins by grapevines in resposne to ultraviolet radiation. *Phytochemistry*, **16**, 1193–1196.

Lyons, M.M., Yu, C.W., Toma, R.B. et al. (2003). Resveratrol in raw and baked blueberries and bilberries. *J. Agric. Food Chem.*, **51**, 5867–5870.

Maier-Salamon, A., Hagenauer, B., Wirth, M. et al. (2006). Increased transport of resveratrol across monolayers of the human intestinal Caco-2 cells is mediated by inhibition and saturation of metabolites. *Pharmaceut. Res.*, **23**, 2107–2115.

Meng, X., Maliakal, P., Lu, H. et al. (2004). Urinary and plasma levels of resveratrol and quercetin in humans, mice, and rats after ingestion of pure compounds and grape juice. *J. Agric. Food Chem.*, **52**, 935–942.

Patel, K.R., Brown, V.A., Jones, D.J. et al. (2010). Clinical pharmacology of resveratrol and Its metabolites in colorectal cancer patients. *Cancer Res.*, **70**, 7392–739.

Pervaiz, S. (2003). Resveratrol: from grapevines to mammalian biology. *FASEB J.*, **17**, 1975–1985.

Pezet, R., Perret, C., Jean-Denis, J.B. et al. (2003). δ-Viniferin, a resveratrol dihydrodimer: one of the major stilbenes synthesized by stressed grapevine leaves. *J. Agric. Food Chem.*, **51**, 5488–5492.

Potrebko, I. and Resurreccion, A.V.A. (2009). Effect of ultraviolet doses in combined ultraviolet-ultrasound treatments on *trans*-resveratrol and *trans*-piceid Contents in sliced peanut kernels. *J. Agric. Food Chem.*, **57**, 7750–7756.

Ragab, A.S., Van Fleet, J., Jankowski, B. et al. (2006). Detection and quantitation of resveratrol in tomato fruit (*Lycopersicon esculentum* Mill.). *J. Agric. Food Chem.*, **54**, 7175–7179.

Romero-Perez, A.I., Ibern-Gomez, M., Lamuela-Raventos, R.M. et al. (1999). Piceid, the major resveratrol derivative in grape juices. *J. Agric. Food Chem.*, **47**, 1533–1536.

Romero-Perez, A.I., Lamuela-Raventos, R.M., Buxaderas, S. et al. (1996). Resveratrol and piceid as varietal markers of white wines. *J. Agric. Food Chem.*, **44**, 1975–1978.

Rudolf, J.R. and Resurreccion, A.V.A. (2005). Elicitation of resveratrol in peanut kernels by application of abiotic stresses. *J. Agric. Food Chem.*, **53**, 10186–10192.

Sale, S., Verschoyle, R.D., Boocock, D. et al. (2004). Pharmacokinetics in mice and growth-inhibitory properties of the putative cancer chemopreventive agent resveratrol and the synthetic analogue trans-3,4,5,49-tetramethoxystilbene. *Brit. J. Can.*, **90**, 736–744.

Sanders, T.H., McMichael, R.W., and Hendrix, K.W. (2000). Occurrence of resveratrol in edible peanuts. *J. Agric. Food Chem.*, **48**, 1243–1246.

Sato, M., Suzuki, Y., Okuda, T. et al. (1997). Contents of resveratrol, piceid, and their isomers in commercially available wines made from grapes cultivated in Japan. *Biosci. Biotech. Biochem.*, **61**, 1800–1805.

Shen, T., Wang, X.N., and Lou, H.X. (2009). Natural stilbenes: an overview. *Nat. Prod. Rep.*, **26**, 916–935.

Sobolev, V.S. and Cole, R.J. (1999). *Trans*-resveratrol content in commercial peanuts and peanut products. *J. Agric. Food Chem.*, **47**, 1435–1439.

Stervbo, U., Vang, O., and Bonnesen, C. (2007). A review of the content of the putative chemo-preventive phytoalexin resveratrol in red wine. *Food Chem.*, **101**, 449–457.

Tinttunen, S. and Lehtonen, P. (2001). Distinguishing organic wines from normal wines on the basis of concentrations of phenolic compounds and spectral data. *Eur. Food Res. Technol.*, **212**, 39–394.

Urpi-Sarda, M., Zamora-Ros, R., Lamuela-Raventos, R. et al. (2007). HPLC-tandem mass spectrometric method to characterize resveratrol metabolism in humans. *Clin. Chem.*, **53**, 292–299.

Vastano, B.C., Chen, Y., Zhu, N.Q. et al. (2000). Isolation and identification of stilbenes in two varieties of *Polygonum cuspidatum*. *J. Agric. Food Chem.*, **48**, 253–256.

Vaz-da-Silva, M., Loureiro, A.I., Falcao, A. et al. (2008). Effect of food on the pharmacokinetic profile of *trans*-resveratrol. *Int. J. Clin. Pharmacol. Ther.*, **46**, 564–570.

Vitaglione, P., Ottanelli, B., Milani, S. et al. (2009). Dietary *trans*-resveratrol bioavailability and effect on CCl4-induced liver lipid peroxidation. *J. Gastroent., Hepatol.*, **24**, 618–622.

Vitaglione, P., Sforza, S., Galaverna, G. et al. (2005). Bioavailability of *trans*-resveratrol from red wine in humans. *Mol. Nutr. Food Res.*, **49**, 495–504.

Vitrac, X., Bornet, A,. Vanderlinde, R. et al. (2005). Determination of stilbenes (δ-viniferin, *trans*-astringin, *trans*-piceid, *cis*- and *trans*-resveratrol, ε-viniferin) in Brazilian wines. *J. Agric. Food Chem.* **53**, 5664–5669.

Vitrac, X., Desmoulière, A., Brouillaud, B. et al. (2003). Distribution of [^{14}C]*trans*-resveratrol, a cancer chemopreventive polyphenol, in mouse tissues after oral administration. *Life Sci.*, **72**, 2219–2233.

Waffo-Téguo, P., Lee, D. et al. (2001). Two new stilbene dimer glucosides in grape (*Vitis vinifera*) cell cultures. *J. Nat. Prod.*, **64**, 136–138.

Walle, T., Hsieh, F., DeLegge, M.H. et al. (2004). High absorption but very low bioavailability of oral resveratrol in humans. *Drug Metab. Dispos.*, **32**, 1377–1382.

Wang, Y., Catana, F., Yang, Y.N. et al. (2002). An LC-MS method for analyzing total resveratrol in grape juice, cranberry juice, and in wine. *J. Agric. Food Chem.*, **50**, 431–435.

Wang, D.G., Hang, T.J., Wu, C.Y. et al. (2005). Identification of the major metabolites of resveratrol in rat urine by HPLC-MS/MS. *J. Chromatogr. B*, **829**, 97–106.

Wenzel, E. and Somoza, V. (2005). Metabolism and bioavailability of *trans*-resveratrol. *Mol. Nutr. Food Res.*, **49**, 472–481.

Wenzel, E., Soldo, T., Erbersdobler, H. et al. (2005). Bioactivity and metabolism of *trans*-resveratrol orally administered to Wistar rats. *Mol. Nutr. Food Res.*, **49**, 482–494.

Zhang, D.L., Li, X.A., Hao, D.X. et al. (2009). Systematic purification of polydatin, resveratrol and anthraglycoside B from *Polygonum cuspidatum* Sieb. et Zucc. *Sep. Purific. Technol.*, **66**, 329–339.

9 Bioavailability and Metabolism of Ellagic Acid and Ellagitannins

Mar Larrosa, María T. García-Conesa, Juan C. Espín, and Francisco A. Tomás-Barberán

CONTENTS

9.1 INTRODUCTION

Ellagitannins are hydrolyzable tannins that release ellagic acid upon hydrolysis. They exhibit various biological activities *in vitro* that have been associated with pharmacological (ellagitannin-containing medicinal plants) and nutritional (ellagitannin-containing foods) effects *in vivo*. The potential health effects are mainly related to the prevention of cardiovascular diseases and cancer. *In vivo* biological effects may be due partially to the potent free-radical scavenging activity that these compounds exert *in vitro*. It is, however, necessary to take into account the fate of ellagitannins in the gastrointestinal tract, their bioaccessibility, bioavailability, metabolism, and tissue distribution of the corresponding metabolites to understand the efficacy and the physiological role of dietary and medicinal ellagitannins. In the present chapter, we review the current knowledge regarding the bioavailability and metabolism of ellagitannins and point out various unresolved issues within these processes in humans that require further research.

9.2 RELEVANCE OF ELLAGITANNINS IN THE DIET

Ellagitannins are present in significant amounts in many berries, including strawberries, raspberries (Zafrilla et al. 2001), blackberries, as well as nuts, such as walnuts (Fukuda et al. 2003), pistachios, cashew nuts, chestnuts, oak acorns (Cantos et al. 2003), and pecans (Villarreal-Lozoya et al. 2007). They are also abundant in pomegranates (Gil et al. 2000) and muscadine grapes (Lee and Talcot 2002) and are important constituents of wood, particularly oak (Glabasnia and Hofmann 2006). Ellagitannins can be incorporated into several food products such as wines and whisky, through migration from oak barrels into the liquid matrix during ageing processes. Ellagic acid, a structural unit of ellagitannins, has been found in several types of honey, and it has been proposed as a floral marker for honey produced from heather (Ferreres et al. 1996). Free ellagic acid and glycoside derivatives, including glucosides, rhamnosides, arabinosides, and the corresponding acetyl esters conjugates, are also present in these food products (Zafrilla et al. 2001).

In an earlier review, it was documented that reliable figures on the ellagitannin dietary burden were not available, but that it would probably not exceed 5 mg/day (Clifford and Scalbert 2000). Since then a number of studies have shown that the ellagitannin content of several food products can be quite high. A glass of pomegranate juice and a 100 g serving of raspberries can each provide as much as 300 mg of ellagitannins, a strawberry serving 70 mg, and four walnuts some 400 mg. As a result, the intake of dietary ellagitannins is almost certainly much higher than previously estimated, especially if some of the ellagitannin-rich foods (Table 9.1) are consumed on a regular basis.

Representative dietary ellagitannins are shown in Figure 9.1. Punicalagin is typical of pomegranate, vescalagin of oak-aged wines and spirits, sanguiin H-6 of strawberry and raspberry, while pedunculagin occurs in walnuts. All of them release ellagic acid upon hydrolysis although other metabolites can also be produced and are distinctive of individual ellagitannins (i.e., gallagic and ter-gallagic acids).

9.3 MODELS TO STUDY BIOAVAILABILITY AND METABOLISM

Several models have been applied to the study of ellagitannin bioavailability and metabolism.

- *In vitro* investigations including the evaluation of metabolism using cell lines and bacteria. These studies can give evidence of the biochemical changes undergone by the ellagitannins in the cell culture or bacteria media and also follows their metabolic fate once they enter cells and are conjugated into the specific metabolites.

TABLE 9.1
Ellagic Acid Content in Various Food Products[a]

Foods	Content	Reference
Fresh Fruits		
Arctic bramble	69–320 mg/100 g f.w.	Törrönen (2009)
Blackberry	1.5–2.0 mg/g d.w.	Clifford and Scalbert (2000)
Blackberry	1090 mg/100 g f.w.	Bakkalbasi et al. (2009)
Cloudberry	315 mg/100 g f.w.	Koponen et al. (2007)
Cloudberry	56–360 mg/100 g f.w.	Törrönen (2009)
Grapes (Muscadine)	36–91 mg/100 g f.w.	Törrönen (2009)
Pomegranates	35–75 mg/100 g f.w. (arils)	Gil et al. (2000)
Raspberry	263–330 mg/100 g f.w.	Koponen et al. (2007)
Raspberry	51–330 mg/100 g f.w.	Törrönen (2009)
Strawberry	77–85 mg/100 g f.w.	Koponen et al. (2007)
Strawberry	25 mg/100 g f.w.	Aaby et al. (2007)
Nuts		
Chestnut	1.6–24.9 mg/kg d.w.	Gonçalves et al. (2010)
Pecan	21–86 mg/g	Malik et al. (2009)
Walnut	802 mg/50g (8 nuts)	Anderson et al. (2001)
Processed Fruits		
Muscadine grape juice	8–84 mg/L	Lee and Talcot (2002)
Pomegranate juice (cv. Wonderful)	1500–1900 mg/L punicalagin	Gil et al. (2000)
Pomegranate juice (cv. Wonderful)	2020–2660 mg/L (ellagitannins and ellagic acid)	Gil et al. (2000)
Pomegranate juice (cv. Mollar)	5700 mg/L (ellagitannins and ellagic acid)	Cerdá et al. (2006)
Raspberry jam	76 mg/100 g f.w.	Koponen et al. (2007)
Strawberry jam	24 mg/100 g f.w.	Koponen et al. (2007)
Wines		
Muscadine grape wine	2–65 mg/L	Lee and Talcot (2002)
Oak aged red wine	9.4 mg/L	Glabasnia and Hofmann (2006)
Oak aged red wine	50 mg/L	Clifford and Scalbert (2000)
Spirits		
Cognac	31–55 mg/L	Clifford and Scalbert (2000)
Whisky (sour mash)	23.8 mg/L	Bakkalbasi et al. (2009)
Whisky	1–2 mg/L	Glabasnia and Hofmann (2006)

[a] f.w.: fresh weight; d.w.: dry weight.

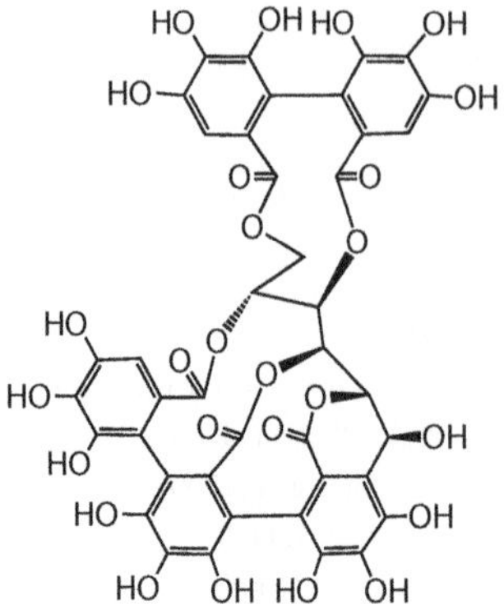

Vescalagin

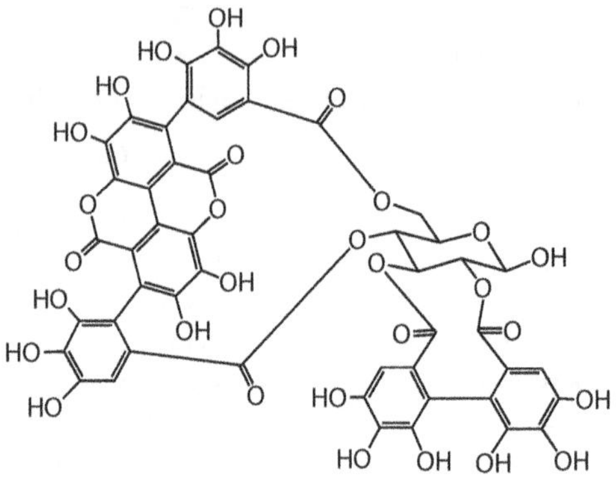

Punicalagin

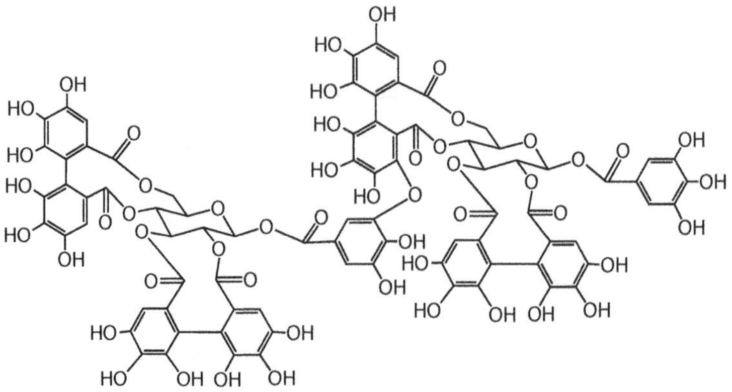

Sanguiin H-6

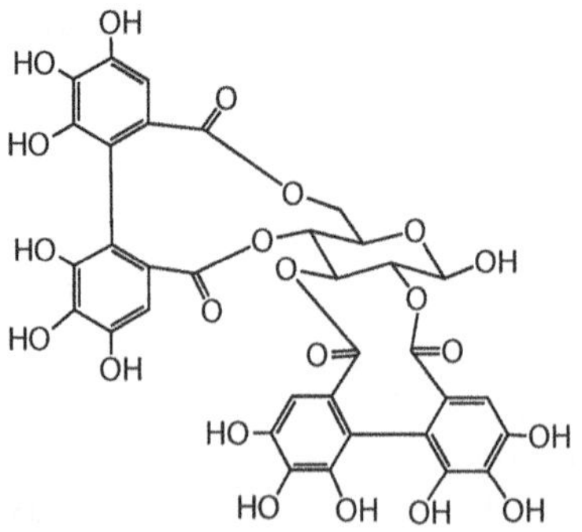

Pedunculagin

FIGURE 9.1 Representative ellagitannins in food products.

- *In vivo* studies with animal models that are essential for toxicological studies and to confirm bioavailability and metabolism in mammals and to evaluate tissue distribution.
- Clinical interventions with human volunteers to determine human bioavailability and metabolism. Using knowledge generated in animal studies, confirming the relevant metabolites as well as their *in vivo* concentrations in humans.

9.3.1 *IN VITRO* UPTAKE AND METABOLISM

The potential beneficial effects of ellagitannins and ellagic acid as antioxidants, and against various diseases, such as atherosclerosis and cancer, have been addressed in numerous *in vitro* studies. However, the number of *in vitro* studies designed to investigate the metabolism of these compounds by the cells is very scarce. In rat explants of oesophagus, fore-stomach, colon, bladder, trachea, lung, and liver incubated with [^{3}H]-ellagic acid, binding of ellagic acid to DNA was observed, with most being found in the oesophagus and the lowest in the lungs. Three metabolites of ellagic acid were identified in the cell culture media, sulfate ester, glucuronide, and glutathione conjugates (Teel et al. 1987). In a study carried out with [^{14}C]-ellagic acid in Caco-2 cells trans-wells, Whitley et al. (2003) observed a fast absorptive transport across the apical membrane with a high accumulation of radioactivity in the cells, whereas the apical to basolateral transport was minimal. As much as 93% of the intracellular radioactivity was detected irreversibly bound to macromolecules (proteins and DNA). Data obtained in experiments in which [^{14}C]-ellagic acid was incubated with Caco-2 cells and human esophageal epithelial cell line HET-1A led to the suggestion of an involvement of the organic anion transporters hOAT1 and hOAT4 in the observed apical transport (Whitley et al. 2006). However, the use of radiolabeled ellagic acid in these experiments did not confirm if ellagic acid was metabolized or not by the cells.

Larrosa et al. (2006a) reported that when incubated in Caco-2 cell media under physiological conditions at ~pH 7.2, punicalagin, the most abundant ellagitannin in pomegranate, was spontaneously hydrolyzed to ellagic acid, which was further metabolized by the cells to dimethyl-ellagic acid, a conversion that involves the active participation of catechol-*O*-methyltransferase. In addition, conjugation with glucuronic acid yielded dimethyl-ellagic acid glucuronide, which was the most abundant metabolite. All the metabolites were detected in both the cell culture medium and within the cells (Larrosa et al. 2006a). Another set of experiments was conducted to investigate metabolism by colonic microbiota (Cerdá et al. 2005a). Ellagic acid, punicalagin, and a walnut extract, containing several ellagitannins, were incubated with human fecal slurries. 3,8-Dihydroxy-dibenzopyranone (urolithin-A) was the main metabolite derived from both ellagic acid and the ellagitannins. Subsequently, it was shown that geraniin, an ellagitannin found in geraniums and other higher plants, was converted to 3,8,9-trihydroxy-dibenzopyranone (urolithin-C),

3,8-dihydroxy-9-methoxy-dibenzopyranone, 3,9-dihydroxy-8-methoxy-dibenzopyranone, and 3,4,8,9,10-pentahydroxy-dibenzopyranone (urolithin-M5) when incubated with rat fecal suspensions (Ito et al. 2008).

9.3.2 In Vivo Studies Using Animal Models

Doyle and Griffiths (1980) described for the first time the conversion of ellagic acid by rat colon microbiota into urolithin-A and an unidentified metabolite that was probably 3-hydroxy-dibenzopyranone (urolithin-B). After ingestion of ellagic acid, 10% of the dose was detected as urolithin-A in feces and urine of normal but not germ-free rats. Unmetabolized ellagic acid was detected in urine and feces of germ-free but not normal rats. In mice, after oral administration of 90 mg/kg of ellagic acid, which is equivalent to a dose of 428 mg for a 70-kg human, very low to not detectable levels of ellagic acid were found in blood, lung, or liver, indicating a poor absorption and/or rapid elimination of the phenolic acid (Smart et al. 1986). In contrast, when rats were fed 0.3 mg of [^{3}H]-ellagic acid/kg body weight, equivalent to a human dose of 3.4 mg for a 70-kg person, both free ellagic acid and sulfate ester, glucuronide and glutathione metabolites were detected in urine, bile, and blood. A rapid absorption of [^{3}H]-ellagic acid occurred, mostly within 2 h of oral administration. Levels in blood, bile, and tissues were low, and absorbed compounds were excreted in urine. After 24 h, more than a half of the administered [^{3}H]-ellagic acid remained in the gastrointestinal tract, while ~19% was excreted with feces and ~22% in urine (Teel and Martin 1988).

The results described with mice by Doyle and Griffiths (1980) and by Smart et al. (1986) are in good agreement and suggest that poor absorption of ellagic acid may lead to very low concentrations in tissues and a high concentration in the gut where it is metabolized by colonic microbiota into urolithins. The poor absorption of ellagic acid is supported by a report of the presence of ellagic acid calculi in the gastrointestinal tract of monkeys and goats whose natural diet contains ellagic acid (Van Tassel 1976). The low bioavailability may be caused by several factors including ionization of ellagic acid at physiological pH as well as formation of poorly soluble complexes with Mg and Ca ions. In addition, extensive binding of ellagic acid to the intestinal epithelium may also diminish absorption (Whitley et al. 2003).

Cerdá et al. (2003a) investigated the bioavailability of pomegranate husk ellagitannins, mainly punicalagin, in rats. These ellagitannins are essentially the same as those found in commercial pomegranate juice (Gil et al. 2000). The rats were given 6% of their diet as ellagitannins, and the experiment was used to evaluate absorption, tissue distribution, and toxicity. Values around 3–6% of the ingested punicalagin were excreted as metabolites in feces and urine. In feces, punicalagin was transformed by the microbiota to hydrolysis products and hydroxy-dibenzopyranone derivatives. In plasma, punicalagin was detected at concentrations of 30 μg/mL (28 μM). The absorption of intact punicalagin by rats and its detection in plasma is especially relevant, as it is one of the largest polyphenols, having a molecular weight of 1084 Da, that has been reported to be absorbed into the circulatory system (Cerdá et al. 2003a; Manach et al. 2005). Free ellagic acid was present in plasma in lower concentrations, 3–5 μg/mL (10–17 μM). The main urinary metabolites were urolithins A, B, and C as aglycones and glucuronide conjugates (Cerdá et al. 2003a). Only 3–6% of the ingested punicalagin was detected as the original compound or derived metabolites in urine and feces, suggesting that the majority of the ellagitannin had accumulated in tissues and/or been converted to undetectable metabolites such as CO_2.

Following daily feeding of punicalagins to rats for a period of 37 days, traces of punicalagin metabolites were detected in liver or kidney but neither punicalagin, ellagic acid, nor other derived metabolites were found in lung, brain, and heart tissues (Cerdá et al. 2003b). In addition to small quantities of punicalagins and ellagic acid, ellagi-tannin metabolites detected in plasma included urolithin-C diglucuronide, urolithin-A glucuronide, gallagic acid, dimethyl-ellagic acid glucuronide, and dimethyl-ellagic acid glucuronide methyl ester. As ellagic acid has two *ortho*-dihydroxy groups, it was anticipated that in the liver, as a consequence of catechol-*O*-methyltransferase activity, one methyl ether group would be introduced per dihydroxyl grouping. These metabo-lites were further conjugated with glucuronic acid which increases water solubility and so facilitates their excretion (Cerdá et al. 2003a).

To verify the ellagic acid accumulation in the digestive tract Whitley and col-leagues (2006) compared the oral and intravenous administration of [14C]-ellagic acid in rats. An accumulation of ellagic acid was found in the gastrointestinal tract when it was given orally. It was also detected in the gastrointestinal tract 0.5 h after intravenous administration, indicating a possible blood-to-enterocyte active trans-port system for ellagic acid. In another study, the intraperitoneal (0.3 mg/mouse/day) and oral (0.8 mg/mouse/day) bioavailability of a pomegranate extract were compared (Seeram et al. 2007). After oral intake, ellagic acid detected in plasma after 30 min had cleared after 2 h. No ellagic acid-derived urolithins were detected in plasma during the 24-h time course. When the pomegranate extract was administered intra-peritoneally, the ellagic acid content of plasma was 10-fold higher, and clearance was more prolonged being extended from 2 to 6 h. The presence of ellagic acid in prostate tissue, small intestine, colon, and liver has also been reported (Seeram et al. 2007) while the bioavailability of ellagic acid has been reported to be improved when it is ingested with phospholipids (Murugan et al. 2009).

Pigs, which represent a very useful model with which to study ellagitannin bio-availability, because of their physiological similarity to the man, have been used in acute and chronic feeding studies with ellagitannin-rich acorns (Espín et al. 2007). Thirty-one ellagitannin metabolites were detected, including ellagic acid conjugates and metabolites derived from its microbial degradation. This study also provided information about the distribution of metabolites in various tissues and biological fluids. Twenty-four hours after acute acorn intake, urolithin metabolites were detected in the distal part of the gastrointestinal tract. Following daily chronic acorn feeding, ellagic acid aglycone was detected in the jejunum along with significant amounts of 3,8,9,10-tetrahydroxy-dibenzopyranone (urolithin-D), urolithin-C isomers, and smaller amounts of urolithin-A and urolithin-B. The main metabolite detected in the wall of the small intestine was urolithin-A and its glucuronide conjugate sug-gesting that intestinal tissue uptake occurs more readily with dihydroxy-urolithins than their tri- and tetra-hyroxy analogs. Only small amounts of urolithin-A and urolithin-B were detected in the colon tissues and its contents. Urolithin-A was the main metabolite in feces, which also contained unmetabolized ellagitannins. These results indicate that ellagitannins release ellagic acid in the distal gastrointestinal tract and that the bacteria present in the small intestine are able to convert ellagic acid into urolithins that are absorbed as their lipophilic character increases. In urine, an extensive variety of glucuronide and diglucuronide urolithin conjugates detected, including urolithin-A and urolithin-C diglucuronides as well as urolithins A, B, C, and D glucuronides. Bile contained a diversity of ellagitannin metabolites including urolithins A and C, urolithin-C methyl ether, urolithin-D, urolithin-D methyl ether, and ellagic acid methyl and dimethyl ether glucuronides and diglucuronides. Only one sulfate metabolite was detected, so glucuronides were the main conjugated metabolites. No ellagitannin metabolites were detected in muscle, adipose (subepidermal and visceral), lung, liver, heart, and kidney (Espín et al. 2007).

In a study carried out with geraniin, six new ellagitannin-derived metabolites were identified in rat urine 48 h after the ingestion (Ito et al. 2008). These metabolites were detected principally as glucuronides of urolithin-C, urolithin-D, 3,8-dihydroxy-9-methoxy-dibenzopyranone, 3,9-dihydroxy-8-methoxy-dibenzopyranone, 3,4,8,9,10-pentahydroxy-dibenzopyranone (urolithin-M5), and 3,8,10-trihydroxy-dibenzopyranone (urolithin-M7). As yet, it is not known which bacterial group(s) converts ellagic acid into urolithins by opening the lactone ring, removing the carboxyl residue and dehydroxylating the individual hydroxyl groups of the ellagic acid nucleus. This is a general metabolic trend in mammals after the intake of dietary ellagitannins, as urolithins are produced by rodents, squirrels, beavers, sheeps, bull calves, pigs, and humans but not birds and beetles (González-Barrio et al. 2011). This multianimal study detected another group of ellagic acid-derived compounds, nasutins (see Figure 9.2). These metabolites, in which the phenolic hydroxyl groups

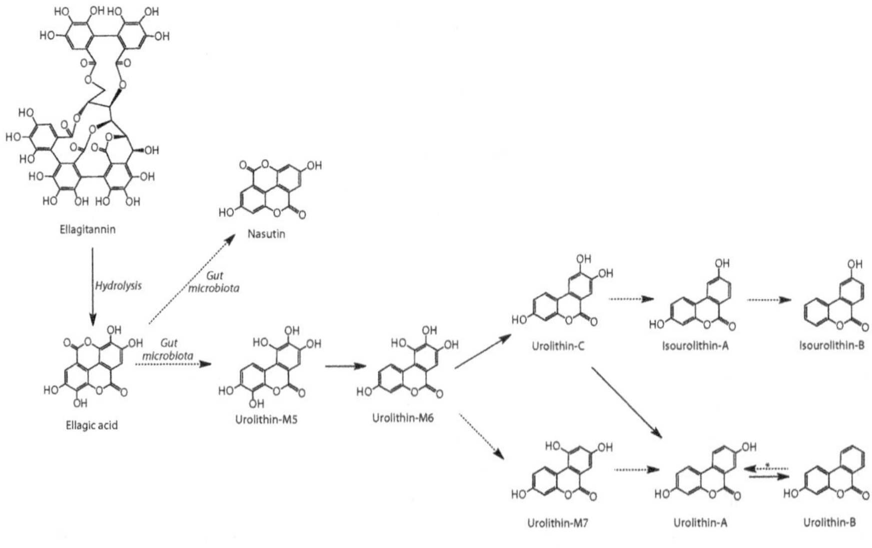

FIGURE 9.2 Metabolism of ellagitannins in the gastrointestinal tract. The asterisk designates the hydroxylation reaction from urolithin-B to urolithin-A observed *in vitro*. (Larrosa, M. et al., *J. Agric. Food Chem.*, **54**, 1611–1620, 2006. With permission.)

of the ellagic acid have been removed without opening and decarboxylation of the lactone ring, were present in castoreum of beavers as well as pig feces and urine (González-Barrio et al. 2011).

9.3.3 HUMAN STUDIES

In humans, two main types of experiments have been carried out: (i) pharmacokinetic studies, in which the absorption of pomegranate polyphenols has been evaluated over a period of several hours after intake, and (ii) bioavailability and metabolism studies in which the metabolites present in plasma and urine are monitored for long periods of time.

In a pharmacokinetic study carried out at the University of California, Los Angeles, ellagic acid was detected at a maximum concentration in plasma (C_{max}) of one human volunteer 1 h postingestion of 180 mL of pomegranate juice (cv. "Wonderful"; containing 25 mg ellagic acid and 318 mg of ellagitannins). The C_{max} was 32 ng/mL (0.1 μM), and the ellagic acid was eliminated from the circulatory system within 4 h (Seeram et al. 2004). In a study carried out in Murcia (Spain), six healthy volunteers consumed 1 L of pomegranate juice (cv. "Mollar" cultivar, 4.4 g of punicalagins, no free ellagic acid). Despite the much higher dose, no ellagic acid was detected in plasma in the 4 h period following juice intake (Cerdá et al. 2004). This difference could simply be due to interindividual variability.

Another human pharmacokinetic study looking at the absorption of ellagic acid has also shown low, but nonetheless, significant absorption of free ellagic acid (less than 1% of the ingested ellagic acid) during the first 2 h after the intake of black raspberries (Stoner et al. 2005), in agreement with the results of Seeram et al. (2004).

In a pharmacokinetic study (Seeram et al. 2006), with a larger number of subjects ($n = 18$), in which 180 mL of pomegranate juice (318 mg punicalagins, 12 mg ellagic acid, 75 mg of other hydrolyzable tannins), the ellagic acid plasma C_{max} was much lower, 0.06 ± 0.01 μM, compared to that in a previous study (Seeram et al. 2004). Urolithins appeared in plasma very early, 30 min after the intake of the pomegranate juice (Seeram et al. 2006). This is unexpected when they usually begin to be detected ~24 h after the ingestion as the parent ellagitannins have to pass through the gastrointestinal tract and reach the colon where they are converted to urolithins by the microbiota (Cerdá et al. 2005a). In a subsequent study in which ellagitannins and ellagic acid were consumed in pomegranate juice, a pomegranate liquid extract, and a pomegranate powder extract, similar plasma ellagic acid C_{max} values were observed, respectively, 0.06 ± 0.01 μM, 0.04 ± 0.01 μM, and 0.02 ± 0.01 μM (Seeram et al. 2008).

González-Barrio et al. (2010) provided *in vivo* evidence of the metabolism of ellagitannins by the colonic microbiota in parallel feeds with healthy human subjects and volunteers who had undergone an ileostomy. Both groups ingested 300 g of raspberries (0.23 g of ellagitannins and 5 mg of ellagic acid), but urolithin-A and urolithin-B glucuronides were detected only in the urine of the healthy subjects with an intact functioning colon in amounts equivalent to 0–8.6% of ellagitannin intake.

Regarding the long-term evaluation of metabolites in plasma and urine, four main studies are of note. In one of these studies, healthy volunteers consumed 1 L of pomegranate juice daily (containing 4.37 g/L punicalagin isomers) on a daily basis for 5 days (Cerdá et al. 2004). No punicalagins appeared in either plasma or urine, but three microbial ellagitannin-derived metabolites were detected, namely urolithin-A, urolithin-C, and urolithin-B glucuronides, in samples from some but not all volunteers. When the metabolites were detected, the plasma C_{max} ranged from 0.5 to 18.6 µM, providing evidence a large inter-individual variability between participants. The same metabolites and their corresponding aglycones were detected in urine excreted after 1 day of juice consumption. Total urinary excretion of urolithins also showed marked person-to-person variability, ranging from 0.7 to 52.7% of the ingested punicalagin. In general, the metabolites found in humans after the pomegranate juice intake are similar to those produced by rats after pomegranate husk intake (Cerdá et al. 2003a).

A further investigation into ellagitannin bioavailability and metabolism was carried out with volunteers who consumed strawberries, raspberries, walnuts, and oak-aged red wine (Cerdá et al. 2005b). These foodstuffs differ in the content and type of ellagitannins; however, it is noteworthy that urolithin-A glucuronide was detected in the urine of all 40 volunteers who participated in the four feeds. Thus, urolithin-A glucuronide was proposed as a candidate biomarker of human exposure to dietary ellagitannins that would be useful in intervention studies with a range of ellagitannin-containing foods (Cerdá et al. 2005a, 2005b). As noted in other studies, there was a large inter-individual variability in the quantity of urolithin-A glucuronide excreted.

Another study on ellagitannin bioavailability was carried out with a group of 15 patients with stable chronic obstructive pulmonary disease (COPD). The volunteers consumed pomegranate juice daily for 5 weeks in a randomized, double blind, placebo-controlled trial (Cerdá et al. 2006). Glucuronides of urolithin-A and urolithin-B were detected in both the plasma and urine and once again a large interindividual variability was observed.

The most recent study was conducted with patients diagnosed with benign prostatic hyperplasia or prostate cancer (González-Sarrías et al. 2010a). Fourteen subjects consumed 35 g of walnuts (202 mg ellagitannins, 8 mg ellagic acid), and 19 volunteers consumed 200 mL of pomegranate juice (265 mg punicalagins, 14 mg ellagic acid) for three days before surgery. Patients were classified into three groups—high, low, and very low urolithin excreters—depending of the amount of urolithin-A glucuronide excreted in urine. The 48% of patients were identified as high excreters, 32% as low excreters, and 21% as very low urolithin excreters. The plasma of the seven high excreters contained urolithin-A glucuronide, in concentrations ranging from 0.05 to 0.2 µM, along with lower levels of urolithin-C glucuronide, urolithin-C methyl ether glucuronide, and dimethyl-ellagic acid glucuronide. Ellagic acid derivatives were detected in only 24% of the prostates analyzed. In the high-urolithin excreter group, urolithin-A glucuronide was in the range of 0.5–2 ng/g tissue and urolithin-B was found in trace quantities. Dimethyl-ellagic acid aglycone was found in four prostate samples from the high excreters and in two prostates of the low excreters (González-Sarrías et al. 2010a).

9.4 BIOAVAILABILITY AND METABOLISM EVENTS IN DIFFERENT BODY SITES

Ellagitannins are large molecules that are usually quite polar and, therefore, according to our general knowledge of pharmacokinetics, they should be either poorly absorbed or not absorbed at all. The available data indicate that dietary ellagitannins are not detected in plasma or in other biological fluids after the intake of ellagitannin-rich food. In a toxicological study in which rats were fed with large amounts of pomegranate ellagitannins for a long period of time, small amounts of punicalagin were detected in plasma and urine (Cerdá et al. 2003a). In other studies carried out with humans and animal models, no ellagitannins were detected in either plasma or urine after the intake of dietary doses of ellagitannin-rich foods (strawberries, walnuts, raspberries, acorns, etc.) (Cerdá et al. 2005a). With high intakes not all ellagitannins are degraded by the colonic microflora as substantial amounts were detected in the feces of rats fed with pomegranate extract and in feces of pigs that had consumed acorns (Cerdá et al. 2005b; Espín et al. 2007).

9.4.1 ABSORPTION OF ELLAGIC ACID

Ellagic acid is already present in most ellagitannin-rich foods and is also produced during food processing and storage. Free ellagic acid is relatively insoluble in aqueous solution and precipitates in juices and liquors (Bakalbassi et al. 2009). Some pharmacokinetic studies show that ellagic acid *per se* can be absorbed 30–60 min after the intake of various foods (Teel and Martin 1988; Seeram et al. 2004, 2006, 2007). This suggests that absorption of ellagic acid begins in the stomach and can be detected in peripheral blood. Seeram et al. (2008) reported that ellagic acid disappears from plasma 2–6 h after ingestion. However, other pharmacokinetic studies carried out with different foods containing similar amounts of free ellagic acid have been unable to detect ellagic acid in plasma (Whitley et al. 2006). Arguably, this indicates that the food matrix (pH, constituents, processing, etc.) may have a crucial impact on ellagic acid absorption in the proximal part of the gastrointestinal tract (Whitley et al. 2006; Seeram et al. 2008). The transporters implicated in ellagic acid absorption in the stomach are unknown and differences between individuals could also be an important factor. The absorption of ellagic acid in distal parts of the gastrointestinal tract appears to be less important. Animal studies show that ellagic acid is released from ellagitannins in the small intestine and absorbed by the cells of the jejunum and ileum where it is readily metabolized yielding methyl-and dimethyl ethers and glucuronide conjugates. These metabolites, but not free ellagic acid, were detected in bile and in peripheral plasma and urine of Iberian pigs that had consumed acorns (Espín et al. 2007).

9.4.2 INTESTINAL MICROBIOTA METABOLISM

One of the main events in ellagitannin metabolism and bioavailability is their microbial-mediated transformation to render a series of hydroxylated dibenzopyran-one derivatives (Cerdá et al. 2005). Among them, the best characterized are urolithin-A and urolithin-B, but intermediates with three and four hydroxyl groups, urolithin-C and urolithin-D, are also produced in the small intestine, absorbed, and excreted in the bile after conversion to methyl ethers and glucuronides (Espín et al. 2007; Ito et al. 2008). Animal experiments show that these metabolites start to be formed in the small intestine, indicating that anaerobic bacteria may be responsible for the transformation of ellagitannins. Ellagitannin metabolism continues along the gastro-intestinal tract, leading to the production of urolithin-A and urolithin-B (Espín et al. 2007). Differences in the profile of these metabolites between human volunteers show that they may be produced by the activity of specific microorganisms present in the gut (González-Sarrías et al. 2010a). If these microbial metabolites, which are more bioavailable than the original ellagitannins or ellagic acid, are the agents responsible for the biological activity associated with consumption of ellagitannin and ellagic acid-rich foods, then it raises the possibility of developing functional foods in which the responsible specific microorganisms could be included together with the parent ellagitannins.

9.4.3 Phase I and Phase II Metabolism

In the gastrointestinal tract and at other sites, principally the liver, ellagic acid, and ellagitannin microbial metabolites are further metabolized either by phase I (hydroxylation) and phase II (methylation, glucuronidation, and sulfation) enzymes to render more soluble metabolites that may accumulate in tissues and also be excreted in urine. A previous *in vitro* study described the hydroxylation of urolithin-B to render urolithin-A, with more possibilities for conjugation to increase excretion (Larrosa et al. 2006b). However, this hypothesis has not yet been confirmed *in vivo*. In a recent gene expression study, it was shown that both urolithin-A and urolthin-B, but particularly urolithin-B, enhanced expression of CYP450 genes in Caco-2 cells by 15–20-fold. This may explain why the di-hydroxy derivative is generally more abundant in tissue such as the liver, as well as in plasma and urine. Phase II metabolites are also produced and methyl ethers, as well as different glucuronide conjugates, are detected in tissues and urine. Sulfated ellagitannins are less abundant in animals and humans than glucuronide conjugates.

9.4.4 Tissue Distribution

To understand the biological activity of ellagitannin and ellagic acid, it is essential to determine which metabolites, and in what concentrations, are present in the different target tissues. In rats, a high accumulation of ellagic acid was detected in the oesophagus and small intestine, even when the ellagic acid was administered intravenously (Whitley et al. 2006). Intraperitoneal administration of pomegranate ellagitannins to mice showed a high concentration ellagic acid in prostate tissue (676 ± 172 ng/g) and to a lesser extent in the intestine, colon, and liver (Seeram et al. 2007). No ellagitannins, ellagic acid, or derived metabolites have been detected in muscle, adipose, heart, lung, or brain tissue, although small amounts of conjugates of the microbial metabolites have been detected in liver and kidney (Cerdá et al. 2003a, 2003b). A study with pigs fed on acorns also revealed a similar distribution of metabolites in systemic tissues (Espín et al. 2007). The same study also showed the presence of large amounts of different conjugates of microbial metabolites in the gall bladder, indicative of enterohepatic circulation, which is probably responsible for the long clearing time of the metabolites. They are detected in urine for as long as 48–76 h after the ellagitannin intake (Cerdá et al. 2005). In humans, urolithin-A glucuronide was detected in prostate tissue at concentrations of 0.5–2 ng/g. Urolithin-B was also present but not in quantifiable amounts (González-Sarrías et al. 2010a).

9.5 THE WHOLE PICTURE OF ELLAGITANNIN BIOAVAILABILITY

Ellagitannins are generally not absorbed. They release ellagic acid in the gut, and this is poorly absorbed in the stomach and small intestine and largely metabolized by unidentified bacteria in the intestinal lumen to produce urolithins. This microbial metabolism starts in the small intestine and the first metabolite produced, urolithin-M-5, which retain five phenolic hydroxyls, is further metabolized by the sequential removal of hydroxyl groups as the hydroxy-dibenzopyranones proceed down the intestinal tract, leading to the appearance of the dihydroxy urolithin-A and monohydroxy urolithin-B in the colon (Figure 9.2). The absorbed metabolites are conjugated with glucuronic acid (one or two units), and methyl ethers can also be produced when *ortho*-dihydroxyl groupings are present. Urolithin-A and urolithin-B conjugates are the main metabolites detected in plasma and urine although some urolithin-C derivatives and ellagic acid-dimethyl ether glucuronide have also been detected but in smaller amounts. Neither urolithins C and D nor ellagic acid derivatives are detected in peripheral plasma, but they are absorbed in the small intestine and transported to the liver where they are further metabolized and excreted in bile into the small intestine, establishing an enterohepatic circulation that is responsible for the long-life of urolithins in plasma and urine. As far as we know, these metabolites do not accumulate in organs or tissues (Cerdá et al. 2003b).

9.6 CONCLUSIONS

The biological activity and physiological effects of ellagitannins and ellagic acid have to be consistent with their bioavailability and metabolism. Thus, results obtained from *in vitro* studies using cultured cells that *in vivo* would not be in direct contact with food ellagitannins should be considered with caution. This applies, for instance, in the case of studies testing activity of ellagitannins on liver or breast cancer cell lines. In cultured cells representing systemic tissues and organs, we should evaluate the bioactivity of the actual metabolites circulating in plasma, urolithins, and their glucuronic acid conjugates, and at the *in vivo* concentrations reached in plasma (~1–5 µM) and in tissues (nM). Ellagitannins and ellagic acid are more relevant in terms of *in vivo* bioactivity in the gastrointestinal tract, where they can be present at significant concentrations. Both *in vitro* and *in vivo* studies show the potential of these polyphenols as chemopreventive agents in several types of cancer. The metabolic transformation of ellagitannins into urolithins is a widespread phenomenon in nature. Urolithin metabolites have been reported to be present in the feces of the squirrel, *Trogopterus xanthippes,* and their hyaluronidase inhibitory activity has been demonstrated (Jeong et al. 2000). They have also been isolated from kidney stones in cattle suffering from the "clover stone" disease that is most likely associated to a large, chronic intake of a clover species (*Trifolium subterraneum*) rich in ellagitannins (Pope 1964). Urolithins A and B have also been isolated from beaver excretions and as such are constituents of castoreum (Lederer 1949). Castoreum is a urine-based fluid from castor sacs (Rosell et al. 2000). Production of urolithins has been described in ruminants, rodents, squirrels, beavers, pigs, and humans (González-Barrio et al. 2011). In addition to hyaluronidase inhibitory activity (Jeong et al. 2000), urolithins may have potential estrogenic/antiestrogenic effects in doses similar to those reported for other well-known phytoestrogenic compounds such as enterolactone, resveratrol, genistein, or daidzein (Larrosa et al. 2006b). Further research is warranted to evaluate the possible role of ellagitannins and ellagic acid as dietary "prophytoestrogens."

Urolithins have anti-inflammatory activity *in vitro* and *in vivo* (González-Sarrías et al. 2010b; Larrosa et al. 2010). *In vitro* urolithin-A, and to a lesser extent urolithin-B, inhibited prostaglandin E_2 (PGE$_2$) production inhibiting translocation of NF-κB to the nucleus (González-Sarrías et al. 2010b). In a rat model of colitis, urolithin-A decreased the inflammation markers, iNOS, cycloxygenase-2, mPGES1 and PGE$_2$, in colonic mucosa, preserved colon architecture, and favorably modulated the gut microbiota (Larrosa et al. 2010). It has also been reported that urolithins showed antiglycative activity in an *in vitro* experimental model at physiological concentrations, becoming good candidates in the control of cardiovascular hyperglycemia-related complications (Verzelloni et al. 2011). Further research is warranted to clarify the biological effects of urolithins and their conjugate derivatives.

ACKNOWLEDGMENTS

This work has been supported by the Spanish MICINN (Consolider Ingenio 2010-Fun-C-Food CSD2007-0063) and Fundación Seneca de la Region de Murcia (grupo de excelencia GERM 06, 04486).

REFERENCES

Aaby, K., Wrolstad, R.E., Ekeberg, D. et al. (2007). Polyphenol composition and antioxidant activity in strawberry purees; Impact of achene level and storage. *J. Agric. Food Chem.,* **55**, 5156–5166.

Anderson, K.J., Teuber, S.S., Gobeille, A. et al. (2001). Walnut polyphenolics inhibit in vitro human plasma and LDL oxidation. *J. Nutr.,* **131**, 2837–2842.

Bakkalbasi, E., Mentes, O., and Artik, N. (2009). Food ellagitannins—Occurrence, effects of processing and storage. *Crit. Rev. Food Sci. Nut.,* **49**, 283–298.

Cantos, E., Espín, J.C., López-Bote, C. et al. (2003). Phenolic compounds and fatty acids from acorns (*Quercus* spp.), the main dietary constituent of free-ranged Iberian pigs. *J. Agric. Food Chem.,* **51**, 6248–6255.

Cerdá, B., Cerón, J.J., Tomás-Barberán, F.A. et al. (2003b). Repeated oral administration of high doses of the pomegranate ellagitannin punicalagin to rats for 37 days is not toxic. *J. Agric. Food Chem.,* **51**, 3493–3501.

Cerdá, B., Espín, J.C., Parra, S. et al. (2004). The potent *in vitro* antioxidant elagitannins from pomegranate juice are metabolised into bioavailable but poor antioxidant hydroxyl-6H-dibenzopyran-6-one derivatives by the colonic microflora of healthy humans. *Eur. J. Nutr.,* **43**, 205–220.

Cerdá, B., Llorach, R., Cerón, J.J. et al. (2003a). Evaluation of the bioavailability and metabolism in the rat of punicalagin, an antioxidant polyphenols from pomegranate juice. *Eur. J. Nutr.,* **42**, 18–28.

Cerdá, B., Periago, P., Espín, J.C. et al. (2005a). Identification of urolithin A as a metabolite produced by human colon microflora from ellagic acid and related compounds. *J. Agric. Food Chem.,* **53**, 5571–5576.

Cerdá, B., Tomás-Barberán, F.A., and Espín, J.C. (2005b). Metabolism of antioxidant and chemopreventive ellagitannins from strawberries, raspberries, walnuts, and oak-aged wine in humans: Identification of biomarkers and individual variability. *J. Agric. Food Chem.,* **53**, 227–235.

Cerdá, B.; Soto, M.C., Albaladejo, M.D. et al. (2006). Pomegranate juice supplementation in COPD: A 5-week randomised, double blind, placebo-controlled trial. *Eur. J. Clin. Nutr.,* **60**, 245–253.

Clifford, M.N. and Scalbert, A. (2000). Ellagitannins—nature, occurrence and dietary burden. *J. Sci. Food Agric.,* **80**, 1118–1125.

Doyle, B. and Griffiths, L.A. (1980). The metabolism of ellagic acid in the rat. *Xenobiotica,* **10**, 247–256.

Espín, J.C., González-Barrio, R., Cerdá, B. et al. (2007b). Iberian pig as a model to clarify obscure points in the bioavailability and metabolism of ellagitannins in humans. *J. Agric. Food Chem.,* **55**, 10476–10485.

Ferreres, F., Andrade, P., Gil, M.I. et al. (1996). Floral nectar phenolics as biochemical markers for the botanical origin of heather honey. *Z. Leb-Unt.–Forsch.,* **202**, 40–44.

Fukuda, T., Ito, H., and Yoshida, T. (2003). Antioxidative polyphenols from walnuts (*Juglans regia* L.). *Phytochemistry,* **63**, 795–801.

Gil, M.I., Tomás-Barberán, F.A., Hess-Pierce, B. et al. (2000). Antioxidant activity of pomegranate juice and its relationship with phenolic composition and processing. *J. Agric. Food Chem.,* **48**, 4581–4589.

Glabasnia, A. and Hofmann, T. (2006). Sensory directed identification of taste-active ellagitannins in American (*Quercus alba* L.) and European oak wood (*Quercus robur* L.) and quantitative analysis in bourbon whiskey and oak-matured red wines. *J. Agric. Food Chem.,* **54**, 3380–3390.

Gonçalves, B., Borges, O., Costa, H.S. et al. (2010). Metabolite composition of chestnut (*Castanea sativum* Mill.) upon cooking: Proximate analysis, fibre, organic acids and phenolics. *Food Chem.,* **122**, 154–160.

González-Barrio, R., Borges, G., Mullen, W. et al. (2010). Bioavailability of anthocyanins and ellagitannins following consumption of raspberries by healthy humans and subjects with an ileostomy. *J. Agric. Food Chem.,* **58**, 3933–3939.

González-Barrio, R., Truchado, P., Ito, H. et al. (2011). UV and MS Identification of urolithins and nasutins, the bioavailable metabolites of ellagitannins and ellagic acid in different mammals. *J. Agric. Food Chem.,* **55**, 1152–1162.

González-Sarrías, A., Giménez-Bastida, J.A., García-Conesa, M.T. et al. (2010a). Occurrence of urolithins, gut microbiota ellagic acid metabolites and proliferation markers expression response in the human prostate gland upon consumption of walnuts and pomegranate juice. *Mol. Nutr. Food Res.,* **54**, 311–322.

González-Sarrías, A., Larrosa, M., Tomás-Barberán, F.A. et al. (2010b). NF-kappaB-dependent anti-inflammatory activity of urolithins, gut microbiota ellagic acid-derived metabolites, in human colonic fibroblasts. *Br. J. Nutr.,* **104**, 503–512.

Ito, H., Iguchi, A., and Hatano, T. (2008). Identification of urinary and intestinal bacterial metabolites of ellagitannin geraniin in rats. *J Agric. Food Chem.*, **56**, 393–400.

Jeong, S.J., Kim, N.Y., Kim, D.H. et al. (2000). Hyaluronidase inhibitory active 6H-dibenzo[b,d] pyran-6-ones from the faeces of *Trogopterus xanthippes*. *Planta Med.*, **66**, 76–77.

Koponen, J.M., Happonen, A.M., Mattila, P.H. et al. (2007). Contents of anthocyanins and ellagitannins in selected foods consumed in Finland. *J. Agric. Food Chem.*, **55**, 1612–1619.

Larrosa, M., González-Sarrías, A., García-Conesa, M.T. et al. (2006b). Urolithins, ellagic acid-derived metabolites produced by human colonic microflora, exhibit estrogenic and antiestrogenic activities. *J. Agric. Food Chem.*, **54**, 1611–1620.

Larrosa, M., González-Sarrías, A., Yáñez-Gascón, M.J. et al. (2010). Anti-inflammatory properties of a pomegranate extract and its metabolite urolithin-A in a colitis rat model and the effect of colon inflammation on phenolic metabolism. *J. Nutr. Biochem.*, **21**, 717–725.

Larrosa, M., Tomás-Barberán, F.A., and Espín, J.C. (2006a). The dietary hydrolysable tannin punicalagin releases ellagic acid that induces apoptosis in human colon adenocarcinoma Caco-2 cells by using the mitochondrial pathway. *J. Nutr. Biochem.*, **17**, 611–625.

Lederer, E. (1949). Chemistry and biochemistry of some mammalian secretions and excretions. *J. Chem. Soc.*, 2115–2125.

Lee, J.H. and Talcott, S.T. (2002). Ellagic acid and ellagitannins affect on sedimentation in muscadine juice and wine. *J. Agric. Food Chem.*, **50**, 3971–3976.

Malik, N.S.A., Perez, J.L., Lombardini, L. et al. (2009). Phenolic compounds and fatty acids composition of organic and conventional grown pecan kernels. *J. Sci. Food Agric.*, **89**, 2207–2213.

Manach, C., Williamson, G., Morand, C. et al. (2005). Bioavailability and bioefficacy of phytonutrients in humans. I. Review of 97 bioavailability studies. *Am. J. Clin. Nutr.*, **81**, 230S–242S.

Mertens-Talcott, S.U., Jilma-Stohlawetz, P., Rios, J. et al. (2006). Absorption, metabolism, and antioxidant effects of pomegranate (*Punica granatum* L.) polyphenols after ingestion of a standardized extract in healthy human volunteers. *J. Agric. Food Chem.*, **54**, 8956–8961.

Murugan, V., Mukherjee, K., Maiti, K. et al. (2009). Enhanced oral bioavailability and antioxidant profile of ellagic acid by phospholipids. *J. Agric. Food Chem.*, **57**, 4559–4565.

Pope, G.S. (1964). Isolation of two benzocoumarins from 'clover stone', a type of renal calculus found in sheep. *Biochem. J.*, **93**, 474–477.

Rosell, F., Johansen, G., and Parker, H. (2000). Eurasian beavers (*Castor fiber*) behavioural response to simulated territorial intruders, *Can J. Zool.*, **78**, 931–935.

Seeram, N.P., Aronson, W.J., Zhang, Y. et al. (2007). Pomegranate ellagitannin-derived metabolites inhibit prostate cancer growth and localize to the mouse prostate gland. *J. Agric. Food Chem.*, **55**, 7732–7737.

Seeram, N.P., Henning, S.M., Zhang, Y. et al. (2006). Pomegranate juice ellagitannin metabolites are present in human plasma and some persist in urine up to 48 hours. *J. Nutr.*, **136**, 2481–2485.

Seeram, N.P., Lee, R., and Heber, D. (2004). Bioavailability of ellagic acid in human plasma after consumption of ellagitannins from pomegranate (*Punica granatum* L.) juice. *Clin. Chim. Acta*, **348**, 63–68.

Seeram, N.P., Zhang, Y., McKeever, R. et al. (2008). Pomegranate juice and extracts provide similar levels of plasma and urinary ellagitannin metabolites in human subjects. *J. Med. Food*, **11**, 390–394.

Smart, R.C., Huang, M.T., Chang, R.L. et al. (1986). Disposition of the naturally occurring antimutagenic plant phenol, ellagic acid, and its synthetic derivatives, 3-*O*-decylellagic acid and 3,3'-di-*O*-methylellagic acid in mice. *Carcinogenesis*, **7**, 1663–1667.

Stoner, G.D., Sardo, C., Apseloff, G. et al. (2005). Pharmacokinetics of anthocyanins and ellagic acid in healthy volunteers fed freeze-dried black raspberries daily for 7 days. *J. Clin. Pharmacol.*, **45**, 1153–1164.

Teel, R.W. and Martin, R.W. (1988). Disposition of the plant phenol ellagic acid in the mouse following oral administration by gavage. *Xenobiotica,* **18**, 397–405.

Teel, R.W.; Martin, R.W., and Allahyari, R. (1987). Ellagic acid metabolism and binding to DNA in organ explant cultures of the rat. *Cancer Lett.,* **36**, 203–211.

Törrönen, R. (2009). Sources and health effects of dietary ellagitannins. In S. Quideau (ed.), *Chemistry and Biology of Ellagitannins*. World Scientific, London, pp. 298–319.

Van Tassel, R. (1973). Bezoars, *Janus* **60**, 241–259.

Verzelloni, E., Pellacani, C., Tagliazucchi, D. et al. (2011). Antiglycative and neuroprotective activity of colon-derived polyphenol catabolites. *Mol. Nutr. Food. Res.*, **55**, S35–S43.

Villarreal-Lozoya, J.E., Lombardini, L., and Cisneros-Zevallos, L. (2007). Phytochemical constituents and antioxidant capacity of different pecan (*Carya illinoinensis*) cultivars. *Food Chem.*, **102**, 1241–1249.

Whitley, A.C., Stoner, G.D., Darby, M.V. et al. (2003). Intestinal epithelial cell accumulation of the cancer preventive polyphenol ellagic acid—extensive binding to protein and DNA. *Biochem Pharmacol.*, **66**, 907–915.

Whitley, A.C., Sweet, D.H. and Walle, T. (2006). Site-specific accumulation of the cancer preventive dietary polyphenol ellagic acid in epithelial cells of the aerodigestive tract. *J. Pharm. Pharmacol.*, **58**, 1201–1209.

Zafrilla, P., Ferreres, F., and Tomás-Barberán, F.A. (2001). Effect of processing and storage on the antioxidant ellagic acid derivatives and flavonoids of red raspberry (*Rubus idaeus*) jams, *J. Agric.Food Chem.*, **49**, 3651–3655.

10 Colon-Derived Microbial Metabolites of Dietary Phenolic Compounds

Anna-Marja Aura

CONTENTS

10.1 INTRODUCTION

Epidemiological studies show that consumption of plant foods is associated with lowered risk of chronic diseases such as cancer and cardiovascular disease (WHO 2003). Dietary recommendations are based partly on this epidemiological evidence. Plant foods consist of cereal grains, vegetables, legumes, fruits, and berries. These plant foods contain (poly)phenolic compounds, which have been discussed in previous chapters of this book.

The human colon contains more than 1000 different bacterial species based on metagenomic analysis (Quin et al. 2010). Both intra- and interindividual variations of colonic microbiota occur as a result of age, stress, disease, and diet (Salminen et al. 1998; Kleessen et al. 2000). Diversity of microbiota of the stool and mucosal community is immense (Eckburg et al. 2005). Microbiota can be divided into non-adherent and adherent communities, which show different activities. Biofilm populations, attached to particulate matter in the lumen, are able to digest polysaccharides, whereas non-adherent communities ferment the liberated oligosaccharides more rapidly (Macfarlane and Macfarlane 2006, 2007; Macfarlane and Dillon 2007).

Intervention studies demonstrate the diversity of metabolites formed after consumption of a diet rich in plant foods. These metabolites include hepatic conjugates and microbial degradation products originating from the colon. Microbial conversions were studied between the 1950s and 1980s using mostly qualitative methodology, and these studies have been well reviewed by Scheline (1978) and Griffiths (1982). The fact that dietary fiber serves as a vehicle for phenolic compounds, coupled with recent *in vivo* and *in vitro* findings, have reignited interest in the importance of the structural transformations occurring in the colon with the topic being reviewed by Aura (2008), Selma et al. (2009), and Del Rio et al. (2010). This chapter aims to update information concerning structural transformations of dietary (poly)phenolic compounds mediated by human gut microbiota. The (poly)phenolic compounds are abundant in foods consumed in Mediterranean countries (fruits, vegetables, and beverages) and Nordic countries (wholegrain cereals and berries). Dietary sources of flavonoids and related phenolic compounds and their fate in the upper gastrointestinal tract have been described in detail in previous chapters. This review will focus on studies concerning the production and biological activities of microbial metabolites as well as future prospects for the application of systems biology.

10.2 RELEASE AND ABSORPTION OF PHENOLIC COMPOUNDS AND THEIR METABOLITES

When flavonoids and related phenolic compounds are consumed in the diet, they are released from the matrix after mastication. The stomach reduces the size of food particle, which further enhances the release of phenolic compounds (Scalbert et al. 2002). Phenolic acids can be absorbed in the free form in the stomach (Zhao et al. 2004; Lafay et al. 2006). Flavonoid glycosides can be deglycosylated in the mouth by microbiota or by oral epithelial cells (Walle et al. 2005). Aglycones can also be formed in the small intestine by the action of membrane-bound lactase phloridzin hydrolase, and they are absorbed passively through the epithelium, or flavonoids can be transported through the epithelium as glycosides by sugar transporters. In the epithelial cells, cytosolic β-glucosidase hydrolyses the glycosides and aglycones are formed after absorption (Day et al. 2000; Nemeth et al. 2003). Generally, absorption is affected by the structure of phenolic compounds (glycosylation, molecular weight, and esterification). Once absorbed, aglycones and phenolic acids undergo conjugation in the ileal epithelium and/or the liver (Scalbert et al. 2002). Subsequently, luminal non-absorbed phenolic compounds and those still bound into the food matrix enter the colon.

In the colon, cleavage of the glycosyl moiety from the phenolic backbone is catalyzed by microbial enzymes resulting in the transient appearance of aglycones prior to further catabolism by the microbiota (Aura et al. 2002; Rechner et al. 2004; Hein et al. 2008). A part of the hepatic metabolites (methylated, sulfated, or glucuronidated conjugates) is returned to the lumen via bile and reabsorbed after deconjugation by the luminal bacteria from the colon (enterohepatic recirculation) (Donovan et al. 1999; Felgines et al. 2003; Natsume et al. 2003).

Deconjugation is catalyzed by fecal microbial enzymes (α-rhamnosidase, β-glucosidase, and β-glucuronidase), the specific activities of which reflect the *in vitro* deconjugation rates of phenolic compounds. Many flavonoid aglycones undergo C-ring-fission, as a result of which hydroxylated aromatic compounds are formed from the A-ring and phenolic acids from the B-ring (Aura et al. 2002, 2005a; Rechner et al. 2004). In contrast to deconjugation, sulfate conjugation of phenolic hydroxyl groups is catalyzed by arylsulfotransferases originating from human intestinal bacteria. This enzyme activity has been detected in human and rat feces (Kim and Kobashi 1986; Kim et al. 1986). Furthermore, lactones can be formed from ellagic acid and plant lignans (Heinonen et al. 2001; Cerda et al. 2005).

Pharmacokinetic studies show that microbial metabolites have up to a 24–48 h residence time in the bloodstream after a single dose of their parent compounds (Sawai et al. 1987; Gross et al. 1996; Nesbitt et al. 1999; Juntunen et al. 2000; Kilkkinen et al. 2003; Kuijsten et al. 2005). After formation, microbial metabolites are absorbed from the colon and may be substrates for liver metabolism, resulting in the production of glucuronidated, methylated, glycinated, and sulfated derivatives (Axelson and Setchell 1981; Adlercreutz et al. 1995; Lampe 2003; Olthof et al. 2003). These hepatic metabolites undergo enterohepatic circulation, contributing to the low diurnal variation of the metabolite concentrations in the blood (Bach Knudsen et al. 2003). Finally, the phenolic microbial metabolites are excreted via urine as hepatic conjugates (Adlercreutz et al. 1995; Seeram et al. 2006).

High individual variation in the extent of metabolism is a reflection of the inter- and intraindividually variable colonic microbiota (Kilkkinen et al. 2002; Lampe 2003; Cerda et al. 2005). Personal differences in the enzymatic conversion activities, intestinal diseases, or medication cause qualitative and quantitative variations of the metabolite profiles. Using NMR, GC-MS, or GCxGC-MS-TOF methods enables a metabolomic approach to profiling of the colonic conversions. These techniques can be applied to samples derived from colon models and urine as well as fecal water from human intervention studies (Aura et al. 2008; Grün et al. 2008; Jacobs et al. 2008; Pettersson et al. 2008).

10.2.1 FLAVONOLS

Deconjugated flavonols, such as quercetin and myricetin, are degraded to phenylacetic acids, reflecting the hydroxylation pattern of the B-ring. Thus, the primary catabolite of quercetin is 3',4'-dihydroxyphenylacetic acid and that of myricetin is 3',5'-dihydroxyphenylacetic acid. Further dehydroxylation results in the formation of 3'-hydroxyphenylacetic acid from both the dihydroxylated derivatives (Griffiths and Smith 1972a; Aura et al. 2002). The early appearance of the maximum concentration of 3',4'-dihydroxyphenylacetic acid and its decline concomitant with the appearance of 3'-hydroxyphenylacetic acid indicates that ring-fission occurs first, followed by dehydroxylation (Aura et al. 2002; Rechner et al. 2004). Other catabolites derived from quercetin include phloroglucinol (1,3,5-trihydroxybenzene), 3,4-dihydroxybenzaldehyde, and 3,4-dihydroxytoluene (4-methylcatechol) (Krishnamurty et al. 1970; Sawai et al. 1987; Winter et al. 1991; Schneider et al. 1999; Rechner et al. 2004). It appears that different microbial species in human fecal material are responsible for degycosylation and ring-fission of flavonols (Krishnamurty et al. 1970; Winter et al. 1991; Schneider et al. 1999; Schneider and Blaut 2000). Hollman and Katan (1998) suggested that *in vivo* *O*-methylation of 3′,4′-dihydroxyphenylacetic acid occurs in the liver after microbial catabolism of quercetin resulting in the formation of the 3′-methoxy-4′-hydroxyphenylacetic acid. This proposal is supported by data showing that while methylated catabolites are found in body fluids (Sawai et al. 1987; Gross et al. 1996) they are not produced by *in vitro* fecal incubations (Aura et al. 2002).

Walle et al. (2001) found that a substantial part (23–81%) of [4-^{14}C]quercetin ingested by humans is also recovered as [^{14}C]carbon dioxide. Mullen et al. (2008a) reported on the *in vivo* bioavailability of [2-^{14}C]quercetin-4′-*O*-glucoside in rats dosed by gavage. Quercetin was found in the proximal gastrointestinal tract as glucuronide, methylated and sulfated derivatives, one hour after ingestion, and only traces were excreted via urine, whereas after 72 h, the radiolabel was associated almost exclusively with ring-fission metabolites such as 3′,4′-dihydroxyphenylacetic acid, 3′-hydroxyphenylacetic acid, 3,4-dihydroxybenzoic acid (protocatechuic acid), benzoic acid, and hippuric acid (Mullen et al. 2008a). In contrast to the data of Walle and coworkers, obtained with [4-^{14}C]quercetin, almost all the radioactivity in the Mullen et al. study, which employed [2-^{14}C]quercetin-4′-*O*-glucoside, was recovered as radiolabeled metabolites and catabolites, indicating that there was no substantial loss as radiolabeled carbon dioxide. The metabolism of quercetin is summarized in Figure 10.1.

10.2.2 ANTHOCYANINS

Anthocyanins have a labile flavylium cation structure, and their microbial metabolism has been investigated only recently. Because of their instability, the detection of anthocyanins from human samples has been challenging. The total amount of anthocyanins in 24 h urine is very low, <0.2% of the ingested dose. Usually the parent anthocyanins predominate in urine, but methylated and glucuronidated derivatives have also been detected (Vitaglione et al. 2007) (see Chapter 2).

The microbial metabolism of anthocyanin glycosides includes deconjugation and ring-fission as described earlier for flavonols (Aura et al. 2005; Keppler and Humpf 2005; Fleschhut et al. 2006). Cyanidin is converted to 3,4-dihydroxy-benzoic acid, malvidin to 3,5-dimethoxy-4-hydroxybenzoic acid (syringic acid), peonidin to 3-methoxy-4-hydroxybenzoic acid (vanillic acid), and pelargonidin to 4-hydroxybenzoic acid. Again, the ring-fission catabolites reflect the substitution pattern of the B-ring of the precursors. Furthermore, an unidentified derivative of the anthocyanin aglycone with a molecular weight 85 mass units higher than that of the corresponding aglycone was detected for anthocyanidins originating from malvidin-, peonidin-, petunidin-, delphinidin-, and cyanidin-3-*O*-glucosides (Aura et al. 2005). 2,4,6-Trihydroxyphenylacetaldehyde, benzaldehyde, 3-methoxy-4-hydroxybenzoic acid, phloroglucinol aldehyde (2,4,6-trihydroxybenzaldehyde), 2,4,6-trihydroxybenzoic acid, and gallic acid produced from anthocyanins *in vitro* are thought to be derived from the A-ring (Keppler and Humpf 2005; Fleschhut et al. 2006). The metabolism of anthocyanins is summarized in Figure 10.2. The

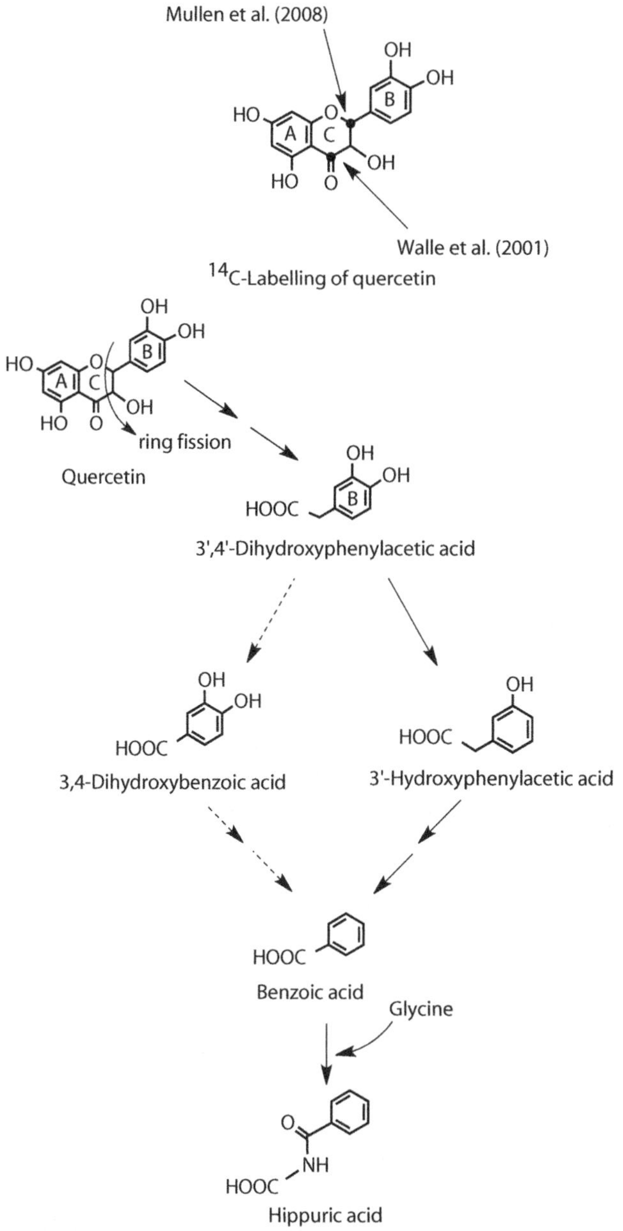

FIGURE 10.1 Metabolites of quercetin originating from colon and tissues (liver, intestinal epithelia) (According to Aura, A.-M. et al., *J. Agric. Food Chem.*, **50**, 1725–1730, 2002; Walle, T., Walle, U.K., and Halushka, P.V. *J. Nutr.*, **131**, 2684–2652, 2001; Mullen, W. et al. *J. Agric Food Chem.*, **56**, 12127–12137, 2008a.) Note that when the ingested [14C]quercetin was labeled at the 2-carbon, lost substantial amounts of radioactivity. With labeled at the 2-carbon, recovered 98% of the label after 72 h with the labeled metabolites being derived from ring B. The feed provides no information of compounds being produced from unlabeled ring A.

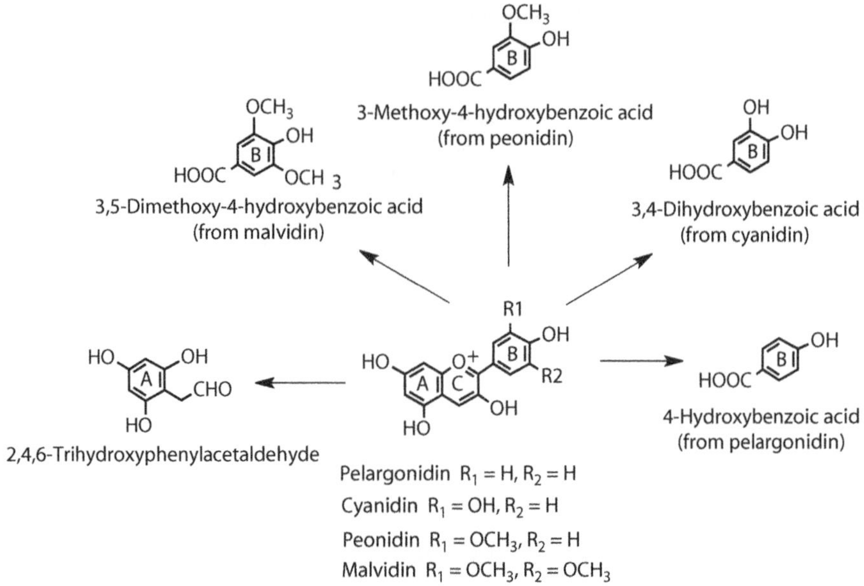

FIGURE 10.2 Microbial degradation of anthocyanins. (According to Aura, A.-M. et al., *Eur. J. Nutr.*, **44**, 133–142, 2005a and Fleschhut, J. et al., *Eur. J. Nutr.*, **45**, 7–15, 2006.)

major *in vivo* metabolite of cyanidin-*O*-glucosides, 3,4-dihydroxybenzoic acid, is excreted in 24-h urine and feces in amounts corresponding to 73% of the ingested anthocyanin (Vitaglione et al. 2007). Thus, results *in vitro* and *in vivo* are in agreement.

Compounds similar to colonic anthocyanin metabolites can be formed at neutral pH. Glycosidic anthocyanins exist in equilibrium between the flavylium cation (red colored and abundant at low pH), blue quinoidal bases, and the carbinol and chalcone pseudobases (noncolored), the three latter compounds being more abundant at pH 7 (Clifford 2000) (see Chapter 2). Anthocyanidins can form a reactive α-diketone, which can readily decompose to phenolic acids and the corresponding aldehydes (Keppler and Humpf 2005). However, under the mildly acidic conditions prevailing in the *in vitro* colon model, phenolic acids were not detected, unless active microbiota were present (Aura et al. 2005). This suggests that both microbial and spontaneous decomposition of anthocyanins can occur in the intestinal environment.

Forrester and Waterhouse (2008) carried out an *in vitro* study incubating an extract of Cabernet Sauvignon grapes with pig microbiota. Delphinidin-, petunidin-, peonidin-, and malvidin-3-*O*-glucosides disappeared within 6 h and three major metabolites accumulated, namely 3-*O*-methylgallic acid, 3,4-dimethoxybenzoic acid, and 2,4,6-trihydroxybenzaldehyde. The microbial metabolites from pigs, therefore, resemble those of humans. Anthocyanin bioavailability has also been investigated in Sprague-Dawley rats after ingestion of wild bilberries. Benzoic acid was the main metabolite in the liver and brain tissues after 4 and 8 week supplementation

(Bò et al. 2010). In addition, anthocyanins and their glucuronides were detected in pig brain tissues after administration of a 2% blueberry diet. The levels were low (femtomoles per gram fresh weight), but nonetheless they are indicative of passage through the blood–brain barrier (Milbury and Kalt 2010). These studies indicate that metabolites, despite their size and hydrophilic nature, may enter the liver and brain tissues and have the potential to be bioactive at these sites.

10.2.3 FLAVAN-3-OLS

Flavan-3-ols can be divided into monomers such as (+)-catechin and (−)-epicatechin and their galloylated derivatives, which occur in high amounts in green tea, and dimeric, oligomeric, and polymeric proanthocyanidins, which are abundant in cocoa, apples, grapes, and, as a consequence, beverages such as cider and wine.

Hydroxyphenylvaleric acids were shown to be formed from (+)-catechin by the *in vitro* action of rabbit intestinal microbiota (Scheline 1970). When two different batches of human microbiota was used in *in vitro* incubations, stereoisomerism did not have a marked effect on the conversion of (+)-catechin and (−)-epicatechin, but the site of ring-fission was dependent on the microbiota. (+)-Catechin and (−)-epicatechin were transformed either to hydroxyphenylpropionic or hydroxyphenylvaleric acid derivatives in two separate experiments using pools of microbiota obtained from different donors of the fecal inocula (Aura et al. 2008). (−)-Epicatechin, (−)-epigallocatechin, and their 3-O-gallates were catabolized extensively by human fecal suspensions, whereas gallates resisted degradation by a rat cecal suspension, suggesting species differences in catabolic activity of the two microbiota (Meselhy et al. 1997). In a recent study, (−)-epicatechin-3-O-gallate incubated *in vitro* with rat cecal microbiota was converted to 1-(3′,4′,5′-trihydroxyphenyl)-3-(2″,4″,6″-trihydroxy)propan-2-ol, 1-(dihydroxyphenyl-3-(2″,4″,6″-trihydroxy)propan-2-ol, and 5-(3′,5′-dihydroxyphenyl)-4-hydroxyvaleric acid (Takagaki and Nanjo 2010). A series of catechin isomers and their gallates were incubated in pig cecal microbiota, and all had a similar qualitative metabolite profile, including phenylpropionic acids, phenylacetic acids, benzoic acids, and phloroglucinol (van't Slot and Humpf 2009).

The first *in vivo* study identified 11 metabolites in human urine after (+)-catechin intake. The major components were 3-(3′-hydroxyphenyl)propionic acid and 5-(3′-hydroxyphenyl)-γ-valerolactone (Das 1971). In a pharmacokinetic study of tea catechins, ring-fission catabolites appeared in significant amounts after 3 h and peaked in plasma at 8–15 h after consumption (Lee et al. 2002). Ring-fission metabolites from tea polyphenols were excreted in urine as di- and trihydroxylated derivatives of phenyl-γ-valerolactone and they accounted for 1.5–16% of the ingested catechins (Meng et al. 2002).

A substantial proportion of ingested green tea flavan-3-ols passes through the small intestine to the large intestine (Stalmach et al. 2009). Roowi et al. (2010) showed that a minimum 40% of the intake of flavan-3-ols from green tea is converted to phenolic acid metabolites in the colon compared with ~8% excretion as flavan-3-ol methyl, glucuronide, and sulfate metabolites, which reflect small intestine absorption. Furthermore, colonic metabolites undergo hepatic metabolism, since urinary metabolites comprise 4-hydroxybenzoic acid, hippuric acid, and 3'-methoxy-4'-hydroxyphenylacetic acid, which do not accumulate in the fecal suspensions. Conversely, 5-(3',4'- dihydroxyphenyl)-γ-valerolactone and 5-(3',4'-dihydroxyphenyl)-γ-valeric acid, which were major (–)-epicatechin metabolites in fecal suspensions, were not excreted in urine (Roowi et al. 2010).

van't Slot et al. (2010) found that flavan-3-ols dimers, but not trimers, were degraded by pig cecal microbiota *in vitro*. When proanthocyanidin dimers were incubated with human colonic microbiota, the main metabolites were 3',4'-dihydroxyphenylacetic acid and 5-(3',4'-dihydroxyphenyl)-γ-valerolactone (Appeldoorn et al. 2009). In this model, the metabolism of dimers was slower than that of monomers (Aura et al. 2008). It should be noted that direct correlations cannot be made between different studies because the experiments did not use the same donors of microbiota. However, the conversion rates can be compared when using the same colon model, when the repeatability of the experiments is enhanced by pooling feces from several donors to obtain a diverse spectrum of microbiota for the inoculating suspensions.

Despite the similarities of microbial metabolite profiles of (–)-epicatechin and proanthocyanidin B_2 (10 common metabolites), five unique metabolites of the dimer B_2 were produced by human fecal microbiota (Stoupi et al. 2010a). These were dimeric intermediates, where the C-ring was cleaved either in the upper or in the lower unit (Stoupi et al. 2010b). Figure 10.3 shows possible routes for microbial metabolism of flavan-3-ol dimers. Stoupi et al. (2010c) also showed in male rats that orally administered ^{14}C-labeled dimer B2 was bioavailable only to a low degree (8–11%; $AUC_{0-24\,h}$) in blood, and that 63% of the label was excreted via urine within 4 days of administration, demonstrating the long residence time of the labeled metabolites prior to final excretion.

Deprez et al. (2000) found that only 9–22% of the label from [14C]proanthocyanidin was incorporated into the metabolite pool after *in vitro* fermentation with human colonic microbiota. The metabolites that were produced in this study included several derivatives of phenylvaleric acid, phenylpropionic acid, phenylacetic acid, and benzoic acids with different patterns of hydroxylation, and the total yields decreased significantly with an increased polymerization (Gonthier et al. 2003a; Rios et al. 2003; Bazzocco et al. 2008). Phenylcarboxylic acid metabolites derived from monomeric or dimeric catechins are also formed from their analogs in apples, cocoa, and almonds (Bazzocco et al. 2008; Urpi-Sarda et al. 2009a, 2009b). Lorach et al. (2009) detected 27 metabolites using metabolomic profiling of urine samples after cocoa intake. These metabolites included alkaloid derivatives and polyphenol metabolites, as well as processing-derived products such as diketopiperazines.

Further evidence linking a high degree of flavan-3-ol polymerization with reduced rates of degradation was obtained by Bazzocco et al. (2008) who showed that Marie Ménard apples and cider proanthocyanidins, with a respective average degree of polymerization 8.2 and 2.2, were broken down more extensively and yielded higher metabolite concentrations than Averolles apples and cider with an average degree of polymerization 71.2 and 7.4. Polymeric flavan-3-ols have a tendency to bind to

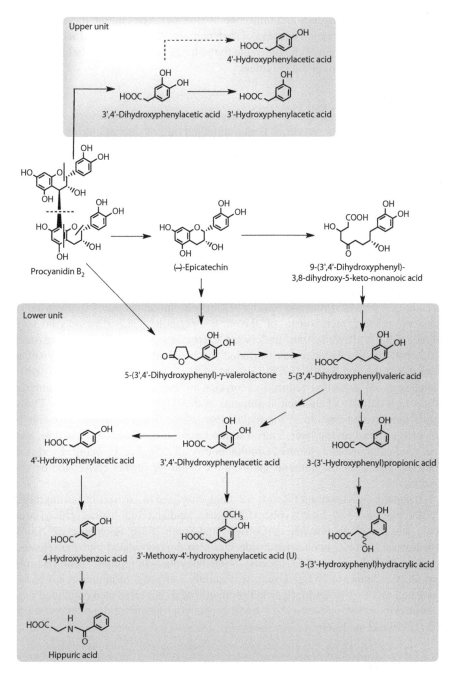

FIGURE 10.3   The proposed microbial metabolism of flavan-3-ol B$_2$ dimer to microbial metabolites. (Based on the data of Appeldoorn, M.M. et al., *J. Agric. Food Chem.*, **57**, 1084–1092, 2009; Roowi, S. et al., *J. Agric. Food Chem.*, **58**, 1296–1304, 2010; and Stoupi, S. et al., *Mol. Nutr. Food Res.*, **54**, 747–759, 2010a.)

proteins, causing astringency and inhibition of enzymes. The binding increases with the degree of polymerization because of high number of interacting hydroxyl groups (Scalbert 1991). This may well contribute to the lower number of metabolites and to the reduced metabolism of the polymers.

10.2.4 FLAVANONES AND FLAVONES

Flavanones are present in citrus fruits and their juices. Processing and storage causes precipitation of flavanones, which influences their fate in the gastrointestinal tract. Maximal concentrations in the plasma correlated well with the soluble flavanone intake, especially hesperetin-7-O-rutinoside, but not with the total flavanones ingested (Vallejo et al. 2010). Furthermore, intake of full-fat yogurt decreased the absorption of flavanones from orange juice (Mullen et al. 2008b). The flavone, apigenin-7-O-glucoside, is absorbed in amounts corresponding to 13% of the dose, and deglycosylation occurs *in vitro* when it is incubated with cytosolic extracts of *Eubacterium ramulus* and *Bacteroides distasonis* and also a microsomal fraction of intestinal mucosa of rats (Hanske et al. 2009).

The importance of the position of hydroxyl groups on the flavonoid skeleton was addressed by Griffiths and Smith (1972b), who found out that 5-hydroxylation enhanced colonic metabolism of flavones. Apigenin (4',5,7-trihydroxyflavone), but not 4',7-dihydroxyflavone, underwent C-ring fission, and likewise genistein (4',5,7-trihydroxyisoflavone) was degraded but not daidzein (4',7-dihydroxyisoflavone) (Griffiths and Smith 1972b). Simons et al. (2010) showed that 4',5,7-trihydroxyflavonoids exhibited reduced absorption in the upper intestine in humans, indicating a delivery of substantial amounts of these flavonoids to the colon. These results suggest that the site of the metabolism of the 5-hydroxylated favonoids is the colon rather than the liver. Thus, the food matrix, the flavonoid structure, and other food components consumed simultaneously all influence flavonoid absorption efficiency and, thus, the amounts entering the colon for further conversion by the microbiota.

Naringenin-7-O-neohesperidoside was deglycosylated to naringenin and degraded to phloroglucinol, 3-(4'-hydroxyphenyl)propionic acid and 3-(3'-hydroxyphenyl)propionic acid by colonic microbiota (Griffiths and Smith 1972b; Rechner et al. 2004). Dihydrochalcone-4-O-neohesperidoside was also deglycosylated and the released aglycone hydrolyzed to 3-(3'-hydroxy-4'-methoxyphenyl)propionic acid (Braune et al. 2005). Apigenin-7-O-glucoside was mainly converted in human microbiota-associated rats to 3-(4'-hydroxyphenyl)propionic acid, but urine also contained naringenin, eriodictyol, phloretin, 3-(3'-hydroxyphenyl)propionic acid, and *p*-coumaric acid (Hanske et al. 2009).

10.2.5 ISOFLAVONES

The isoflavones daidzein and genistein occur in soy foods as glycosides although the aglycones are found in some processed products (see Chapter 5). Setchell et al. (2002a) detect urinary isoflavones as deglycosylated sulfates and glucuronides, implying the involvement of intestinal metabolism in their bioavailability, while Xu et al. (1995) concluded that isoflavone bioavailability is also related to the extent to which they are degraded by the gut microbiota.

In vitro anaerobic incubation with human feces has shown that the half-lives of genistein and daidzein are ~3 and 7.5 h, respectively (Setchell et al. 2002b). Two routes of metabolism are recognized for daidzein, depending on subjects and their gut microbiota. Some subjects produce equol [7-hydroxy-3-(4'-hydroxyphenyl)chroman] via dihydrodaidzein and tetrahydrodaidzein, and others produce *O*-desmethylangolensin via 2'-dehydro-*O*-desmethylangolensin (Figure 10.4). Thus, there are two sub-populations: those who have the microbiota capable of synthesizing equol and those who lack the microbes to do so (Joannou et al. 1995; Setchell et al. 2002b).

Transepithelial transport of equol was studied *in vitro* using CaCo-2 cells. Free and conjugated equol were followed in apical medium, cells, and in basolateral medium for 8 h. Apical medium showed a decline of free and an increase of conjugated equol, CaCo-2 cells contained only conjugated equol, and the basolateral medium showed a transient appearence of free equol and a steady increase of conjugated equol (Walsh and Failla 2009). Thus, *in vivo* most equol can be expected to appear in conjugated derivatives.

Microbial metabolism of genistein is different from that of daidzein. Genistein is reduced to dihydrogenistein, which is further metabolized to 6'-hydroxy-*O*-desmethylangolensin (Joannou et al. 1995). *In vitro* incubations with human fecal and rat cecal microbiota revealed formation of 2-(4'-hydroxyphenyl)propionic acid from 6'-hydroxy-*O*-desmethylangolensin, indicating C-ring fission (Coldham et al. 2002) (Figure 10.5). Heinonen and coworkers (2004a) identified four novel

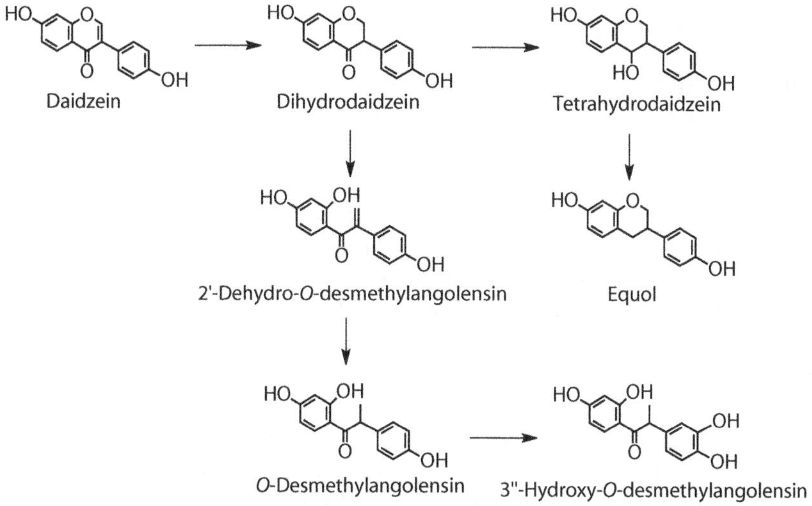

FIGURE 10.4 Microbial metabolism of the isoflavone genistein. (According to Joannou, G.E. et al., *J. Steroid Biochem. Mol. Biol.*, **54**, 167–184, 1995; and Setchell, K.D., Brown, N.M., and Lydeking-Olsen, E., *J. Nutr.*, **132**, 3577–3584, 2002b.)

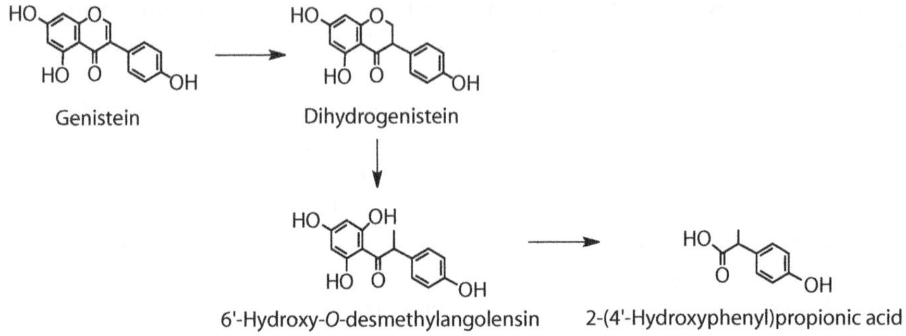

Genistein   Dihydrogenistein

6'-Hydroxy-O-desmethylangolensin   2-(4'-Hydroxyphenyl)propionic acid

FIGURE 10.5 Microbial metabolism of isoflavonoid daidzein. (According to Joannou, G.E. et al., *J. Steroid Biochem. Mol. Biol.*, **54**, 167–184, 1995; and Coldham, N.G. et al., *Xenobiotica*, **32**, 45–62, 2002.)

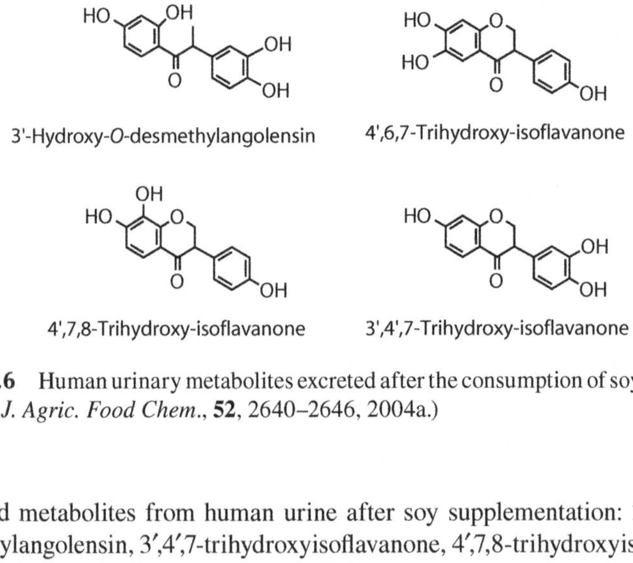

3'-Hydroxy-O-desmethylangolensin   4',6,7-Trihydroxy-isoflavanone

4',7,8-Trihydroxy-isoflavanone   3',4',7-Trihydroxy-isoflavanone

FIGURE 10.6 Human urinary metabolites excreted after the consumption of soy. (Heinonen, S.-M. et al., *J. Agric. Food Chem.*, **52**, 2640–2646, 2004a.)

isoflavonoid metabolites from human urine after soy supplementation: 3″-hydroxy-O-desmethylangolensin, 3',4',7-trihydroxyisoflavanone, 4',7,8-trihydroxyisoflavanone, and 4',6,7-trihydroxyisoflavanone (Figure 10.6).

Red clover, which contains the isoflavones biochanin A and formononetin, is consumed by cows in Finland and, as a consequence, high levels of equol have been detected in Finnish milk that is sold for human consumption (Hoikkala et al. 2007). In humans, urinary excretion indicates that formononetin is readily metabolized to daidzein, which can be converted to O-desmethylangolensin and equol, as well as being metabolized to dihydroformononetin and angolensin. Biochanin A is metabolized to dihydrobiochanin A and 6'-hydroxyangolensin and is also converted, via genistein, to 6'-hydroxy-O-desmethylangolensin (Figure 10.7) (Heinonen et al. 2004b). This explains the similarity of these microbial metabolites to those of daidzein and genistein.

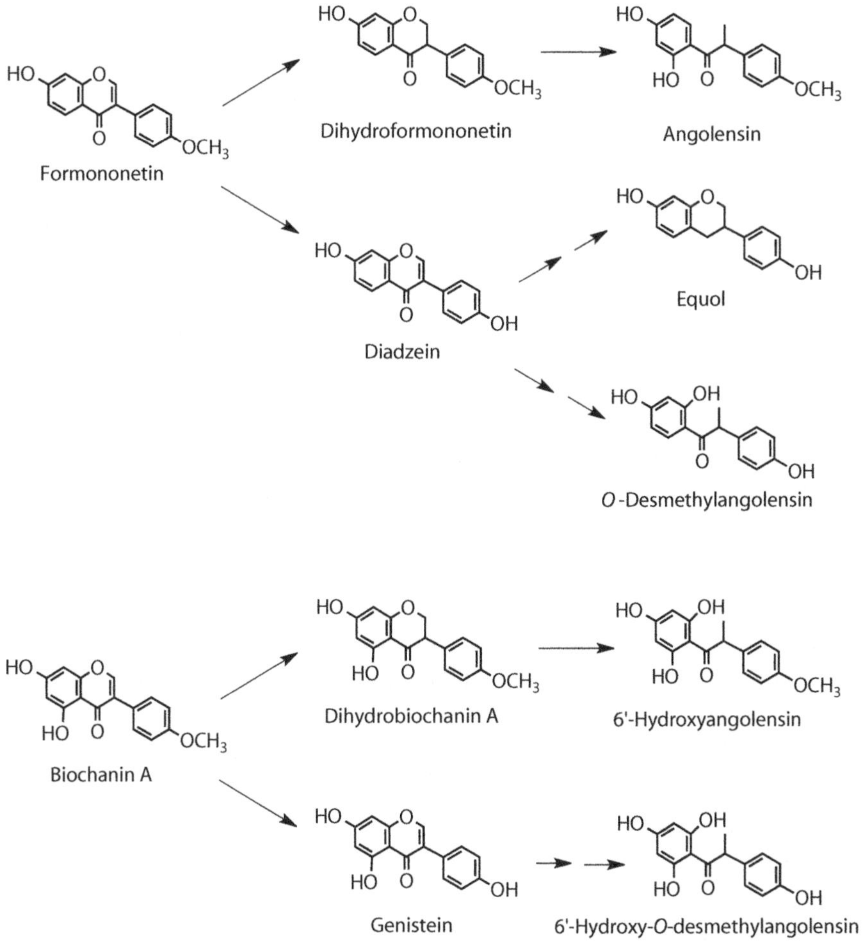

FIGURE 10.7 Human metabolism of the red clover isoflavones formononetin and biochanin A. (After Heinonen, S.-M., Wähälä, K., and Adlercreutz, H., *J. Agric. Food Chem.*, **52**, 6802–6809, 2004b.)

10.2.6 PHENOLIC ACIDS

Phenolic acids are present in a number of foods with especially high amounts occurring in whole grains, berries, and wine. The most abundant phenolic acid in cereals is the hydroxycinnamate ferulic acid. Since a high proportion of the ferulic acid in rye is bound to arabinoxylans by ester bonds, its bioavailability requires the presence of degradative esterases in intestinal tissues and microbiota (Kroon et al. 1997; Plumb et al. 1999; Andreasen et al. 2001a, 2001b; Rondini et al. 2004). In addition, Andreasen et al. (2001a) have demonstrated the release of sinapic acid and *p*-coumaric acid from rye and wheat bran by human colonic esterases. Hydrolysis by intestinal esterases is probably the major route for release of soluble, conjugated hydroxycinnamic acids *in vivo.* Such free phenolic acids can then be absorbed across the gastrointestinal barrier and enter the peripherial circulation of mammals (Andreasen et al. 2001b).

When eight postmenopausal women consumed rye bread, daily urinary excretion of ferulic acid was 47% of daily intake (Harder et al. 2004). This bioavailability of ferulic acid is also affected by its further metabolism to other phenolic acids. Bioprocessing of wheat bran increases both release and metabolism of phenolic acids. An experiment was performed using wholegrain wheat breads fortified with either bioprocessed (by enzymatic and yeast fermentation) or native wheat bran. Bioaccessibility in the upper intestine and colonic conversions were studied in human *in vitro* gastrointestinal models TIM-1 and TIM-2, which are computer-controlled simulations of the intestinal tract, including regulation of pH at different stages, additions of enzymes, and finally human microbiota. The models also include peristalsis and absorption through membranes. The bioaccessibility of ferulic acid, *p*-coumaric acid, and sinapic acid was increased five-fold if the bran in the bread had undergone bioprocessing. Since the contents of *p*-coumaric acid and sinapic acid were low and their release was mainly in the TIM-1 model, the microbial conversion products were better connected to matrix bound ferulic acid (Mateo Anson et al. 2009). The major colonic metabolites in the TIM-2 model from the wheat bran fortified breads were 3-(3′-hydroxyphenyl)propionic acid and 3-phenylpropionic acid independent of the bioprocessing; however, the release and conversion were enhanced by enzymatic and yeast fermentation processes of the wheat bran (Mateo Anson et al. 2009). A similar pattern was observed when 300 g of the same breads were digested by eight healthy volunteers after which plasma and urine samples were collected for 24 h after intake. In addition to ferulic, sinapic, and coumaric acids, 3,4-dimethoxybenzoic acid was detected as an early metabolite from the whole grain breads. Bioavailability of the phenolic acids was increased 2–3-fold by bioprocessing, and 3-(3′-hydroxyphenyl)propionic acid and 3-phenylpropionic acid were detected as major metabolites from the matrix-bound phenolics entering the colon (Mateo Anson et al. 2011).

Additional conversions of hydroxycinnamic acids can be observed *in vitro.* Formation of 3-(3′-hydroxyphenyl)propionic acid and small amounts of benzoic acid occurred when caffeic acid and its esters, chlorogenic acid and caftaric acid, were used a substrates in an *in vitro* human colon model (Gonthier et al. 2006). The side-chain shortening of phenylpropionic acid occurs via β-oxidation. Caffeic acid can also be decarboxylated and reduced to 4-ethylcatechol by human fecal microbiota (Peppercorn and Goldman 1971).

Ferulic acid dimers are another form of released phenolic acids from cereal matrices (Andreasen et al. 2000). Two ferulic acid moieties can be bound to each other either via either an 8-*O*-4- or a 5-5-linkage. Dehydrodiferulic acid 8-*O*-4- and 5-5-derivatives were incubated with human fecal microbiota. The 8-*O*-4-derivative was shown to degrade transiently to monomeric ferulic acid, which was further metabolized to 3-(3′,4′-dihydroxyphenyl)propionic acid, 3′,4′-dihydroxyphenyl acetic acid, 3-phenylpropionic acid, and benzoic acid. An alternative route to benzoic acid involved metabolism to 3-(4′-hydroxy-3′-methoxyphenyl)pyruvic acid (Figure 10.8). In contrast, the 5-5-diferulate derivative was subject only to demethylation and/or side-chain reduction as shown in Figure 10.9 (Braune et al. 2009).

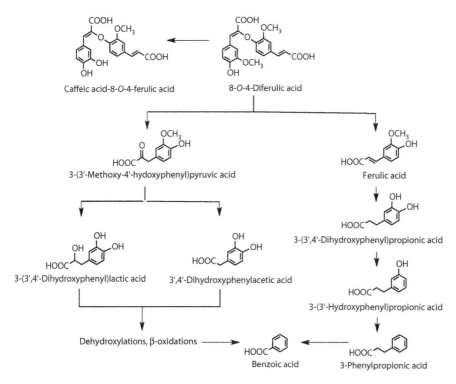

FIGURE 10.8 Metabolism of 8-*O*-4 ferulic acid dimer. (From Gonthier, M.P. et al., *Biomed. Pharmacotherapy*, **60**, 536–540, 2006; Braune, A. et al., *J. Agric. Food Chem.*, **57**, 3356–3362, 2009; and Mateo Anson, N. et al., *J. Agric Food Chem.*, **57**, 6148–6155, 2009.)

Feeding studies with chlorogenic acid (5-*O*-caffeoylquinic acid) revealed the appearance of hydroxylated cinnamic acids, benzoic acid, and hippuric acid in plasma and urine of rats. Hippuric acid is formed by glycination of benzoic acid in the liver (Gonthier et al. 2003b). In humans, half of the ingested chlorogenic acid was excreted in urine as hippuric acid (Olthof et al. 2003), while 57% of the ingested chlorogenic acid was detected in rat plasma and urine as microbial metabolites (Gonthier et al. 2003b). Coffee is an especially rich source of caffeoylquinic acids, and Chapter 6 contains information on recent research on their absorption in the small and large intestine, metabolism, and urinary excretion in healthy humans and volunteers with an ileostomy by Stalmach et al. (2009, 2010).

In rats, cranberry phenolic acids are excreted in urine as methylated, sulfated, and glucuronidated conjugates of hydroxyphenylacetic acid and hydroxypropionic acid. The degree of conjugation ranged from 65 to 100% depending on the phenolic acid (Prior et al. 2010). Nurmi and coworkers (2009) followed urinary excretion of phenolic acids after ingestion of a bilberry–lingonberry purée for a 48-h period. Early excretion of 3,4-dihydroxybenzoic acid was observed, together with caffeic acid, *p*-coumaric acid, and ferulic acid. A low-level excretion of isoferulic acid and dehydroisoferulic acid was also detected. 3′-Methoxy-4′-hydroxyphenylacetic

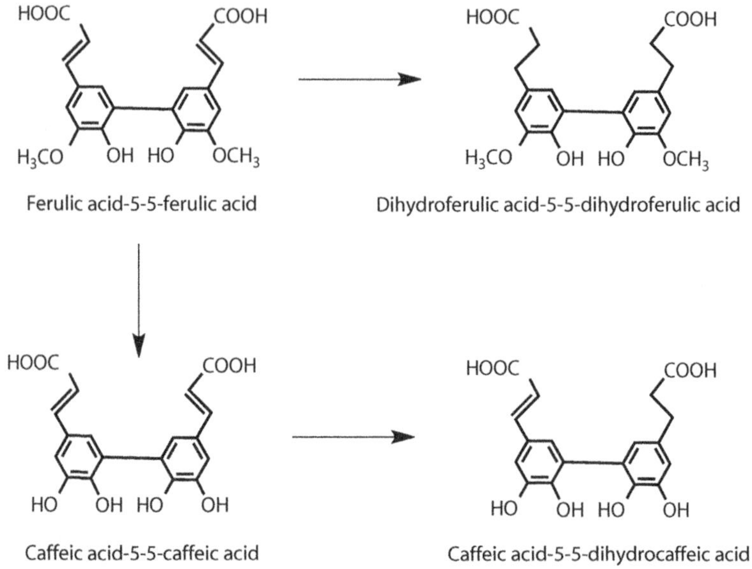

FIGURE 10.9 Metabolism 5-5-ferulic acid dimer modified. (From Braune, A. et al., *J. Agric. Food Chem.*, **57**, 3356–3362, 2009.)

acid (homovanillic acid) and 3-methoxy-4-hydroxybenzoic acid (vanillic acid) were the most abundant metabolites, peaking after 6 h. 3-(3′,4′-Dihydroxyphenyl) propionic and the corresponding acetic acid appeared 12 h of ingestion, and at the same time, their monohydroxylated 3′-derivatives maintained steady profiles (Nurmi et al. 2009). The metabolic pattern of bilberry–lingonberry purée reflects the presence of bilberry anthocyanins and lignonberry phenolic acids and flavonols.

A berry mixture containing bilberries, lingonberries, blackcurrants, and chokeberries was ingested in a placebo-controlled human intervention study. (Poly)phenol intake was 837 mg/day, and quercetin, *p*-coumaric acid, 3′-hydroxyphenylacetic acid, caffeic acid, 3,4-dihydroxybenzoic acid, 3-methoxy-4-hydroxybenzoic acid, 3′-methoxy-4′-hydroxyphenylacetic acid, and 3-(3′-hydroxyphenyl)propionic acid increased significantly in the plasma of the berry group compared with that of the control group. The urinary excretion of quercetin, *p*-coumaric acid, and 3′-hydroxyphenylacetic acid was increased likewise in the berry group (Koli et al. 2010).

10.2.7 Ellagitannins

Ellagitannins are large molecules that are not absorbed but are subject to metabolism by the colonic microflora. Doyle and Griffiths (1980) showed that in rats 10% of ellagic acid intake was excreted in urine and feces along with an unidentified compound. Later it was shown that the degradation pathway of ellagitannins by human and animal microbiota involves hydrolysis to ellagic acid that is converted to a series of lactones, called urolithins (3,8-dihydroxy-6H-dibenzopyran-6-one and

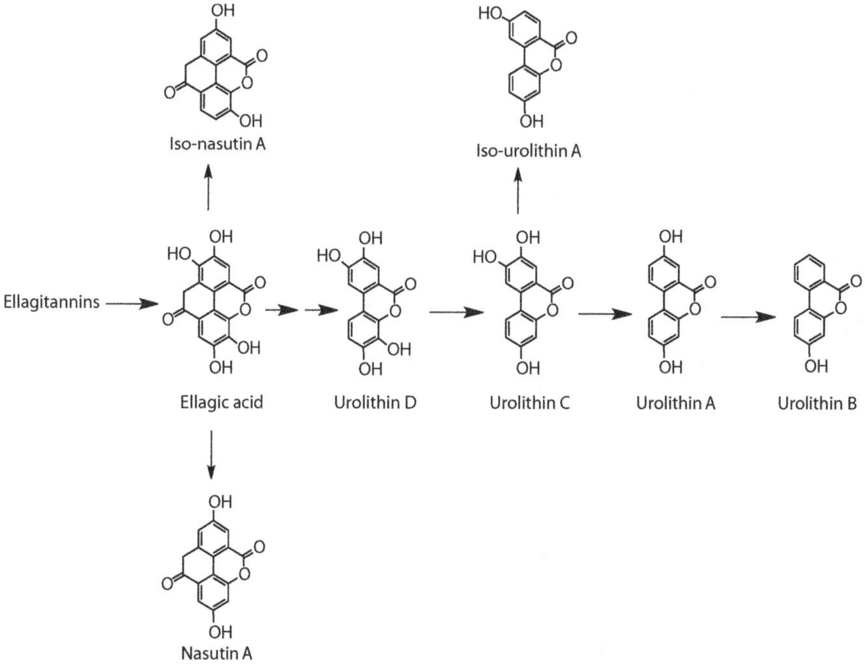

FIGURE 10.10 Proposed route for the conversion of an ellagitannin, punicalagin, to ellagic acid, urolithins and nasutins. (Based on the findings of Cerda, B. et al., *Eur. J. Nutr.*, **43**, 205–220, 2004; Seeram, N.P. et al., *J. Nutr.*, **136**, 2481–2485, 2006; and González-Barrío, R. et al., *J. Agric. Food Chem.*, **59**, 1152–1162, 2011.)

its hydroxylated analogs). An assortment of urolithins, along with nasutins, is produced by the routes illustrated in Figure 10.10 (Cerda et al. 2004; Seeram et al. 2006; González-Barrío et al. 2011). Urolithins occur in human plasma as glucuronides, and sulfates have also been detected (González-Barrío et al. 2011). The total human urinary excretion of urolithin B derivatives ranged from 2.8 to 16.6% of the ingested ellagitannins after a single dose of strawberries, red raspberries, walnuts, or oak-aged red wine. There are large individual differences dividing volunteers into high- and low-urolithin producers (Cerda et al. 2005; González-Barrío et al. 2010) (see Chapter 9). Ellagitannins are reported to modulate colonic microbiota by inhibiting pathogenic clostridia and *Staphylococcus aureus*, without affecting to any extent lactobacilli and bifidobacteria (Bialonska et al. 2009).

10.2.8 PLANT LIGNANS

In epidemiological studies, a diet rich in plant lignans, which occur principally in wholegrain, dietary fiber, fruits, and vegetables, is associated with high *in vivo* concentrations of enterolactone in Finnish and Dutch men and women (Kilkkinen et al. 2003; Milder et al. 2007). The conversion of plant lignans to enterolactone has been observed in intervention studies with rats (Aura et al. 2006), humans (Juntunen et al.

2000; Mazur et al. 2000; Knust et al. 2006), and cannulated pigs (Bach Knudsen et al. 2003).

Microbial metabolism of plant lignans to the mammalian lignans enterodiol and enterolactone is the first and the most studied conversion process. The metabolism of plant lignans involves both mammalian (glucuronidation and to a lesser degree sulfation) and gut microbial processes (Rowland et al. 2003). Despite the structural diversity of plant lignans from different sources, they undergo conversion to enterodiol and enterolactone by human gut microbiota (Heinonen et al. 2001; Penalvo et al. 2005a, 2005b; Eeckhaut et al. 2008). In the colon, secoisolariciresinol, matairesinol, pinoresinol, and lariciresinol are all metabolized to enterolactone, although syringaresinol is probably transformed via another route. After consumption of rye, conversion of medioresinol, and other as yet unidentified lignans, to enterolactone also contribute to enterolignan exposure (Peñalvo et al. 2005a).

Enterolactone production *in vitro* is slow compared with synthesis of enterodiol (Aura et al. 2006; Eeckhaut et al. 2008). The ratio of enterolactone- and enterodiol-producing bacteria is 1:2000 as shown by culture-based and 16S rRNA-targeted molecular methods. These observations indicate that enterodiol-synthesizing bacteria are dominant while enterolactone-forming microbiota are a minor population (Clavel et al. 2005). Interestingly, several research groups have shown a negative association between frequent bowel movements and enterolactone levels (Kilkkinen et al. 2001; Johnsen et al. 2004; Milder et al. 2007). Human subjects with more rapid colonic transit may have less time to convert plant lignans to enterolactone than subjects whose colon motility is slow and the subsequent transit time of the colonic contents more prolong.

Sesaminol triglucoside and sesamin are the main lignans of sesame (*Sesamum indicum* L.). Analysis of human urine has shown that these lignans are converted to enterolactone (Peñalvo et al. 2005b) and glucuronidated/sulfated catechol metabolites (Moazzami et al. 2007; Jan et al. 2010). The catechol metabolites are also detectable in liver, lung, kidney, heart, and brain tissues of rats at 40- to 100-fold lower concentrations than in plasma (Jan et al. 2010). Furthermore, deuterium-labeled polymeric lignin and a monomeric 7′-hydroxy-matairesinol from the coniferous tree *Picea abies* are converted to enterolactone in rats (Saarinen et al. 2000; Begum et al. 2004). Figure 10.11 summarizes the structures of plant lignans present in cereal grains and sesame seeds and their microbial metabolites, enterodiol and enterolactone.

Stereoisomerism appears to have a role in the metabolism of plant lignans as (−)-secoisolariciresinol is converted to (−)-enterolactone, whereas (−)-secoisolariciresinol diglucoside is converted to (+)-secoisolariciresinol, which is metabolized to (+)-enterolactone (Saarinen et al. 2002). Secoisolariciresinol is demethylated and hydroxylated before conversion to enterodiol and enterolactone (Blaut and Clavel 2007; Eeckhaut et al. 2008). Additional hydroxylation and conjugation patterns diversify enterolignan derivatives in the circulation (Jacobs et al. 1999). Secoisolariciresinol and its anhydro-derivative are demethylated by *Peptostreptococcus productus*, *Eubacterium limosum*, and *Clostridium methoxybenzovorans*, and the demethylated products are efficiently dehydroxylated to enterodiol by *Eggerthella lenta* (Struijs

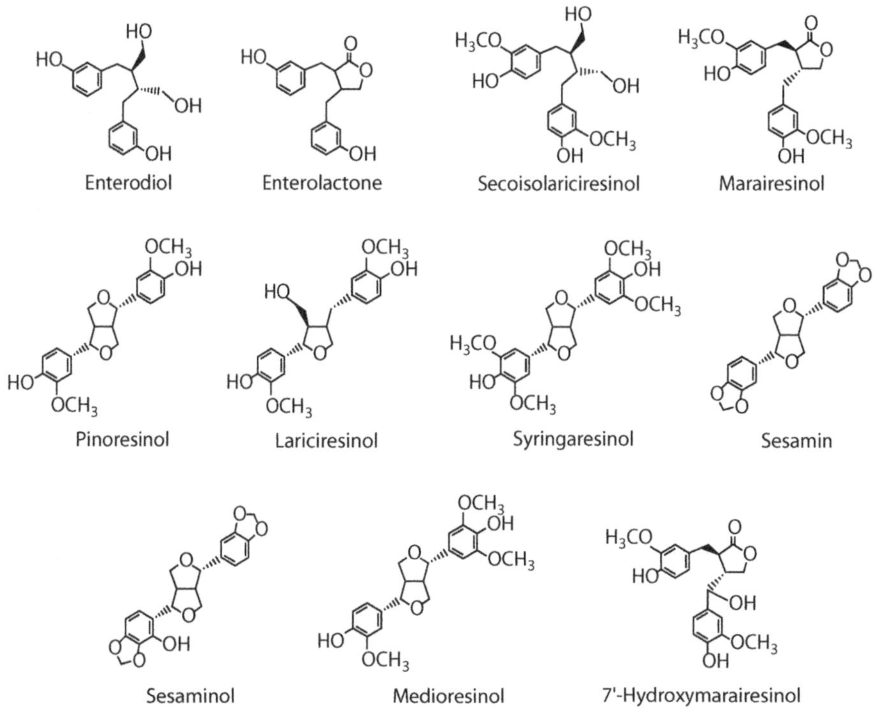

FIGURE 10.11 Structures of plant lignans present in cereal grains and sesame seeds, and their microbial conversion products enterodiol and enterolactone. (According to Heinonen, S. et al., *J. Agric. Food Chem.*, **49**, 3178–3186, 2001; Peñalvo, J.L. et al., *J. Agric. Food Chem.*, **53**, 9342–9347, 2005a; and Peñalvo, J.L. et al., *J. Nutr.*, **135**, 1056–1062, 2005b.)

et al. 2009). *Ruminococcus* (*R.*) sp. END-1 is capable of oxidizing (−)-enterodiol to (−)-enterolactone via the intermediate enterolactol. Furthermore, arctiin and secoisolariciresinol diglucoside are, respectively, converted to (−)-dihydroxyenterolactone and (+)-dihydroxyenterodiol by the bacterium. When *Eggerthella* sp. SDG-2 is coincubated with *Ruminococcus*, further dehydroxylation occurs, and (−)-enterolactone and (+)-enterodiol are synthesized (Jin and Hattori 2009). Thus, enterolactone conversion is a complex phenomenon involving several precursors, several bacterial species, different intermediary metabolites, and diverse conjugation patterns.

10.3 SIGNIFICANCE OF PHENOLIC METABOLITES FOR HUMAN HEALTH

The complex microbial community provides a high metabolic potential and interaction with the host via the colonic epithelium. The metabolite profiles show considerable inter- and intra-individual differences. The metabolite concentrations are more variable than the metabolite compositions among subjects, suggesting that different colonic communities share general metabolic activities, which convert food components to specific metabolite profiles (Jacobs et al. 2009). Thus, a large number of different phenolic precursors are converted into a relatively small number of colonic phenolic acids and lactones, which comprise the predominant metabolites.

The daily flux of water through the colonic epithelium is estimated to be more than 1 L calculated from the daily volumes of the chyme entering the cecum (~1500 mL) and that of the fecal excretion (~200 mL) (Guyton and Hall 1996). As the with metabolites decribed earlier, their hepatic conjugates are found in plasma and urine. They therefore circulate through the body and may exhibit both local and systemic effects. Phenolic metabolite levels in plasma range from low to high nanomolar concentrations (Sawai et al. 1987; Kilkkinen et al. 2001; Kern et al. 2003; Johnsen et al. 2004; Kuijsten et al. 2006), whereas urinary excretion is in the micromolar range (Gross et al. 1996; Knust et al. 2006). In peripheral tissues, the concentrations are anticipated to be even lower.

Local effects at the colonic epithelia are related to reduced instances of inflammation and colon cancer. Colonic microbiota are intimately associated with the colonic epithelium, preventing the invasion of pathogenic bacteria and interacting with the immune system via lymphocytes in the epithelia and mucosa (Salminen et al. 1998). Fecal water is an important mediator, since it is the carrier of the metabolites through the colonic epithelium and is, thus, closely connected to its immunoresponses. Human fecal water contains a mixture of phenolic acids, including phenylacetic acid, phenylpropionic acid, benzoic acid, and hydroxycinnamic acid derivatives (Jenner et al. 2005). According to recent data from NMR-based metabolite profiling, fecal water from vegetarians contains acetic acid, butanoic acid, propanoic acid, glutamic acid, and alanine as major metabolites, and smaller amounts of glycine and fumaric acid, which are known to have anti-inflammatory and antitumorigenic properties. Fecal water of vegetarians was able to inhibit prostaglandin E2 production in human colon HT-29 cell line. Thus, vegetarians have a metabolite profile, suggesting a lower risk of chronic diseases (Pettersson et al. 2008). NMR-based metabolite profiling of fecal water has also been used to investigate potential biomarkers of colon cancer (Monleón et al. 2009; Pearson et al. 2009). Kuijsten et al. (2006) found a substantial reduction of colorectal adenoma risk among subjects with a high plasma concentration of enterolignans, particularly enterodiol. However, the association between plasma enterodiol and colorectal cancer risk was less apparent among women, who were regular smokers and had a high body-mass index (Kuijsten et al. 2008).

It has also been observed that anti-inflammatory potential of blueberry (poly) phenolics is mediated by their colonic metabolites rather than by the berry (poly) phenolics *per se* (Russell et al. 2007). Microbial phenolic acid metabolites have been shown to inhibit prostaglandin production in colon fibroblast cells and reduce expression of cytokines, malonyldialdehyde, and oxidative-DNA damage in distal colonic mucosa (Larrosa et al. 2009). Urolithin A may have a role in cancer prevention as it is able to inhibit Wnt signalling, which is activated in 90% of colon cancer cases (Sharma et al. 2010). Futhermore, microbial metabolites inhibit proinflammatory cytokine production from lipopolysaccharide-stimulated peripheral blood mononuclear cells (Monagas et al. 2009). Enterodiol and enterolactone modulate immunoresponse in human peripheral blood lymphocytes after lipopolysaccharide and monoclonal antibody stimulation. This effect is mediated by Nuclear Factor-κB signaling (Corsini et al. 2010). Jan et al. (2010) demonstrated a reduction of cytokines in lipopolysaccharide-stimulated murine macrophages by catechol metabolites derived from sesaminol triglucoside.

Karlsson and coworkers (2005) showed that phenolic acid metabolites at physiologically relevant concentrations of 0.01–417 μM, inhibited cyclooxygenase-2 and affected its protein level in the colonic HT-20 cell line. This enzyme plays a major role in regulation of inflammation. Gao et al. (2006) showed that 3′,4′-dihydroxyphenylacetic acid had antiproliferative activity against colon cancer cells, and it was more inhibitory against colon cancer HCT116 cells than against a normal intestinal cell line (IEC6). Similarly, ellagic acid and punigalagin, a pomegranate ellagitannin, which is converted to ellagic acid in the colon (see Chapter 9), were able to induce apoptosis in colon cancer Caco-2 cell line but not in normal colon cells (Larrosa et al. 2006a). Urolithins also inhibit cell proliferation and induce apoptosis in colon cancer HT-29 cells (Kasimsetty et al. 2010).

The local effects of colonic metabolites can also be exerted on intestinal epithelia and the liver due to enterohepatic circulation. Some studies have been conducted *in vivo* with animals and in cell lines *in vitro*. Urolithins A and B were shown to modulate phase I and II detoxifying enzymes in CaCo-2 cells. Expression and activity of phase I and phase II enzymes (including cytochrome P450 1A1 and glucuronosyltransferase 1A10) were induced and several sulfotransferases inhibited, which favored the formation of glucuronides over sulfate conjugates. However, the *in situ* introduction of urolithins, or a pomegranate extract, to rats showed CYP1A1 induction only in a buffer matrix, but not in oil or when the components were introduced in feed (Gonzáles-Sarrías et al. 2009).

The anthocyanin catabolite protocatechuic acid (3,4-dihydroxybenzoic acid) has been shown to modulate cytochrome P450 and phase II enzymes in mouse liver and kidney (Krajna-Kuzniak et al. 2005), and it has a protective effect *in vivo* on tert-butyl-hyperoxide-induced rat hepatotoxicity (Liu et al. 2002). Protocatechuic acid also has an apoptotic effect in human breast, lung, liver, cervix, and prostate cancer cells, lowers interleukins in a concentration-dependent manner, and suppresses vascular endothelial growth factor production at concentrations of 2–8 μM (Yin et al. 2009). It remains to be determined if these *in vitro* effects of protocatechuic acid are apparent *in vivo* when the metabolite is produced in the colon after intake of dietary cyanidin-based anthocyanins.

The term phytoestrogen is used for lignans and isoflavones because of their structural similarity to estradiol-17β (Cassidy et al. 2000; Adlercreutz 2002; Setchell et al. 2005). Their intake is associated with a decreased risk of hormone-related cancers (Adlercreutz 2002; Magee and Rowland 2004; Lof and Weiderpass 2006). Equol has affinity for both estrogen receptors, ERα and ERβ, and it possesses high estrogenic activity, possibly explaining hormone-dependent responses of soy products (Setchell et al. 2002b). A higher concentration of plasma enterolactone is associated with a lower risk of breast cancer (Webb and McCullough 2005; Touillaud et al. 2007). Urolithins have shown estrogenic and antiestrogenic activities as potential endocrine-disrupting molecules, as they were able to enter estrogen sensitive MCF-7 breast cancer cells and bind to ERα and ERβ (Larrosa et al. 2006b). Jan et al. (2010) demonstrated estrogenic activity of sesaminol metabolites by ligand-dependent transcriptional activation of estrogen receptors. Components affecting the cytochrome P450 CYP1B1 enzyme in a prostate cell line are considered chemopreventive and urolithins are able to decrease activity and expression in 22Rv1 prostate cancer cells (Kasimsetty et al. 2009).

Other beneficial effects of phenolic metabolites of relevance to human health include elevated plasma enterolactone levels being associated with a lower risk of acute coronary events (Vanharanta et al. 1999) and isoflavone intake and equol production being linked with prevention of osteoporosis and maintaining good bone health (Setchell et al. 2002b). Low and in some instances sub-mM concentrations of several colonic-derived phenolics and lactones have also been shown *in vitro* to have antiglycative effects and to protect human neurons from midoxidative stress (Verzelloni et al. 2011).

10.4 FUTURE PROSPECTS

Dietary components undergo several transformations when they pass through the digestive tract. The colon and its microbiota have a crucial role in these processes. Simple phenolic compounds, derived from dietary flavonoids and related compounds, have been detected in plasma, serum, and urine. A large number of different (poly)phenolic substrates are converted by the colonic microbiota to a comparatively small number of phenolic acid metabolites and lactones. The metabolites form a pool, the elucidation of which has recently been facilitated by novel profiling techniques. As a consequence, it is now becoming possible to take into account the synergistic effects mediated by the entire dietary phytochemical pool, instead of merely studying the impact of individual compounds. Developments in data processing will facilitate the discovery of correlations between the metabolite diversity in the circulation and *in vivo* biomarkers of human homeostasis, when subjects are exposed to a defined dietary regime. Metabolomics and nutrigenomics provide tools to investigate the regulation of biological processes through associations with signal pathways. In the near future, systems biology is likely to facilitate bridge building between food chemistry and biomarkers of exposure and human health.

REFERENCES

Adlercreutz, H. (2002). Phytoestrogens and cancer. *Lancet Oncol.*, **3**, 364–373.

Adlercreutz, H., van der Wildt, J., Kinzel, J. et al. (1995). Lignan and isoflavonoid conjugates in human urine. *J. Steroid Biochem. Mol. Biol.*, **52**, 97–103.

Andreasen, M.F., Christensen, L.P., Meyer, A.S. et al. (2000). Content of phenolic acids and ferulic acid dehydrodimers in 17 rye (*Secale cereale* L.) varieties. *J. Agric. Food Chem.*, **48**, 2837–2842.

Andreasen, M.F., Kroon, P.A., Williamson, G. et al. (2001a). Intestinal release and uptake of phenolic antioxidant diferulic acids. *Free Radic. Biol. Med.*, **31**, 304–314.

Andreasen, M.F., Kroon, P.A., Williamson, G. et al. (2001b). Esterase activity able to hydro-lyze dietary antioxidant hydroxycinnamates is distributed along the intestine of mammals. *J. Agric. Food Chem.*, **49**, 5679–5684.

Appeldoorn, M.M., Vincken, J.-P., Aura, A.-M. et al. (2009). Procyanidin dimers are metabolised by human microbiota with 2-(3,4-dihydroxyphenyl)-acetic acid and 5-(3,4-dihydroxyphenyl)-γ-valerolactone as the major metabolites. *J. Agric. Food Chem.*, **57**, 1084–1092.

Aura, A.-M. (2008). Microbial metabolism of dietary phenolic compounds in the colon. *Phytochem. Rev.*, **7**, 407–429.

Aura, A.-M., Martin-Lopez, P., O'Leary, K.A. et al. (2005). *In vitro* metabolism of anthocya-nins by human gut microflora. *Eur. J. Nutr.*, **44**, 133–142.

Aura, A.-M., Mattila, I., Seppänen-Laakso, T. et al. (2008). Microbial metabolism of catechin stereoisomers by human faecal microbiota: Comparison of targeted analysis and a non-targeted metabolomics method. *Phytochem. Lett.*, **1**, 18–22.

Aura, A.-M., Oikarinen, S., Mutanen, M. et al. (2006). Suitability of a batch *in vitro* fer-mentation model using human faecal microbiota for prediction of conversion of flax-seed lignans to enterolactone with reference to an *in vivo* rat model. *Eur. J. Nutr.*, **45**, 45–51.

Aura, A.-M., O'Leary, K.A., Williamson, G. et al. (2002). Quercetin derivatives are deconju-gated and converted to hydroxyphenylacetic acids but not methylated by human fecal microflora *in vitro*. *J. Agric. Food Chem.*, **50**, 1725–1730.

Axelson, M. and Setchell, K.D.R. (1981). The excretion of lignans in rats—evidence for an intestinal bacterial source for this new group of compounds. *FEBS Lett.*, **123**, 337–342.

Bach Knudsen, K.E., Serena, A., Kjaer, A.K. et al. (2003). Rye bread in the pigs enhances the formation of enterolactone and increases its levels in plasma, urine and feces. *J. Nutr.*, **133**, 1368–1375.

Bazzocco, S., Mattila, I., Guyot, S. et al. (2008). Factors affecting the conversion of apple polyphenols to phenolic acids and fruit matrix to short-chain fatty acids by human faecal microbiota *in vitro*. *Eur. J. Nutr.*, **47**, 442–452.

Begum, A.N., Nicolle, C., Mila, I. et al. (2004). Dietary lignins are precursors of mammalian lignans in rats. *J. Nutr.*, **134**, 120–127.

Bialonska, D., Kasimsetty, S.G., Schrader, K.K. et al. (2009). The effect of pomegranate (*Punica granatum* L.) byproducts and ellagitannins on the growth of human gut bacte-ria. *J. Agric. Food Chem.*, **57**, 8344–8349.

Blaut, M. and Clavel, T. (2007). Metabolic diversity of the intestinal microbiota: Implications for health and disease. *J. Nutr.*, **137**, 751S–755S.

Bò, C.D., Ciappellano, S., Klimis-Zacas, D. et al. (2010). Anthocyanin absorption, metabolism, and distribution from a wild bilberry-enriched diet (*Vaccinum angustifolium*) is affected by diet duration in the Sprague-Dawley rat. *J. Agric. Food Chem.*, **58**, 2491–2497.

Braune, A., Bunzel, M., Yonekura, R. et al. (2009). Conversion of dehydrodiferulic acids by human intestinal microbiota. *J. Agric. Food Chem.*, **57**, 3356–3362.

Braune, A., Engst, W., and Blaut, M. (2005). Degradation of neohesperidin dihydrochalcone by human intestinal bacteria. *J. Agric. Food Chem.*, **53**, 1782–790.

Cassidy, A., Hanley, B., and Lamuela-Raventós, R.M. (2000). Isoflavones, lignans and stil-benes—origins, metabolism and potential importance to human health. *J. Sci. Food Agric.*, **80**, 1044–1062.

Cerda, B., Espin, J.C., Parra, S. et al. (2004). The potent *in vitro* antioxidant ellagitannins from pomegranate juice are metabolised into bioavailable but poor antioxidant hydroxy-6H-dibenzopyran-6-one derivatives by the colonic microflora of healthy humans. *Eur. J. Nutr.*, **43**, 205–220.

Cerda, B., Tomás-Barberán, F.A., and Espin, J.C. (2005). Metabolism of antioxidant and che-mopreventive ellagitannins from strawberries, raspberries, walnuts, and oak-aged wine in humans: Identification of biomarkers and individual variability. *J. Agric. Food Chem.*, **53**, 227–235.

Clavel, T., Henderson, G., Alpert, C.-A. et al. (2005). Intestinal bacterial communities that produce active estrogen-like compounds enterodiol and enterolactone in humans. *Appl. Environ. Microbiol.*, **71**, 6077–6085.

Clifford, M.N. (2000). Anthocyanins—nature, occurrence and dietary burden. *J. Sci. Food. Agric.*, **80**, 1063–1072.

Coldham, N.G., Darby, C., Hows, M. et al. (2002). Comparative metabolism of genistein in human and rat gut microflora: Detection and identification of the end-products of metab-olism. *Xenobiotica*, **32**, 45–62.

Corsini, E., Agli, M.D., Facchi, A. et al. (2010). Enterodiol and enterolactone modulate the immune response by acting on nuclear factor-κB (NK-κB) signalling. *J. Agric. Food Chem.*, **58**, 6678–6684.

Das, N.P. (1971). Studies on flavonoid metabolism. Absorption and metabolism of (+)-cat-echin in man. *Biochem. Pharmacol.*, **20**, 3435–3445.

Day, A.J., Cañada, F.J., Díaz, J.C. et al. (2000). Dietary flavonoid and isoflavone glycosides are hydrolysed by the lactase site of lactase phlorizin hydrolase. *FEBS Lett.*, **468**, 166–170.

Del Rio, D., Costa, L.G., Lean, M.E.J. et al. (2010). Polyphenols and health: What compounds are involved. *Nutr. Met. Cardiovasc. Dis.*, **20**, 1–6.

Deprez, S., Brezillon, C., Rabot, S. et al. (2000). Polymeric proanthocyanidins are catabolized by human colonic microflora into low-molecular-weight phenolic acids. *J. Nutr.*, **130**, 2733–2738.

Donovan, J.L., Bell, J.R., Kasim-Karakas, S. et al. (1999). Catechin is present as metabolites in human plasma after consumption of red wine. *J. Nutr.*, **129**, 1662–1668.

Doyle, B. and Griffiths, L.A. (1980). The metabolism of ellagic acid in the rat. *Xenobiotica*, **10**, 247–256.

Eckburg, P.B., Bik, E.M., Bernstein, C.N. et al. (2005). Diversity of the human intestinal microbial flora. *Science*, **308**, 1635–1638.

Eeckhaut, E., Struijs, K., Possemiers, S. et al. (2008). Metabolism of lignan macromolecule into eneterolignans in the gastrointestinal lumen as determined in the simulator of the human intestinal microbial ecosystem. *J. Agric. Food Chem.*, **56**, 4806–4812.

Felgines, C., Talavéra, S., Gonthier, M.-P. et al. (2003). Strawberry anthocyanins are recovered in urine as glucuro- and sulfoconjugates in humans. *J. Nutr.*, **133**, 1296–1301.

Fleschhut, J., Kratzer, F., Rechkemmer, G. et al. (2006). Stability and biotransformation of various dietary anthocyanins *in vitro*. *Eur. J. Nutr.*, **45**, 7–15.

Forrester, S.C. and Waterhouse, A.L. (2008). Identification of Cabernet Sauvignon anthocy-anin gut microflora metabolites. *J. Agric. Food Chem.*, **56**, 9299–9304.

Gao, K., Xu, A., Krul, C. et al. (2006). Of the major phenolic acids formed during human microbial fermentation of tea, citrus, and soy flavonoid supplements, only 3,4-dihy-droxyphenylacetic acid has antiproliferative activity. *J. Nutr.*, **136**, 52–57.

Gonthier, M.-P., Donovan, J.L., Texier, O. et al. (2003a). Metabolism of dietary procyanidins in rats. *Free Rad. Biol. Med.*, **35**, 837–844.

Gonthier, M.-P., Verny, M.-A., Besson, C. et al. (2003b). Chlorogenic acid bioavailabil-ity largely depends on its metabolism by the gut microflora in rats. *J. Nutr.*, **133**, 1853–1859.

Gonthier, M.P., Remesy, C., Scalbert A. et al. (2006). Microbial metabolism of caffeic acid and its esters chlorogenic and caftaric acids by human faecal microbiota *in vitro*. *Biomed. Pharmacotherapy*, **60**, 536–540.

González-Barrio, R., Borges, G., Mullen, W., et al. (2010). Bioavailability of raspberry anthocyanins and ellagitannins following consumption of raspberries by healthy humans and subjects with an ileostomy. *J. Agric. Food Chem.* **58**, 3933–3939.

González-Barrío, R., Truchado, P., Ito, H. et al. (2011). UV and MS identification of urolithins and nasutins, the bioavailable metabolites of ellagitannins and ellagic acid in different mammals. *J. Agric. Food Chem.*, **59**, 1152–1162.

González-Sarrías, A., Azorín-Ortuño, M., Yáñez-Gascón, M.-J. et al. (2009). Dissimilar *in vitro* and *in vivo* effects of ellagic acid and its microbiota-derived metabolites, urolithins, on the cytochrome P450 1A1. *J. Agric. Food Chem.*, **57**, 5623–5632.

Griffiths, L.A. (1982). Mammalian metabolism of flavonoids. In J.B. Harborne and T.J. Marby (eds.), *The Flavonoids: Recent Advances in Research.* Chapman and Hall, London, pp. 681–718.

Griffiths, L.A. and Smith, G.E. (1972a). Metabolism of myricetin and related compounds in the rat metabolite formation *in vivo* and by intestinal microflora *in vitro. Biochem. J.*, **130**, 141–151.

Griffiths, L.A. and Smith, G.E. (1972b). Metabolism of apigenin and related compounds in the rat. Metabolite formation *in vivo* and by intestinal microflora *in vitro. Biochem. J.*, **128**, 901–911.

Gross, M., Pfeiffer, M., Martini, M. et al. (1996). The quantitation of metabolites of quercetin flavonols in human urine. *Cancer Epidemiol. Biomark. Prev.*, **5**, 711–720.

Guyton, A.C. and Hall, J.E. (1996). Gastrointestinal physiology. In A.C. Guyton and J.E. Hall (eds.), *Textbook of Medical Physiology*, 9th ed. W.B. Saunders Company, Philadelphia, pp. 793–813.

Grün, C.H., van Dorsten F.A., Jacobs, D.M. et al. (2008). GC-MS methods for metabolite profiling of microbial fermentation products of dietary polyphenols in human and *in vitro* intervention studies. *J. Chromogr. B*, **871**, 212–219.

Hanske, L., Loh, G., Sczesny, S. et al. (2009). The bioavailability of apigenin-7-glucoside is influenced by human intestinal microbiota in rats. *J. Nutr.*, **139**, 1095–1102.

Harder, H., Tetens, I., Let, M.B. et al. (2004). Rye bread intake elevates urinary excretion of ferulic acid in humans, but does not affect the susceptibility of LDL to oxidation. *Eur. J. Nutr.*, **43**, 230–236.

Hein, E.-M., Rose, K., van't Slot, G. et al. (2008). Deconjugation and degradation of flavonol glycosides by pig cecal microbiota characterized by fluorescence in situ hybridization (FISH). *J. Agric. Food Chem.*, **56**, 2281–2290.

Heinonen, S., Nurmi, T., Liukkonen, K. et al. (2001). *In vitro* metabolism of plant lignans: New precursors of mammalian lignans enterolactone and enterodiol. *J. Agric. Food Chem.*, **49**, 3178–3186.

Heinonen, S.-M., Wähälä, K., Liukkonen, K.-H. et al. (2004a). Studies of the *in vitro* intestinal metabolism of isoflavones aid in the identification of their urinary metabolites. *J. Agric. Food Chem.*, **52**, 2640–2646.

Heinonen, S.-M., Wähälä, K., and Adlercreutz, H. (2004b). Identification of urinary metabolites of the red clover isoflavones formononetin and biochanin A in human subjects. *J. Agric. Food Chem.*, **52**, 6802–6809.

Hoikkala, A., Mustonen, E., Saastamoinen, I. et al. (2007). High levels of equol in organic skimmed Finnish cow milk. *Mol. Nutr. Food Res.*, **51**, 782–786.

Hollman, P.C.H. and Katan, M.B. (1998). Absorption, metabolism and bioavailability of flavonoids. In C. Rice-Evans and L. Packer (eds.), *Flavonoids in Health and Disease*, Marcel Dekker, New York, pp. 483–522.

Jacobs, D.M., Deltimple, N., van Velzen, E. et al. (2008). ^{1}H-NMR metabolite profiling of feces as a tool to assess the impact of nutrition on the human microbiome. *NMR Biomed.*, **21**, 615–626.

Jacobs, D.M., Gaudier E., van Duynhoven, J. et al. (2009). Non-digestible food ingredients, colonic microbiota and the impact on gut health and immunity: A role for metabolomics. *Curr. Drug Metabol.,* **10**, 41–54.

Jacobs, E., Kulling, S.E. and Metzler, M. (1999). Novel metabolites of the mammalian lignans enterolactone and enterodiol in human urine. *J. Steroid Biochem. Mol. Biol.,* **68**, 211–218.

Jan, K.-C., Ku, K.-L., Chu, Y.-H. et al. (2010). Tissue distribution and elimination of estrogenic and anti-inflammatory catechol metabolites from sesaminol triglucoside in rats. *J. Agric Food Chem.,* **58**, 7693–7700.

Jenner, A.M., Rafter, J. and Halliwell, B. (2005). Human fecal water contents: The extent of colonic exposure to aromatic compounds. *Free Radic. Biol. Med.,* **38**, 763–772.

Jin, J.-S. and Hattori, M. (2009). Further studies on a human intestinal bacterium *Ruminococcus* sp. END-1 for transformation of plant lignans to mammalian lignans. *J. Agric. Food Chem.,* **57**, 7537–7542.

Joannou, G.E., Kelly, G.E., Reeder, A.Y. et al. (1995). A urinary profile study of dietary phytoestrogens. The identification and mode of metabolism of new isoflavonoids. *J. Steroid Biochem. Mol. Biol.,* **54**, 167–184.

Johnsen, N.F., Hausner, H., Olsen, A. et al. (2004). Intake of whole grains and vegetables determines the plasma enterolactone concentration of Danish women. *J. Nutr.,* **134**, 2691–2697.

Juntunen, K.S., Mazur, W.M., Liukkonen, K.H. et al. (2000). Consumption of wholemeal rye bread increases serum concentrations and urinary excretion of enterolactone compared with consumption of white wheat bread in healthy Finnish men and women. *Br. J. Nutr.* **84**, 839–846.

Karlsson, P.C., Huss, U., Jenner, A. et al. (2005). Human fecal water inhibits COX-2 in colonic HT-29 cells: Role of phenolic compounds. *J. Nutr.,* **135**, 2343–2349.

Kasimsetty, S.G., Bialonska, D., Reddy, M.K. et al. (2009). Effects of pomegranate chemical constituents/Intestinal microbial metabolites on CYP1B1 in 22Rv1 prostate cancer cells. *J. Agric. Food Chem.,* **57**, 10636–10644.

Kasimsetty, S.G., Bialonska, D., Reddy, M.K. et al. (2010). Colon cancer chemopreventive activities of pomegranate ellagitannins and urolithins. *J. Agric. Food Chem.,* **58**, 2180–2187.

Keppler, K. and Humpf, H.U. (2005). Metabolism of anthocyanins and their phenolic degradation products by the intestinal microflora. *Bioorg. Med. Chem.,* **13**, 5195–5205.

Kern, S.M., Bennett, R.N., Mellon, F.A. et al. (2003). Absorption of hydroxycinnamates in humans after high-bran cereal consumption. *J. Agric Food Chem.,* **51**, 6050–6055.

Kilkkinen, A., Pietinen, P., Klaukka, T. et al. (2002). Use of oral antimicrobials decreases serum enterolactone concentration. *Am. J. Epidemiol.,* **155**, 472–477.

Kilkkinen, A., Stumpf, K., Pietinen, P. et al. (2001). Determinants of serum enterolactone concentration. *Am. J. Clin. Nutr.,* **73**, 1094–1100.

Kilkkinen, A., Valsta, L.M., Virtamo, J. et al. (2003). Intake of lignans is associated with serum enterolactone concentration in Finnish men and women. *J. Nutr.,* **133**, 1830–1833.

Kim, D.-H. and Kobashi, K. (1986). The role of intestinal flora in metabolism of phenolic sulfate esters. *Biochem. Pharmacol.,* **35**, 3507–3510.

Kim, D.-H., Konishi, L., and Kobashi, K. (1986). Purification, characterization and reaction mechanism of novel arylsulfotransferase obtained from an anaerobic bacterium of human intestine. *Biochim. Biophys Acta,* **872**, 33–41.

Kleessen, B., Bezirtzoglou, E., and Mättö, J. (2000). Culture-based knowledge on biodiversity, development and stability of human gastrointestinal microflora. *Microb. Ecol. Health Disease,* 412 (Suppl 2), 53–63.

Knust, U., Spiegelhalder, B., Strowitzki, T. et al. (2006). Contribution of lignan intake to urine and serum enterolignan levels in German females: A randomised controlled intervention trial. *Food Chem. Toxicol.,* **44**, 1057–1064.

Koli, R., Erlund, I., Jula, A. et al. (2010). Bioavailability of various polyphenols from a diet containing moderate amounts of berries. *J. Agric. Food Chem.*, **58**, 3927–3932.

Krajka-Kuzniak, V., Szaefer, H., and Baer-Dubowska, W. (2005). Modulation of cytochrome P450 and phase II enzymes by protocatechuic acid in mouse liver and kidney. *Toxicology*, **216**, 24–31.

Krishnamurty, H.G., Cheng, K.J., Jones, G.A. et al. (1970). Identification of products by the anaerobic degradation of rutin and related flavonoids by *Butyrovibrio* sp. C3. *Can. J. Microbiol.*, **16**, 759–767.

Kroon, P.A., Faulds, C.B., Ryden, P. et al. (1997). Release of covalently bound ferulic acid from fiber in the human colon. *J. Agric. Food Chem.*, **45**, 661–667.

Kuijsten, A., Arts, I.C.W., Hollman, P.C.H. et al. (2006). Plasma enterolignans are associated with lower colorectal adenoma risk. *Cancer Epidemiol. Biomarkers Prev.*, **15**, 1132–1136.

Kuijsten, A., Arts, I.C.W., Vree, T.B. et al. (2005). Pharmacokinetics of enterolignans in healthy men and women consuming a single dose of secoisolariciresinol diglucoside. *J. Nutr.*, **135**, 795–801.

Kuijsten, A., Hollman, P.C., Boshuizen, H.C. et al. (2008). Plasma enterolignan concentrations and colorectal cancer risk in a nested case-control study. *Am. J. Epidemiol.*, **167**, 734–742.

Lafay, S., Gil-Izquierdo, A., Manach, C. et al. (2006). Chlorogenic acid is absorbed in its intact form in the stomach of rats. *J. Nutr.*, **136**, 1–6.

Lampe, J.W. (2003). Isoflavonoid and lignan phytoestrogens as dietary biomarkers. *J. Nutr.*, **133**, 956S–964S.

Larrosa, M., Gonzáles-Barrio, R., Garcia-Conesa, M.T. et al. (2006b). Urolithins, ellagic acid-derived metabolites produced by human colonic microflora, exhibit estrogenic and anti-estrogenic activities. *J. Agric. Food Chem.*, **54**, 1611–1620.

Larrosa, M., Luceri, C., Vivoli, E. et al. (2009). Polyphenol metabolites form colonic micro-biota exert anti-inflammatory activity on different inflammation models. *Mol. Nutr. Food Res.*, **53**, 1044–1054.

Larrosa, M., Tomás-Barberán, F.A., and Espin, J.C. (2006a). The dietary hydrolysable tannin punicalagin releases ellagic acid that induces apoptosis in human colon adenocarcinoma Caco2 cells by using the mitochondrial pathway. *J. Nutr. Biochem.*, **17**, 611–625.

Lee, M.-J., Maliakal, P., Chen, L. et al. (2002). Pharmacokinetics of tea catechins after ingestion of green tea and (–)-epicatechin-3-gallate by humans: Formation of different metabolites and individual variability. *Cancer Epidemiol. Biomark. Prev.*, **11**, 1025–1032.

Liu, C.-L., Wang, J.-M., Chu, C.-Y. et al. (2002). *In vivo* protective effect of protocatechuic acid on tert-butylhydroperoxide-induced rat hepatotoxicity. *Food Chem. Toxicol.*, **40**, 635–641.

Llorach, R., Urpi-Sarda, M., Jauregui, O. et al. (2009). An LC-MS-based metabolomics approach for exploring urinary metabolome modifications after cocoa consumption. *J. Proteome. Res.*, **8**, 5060–5068.

Lof, M. and Weiderpass, E. (2006). Epidemiologic evidence suggests that dietary phytoestrogen intake is associated with reduced risk of breast, endometrial, and prostate cancer. *Nutr. Res.*, **26**, 609–619.

Macfarlane, G.T. and Macfarlane, S. (2007). Models for intestinal fermentation: Association between food components, delivery systems, bioavailability and functional interactions in the gut. *Curr. Opinion Biotechnol.*, **18**, 156–162.

Macfarlane, S. and Dillon, J.F. (2007). Microbial biofilms in the human gastrointestinal tract. *J. Appl. Microbiol.*, **102**, 1187–1196.

Macfarlane, S. and Macfarlane, G.T. (2006). Composition and metabolic activities of bacterial biofilms colonizing food residues in the human gut. *Appl. Environ. Microbiol.*, **72**, 6204–6211.

Magee, P.J. and Rowland, I.R. (2004). Phytoestrogens, their mechanism of action: Current evidence for a role in breast and prostate cancer. *Br. J. Nutr.,* **91**, 513–531.

Mateo Anson, N., Aura, A.-M., Selinheimo, E. et al. (2011). Bioprocessing of wheat bran in whole wheat bread increases the bioavailability of phenolic acids in men and exerts anti-inflammatory effects *ex vivo*. *J. Nutr.,* **141**, 137–143.

Mateo Anson, N., Selinheimo, E., Havenaar, R. et al. (2009). Bioprocessing of wheat bran improves *in vitro* bioaccessibility and colonic metabolism of phenolic compounds. *J. Agric Food Chem.,* **57**, 6148–6155.

Mazur, W.M., Uehara, M., Wähälä, K. et al. (2000). Phyto-oestrogen content of berries, and plasma concentrations and urinary excretion of enterolactone after a single strawberry-meal in human subjects. *Br. J. Nutr.,* **83**, 381–387.

Meng, X., Sang, S., Zhu, N. et al. (2002). Identification and characterization of methylated and ring-fission metabolites of tea catechins formed in humans, mice, and rats. *Chem. Res. Toxicol.,* **15**, 1041–1050.

Meselhy, M.R., Nakamura, N., and Hattori, M. (1997). Biotransformation of (–)-epicatechin-3-*O*-gallate by human intestinal bacteria. *Chem. Pharm. Bull.,* **45**, 888–893.

Milbury, P.E. and Kalt, W. (2010). Xenobiotic metabolism and berry flavonoid transport across the blood-brain barrier. *J. Agric. Food Chem.,* **58**, 3950–3956.

Milder, I.E.J., Kuijsteen, A., Arts, I.C.W. et al. (2007). Relation between plasma enterodiol and enterolactone and dietary intake of lignans in a Dutch endoscopy-based population. *J. Nutr.,* **137**, 1266–1271.

Moazzami, A.A., Anderson, R., and Kamal-Eldin, A. (2007). Quantitative NMR analysis of a sesamin catechol metabolite in human urine. *J. Nutr.,* **137**, 940–944.

Monagas, M., Khan, N., Andrés-Lacueva, C. et al. (2009). Dihydroxylated phenolic acids derived from microbila metabolism reduce lipopolysaccharide-stimulated cytokine secretion by human peripheral blood mononuclear cells. *Br. J. Nutr.,* **102**, 210–206.

Monleón, D., Morales, J.M., Barrasa, A. et al. (2009). Metabolite profiling of fecal water extracts from human colorectal cancer. *NMR Biomed.,* **22**, 342–348.

Mullen, W., Rouanet, J.-M., Auger, C. et al. (2008a). Bioavailability of [2-^{14}C]quercetin-4′-glucoside in rats. *J. Agric Food Chem.,* **56**, 12127–12137.

Mullen, W., Archeveque, M.-A., Edwards, C.A. et al. (2008b). Bioavailability and metabolism of orange juice flavanones in humans: Impact of a full-fat yogurt. *J. Agric. Food Chem.,* **56**, 11157–11164.

Natsume, M., Osakabe, N., Oyama, M. et al. (2003). Structures of (–)-epicatechin glucuronide identified from plasma and urine after oral ingestion of (–)-epicatechin: Differences between human and rat. *Free Radic. Biol. Med.,* **34**, 840–849.

Nemeth, K., Plumb, G.W., Berrin, J.G. et al. (2003). Deglycosylation by human intestinal epithelial cell β-glucosidases is a critical step in the absorption and metabolism of dietary flavonoid glycosides in humans. *Eur. J. Nutr.,* **42**, 29–42.

Nesbitt, P.D., Lam, Y., and Thompson, L.U. (1999). Human metabolism of mammalian lignan precursors in raw and processed flaxseed. *Am. J. Clin. Nutr.,* **69**, 549–555.

Nurmi, T., Mursu, J., Heinonen, M. et al. (2009). Metabolism of berry anthocyanins to phenolic acids in humans. *J. Agric. Food Chem.,* **57**, 2274–2281.

Olthof, M.R., Hollman, P.C.H., Buijsman, M.N.C.P. et al. (2003). Chlorogenic acid, quercetin-3-rutinoside and black tea phenols are extensively metabolized in humans. *J. Nutr.,* **133**, 1806–1814.

Pearson, J.R., Gill, C.I.R., and Rowland I.R. (2009). Diet, fecal water, and colon cancer—development of a biomarker. *Nutr. Rev.,* **67**, 509–526.

Peñalvo, J.L., Haajanen, K.M., Botting, N. et al. (2005a). Quantification of lignans in food using isotope dilution gas chromatography/mass spectrometry. *J. Agric. Food Chem.,* **53**, 9342–9347.

Peñalvo, J.L., Heinonen, S.-M., Aura, A.-M. et al. (2005b). Dietary sesamin is converted to enterolactone in humans. *J. Nutr.,* **135**, 1056–1062.

Peppercorn, M.A. and Goldman, P. (1971). Caffeic acid metabolism by bacteria of the human gastrointestinal tract. *J. Bacteriol.,* **108**, 996–1000.

Pettersson, J., Karlsson P.C., Choi, Y.H. et al. (2008). NMR metabolomic analysis of fecal water from subjects on a vegetarian diet. *Biol. Pharm Bull.,* **31**, 1192–1198.

Plumb, G.W., Garcia-Conesa, M.T., Kroon, P.A. et al. (1999). Metabolism of chlorogenic acid by human plasma, liver, intestine and gut microflora. *J. Sci. Food Agric.,* **79**, 390–392.

Prior, R.L., Rogers, T., Khanal, R.C. et al. (2010). Urinary excretion of phenolic acids in rats fed cranberry. *J. Agric. Food Chem.,* **58**, 3940–3949.

Quin, J., Li, R., Raes J., Arumugam, M. et al. (2010). A human gut microbial gene catalogue established by metagenomic sequencing. *Nature,* **464**, 59–65.

Rechner, A.R., Smith, M.A., Kuhnle, G. et al. (2004). Colonic metabolism of dietary polyphenols: Influence of structure on microbial fermentation products. *Free Radic. Biol. Med.* 36, 212–225.

Rios, L.Y., Gonthier, M.-P., Rémesy, C. et al. (2003). Chocolate intake increases urinary excretion of polyphenol-derived phenolic acids in healthy human subjects. *Am. J. Clin. Nutr.,* **77**, 912–918.

Rondini, L., Peyrat-Maillard, M.-N., Marsset-Baglieri, A. et al. (2004). Bound ferulic acid from bran is more bioavailable than the free compound in rat. *J. Agric. Food Chem.,* **52**, 4338–4343.

Roowi, S., Stalmach, A., Mullen, W. et al. (2010). Green tea flavan-3-ols: Colonic degradation and urinary excretion of catabolites by humans. *J. Agric. Food Chem.,* **58**, 1296–1304.

Rowland, I., Faughnan, M., Hoey, L. et al. (2003). Bioavailability of phytoestrogens. *Br. J. Nutr.,* **89**, S45–S58.

Russell, W.R., Labat, A., Scobbie, L. et al. (2007). Availability of blueberry phenolics for microbial metabolism in the colon and potential inflammatory implications. *Mol. Nutr. Food Res.,* **51**, 726–731.

Saarinen, N.M., Smeds, A., Mäkelä, S.I. et al. (2002). Structural determinants of plant lignans for the formation of enterolactone *in vivo. J. Chromatogr. B,* **777**, 311–319.

Saarinen, N.M., Wärri, A., Mäkelä, S.I. et al. (2000). Hydroxymatairesinol, a novel enterolactone precursor with antitumor properties from coniferous tree (*Picea abies*). *Nutr. Cancer,* **36**, 207–216.

Salminen, S., Bouley, C., Boutron-Ruault, M.-C. et al. (1998). Functional food science and gastrointestinal physiology and function. *Br. J. Nutr.,* **80**, S147–S171.

Sawai, Y., Kohsaka, K., Nishiyama, Y. et al. (1987). Serum concentrations of rutoside metabolites after oral administration of a rutoside formulation to humans. *Drug Res.,* **37**, 729–732.

Scalbert, A. (1991). Antimicrobial properties of tannins. *Phytochemistry,* **30**, 3875–3883.

Scalbert, A., Morand, C., Manach, C. et al. (2002). Absorption and metabolism of polyphenols in the gut and impact on health. *Biomed. Pharmacother.,* **56**, 276–282.

Scheline, R.R. (1970). The metabolism of (+)-catechin to hydroxyphenylvaleric acids by the intestinal microflora. *Biochim. Biophys. Acta,* **222**, 228–230.

Scheline, R.R. (ed.). (1978). *Mammalian Metabolism of Plant Xenobiotics.* Academic Press, London.

Schneider, H. and Blaut, M. (2000). Anaerobic degradation of flavonoids by *Eubacterium ramulus. Arch. Microbiol.,* **173**, 71–75.

Schneider, H., Schwiertz, A., Collins, M.D. et al. (1999). Anaerobic transformation of quercetin-3-glucoside by bacteria from the human intestinal tract. *Arch. Microbiol.,* 171, 81–91.

Seeram, N.P., Henning, S.M., Zhang, Y. et al. (2006). Pomegranate juice ellagitannin metabolites are present in human plasma and some persist in urine for up to 48 h. *J. Nutr.,* **136**, 2481–2485.

Selma, M.V., Espín, J.C., and Tomás-Barberán, F.A. (2009). Interaction between phenolics and gut microbiota: Role in human health. *J. Agric Food Chem.,* **57**, 6485–6501.

Setchell, K.D., Brown, N.M., and Lydeking-Olsen, E. (2002b). The clinical importance of the metabolite equol—a clue to the effectiveness of soy and its isoflavones. *J. Nutr.,* **132**, 3577–3584.

Setchell, K.D., Brown, N.M., Zimmer-Nechemias, L. et al. (2002a). Evidence for lack of absorption of soy isoflavone glycosides in humans, supporting the crucial role intestinal metabolism for bioavailability. *Am. J. Clin. Nutr.,* **76**, 447–453.

Setchell, K.D., Clerici, C., Lephart, E.D. et al. (2005). *S*-Equol, a potent ligand for estrogen receptor β, is the exclusive enantiomeric form of the soy isoflavone metabolite produced by human intestinal bacterial flora. *Am. J. Clin. Nutr.,* **81**, 1072–1079.

Sharma, M., Li, L., Celver, J., Killian, C. et al. (2010). Effects of ellagitannin extracts, ellagic acid, and their colonic metabolite, urolithin A, on Wnt signaling. *J. Agric Food Chem.,* **58**, 3965–3969.

Simons, A.L., Renouf, M., Murphy, P.A. et al. (2010). Greater apparent absorption of flavonoids is associated with lesser human fecal flavonoid disappearance rates. *J. Agric Food Chem.,* **58**, 141–147.

Stalmach, A., Mullen, W., Barron, D. et al. (2009). Metabolite profiling of hydroxycinnamate derivatives in plasma and urine after ingestion of coffee by humans: Identification of biomarkers of coffee consumption. *Drug Metab. Dispos.* **37**, 1749–1758.

Stalmach, A., Steiling, H., Williamson, G. et al. (2010). Bioavailability of chlorogenic acids following acute ingestion of coffee by humans with an ileostomy. *Arch. Biochem. Biophys.,* **501**, 98–105.

Stalmach, A., Troufflard, S., Serafini, M. et al. (2009). Absorption, metabolism and excretion of Choladi green tea flavan-3-ols by humans. *Mol. Nutr. Food Res.,* **53**, S44–S53.

Stoupi, S., Williamson, G., Drynan, J.W. et al. (2010a). A comparison of the *in vitro* transformation of (–)-epicatechin and proanthocyanidin B2 by human faecal microbiota. *Mol. Nutr. Food Res.,* **54**, 747–759.

Stoupi, S., Williamson, G., Drynan, J.W. et al. (2010b). Procyannidin B2 catabolism by human fecal microflora: Partial characterization of 'dimeric' intermediates. *Arch. Biochem. Biophys.,* **501**, 73–76.

Stoupi, S., Williamson, G., Viton, F. et al. (2010c). *In vivo* bioavailability, absorption, excretion, and pharmacokinetics of [^{14}C]procyanidin B2 in male rats. *Drug Metab. Dispos.,* **38**, 287–291.

Struijs, K., Vincken, J.-P., and Gruppen, H. (2009). Bacterial conversion of secoisolariciresinol and anhydrosecoisolariciresinol. *J. Appl. Microbiol.,* **107**, 308–317.

Takagaki, A. and Nanjo, F. (2010). Metabolism of (–)-epigallocatechin gallate by rat intestinal flora. *J. Agric Food Chem.,* **58**, 1313–1321.

Touillaud, M.S., Thiébaut, A.C.M., Fournier, A. et al. (2007). Dietary lignan intake and postmenopausal breast cancer risk by estrogen and progesterone receptor status. *J. Natl. Cancer Inst.,* **99**, 475–486.

Vallejo, F., Larrosa M., Escudero, E. et al. (2010). Concentration and solubility of flavones in orange beverages affect their bioavailability in humans. *J. Agric Food Chem.,* **58**, 6516–6524.

Vanharanta, M., Voutilainen, S., Lakka, T.A. et al. (1999). Risk of acute coronary events according to serum concentrations of enterolactone: A prospective population-based case-control study. *Lancet,* **354**, 2112–2115.

van 't Slot, G. and Humpf, H.-U. (2009). Degradation and metabolism of catechin, epicatechin-3-gallate (EGCG), and related compounds by the intestinal microbiota in the pig cecum model. *J. Nutr.*, **57**, 8041–8048.

van 't Slot, G., Mattern, W., Rzeppa, S. et al. (2010). Complex flavonoids in cocoa synthesis and degradation by intestinal microbiota. *J. Agric. Food Chem.*, **58**, 8879–8886.

Verzelloni, E., Pellacani, C., Tagliazucchi, D. et al. (2011). Antiglycative and neuroprotective activity of colon-derived polyphenol catabolites. *Mol. Nutr. Food Res.*, **55** (Suppl 1), 535–543.

Vitaglione, P., Donnarumma, G., Napolitano, A. et al. (2007). Protocatechuic acid is the major human metabolite of cyanidin-glucosides. *J. Nutr.*, **137**, 2043–2048.

Walle, T., Browning, A.M., Steed, L.L. et al. (2005). Flavonoid glycosides are hydrolyzed and thus activated on the oral cavity in humans. *J. Nutr.*, **135**, 48–52.

Walle, T., Walle, U.K. and Halushka, P.V. (2001). Carbon dioxide is the major metabolite of quercetin in humans. *J. Nutr.*, **131**, 2684–2652.

Walsh, K.R. and Failla, M.L. (2009). Transport and metabolism of equol by CaCo-2 human intestinal cells. *J. Agric. Food Chem.*, **57**, 8297–8302.

Webb, A.L. and McCullough, M.L. (2005). Dietary lignans: Potential role in cancer prevention. *Nutr. Cancer*, **51**, 117–131.

World Health Organization. (2003). Diet, nutrition and the prevention of chronic diseases. *WHO Technical Report Series 916*, Geneva, 149 pp.

Urpi-Sarda, M., Garrido, I., Mongas, M. et al. (2009b). Profile of plasma and urine metabolites after the intake of almonds [*Prunus dulcis* (Mill.) D.A. Webb] polyphenols in humans. *J. Agric. Food Chem.*, **57**, 10134–10142.

Urpi-Sarda, M., Mongas, M., Khan, N. et al. (2009a). Epicatechin, proanthocyanidins, and phenolic microbial metabolites after cocoa intake in humans and rats. *Anal. Bioanal. Chem.*, **394**, 1545–1556.

Xu, X., Harris, K.S., Wang, H.-J. et al. (1995). Bioavailability of soybean isoflavones depends upon gut microflora in women. *J. Nutr.*, **125**, 2307–2315.

Yin, M.-C., Lin, C.-C., Wu, H.-C. et al. (2009). Apoptotic effects of protocatechuic acid in human breast, lung, liver, cervix and prostate cancer cells: Potential mechanisms of action. *J. Agric Food Chem.*, **57**, 6468–6473.

Zhao, Z., Egashira, Y., and Sanada, H. (2004). Ferulic acid is quickly absorbed from rat stomach as the free form and then conjugated mainly in liver. *J. Nutr.*, **134**, 3083–3088.

11 Synthesis of Dietary Phenolic Metabolites and Isotopically Labeled Dietary Phenolics

Denis Barron, Candice Smarrito-Menozz, René Fumeaux, and Florian Viton

CONTENTS

11.1 INTRODUCTION

Among the various classes of naturally occurring plant secondary metabolites, flavonoid and related phenolic compounds are well represented in food. These include flavan-3-ols and their oligomeric procyanidin forms, flavonols, anthocyanins, flavanones, isoflavones, lignans, and hydroxycinnamic acids. The bioavailability of these compounds after ingestion, their absorption, metabolism, and excretion, has now become one of the major topics in food research. Once absorbed in the upper gastrointestinal tract, dietary phenols are almost invariably conjugated enzymatically and, thus, are detected in biological fluids in the form of a set of glucuronidated, sulfated, and/or methylated derivatives. However, significant amounts are not absorbed and continue their journey inside the digestive track to the colon, where they are degraded by the microflora. Examples of the metabolites generated by the colon microflora are displayed in Figure 11.1. The metabolites produced in the colon can also be conjugated by phase II enzymes after absorption. The regular advances in the knowledge of the absorption and the metabolism of dietary phenols have been largely linked to the parallel development of sophisticated HPLC-MS analytical techniques for their detection in biological fluids and tissues. Nevertheless, these tools remain largely analytical, and limitations were rapidly faced by the scientific community when quantitative data were required. This led to a renewed interest in the synthesis of dietary (poly)phenolic metabolites and their conjugates as analytical standards. These compounds are also invaluable in studies assessing bioactivity of *in vivo* phenolic metabolites in various *in vitro* models.

The number of synthesis studies has increased considerably in recent years and the topic has been covered in two previous reviews by our group (Barron et al. 2003; Barron 2008). The present review will focus on the literature published since 2005. The chemical reagents, which are commonly used in sulfation and glucuronidation reactions, are summarized in Figure 11.2, together with a set of frequently encountered protected, partially deprotected, and fully deprotected glucuronyl substituents.

One of the most reliable and sensitive method of quantification is based on the addition of a known amount of stable-isotope analog of the analyte of interest, in a given volume of sample. This technique is known as SIDA (stable isotope dilution assay) or SID-MS as only mass spectrometry allows distinct quantification of isotopes. Since the analyte and its isotopologues possess almost identical chemical and physical properties, the isotopic ratio remains stable as they are affected equally by variations in extraction, sample preparation, injection, and/or instrument parameters. Furthermore, SID minimizes the interferences of the matrix with the ionization process. This approach is particularly well suited for the quantification of analytes in low concentrations such as metabolites of dietary phenolics in biological fluids. However, the use of stable isotopes supposes a precise targeting of the analytes to be quantified and, thus, a deep knowledge of the metabolism of the absorbed compounds. This is rarely the case for complex dietary phenolics whose fate in living organisms is usually unclear. In these circumstances,

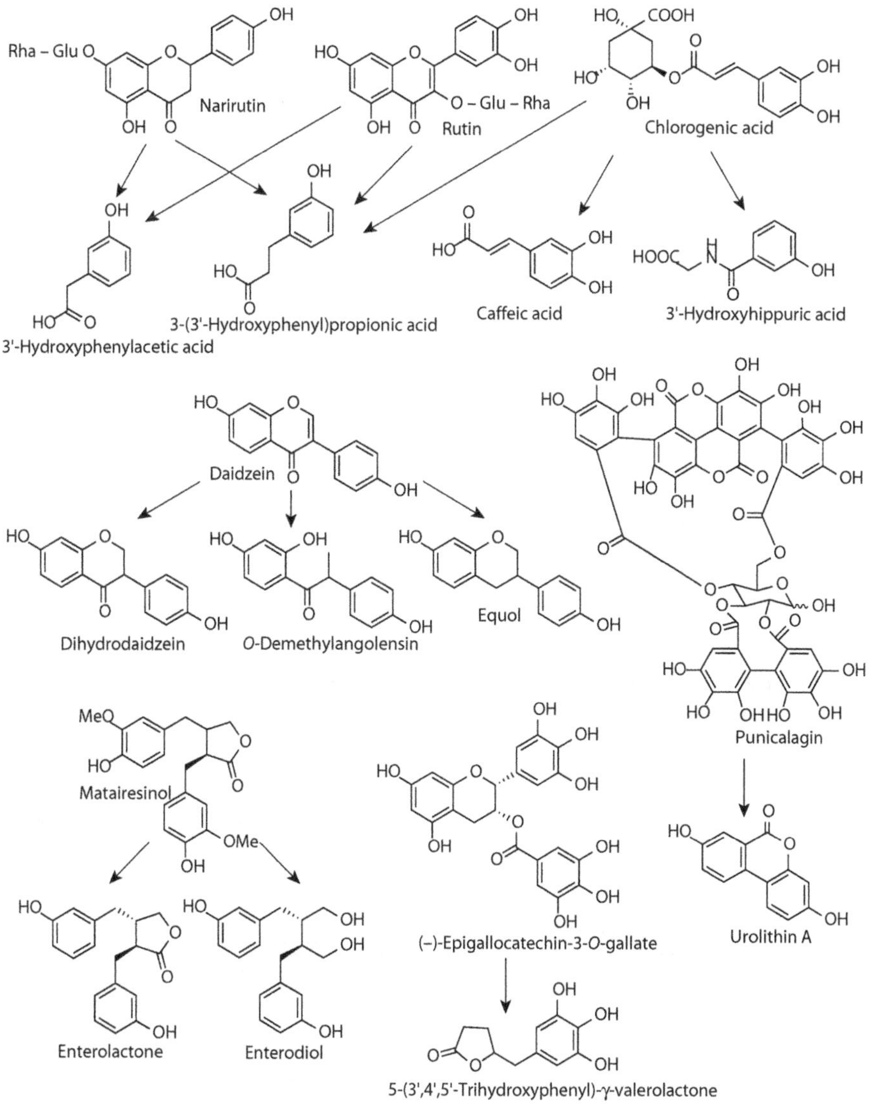

FIGURE 11.1 A selection of phenolic metabolites resulting from colonic fermentations.

the use of radiolabeled compounds is of value, especially to study organ or tissue distribution.

The synthesis of isotopically labeled dietary phenolics has been discussed in detail in a sister publication (Barron et al. 2012). Only the preparation of compounds, which were not covered in this paper, is covered in the present review. However, the reader will find in this chapter tabular lists of previously reported labeled compounds with supporting references.

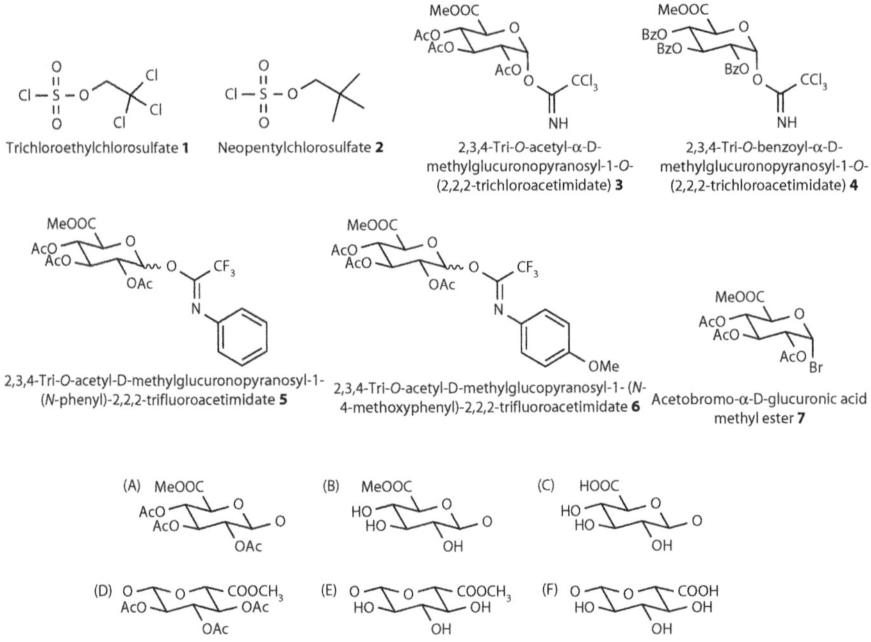

FIGURE 11.2 Frequently encountered sulfating reagents **1** and **2**, glucuronic acid donors **3–6**, and glucuronyl residues A–F.

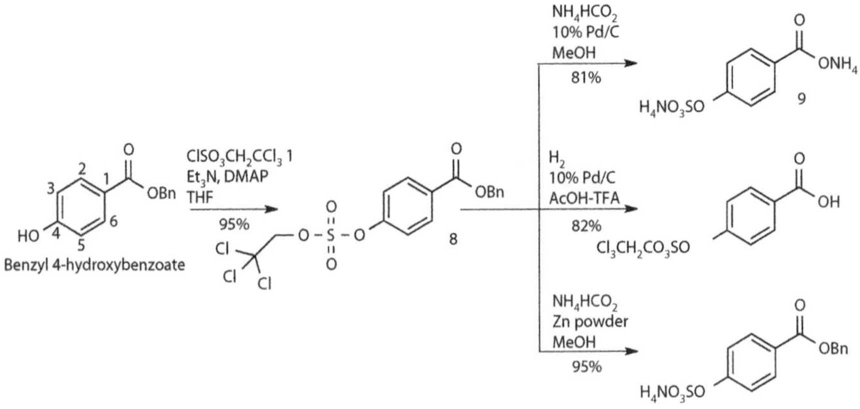

SCHEME 11.1 The synthesis of 4-sulfonyloxy benzoic acids using trichloroethylchlorosulfate as a reagent.

11.2 BENZOIC ACIDS

Almost all the benzoic acids resulting from mammalian metabolism of dietary phenols can be acquired from commercial sources. Although this was not related to a metabolic study, p-hydroxybenzoic acid was used in research with a new sulfating reagent, trichloroethylchlorosulfate (Scheme 11.1) (Liu et al. 2004). The novelty and the attraction of the use of this reagent is that one of the acidic functions of the sulfate is protected and so leads to a nonpolar protected sulfate conjugate, which is readily purified. It is noteworthy that while phosphate protection has been routinely used in the synthesis of phosphate esters for many decades, to date, a similar approach has not been applied to sulfate esters. Furthermore, the sulfate protective group and its removal were shown to be compatible with the presence of a benzyl protective group on the carboxylic acid. Thus, benzyl 4-hydroxybenzoate was sulfated to sulfate ester **8** with trichloroethylchorosulfate **1**. Catalytic transfer hydrogenolysis of **8** led to simultaneous deprotection of the sulfate and carboxylic acid groups, yielding 4-sulfonyloxybenzoic acid **9** as ammonium salt in good yield. On the other hand, it was shown that by manipulating the deprotection conditions, either the trichloroethyl or the benzyl groups could be selectively eliminated (Scheme 11.1), further documenting the versatility of the sulfation reagent.

There is a procedure for preparing of N-(3-hydroxybenzoyl)-, N-(4-hydroxybenzoyl)-, and N-(vanilloyl)-glycines (Van Brussel et al. 1978). The method is based on the activation of the carboxylic acid group in the form of their N-hydroxysuccinimide esters **10–12**, followed by a coupling with glycine (Scheme 11.2). Neither the activation nor the coupling steps require any protection of the phenolic groups of the benzoyl moiety. However, this procedure turned out not to be applicable to the synthesis of N-(3,4-dihydroxybenzoyl)-glycine (Hanselaer et al. 1983), presumably because of the sensitivity of the catechol group to oxidation. To overcome this problem a new synthetic route was developed (Schnabelrauch et al. 2000). The methoxycarbonyl group was used as a protector of the phenolic hydroxyls, and the carboxylic acid function was activated in the form of a chloride (Scheme 11.3). After conjugation of compound **13** with glycine, compound **14** was deprotected under alkaline conditions to give N-(protocatechuyl)-glycine.

11.3 PHENYLACETIC ACIDS, TYROSOL, 3′-HYDROXYTYROSOL, AND THEIR CONJUGATES

Most phenylacetic acids resulting from the metabolism of dietary phenols are available commercially. This is the case of homoprotocatechuic acid (3′,4′-dihydroxyphenylacetic acid), homovanillic acid (3′-methoxy-4′-hydroxyphenylacetic acid), homoisovanillic acid (3′-hydroxy-4′-methoxyphenylacetic acid), and 3′-hydroxyphenylacetic acid. However, the colonic fermentation of dietary phenols may generate phenylacetic acids with unusual substitution pattern. The basics of the preparation of phenylacetic acids are therefore still of interest. Phenylacetic acids are also useful intermediates in the synthesis of isoflavones, and as a consequence, many details about their preparation can be found in papers related to the synthesis of these flavonoids, including isotopically labeled derivatives (Whalley et al. 2000; Oldfield et al. 2004). 4′-Methoxyphenylacetic acid and 4′-hydroxyphenylacetic acid have been synthesized from 4-methoxybenzyl chloride (Kuwajima et al. 1998) (Scheme 11.4).

4-Methoxybenzyl chloride was converted to 4-methoxybenzylmagnesium chloride and carboxylated with CO_2 to 4′-methoxyphenylacetic acid, which was demethylated to 4′-hydroxyphenylacetic acid. After methylation of the carboxylic acid function and reduction of the methyl ester, 4′-hydroxyphenylacetic acid also

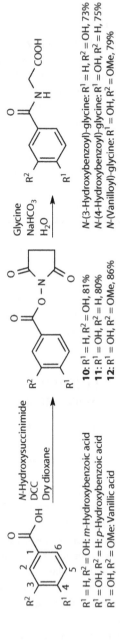

SCHEME 11.2 Syntheses of *N*-hydroxybenzoyl- and *N*-vanilloyl-glycines.

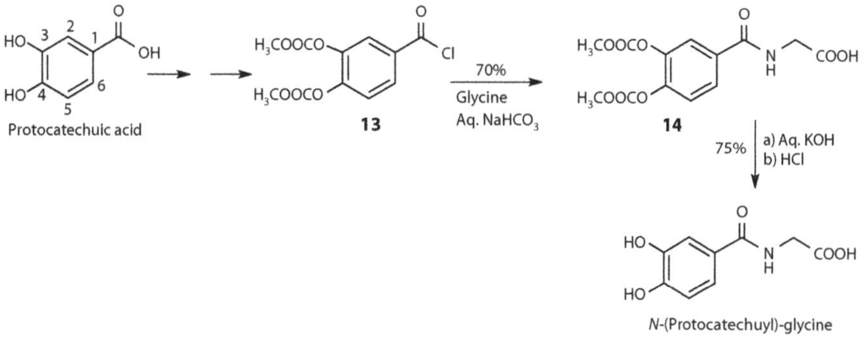

SCHEME 11.3 Preparation of *N*-(protocatechuyl)-glycine.

provided 4′-hydroxyphenylethanol (tyrosol). Tyrosol and its dihydroxylated analog 3′-hydroxytyrosol (Scheme 11.5) are the major phenolic constituents of olive oil. The metabolism of 3′-hydroxytyrosol may involve methylation at position 3′ to give homovanillic alcohol. Tyrosol, hydroxytyrosol, and homovanillic alcohol can all be further subjected to glucuronidation. The preparative synthesis of the glucuronides of olive oil phenolic metabolites has been carried out using biocatalysis with porcine liver microsomes and uridine-5′-diphosphoglucuronic acid (UDPGA) as the glucuronyl donor (Khymenets et al. 2006). Milligram amounts of conjugates (5–30 mg depending on the substrate) were obtained with a 5–88% yield. Interestingly, the glucurono-conjugation could either affect the phenolic groups or the alcoholic hydroxyls albeit in lower yield (Scheme 11.5).

A chemical synthesis of hydroxytyrosol glucuronides has also been published (Lucas et al. 2009). 3′-Hydroxytyrosol was prepared by reduction of 3′,4′-dihydroxyphenylacetic acid, and selective acetylation of the primary alcohol was carried out by an immobilized lipase to give **15** in excellent yield (Scheme 11.6). Glucuronidation was carried out directly on the unprotected catechol ring, using methyl 2,3,4-tri-*O*-acetyl-1-*O*-(trichloroacetinidoyl)-α-D-glucuronate **3** as glucuronyl donor. As expected, this produced a mixture of 3′- and 4′-protected glucuronides **16** and **17**, which were deprotected using aqueous Na₂CO₃, and this yielded the 3′- and 4′-*O*-β-D-glucuronides of 3′-hydroxytyrosol. Surprisingly, no attempt was made to separate the two glucuronide conjugates.

In the preparation of the 1-*O*-β-D-glucuronide of 3′-hydroxytyrosol (Scheme 11.7), the catechol group of hydroxytyrosol was first protected as its acetal to give **18**. Compound **18** was efficiently glucuronidated by reagent **4**, producing compound **19**. Treatment of **19** with sodium carbonate removed the benzoyl protective groups of the glucuronide moiety. A second treatment with trifluoroacetic acid eliminated the acetal group, to give hydroxytyrosol-1-*O*-β-D-glucuronide in good yield (Lucas et al. 2009).

3′,4′-Dihydroxyphenylacetic acid and 3′-methoxy-4′-hydroxyphenylacetic acid are metabolites of the important neurotransmitters dopamine and serotonine. Sulfation and glucuronidation of both acids are known to occur in the brain. The enzymatic syntheses of the sulfate conjugates (mediated by a rat liver fraction) and

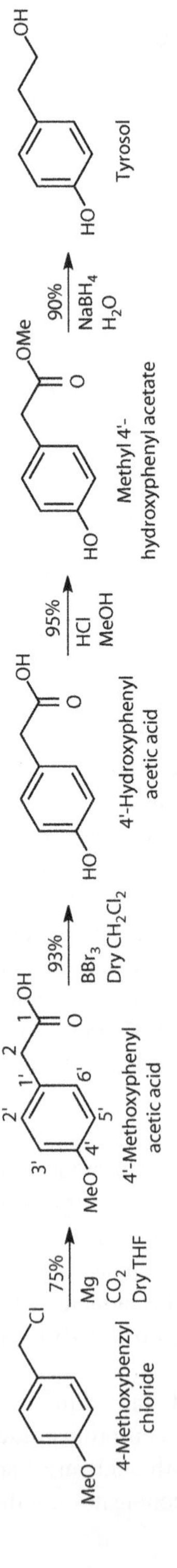

SCHEME 11.4 Syntheses of 4′-methoxyphenylacetic acid, 4′-hydroxyphenylacetic acid and tyrosol.

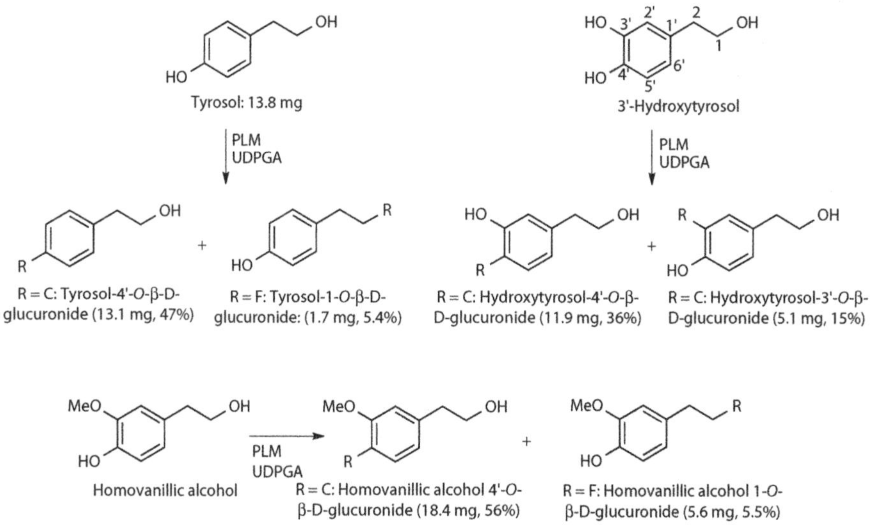

SCHEME 11.5 The biocatalyzed synthesis of tyrosol, 3'-hydroxytyrosol, and homovanillic alcohol glucuronides. PLM – porcine liver microsomes; UDPG - uridine-5'-diphosphoglucuronic acid. R = C or F refers to glucuronyl residues in Figure 11.2.

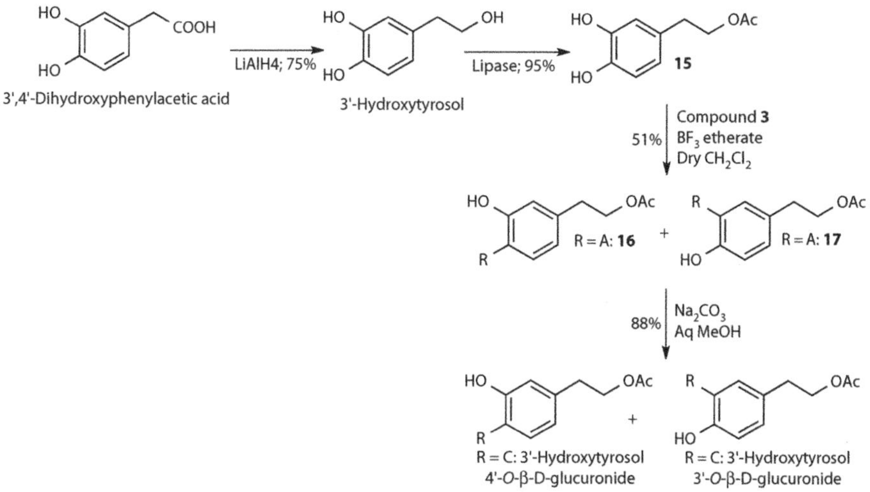

SCHEME 11.6 The chemical syntheses of the 3'- and 4'-O-β-D-glucuronides of 3'-hydroxytyrosol. R = C refers to glucuronyl residue in Figure 11.2.

of the glucuronide conjugates (catalyzed by rat liver microsomes) have been investigated (Uutela et al. 2009). The conjugates were produced in sufficient quantity to be used as analytical standards for the identification of the conjugates present in rat brain microdialysates. The method, however, did not allow the isolation of preparative amounts of the metabolites.

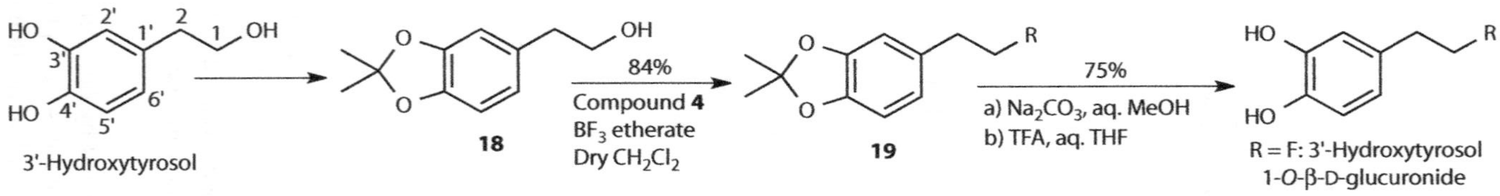

SCHEME 11.7 Chemical synthesis of 3'-hydroxytyrosol-1-O-β-ᴅ-glucuronide. R = F refers to glucuronyl residue in Figure 11.2.

11.4 PHENYLPROPIONIC ACIDS AND THEIR CONJUGATES

Most phenylpropionic acids of metabolic significance are available commercially, and they can easily be prepared either by hydrogenolysis of their corresponding cinnamic acid (see Sharma et al. 2003; Fumeaux et al. 2010; and references cited therein), or from their benzaldehyde precursors by the Knoevenagel-Doebner reaction. This very classical route was optimized by using microwave-assisted synthesis (Sharma et al. 2003). 3-(3'-Hydroxyphenyl)propionic acid was obtained in satisfactory yield using a two-step method (Scheme 11.8). In contrast, no 3-(4'-hydroxy-3'-methoxyphenyl)propionic acid product was formed when using vanillin as precursor.

Hydroxyphenylpropionic acid glucuronides and sulfates have been recently prepared in our laboratory by hydrogenolysis of the corresponding hydroxycinnamic acid glucuronides or direct sulfation of free hydroxyphenylpropionic acids (Scheme 11.9) (Fumeaux et al. 2010). Glucuronidation was carried out on protected ethyl cinnamates using the very efficient glucuronic acid donor 2,3,4-tri-O-acetyl-D-methylglucuronopyranosyl-(N-phenyl)-2,2,2-trifluoroacetimidate **5** (Scheme 11.9). Saponification followed by hydrogenolysis of the exocyclic double bond afforded the glucuronide conjugates of 3-(4'-hydroxyphenyl)-, 3-(3'-hydroxyphenyl)-, 3-(4'-hydroxy-3'-methoxyphenyl)-, and 3-(3'-hydroxy-4'-methoxyphenyl)-propionic acids.

The direct sulfation of hydroxyphenylpropionic acids using sulfur trioxide-trimethylamine complex yielded the corresponding sulfate conjugates (Scheme 11.9). Since the sulfation reaction was carried out in a high salt environment, the final sulfate esters still contained significant amounts of inorganic salts that were difficult to remove completely even after chromatographic purification. Since these inorganic salts were not UV-absorbing, the determination of the purity of the products could not be estimated solely on the basis of their HPLC-UV chromatogram. Acid hydrolysis of aliquots of the sulfate conjugates, followed by UV-quantification of the released aglycone allowed the exact quantification of the amount of conjugate present in the samples, and the calculation of the sulfation yields that are listed in Scheme 11.9 (Fumeaux et al. 2010).

The direct conjugation of 3-(3',4'-dihydroxyphenyl)propionic acid (dihydrocaffeic acid) would have led to a mixture of 3'- and 4'-conjugates. Furthermore, significant degradation of the substrate could take place mainly because of the instability of the catechol moiety under the basic conditions of sulfation. On the other hand, designing a protection strategy for the commercially available 3-(3',4'-dihydroxyphenyl)propionic acid would have been meaningless, as in the absence of a conjugated moiety, the two phenolic hydroxyls would have similar reactivities. For this reason, we decided to introduce the sulfate or glucuronide conjugations at the hydroxycinnamic level, to take advantage of the higher reactivity of the 4'-hydroxyl. Nevertheless, the regioselective preparation of 3'- and 4'-conjugates of 3-(3',4'-dihydroxyphenyl)propionic acids still required a complex protection strategy. Suitable protection was introduced as early as the benzaldehyde level (Scheme 11.10). The common precursor of the 4'-glucuronide and sulfate was ethyl 3-(3'-benzyloxy-4'-hydroxyphenyl)-2-propenoate, prepared by the Horner–Wadsworth–Emmons reaction of 3-benzylprotocatechualdehyde with ethyl (triphenylphosphoranylidene)acetate.

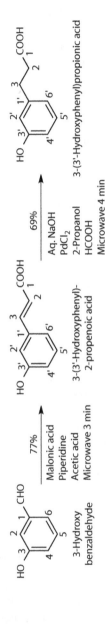

SCHEME 11.8 Microwave-assisted synthesis of 3-(3'-hydroxyphenyl)propionic acid.

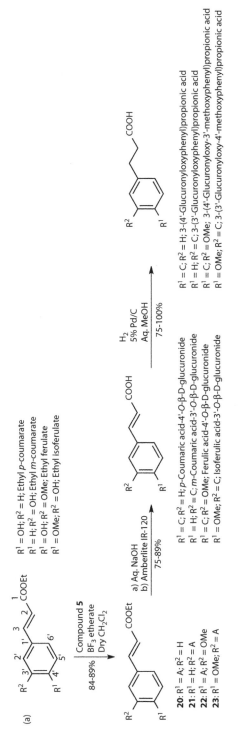

SCHEME 11.9 The syntheses of (a) hydroxycinnamic and hydroxyphenylpropionic acid glucuronides; and (b) hydroxyphenylpropionic acid sulfates. R^1 = A or C refers to glucuronyl residues in Figure 11.2.

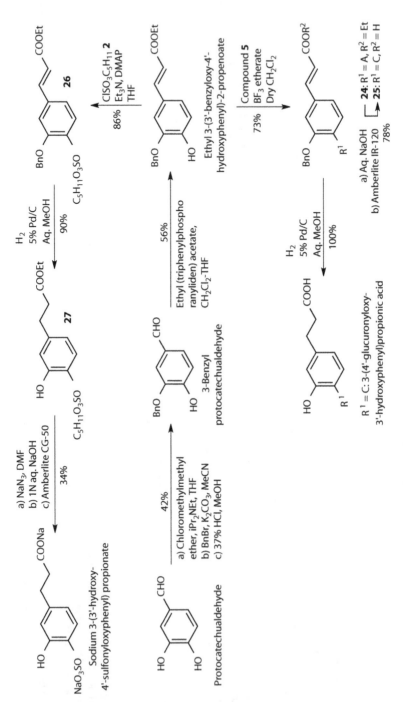

SCHEME 11.10 The syntheses of 3-(4′-glucuronyloxy-3′-hydroxyphenyl)propionic acid (dihydrocaffeic acid-4′-glucuronide) and sodium 3-(3′-hydroxy-4′-sulfonyloxyphenyl)propionate (sodium salt of dihydrocaffeic acid-4′-sulfate). R^1 = A or C refers to glucuronyl residue in Figure 11.2.

Ethyl 3-(3'-benzyloxy-4'-hydroxyphenyl)-2-propenoate was first glucuronidated using the efficient glucuronic acid donor **5**, and all the ester protective groups of the resulting conjugate **24** were removed by alkaline hydrolysis, to give 3'-benzylcaffeic acid **25**. A single-step catalytic hydrogenolysis of **25** resulted in debenzylation and reduction of the exocyclic double bond, to give the desired 3-(4'-glucuronyloxy-3'-hydroxyphenyl)propionic acid (dihydrocaffeic acid-4'-O-glucuronide).

For the preparation of 3-(3'-hydroxy-4'-sulfonyloxyphenyl)propionic acid, we considered the recently developed reagent neopentyl chlorosulfate **2**. Similar to trichloroethylchlorosulfate (see Section 11.2), this reagent is compatible with the use of a number of protective group including a benzyl moiety. Furthermore, purification of the sulfate conjugate could be advantageously carried out before the removal of the sulfate protective neopentyl group. At this stage, the conjugate still retains a reduced hydrophilicity, which simplifies its purification. Thus, ethyl 3-(3'-benzyloxy-4'-hydroxyphenyl)-2-propenoate was regioselectively sulfated with neopentyl chlorosulfate **2** to give the ethyl benzylcaffeate conjugate **26** (Scheme 11.10). A single-step hydrogenolysis of **26** removed the benzyl group and reduced the exocyclic double bond, leading to compound **27**, which was treated with sodium azide (sulfate deprotection) and saponified (removal of the ethyl ester) to give, after deprotection, the sodium salt of 3-(3'-hydroxy-4'-sulfonyloxyphenyl)propionate (dihydrocaffeic acid-4'-sulfate).

Alternatively (Scheme 11.11), 4-benzylprotocatechualdehyde was sulfated in excellent yield by neopentyl chlorosulfate **2**, affording aldehyde **28**, which was submitted to the Horner–Wadsworth–Emmons reaction to give the ethyl cinnamate derivative **29**. Hydrogenolysis of **29** resulted in debenzylation and the reduction of the exocyclic double bond to give **30**. Finally, the three-step deprotection of compound **30** furnished the sodium salt of 3-(4'-hydroxy-3'-sulfonyloxyphenyl)propionic acid (dihydrocaffeic acid-3'-sulfate) (Fumeaux et al. 2010).

For the synthesis of 3-(3'-glucuronyloxy-4'-hydroxyphenyl)propionic acid (dihydrocaffeic acid-3'-glucuronide), we combined the use of the acetate and the benzyl protective groups (Scheme 11.12). Protocatechualdehyde was acetylated to 3,4-diacetyloxybenzaldehyde, which was regioselectively deacetylated at position 4 using an acetate exchange reaction with thiophenol. The resulting

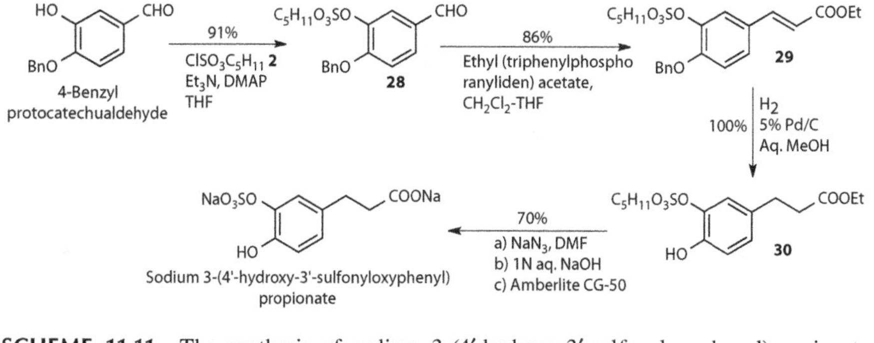

SCHEME 11.11 The synthesis of sodium 3-(4'-hydroxy-3'-sulfonyloxyphenyl)propionate (sodium salt of dihydrocaffeic acid-3'-sulfate).

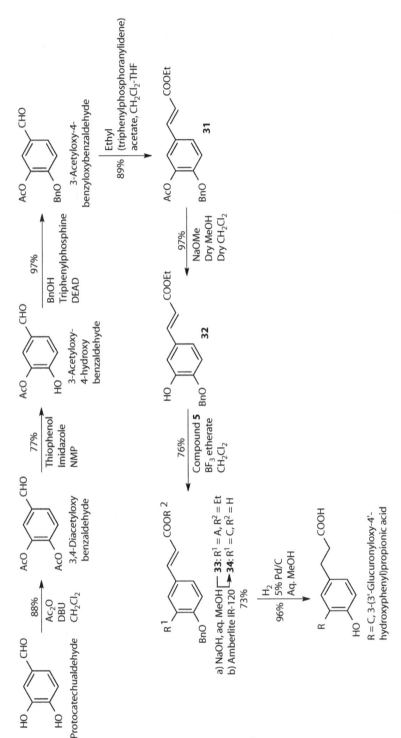

SCHEME 11.12 The preparation of 3-(3′-glucuronyloxy-4′-hydroxyphenyl)propionic acid (dihydrocaffeic acid 3′-glucuronide). R = C refers to glucuronyl residue in Figure 11.2.

3-acetyloxy-4-hydroxybenzaldehyde was benzylated under Mitsunobu conditions, which are compatible with the stability of the acetate group, to give 3-acetyloxy-4-benzyloxybenzaldehyde. This compound was converted to its corresponding ethyl cinnamate **31** by a Horner–Wadsworth–Emmons reaction with ethyl (triphenylphosphoranylidene)acetate. Hydrolysis of the acetate with sodium methylate followed by glucuronidation of the free 3′-hydroxyl group using compound **5** as glucuronyl donor yielded compound **33**. Compound **33** was submitted to a two-step deprotection and reduction sequence, affording 3-(3′-glucuronyloxy-4′-hydroxyphenyl)propionic acid (Fumeaux et al. 2010).

11.5 HYDROXYCINNAMIC ACID CONJUGATES

The first method for the synthesis of ferulic acid 4′-O-β-D-glucuronide was reported by Galland et al. (2008). Methyl 2,3,4-tri-O-acetyl-1-O-(trichloroacetimidoyl)-α-D-glucuronate **3** was used as glycosylation agent (Scheme 11.13). To avoid the formation of the acyl glucuronide, glucuronidation was carried out on methyl ferulate to give compound **35** in good yield. Compound **35** was first deacetylated to **36**. Finally, stirring of compound **36** with sodium hydroxide in aqueous methanol demethylated the two carboxyl groups to give ferulic acid 4′-O-β-D-glucuronide.

This method has been improved by replacing of the two-step deprotection sequence with a single-step procedure, which was applied to a wider range of hydroxycinnamoyl glucuronides (Scheme 11.9A) (Fumeaux et al. 2010). In addition, caffeic acid-3′- and 4′-sulfates were prepared, respectively, from the sulfated aldehyde **28**, and by regioselective sulfation of acetylcaffeic acid ethyl ester (Scheme 11.14). The first step of the preparation of caffeic acid-3′-sulfate was the debenzylation of **28** to aldehyde **37**. The reaction had to be carried out under catalytic transfer hydrogenolysis conditions using cyclohexene as hydrogen donor to avoid the reduction of the aldehyde function. Aldehyde **37** was subsequently submitted to the Horner–Wadsworth–Emmons reaction to give the ethyl cinnamate derivative **38**. Finally, the two steps deprotection of compound **38** furnished sodium 4′-hydroxy-3′-sulfonyloxycinnamate (Fumeaux et al. 2010).

Compound **39** was prepared in quantitative yield from 3,4-diacetyloxybenzaldehyde by the Horner–Wadsworth–Emmons reaction (Scheme 11.14). The deacetylation of acetylated ethyl caffeate **39** turned out not to be regioselective, and a mixture of the monoacetyl derivatives **40** and **41** was obtained. However, it was anticipated that, under the alkaline conditions of the sulfation reaction, transesterification reactions would take place with a tendency for the acetyl groups to migrate to the phenolic hydroxyl of lower acidity, that is, to the 3′-position. Indeed, direct sulfation of the mixture of **40** and **41** led to the formation of the 3′-acetylated-4′-sulfated derivative **42** in 80% yield. The latter compound was deprotected to sodium 3′-hydroxy-4′-sulfonyloxycinnamate in fair yield.

The preparation of caffeic acid glucuronides has also been described, although this could not be achieved in a regioselective route. These compounds have been obtained by glucuronidation of either methyl caffeate (Galland et al. 2008) or a mixture of the monoacetylated ethyl caffeates **40** and **41** (Fumeaux et al. 2010). In both cases, a mixture of the 3′- and 4′-conjugates was obtained from which the individual isomers were isolated chromatographically.

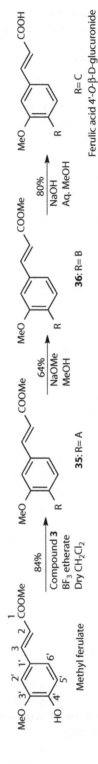

SCHEME 11.13 Synthesis of ferulic acid-4'-O-β-D-glucuronide. R = A, B, or C refers to glucuronyl residues in Figure 11.2.

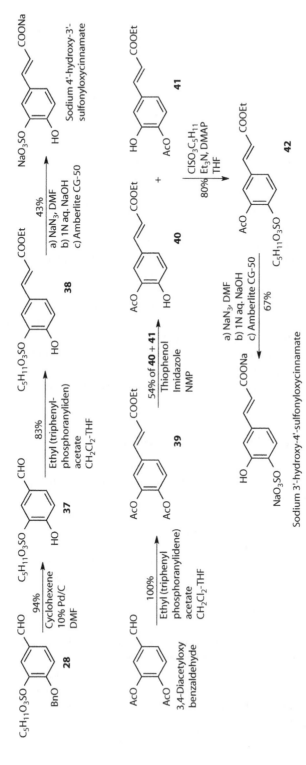

SCHEME 11.14 The syntheses of caffeic acid 3'- and 4'-sulfates.

The preparation of *N-p*-coumaroyl- and *N*-feruloyl-glycines has been carried out (Scheme 11.15) using the same method that was applied to the synthesis of *N*-hydroxybenzoylglycines (Section 11.2) (van Brussel et al. 1978). In the preparation of *N*-caffeoylglycine, protection of the catechol unit by ethoxycarbonyl groups was mandatory (Hanselaer et al. 1983). The deprotection of compound **46** was mediated by hydrazine (Scheme 11.15). Among the three glycine conjugates, only feruloylglycine was detected in the plasma of human volunteers after ingestion of a polyphenol-rich fruit juice (Borges et al. 2010; Mullen et al. 2010). Feruloylglycine was also detected in human urine after administration of a horsetail extract (Graefe and Veit 1999), cereal consumption (Kern et al. 2003), or ingestion of coffee (Stalmach et al. 2009). This prompted the development of alternative methods of synthesis of glycine conjugates. Stalmach et al. (2009) have prepared feruloyl- and isoferuloylglycines by coupling the free ferulic and isoferulic acids with methyl glycinate hydrochloride in the presence of *N*-methyl morpholine (NMM) as a base (Scheme 11.16). 1-Hydroxybenzotriazole (HOBt) was used to activate the carboxylic acid functions of ferulic and isoferulic acids. *O*-(benzotriazol-1-yl)-*N,N,N',N'*-tetramethyluronium tetrafluoroborate (TBTU) was used as peptide coupling agent.

A method optimized for the preparation of glycine-conjugated bile acids (Momose et al. 1997) was applied to the synthesis of *N*-feruloylglycine by Kern et al. (2003), but unfortunately, the reported experimental data are incomplete. The method is close to the one mentioned earlier except that triethylamine was used as a base and diethyl phosphorocyanidate was employed as both a carboxylic acid activator and an amide formation reagent. Interested readers should consult the Momose et al. (1997) for further details.

11.6   FERULOYLQUINIC ACID CONJUGATES

Classical approaches in the synthesis of hydroxycinnamoyl quinic acids are based on the esterification of quinic acid by a hydroxycinnamic acid, which implies a sophisticated protection strategy. Application of this route to the preparation of sulfated and glucuronidated conjugates is expected to be even more difficult because of the further complexity brought about by the necessity to protect the sugar/sulfate moeity. We found it more convenient to develop an alternative strategy for the elaboration of the hydroxycinnamoyl quinic acid core. The key reaction was a Knoevenagel condensation of a properly substituted aldehyde with a malonate ester of quinic acid (Menozzi-Smarrito et al. 2008). Since no protection is required at this step, the suitable sulfate or glucuronide conjugation could be introduced, for example, on the aldehyde fragment before the condensation.

This approach was applied in our group to the preparation of the 4'-sulfate ester and 4'-*O*-β-D-glucuronide of 5-*O*-feruloylquinic acid (Scheme 11.17) (Menozzi-Smarrito et al. 2011). Vanillin (3-methoxy-4-hydroxybenzaldehyde) was either sulfated to the protected sulfate **47** or glucuronidated to the protected glucuronide **50**. The sulfated vanillin **47** was submitted to a Knoevenagel condensation with 5-*O*-malonylquinic acid **48**, to give protected sulfated 5-*O*-feruloylquinic acid **49**. Finally, the sulfate deprotection of compound **49** yielded the sodium salt of 5-*O*-feruloylquinic

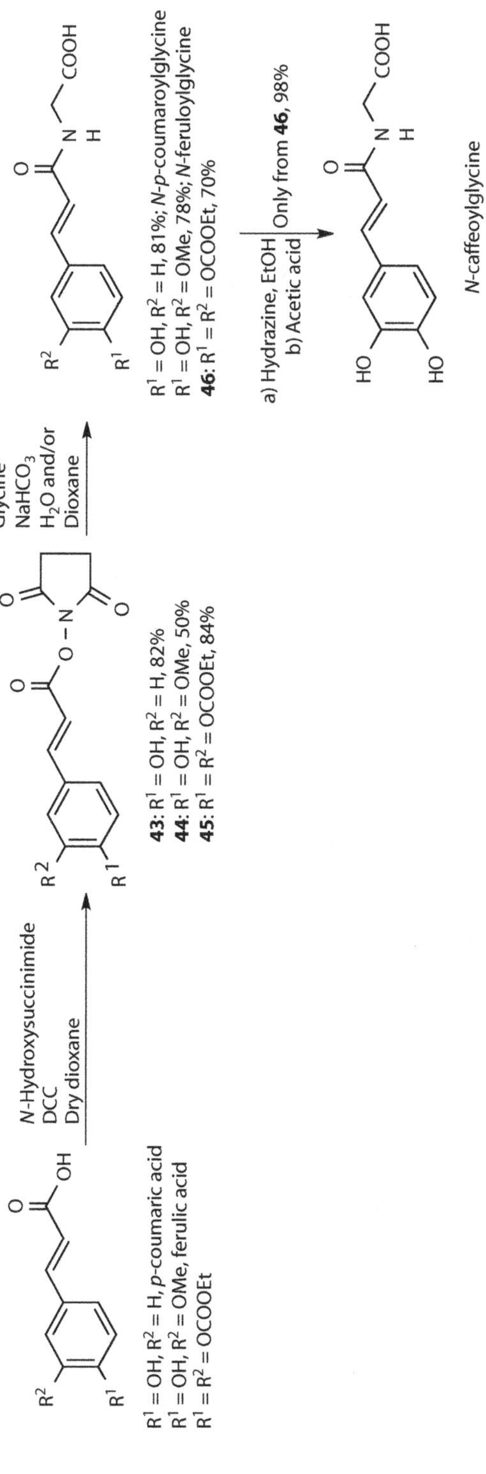

R¹ = OH, R² = H, *p*-coumaric acid
R¹ = OH, R² = OMe, ferulic acid
R¹ = R² = OCOOEt

N-Hydroxysuccinimide
DCC
Dry dioxane

43: R¹ = OH, R² = H, 82%
44: R¹ = OH, R² = OMe, 50%
45: R¹ = R² = OCOOEt, 84%

Glycine
NaHCO₃
H₂O and/or
Dioxane

R¹ = OH, R² = H, 81%; *N-p*-coumaroylglycine
R¹ = OH, R² = OMe, 78%; *N*-feruloylglycine
46: R¹ = R² = OCOOEt, 70%

a) Hydrazine, EtOH
b) Acetic acid — Only from **46**, 98%

N-caffeoylglycine

SCHEME 11.15 The preparations of *N-p*-coumaroyl-, *N*-feruloyl-, and *N*-caffeoyl-glycines.

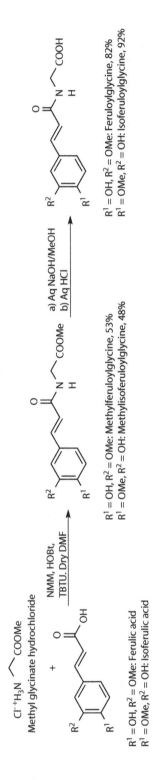

SCHEME 11.16 Alternative syntheses of *N*-feruloyl- and *N*-isoferuloyl-glycines.

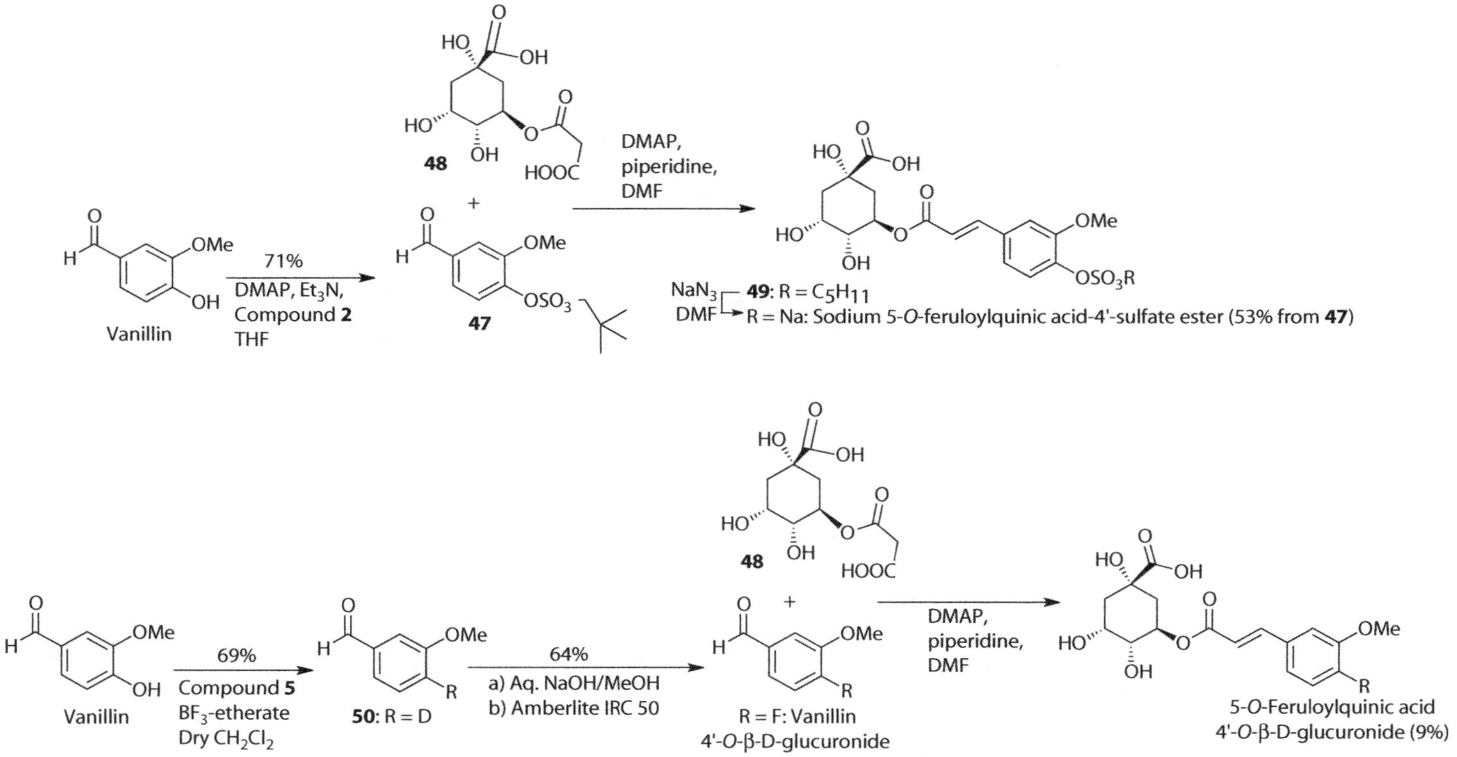

SCHEME 11.17 The syntheses of 5-*O*-feruloylquinic acid 4'-sulfate and 5-*O*-feruloylquinic acid-4'-*O*-β-D-glucuronide. R = F refers to glucuronyl residue in Figure 11.2.

acid-4′-sulfate ester as the major product, together with traces of the 4′-sulfate ester analogs of 3-*O*- and 4-*O*-feruloylquinic acid. Deprotection of glucuronyl derivative **50** was performed at the aldehyde level. The Knoevenagel condensation of the resulting vanillin-4′-*O*-β-D-glucuronide with compound **48** afforded the desired 4′-*O*-β-D-glucuronide of 5-*O*-feruloylquinic acid as major conjugated product.

11.7 PHLORETIN GLUCURONIDE

The incubation of phloretin, the aglycone of the apple dihydrochalcone glucoside phloridzin, with UDP-glucuronyl transferase and UDPGA, yielded two glucuronide conjugates of which only one had the same retention time as the glucuronide present in the ileal fluid. Isolation of this conjugate by analytical HPLC allowed the acquisition of a ¹H NMR spectrum, which identified it as phloretin-2′-*O*-β-D-glucuronide (Kahle et al. 2007). This enzymatic procedure was, however, not suitable for the synthesis of preparative amounts of the conjugate.

11.8 TEA FLAVAN-3-OL–DERIVED LACTONES

In the case of the lactones derived from the fermentation of green tea catechins, the nomenclature which is known by most plant phenol researchers is based on γ-valerolactone (i.e., position 1 of the lactone ring is the C=O), while the official chemical nomenclature is based on furanone (i.e., position 1 of the lactone ring is the oxygen). Thus, depending on the nomenclature used, the carbon bearing the hydroxyphenylmethyl chain can either be 4 or 5, which may introduce confusion (Figure 11.3). In this review, we shall use the common nomenclature based on γ-valerolactone, except when the enantioselective synthesis is discussed.

The colonic fermentation of tea catechins leads to the formation of lactone derivatives including 5-(3′,4′-dihydroxyphenyl)- and 5-(3′,4′,5′-trihydroxyphenyl)-γ-valerolactones. Both compounds have been isolated from human urine after treatment by deconjugating enzymes (Li et al. 2000) and a method for their chemical synthesis has been reported by Lambert et al. (2005). In the preparation of 5-(3′,4′-dihydroxyphenyl)-γ-valerolactone, the key step was the availability of compound **51** as precursor of the lactone ring (Scheme 11.18). This compound was synthesized from guaiacol, which was acetylated and iodinated in preparation for the

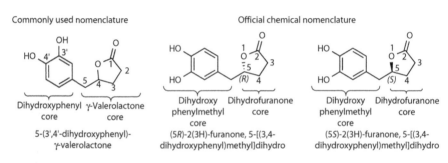

FIGURE 11.3 The numbering of lactone metabolites.

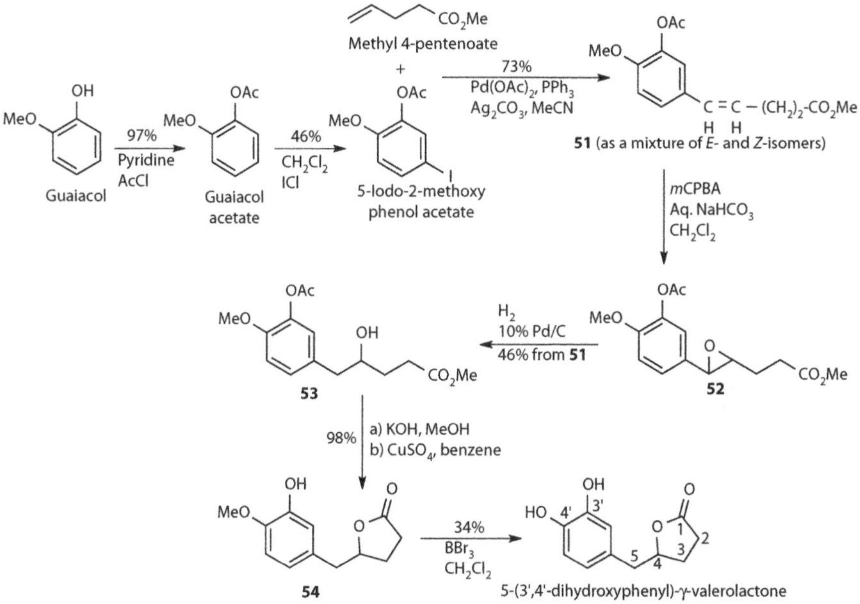

SCHEME 11.18 The chemical syntheses of tea catechins metabolite 5-(3′,4′-dihydroxyphenyl)-γ-valerolactone.

introduction of the side chain. 5-Iodo-2-methoxyphenol acetate was submitted to a Heck reaction with methyl 4-pentenoate, to give **51**. Compound **51** was epoxidized with *m*-chloroperoxybenzoic acid, and the resulting epoxide **52** was opened (cleavage of the benzylic C–O bond) by hydrogenolysis to **53**. Cyclodehydration of **53** to lactone **54** was carried out after ester hydrolysis followed by refluxing in benzene over anhydrous $CuSO_4$. Finally, demethylation of **54** with boron tribromide yielded the desired 5-(3′,4′-dihydroxyphenyl)-γ-valerolactone.

In the synthesis of (−)-5-(3′,4′,5′-trihydroxyphenyl)-γ-valerolactone, a different procedure was used to introduce the side chain (Scheme 11.19). The commercially available 3,4,5-trimethoxybenzaldehyde was reacted with vinylmagnesium bromide to give the alcohol **55**, which was reacted with triethylorthoacetate, and the resultant ether submitted to a Johnson-Claisen rearrangement to give compound **56** (isolated as the *E*-isomer). The remainder of the scheme followed a similar procedure as the synthesis of 5-(3′,4′-dihydroxyphenyl) γ-valerolactone (see Scheme 11.19).

Taking into account the presence of a stereogenic center at position 5 (official chemical nomenclature), these lactones can exist as (5*R*)- and (5*S*)-enantiomers. However, it is only very recently that the enantioselective synthesis of the lactone metabolites of tea catechins has been described. In the synthesis of the (5*R*)-enantiomer (Hamada et al. 2010), the key precursor was the (*S*)-benzyl glycidyl ether (Scheme 11.20). Furthermore, the hydroxylation of the side chain was carried out by coupling of the precursor with a protected aryl bromide **61** instead of the olefin epoxidation shown in Scheme 11.18. This yielded compound **62**, which was debenzylated by catalytic hydrogenolysis to give the alcohol **63**. The alcohol was first oxidized to its corresponding

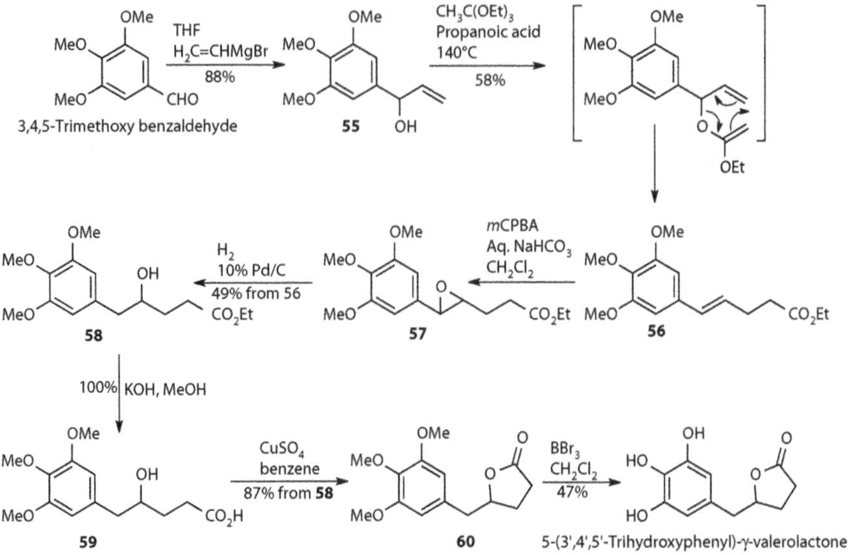

SCHEME 11.19 The chemical syntheses of tea catechins metabolite 5-(3′,4′,5′-trihydroxy-phenyl)-γ-valerolactone.

aldehyde by Swern oxidation (oxalyl choride, DMSO, and triethylamine). The Wittig olefination of the aldehyde yielded the α,β-unsaturated ester **64**. Compound **64** was deacetylated to **65,** which was reduced to **66**. The desilylation and lactonization of **66** were carried out in one step by treatment with *p*-toluenesulfonic acid, yielding the desired (*R*)-5-(3,4-dihydroxyphenyl)-γ-valerolactone (Scheme 11.20). A similar synthesis of the (*S*)-isomer was developed by the same group (Nakano et al. 2008). However, the optical rotations recorded for either the (*R*)- and (*S*)-isomers ($[\alpha]_D = -16.7°$ and $+39.6°$, respectively) differed from those reported in the literature for the metabolite obtained by fermentation of (−)-epicatechin-3-*O*-gallate with enteric bacteria ($[\alpha]_D = -8.6°$) (Meselhy et al. 1997). This suggests that epimerization could have taken place during the fermentation.

11.9 LIGNAN METABOLITES

In the synthesis of enterolactone and various aromatic polyhydroxylated analogs, the aromatic substitution is dictated by an aromatic benzaldehyde on the one hand (precursor of ring A) and by a suitable substituted benzyl bromide on the other (precursor of ring B) (Pelter et al. 1983; Jacobs and Metzler 1999). Thus, the tandem addition of 3-methoxybenzaldehyde bis(phenylthio) acetal, 2-buten-4-olide, and 3-methoxybenzyl bromide furnished the *trans*-adduct **67** (Scheme 11.21). Displacement of the phenylthio groups by a heavy metal (Raney nickel) yielded dimethyl enterolactone, which was demethylated yielding enterolactone using boron tribromide. Enterolactone was reduced to enterodiol by lithium aluminum hydride (Jacobs and Metzler 1999).

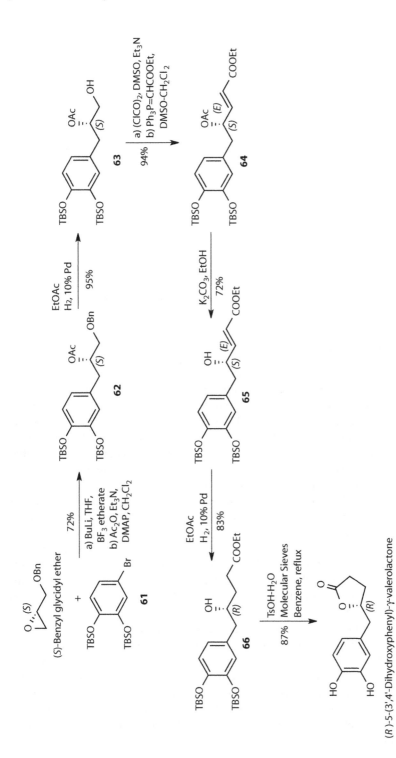

(R)-5-(3',4'-Dihydroxyphenyl)-γ-valerolactone

SCHEME 11.20 The enantioselective synthesis of (R)-5-(3,4-dihydroxyphenyl)-γ-valerolactone.

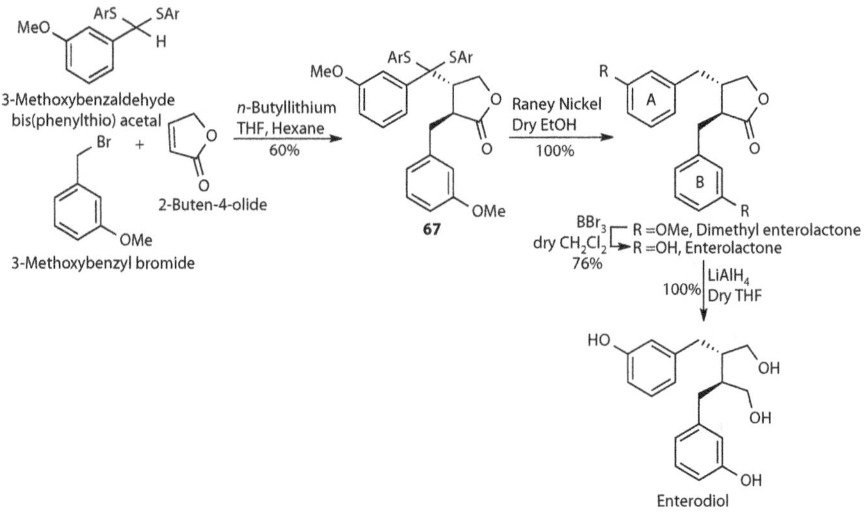

SCHEME 11.21   The syntheses of lignan metabolites enterolactone and enterodiol.

11.10   UROLITHINS AND THEIR CONJUGATES

Urolithins are not only the product of the colonic fermentation of ellagitannins, but they are also part of the bioactives of Shilajit, an organic exudate from steep rocks that is used in Ayurveda medicine (Indian traditional medicine). Furthermore, they are also encountered sequestered in the organs of some herbivores including beavers (González-Barrio et al. 2011). This is presumably related to the diet of the beavers as they consume tree bark that is rich in ellagitannins. The synthesis of urolithins has attracted interest since the late 1980s (Ghosal et al. 1989). This involved the reaction of a bromobenzoic acid with resorcinol under a modified Hurtley condensation. The method is now the classical route to urolithins (Scheme 11.22) and has been used in a number of recent studies including that of Bialonska et al. (2009).

An alternative method had been developed based on the Suzuki coupling of (2,4-dimethoxyphenyl)boronic acid with either 2-bromobenzamide (Alo et al. 1991) or 2-bromobenzoic acid (Sun et al. 2006). After deprotection of the diaryl derivative **69** using boron tribromide, the cyclization into urolithin A was carried out in acetic acid (Scheme 11.23).

The glucuronide of urolithin B was recently prepared using methyl 2,3,4-tri-*O*-acetyl-1-*O*-(trichloroacetimidoyl)-α-D-glucuronate **3** as glucuronidation reagent (Lucas et al. 2009). The protected glucuronide **68** was obtained in a high yield of 95% (Scheme 11.22). However, the alkaline conditions, such as aqueous NaOH or KOH, required for its deprotection were known to induce the opening of the urolithin lactone. To solve this problem, the deprotection was carried out with Na_2CO_3 in aqueous methanol. This afforded the desired urolithin B glucuronide with a good yield of 83%.

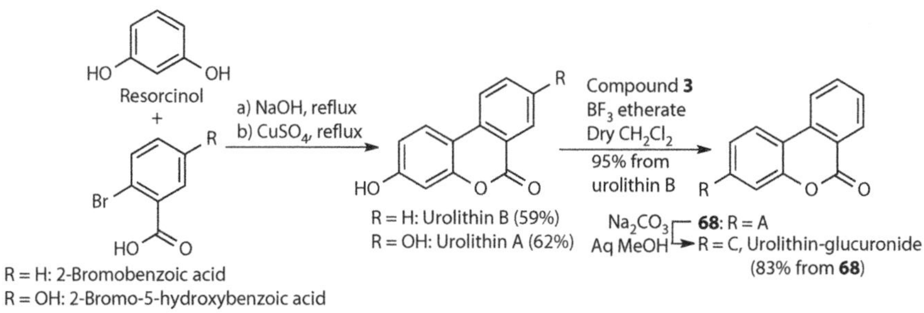

SCHEME 11.22 The syntheses of urolithins A, B and glucuronide conjugate. R = A or C refers to glucuronyl residue in Figure 11.2.

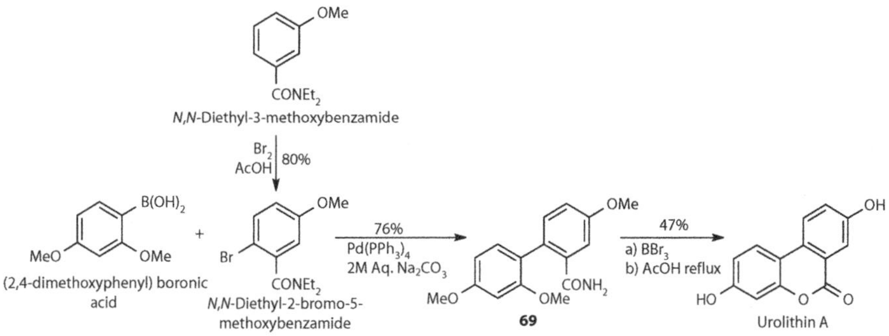

SCHEME 11.23 The synthesis of urolithin A via Suzuki coupling.

11.11 FLAVANONE CONJUGATES

The synthesis of flavanone glucuronides caused specific problems because of the instability of the flavanone ring in the alkaline conditions required for deprotection of the glucuronyl moiety. We recently solved this problem by using an esterase as a biocatalyst (Boumendjel et al. 2009). After glucuronidation of persicogenin (the flavanone of tarragon spice), the protected glucuronide **70** was deprotected using a two-step sequence. The first step with dry zinc acetate in refluxing dry methanol removed all the acetates to give the methyl ester **72**. Treatment of **72** with pig liver esterase in phosphate buffer pH 7.2 ensured the hydrolysis of the methyl ester while preserving the integrity of the flavanone nucleus, to give persicogenin-3′-O-β-D-glucuronide (Scheme 11.24).

This method was also applied to the preparative synthesis of hesperetin-3′-O-β-D-glucuronide (Scheme 11.24). To direct the glucuronidation at position 3′, protection of the 7-hydroxyl was required. To this end, benzylation of hesperetin afforded a 74% yield of 7-O-benzylhesperetin. The direct glucuronidation of this compound using trifluoroacetimidate **5** yielded the protected glucuronide **71** in 59% yield. The removal of the 7-benzyl protective group was performed efficiently

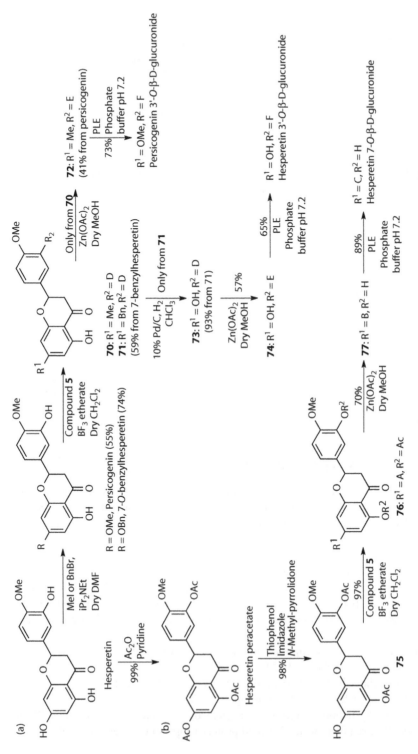

SCHEME 11.24 Efficient syntheses of (a) flavanone-3'-O-β-D-glucuronides; and (b) hesperetin-7-O-β-D-glucuronide. R = A, B, C, D or F refers to glucuronyl residue in Figure 11.2.

by hydrogenolysis without affecting the flavanone nucleus. The resulting compound **73** was submitted to the abovementioned deprotection sequence, to give hesperetin-3'-*O*-β-D-glucuronide.

We synthesized gram amounts of hesperetin-7-*O*-β-D-glucuronide for biological testing by using peracetylated hesperetin as precursor, which was selectively deacetylated at position 7 to produce **75**, which was glucuronidated to **76** and submitted to the double glucuronide deprotection sequence to give hesperetin-7-*O*-β-D-glucuronide (Scheme 11.24). The deacetylation step was performed with a good yield of 70%. All the other steps of the sequence were achieved with high recoveries.

As an alternative deprotection method, the group of Olivier Dangles used the aqueous Na_2CO_3 method previously developed in the synthesis of urolithin B glucuronide (Section 11.10). The deprotection was carried out under nitrogen atmosphere at 0°C, and its progress was followed every hour by HPLC (Khan et al. 2010).

Access to naringenin-3'-*O*-β-D-glucuronide was achieved after monobenzoylation of naringenin (Khan et al. 2010). Because of the higher reactivity of the 7-hydroxyl, the 7-benzoylated compound was the sole product of esterification with one equivalent of benzoyl chloride (Scheme 11.25). Glucuronidation of 7-*O*-benzoyl-naringenin using compound **3** as glucuronide donor gave **78** in fair yield, which was deprotected to naringenin-3'-*O*-β-D-glucuronide. The preparation of naringenin-7-*O*-β-D-glucuronide started from fully benzoylated naringenin. This compound was monodebenzoylated at position 7 to give **79**, which was glucuronidated to **80** and deprotected to yield naringenin-7-*O*-β-D-glucuronide (Scheme 11.25).

11.12 ISOFLAVONE CONJUGATES

The synthesis of isoflavone glucuronides is facilitated by the significant reactivity differences between the 5-, 7-, and 4'-hydroxyls, associated to a good stability of the flavones nucleus—in contrast to flavanones. The original method of synthesis of daidzein-7-*O*-β-D-glucuronide (Needs and Williamson 2001) has been optimized, mainly by replacing acetobromoglucuronic acid methyl ester with the much more efficient methyl 2,3,4-triacetyl-D-glucopyranosyl 1-(*N*-4-methoxyphenyl)-2,2,2-trifluoroacetimidate **6** (Al-Maharik and Botting 2006, 2008). To improve the solubility of daidzein in organic solvents, a protection strategy was adopted. In the presence of an excess of potassium *tert*-butoxide, the 7,4'-diphenolate of daidzein was formed. However, the 4'-phenolate ion was expected to be a better nucleophile than the 7-phenolate ion, due to the conjugation of the latter with the 4-carbonyl. Esterification of daidzein with hexanoyl chloride under these conditions yielded 4'-hexanoyldaidzein in a good yield (Scheme 11.26). Glucuronidation of 4'-hexanoyl-daidzein followed by deprotection of compound **81** in alkaline conditions afforded daidzein-7-*O*-β-D-glucuronide. When a similar 4'-monoprotection route was applied to genistein, the glucuronidation step yielded a complex mixture of compounds. This prompted to protect all the hydroxyl groups by treatment with hexanoyl chloride to give 5,7,4'-trihexanoyl-genistein. The 7-hexanoyl group, being *para* to the electron withdrawing carbonyl, is the most electrophilic ester and, consequently, was selectively transferred to thiophenol in the presence of imidadole, yielding

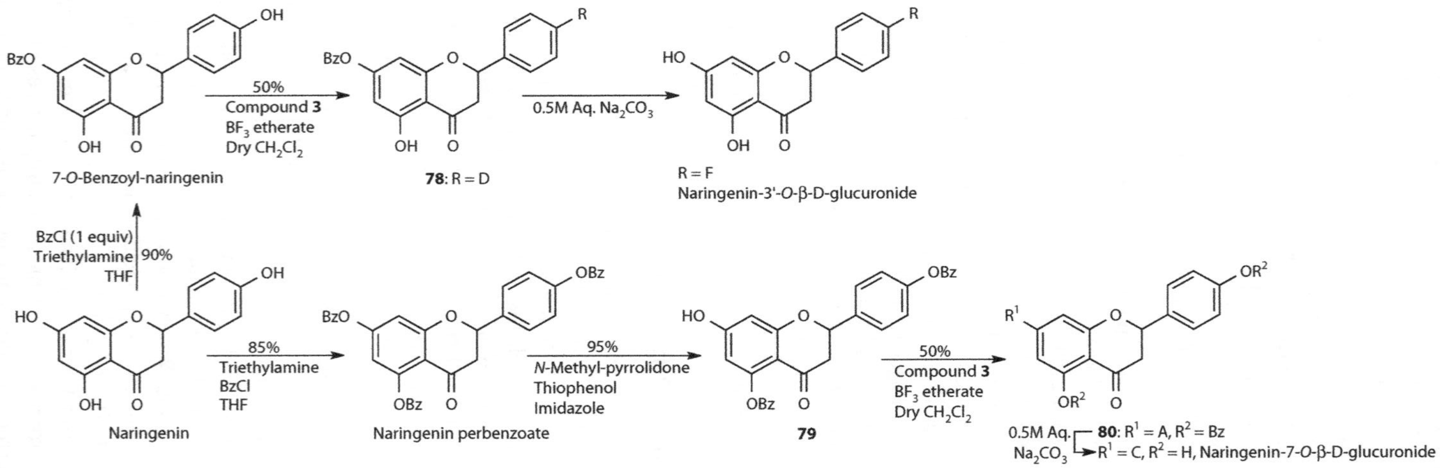

SCHEME 11.25 The syntheses of naringenin-3'- and 7-*O*-β-D-glucuronides. R = A, C, or F refers to glucuronyl residue in Figure 11.2.

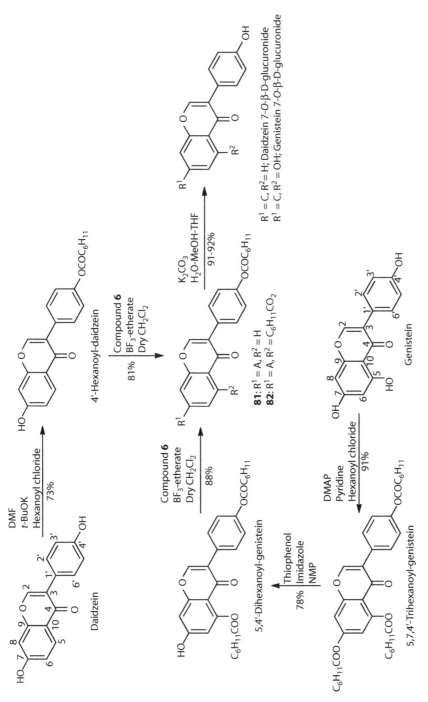

SCHEME 11.26   The syntheses of daidzein- and genistein-7-O-β-D-glucuronides. R = C refers to glucuronyl residue in Figure 11.2.

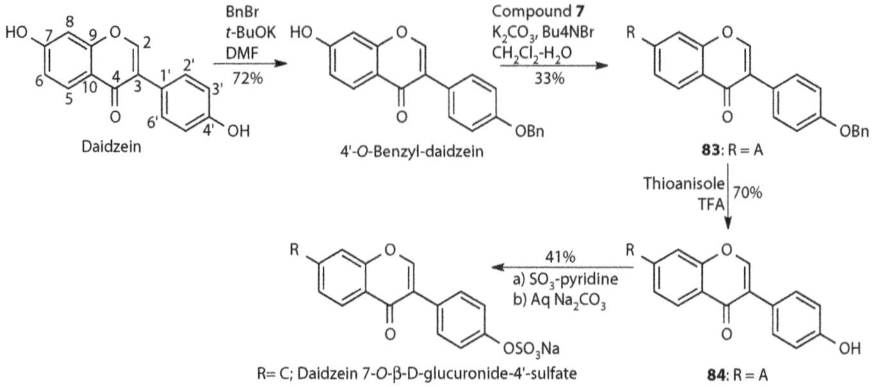

SCHEME 11.27 First synthesis of a mixed sulfate/glucuronide isoflavone. R = C refers to glucuronyl residue in Figure 11.2.

5,4′-dihexanoyl genistein. The two last steps used the same approach as employed for the daidzein-7-O-β-D-glucuronide, affording genistein-7-O-β-D-glucuronide. Later on, a similar approach has been used for the synthesis of glycitein-7-O-β-D-glucuronide (Al-Maharik and Botting 2008).

The preparation of the 7,4′-disulfate conjugates of genistein, daidzein, and their human metabolites dihydrodaidzein, dihydrogenistein, and equol has been carried out using pyridine sulfur trioxide complex as a reagent (Soidinsalo and Wähälä 2004). The synthesis of daidzein-7-O-β-D-glucuronide-4′-sulfate, which has been detected in human urine, was noteworthy. This study was the first report on the preparation of a mixed sulfate/glucuronide conjugate (Soidinsalo and Wähälä 2007). The first step was the 4′-benzyl protection of daidzein in the presence of an excess of potassium *tert*-butoxide (Scheme 11.27), as mentioned earlier in the synthesis of 4′-hexanoyldaidzein. This yielded 4′-O-benzyldaidzein, which was glucuronidated by the commercially available reagent acetobromo-α-D-glucuronic acid methyl ester **7** in phase transfer conditions. Attempts to debenzylate the resulting compound **83** using Pd/C and H$_2$ resulted in partial reduction of the 2,3-double bond. In contrast, debenzylation with thioanisole/trifluoroacetic acid yielded the 4′-deprotected product **84**. Introduction of the sulfate group at position 4′ was mediated by sulfur trioxide-pyridine complex. Finally, the mixed sulfated/glucuronidated product was fully deprotected in alkaline conditions to give daidzein-7-O-β-D-glucuronide-4′-sulfate.

11.13 FLAVONOL CONJUGATES

Improved synthetic methods for the preparation of the major human metabolites of quercetin have been published (Needs and Kroon 2006). In the synthesis of quercetin-3-O-β-D-glucuronide, the starting compound was 7,4′-di-O-benzylquercetin (Scheme 11.28). Interestingly, when this compound was submitted to glucuronidation using acetobromo-α-D-glucuronic acid methyl ester **7**, the conjugation was directed to position 3. This yielded the benzylquercetin glucuronide **85** and its 3′-acetylated

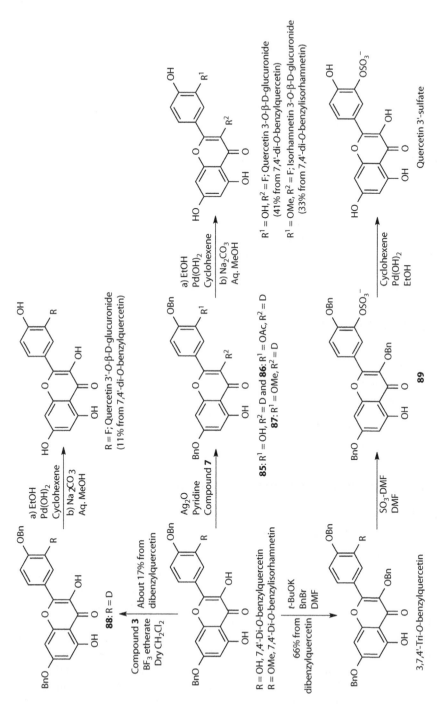

SCHEME 11.28 Preparation of metabolically important flavonol glucuronides and sulfates. R = D or F refers to glucuronyl residue in Figure 11.2.

analogue **86**. This was possibly due to the migration during the reaction of one of the acetyl groups of glucuronide donor **7** to the free 3′-hydroxyl of 7,4′-di-*O*-benzylquercetin. However, this did not decrease the yield of the final conjugate, since after debenzylation and saponification, the 3′-acetyl was eliminated together with the protective groups of the glucuronide moiety, to give quercetin-3-*O*-glucuronide with a 41% overall yield from 7,4′-di-*O*-benzylquercetin. A similar procedure was used for the preparation of isorhamnetin-3-*O*-β-D-glucuronide, starting from 7,4′-di-*O*-benzylisorhamnetin (33% overall yield). This time, of course, no 3′-acetylation took place because of the presence of the 3′-methoxy group.

When 7,4′-di-*O*-benzylquercetin was conjugated using glucuronidating agent **3** instead of **5**, glucuronidation occurred at the 3′ position and compound **88** was obtained, albeit in a low yield. **88** was submitted to debenzylation and saponification, yielding quercetin-3′-*O*-β-D-glucuronide with an 11% overall yield from 7,4′-di-*O*-benzylquercetin (Scheme 11.28).

Finally, the preparation of quercetin-3′-sulfate required the protection of the 3-hydroxyl. Thus 7,4′-di-*O*-benzylquercetin was monobenzylated with 1.6 parts of potassium *tert*-butoxide and benzyl bromide, to give 3,7,4′-tri-*O*-benzylquercetin. The latter compound was sulfated with sulfur trioxide dimethylformamide complex to **89** which was debenzylated to quercetin-3′-sulfate (Scheme 11.28).

11.14 FLAVAN-3-OL CONJUGATES

The direct glucuronidation of catechin using glucuronide donor **7**, followed by deprotection in alkaline conditions, afforded a mixture of the five possible catechin glucuronides (Gonzalez-Manzano et al. 2009). Similar direct sulfation of (+)-catechin or (−)-epicatechin using sulfur trioxide-trimethylamine complex afforded a mixture of mono- and disulfate conjugates. All the conjugates were purified by semipreparative HPLC and analyzed by HPLC-DAD-ESI MS (Gonzalez-Manzano et al. 2009).

11.15 STABLE ISOTOPE LABELED COMPOUNDS

11.15.1 OVERVIEW OF THE KNOWN STABLE ISOTOPICALLY LABELED DIETARY PHENOLS

Deuterium labeling is probably one of the easiest labeling methods since it can be performed on the native molecule, without requiring complex synthetic approaches (Junk et al. 1997). In fact, the activated hydrogen atoms of the compound (e.g., aromatic protons *ortho* and *para* to phenolic hydroxyls, benzylic hydrogens…) are easily exchanged with deuterium. The main drawback, however, is that part of the introduced deuterium labels can be back exchanged with hydrogen when placed in various experimental conditions. Hence, it is of the utmost importance to ensure that the introduced deuterium label will be stable in the experimental context of the studies in which they will be involved. For example, when using GC-MS as the analytical technique, the deuterium label must be stable when treated with silylation reagents (Leppala et al. 2004). Classically, the deuteration of dietary phenols is a two-step procedure involving (i) the polydeuteration of the compound of interest and (ii) the back-exchange of the labile deuterium atoms to hydrogen, ensuring better stability of the final deuterated compound. However, for high precision quantification studies, no unlabeled analog must be present in the deuterium labeled standard, and high isotopic purities are crucial. The higher the number of stable deuterium atoms introduced in the molecule, the more accurate the quantification. In fact, this is the best chance to avoid the MS overlapping of the peak of the stable isotope standard with the M + 1 and M + 2 peaks of the heavier isotopes of the analyte. When preparing deuterium-labeled polyphenols, the presence of at least three to five deuterium atoms is preferable so that when analyzed by MS there is a difference of at least three to five units in the m/z value of the molecular ion (Rasku et al. 1999). A list of previously prepared deuterium-labeled dietary phenolics is presented in Table 11.1. All the synthetic details can be found in our sister review (Barron et al. 2012).

To circumvent the potential instability of deuterium labeled compounds, ^{13}C-labeled analogs have been introduced. In contrast to deuterium, ^{13}C cannot be exchanged to ^{12}C under normal experimental conditions. However, the ^{13}C-label is not easily introduced in the compound of interest, except via O-methylation. This implies the design of elaborated synthesis procedures.

For optimum quantification, the studies involving ^{13}C-labeled compounds have more or less the same requirement as deuterium label studies, that is, a pure standard with enough mass difference to avoid interferences with the natural heavy isotopes of the analyte. The previously prepared ^{13}C-labeled dietary phenols are summarized in Table 11.2 with synthetic details again being found in Barron et al. (2012).

11.15.2 PHENYLACETIC ACIDS AND TYROSOL

The procedures for the syntheses of 4'-methoxyphenylacetic acid, 4'-hydroxyphenylacetic acid, and 3'-hydroxytyrosol were adapted for the preparation of [1-$^{13}C_1$]-4'-methoxyphenylacetic acid, [1-$^{13}C_1$]-4'-hydroxyphenylacetic acid, and [1-$^{13}C_1$]-4'-hydroxyphenylethanol (^{13}C-tyrosol) (Kuwajima et al. 1998). The label was introduced with $^{13}CO_2$, generated by the action of concentrated H_2SO_4 on Ba$^{13}CO_3$ (Scheme 11.29).

11.16 RADIOLABELED COMPOUNDS

11.16.1 OVERVIEW OF THE KNOWN RADIOLABELED DIETARY PHENOLICS

Again, the easiest prepared radiolabeled phenols are the tritiated derivatives. Tritium-labeled compounds share several of the advantages (easy introduction of the label on the native molecule) and drawbacks (label instability) of their deuterated counterparts. However, when measuring tritium, only small amount of radioactivity is required. Hence, the introduction of multiple labels is not necessary, contrarily to deuterium-labeled compounds. The few recently synthesized tritiated phenols are listed in Table 11.3.

The most stable radiolabeled phenolics are ^{14}C-labeled which, like their ^{13}C derivatives, have the disadvantage of requiring complex synthetic methods and equipment

TABLE 11.1
Deuterium Labeled Derivatives of Common Dietary Phenols

Compound	Isotopic Purity	Reference
$[4\alpha\text{-}^2H_1]$- and $[4\beta\text{-}^2H_1]$-Catechins as a mixture	Not reported	Pierre et al. (1997)
$(-)\text{-}[4\text{-}^2H_1]$-Epicatechin	Not reported	Peng et al. (2009)
$(-)\text{-}[4\text{-}^2H_1]$-Epigallocatechin-3-$O$-gallate	Not reported	Kohri et al. (2001)
$[4'\text{-}^2H_1]$-Theaflavin	Not reported	Peng et al. (2009)
$[4\text{-}^2H_1]$-Procyanidin B3 as mixture with a diastereomer	Not reported	Pierre et al. (1997)
$[4\text{-}^2H_1]$-Procyanidin B4	Not reported	Pierre et al. (1997)
$[8,3',5'\text{-}^2H_3]$-Daidzein	93%	Rasku et al. (1999)
$[3',5',6,8\text{-}^2H_4]$-Daidzein	>85%	Hakala and Wähälä (2007)
$[6,8,2',3',5',6'\text{-}^2H_6]$-Daidzein	86 and 90%	Rasku and Wähälä (2000a), Hakala and Wähälä (2007)
$[6,8,2',3',5',6'\text{-}^2H_6]$-Daidzein-7,4'-disulfate	>90%	Soidinsalo et al. (2006)
$[8,3',5'\text{-}^2H_3]$-Daidzein	91.6%	Lewis and Wähälä (1998)
$[2',3',5',6'\text{-}^2H_4]$-Genistein	>90%	Soidinsalo et al. (2006)
$[2',3',5',6'\text{-}^2H_4]$-Genistein-7,4'-disulfate	>90%	Soidinsalo et al. (2006)
$[8,3',5'\text{-}^2H_3]$-Formononetin	91%	Rasku et al. (1999)
$[2',3',5',6'\text{-}^2H_4]$-Biochanin A	70%	Rasku et al. (1999)
$[5,8,2',3',5',6'\text{-}^2H_6]$-Glycitein	>90%	Soidinsalo et al. (2006)
$[5,8,2',3',5',6'\text{-}^2H_6]$-Glycitein-7,4'-disulfate	>90%	Soidinsalo et al. (2006)
$[6,8,3',5'\text{-}^2H_4]$-Dihydrodaidzein	94%	Rasku et al. (1999)
$[3,6,8,3',5'\text{-}^2H_5]$-Dihydrodaidzein	93%	Rasku et al. (1999)
$[3,2',3',5',6'\text{-}^2H_5]$-Dihydrogenistein	66%	Rasku et al. (1999)
$[2,3',3'',5''\text{-}^2H_5]$-$O$-Demethylangolensin	>90%	Hakala and Wähälä (2007)
$[2,3,6,2',5',6'\text{-}^2H_6]$-Matairesinol	>90%	Hakala and Wähälä (2007)
$[2,4,5,6,2',4',5',6'\text{-}^2H_8]$-Enterolactone	>99%	Leppala and Wähälä (2004)
$[2,4,5,6,9,9,2',4',5',6'\text{-}^2H_{10}]$-Enterodiol	>95%	Leppala and Wähälä (2004)
$[3,3',5'\text{-}^2H_3]$-Apigenin	94%	Rasku and Wähälä (2000b)
$[3,6,8,3',5'\text{-}^2H_5]$-Apigenin	>90%	Hakala and Wähälä (2007)
$[3,2',5',6'\text{-}^2H_4]$-Luteolin	86%	Rasku and Wähälä (2000b)
$[2',5',6'\text{-}^2H_3]$-Quercetin	95%	Rasku and Wähälä (2000b)
$[8,2',5',6'\text{-}^2H_4]$-Fisetin	93%	Rasku and Wähälä (2000b)
$[3,3',5'\text{-}^2H_3]$-p-Coumaric acid	Not reported	Sakakibara et al. (2007)
$[3,2',5'\text{-}^2H_3]$-Caffeic acid	Not reported	Sakakibara et al. (2007)
$[3'\text{-}OC^2H_3]$-Ferulic acid	Not reported	Sakakibara et al. (2007)
$[2',5'\text{-}^2H_2, 3'\text{-}OC^2H_3]$-Ferulic acid	Not reported	Sakakibara et al. (2007)
$[2',3',5',6'\text{-}^2H_4]$-$trans$-Resveratrol	96%	Gabriele et al. (2008)

TABLE 11.2
^{13}C Labeled Derivatives of Common Dietary Phenols

Compound (Purity)	Reference
(±)-[4-^{13}C]-Catechin	Nay et al. (2002)
(+)-[4-^{13}C]-Catechin (>99%)	Nay et al. (2001)
(−)-[4-^{13}C]-Epicatechin (>99%)	Nay et al. (2001)
(−)-[4-^{13}C]-Procyanidin B3 (100%)	Arnaudinaud et al. (2001)
[2-^{13}C]-Daidzein (>99%)	Baraldi et al. (1999)
[4-^{13}C]-Daidzein	Whalley et al. (2000)
[3,4,8-^{13}C$_3$]-Daidzein (96%)	Oldfield et al. (2004)
[2,3,4-^{13}C$_3$]-Daidzein	Al-Maharik and Botting (2010)
[4-^{13}C]-Genistein	Whalley et al. (2000)
[3,4,1′-^{13}C$_3$]-Genistein	Oldfield et al. (2007)
[2,3,4-^{13}C$_3$]-Genistein	Al-Maharik and Botting (2010)
[4-^{13}C]-Biochanin A	Whalley et al. (2000)
[2,3,4-^{13}C$_3$]-Biochanin A	Al-Maharik and Botting (2010)
[2-^{13}C]-Formononetin	Baraldi et al. (1999)
[4-^{13}C]-Formononetin	Whalley et al. (2000)
[2,3,4-^{13}C$_3$]-Formononetin	Al-Maharik and Botting (2010)
[2,3,4-^{13}C$_3$]-Glycitein	Zhang and Botting (2004)
[2,3,4-^{13}C$_3$]-Equol	Al-Maharik and Botting (2010)
[1,2,3-^{13}C$_3$]-O-Demethyl angolensin	Al-Maharik and Botting (2010)
[7,7′,9′-^{13}C$_3$]-Enterolactone	Fryatt et al. (2005)
[7,7′,9′-^{13}C$_3$]-Enterodiol	Fryatt et al. (2005)
[7,7′,9′-^{13}C$_3$]-Matairesinol	Fryatt et al. (2005)
[7,7′,9′-^{13}C$_3$]-Secoisolariciresinol	Fryatt et al. (2005)
[7,8,9-^{13}C$_3$]-Medioresinol	Haajanen and Botting (2006)
[7,8,9-^{13}C$_3$]-Sesamin	Haajanen and Botting (2006)
[2-^{13}C]-Quercetin-4′-O-β-glucoside (≥ 97%)	Caldwell et al. (2000)
[2-^{13}C]-Quercetin-3-O-β-D-glucoside (95%)	Caldwell et al. (2006)
[1′,2′,3′,4′,5′,6′-^{13}C$_6$]-p-Coumaric acid (99%)	Ji et al. (2005)
[1,2,3-^{13}C$_3$]-Caffeic acid	Robbins and Schmidt (2004)
[1′,2′,3′,4′,5′,6′-^{13}C$_6$]-Caffeic acid (97%)	Ji et al. (2005)
[1′,2′,3′,4′,5′,6′-^{13}C$_6$]-Ferulic acid (98%)	Ji et al. (2005)
[1,2-^{13}C$_2$]-Sinapic acid	Robbins and Schmidt (2004)

for their preparation. In addition, the use of ^{14}C-labeled precursors is usually restricted to animal studies. Nevertheless, application to human studies is expected to increase in the future with the development of high sensitive detection methods such as accelerator mass spectrometry, allowing the administration of low radioactive ^{14}C doses (Lappin and Garner 2005; Lappin and Seymour 2010). Examples of synthesized ^{14}C-labeled phenolics are displayed in Table 11.4.

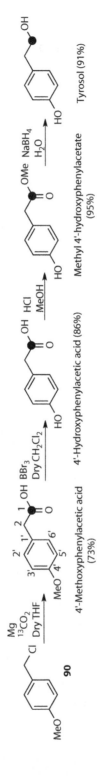

SCHEME 11.29 Syntheses of $[1-^{13}C_1]-4'$-methoxyphenylacetic acid- and $[1-^{13}C_1]-4'$-hydroxyphenylacetic acid, as well of $[1-^{13}C_1]-4'$-hydroxyphenylethanol.

TABLE 11.3
Tritium-Labeled Derivatives of Common Dietary Phenolics

	Radiochemical Purity	Specific Activity	Reference
(−)-[4-³H]-Epigallocatechin-3-O-gallate	99.5%	13 Ci/mmol	Kohri et al. (2001)
(−)-[3-³H]-Epicatechin-3-O-gallate	92.2% (HPLC) 94.7% (TLC)	0.97 Ci/mmol	Takizawa et al. (2004)
[³H]-Puerarin	Not reported	79 mCi/mmol	Lee et al. (2007)

TABLE 11.4
¹⁴C-Labeled Derivatives of Common Dietary Phenolics

Compound	Specific Activity	Reference
[2-¹⁴C₁]-4′-Hydroxyphenylacetic acid	10 mCi/mmol	Komeno et al. (1982)
[2-¹⁴C₁]-4′-Hydroxyphenylethanol	–	Schiefer and Kindl (1971)
(−)-[2-¹⁴C]-Procyanidin B2	1.34 mCi/mmol	Viton et al. (2008)
[2-¹⁴C]-Hispidulin	0.9 mCi/mmol	Kavvadias et al. (2004)
[α-¹⁴C]-5-O-Caffeoylquinic acid	10.8 mCi/mmol	DeBardeleben and Teng (1970)
[β-¹⁴C]-Resveratrol	0.04 mCi/mmol	Zeng et al. (2004)
[3-¹⁴C]-Hesperetin	1.96 mCi/mmol	Honohan et al. (1976)
[2-¹⁴C]-Quercetin-4′-O-β-D-glucoside	3.75 mCi/mmol	Mullen et al. (2008)

Compound	Radiochemical Purity	Reference
[1′,2′,3′,4′,5′,6′-¹⁴C₆]-p-Coumaric acid	94.0%	Ji et al. (2005)
[2-¹⁴C₁]-Caffeic acid	99.0%	Ji et al. (2005)
[1′,2′,3′,4′,5′,6′-¹⁴C₆]-Caffeic acid	98.0%	Ji et al. (2005)
[2-¹⁴C₁]-Ferulic acid	98.7%	Ji et al. (2005)
[1′,2′,3′,4′,5′,6′-¹⁴C₆]-Ferulic acid	99.8%	Ji et al. (2005)

When the labeling of complex polyphenols such as procyanidins is required, bio-labeling experiments can be considered. The compound of interest is produced by cell cultures or plant organs after administration of suitable radiolabeled precursors. Selected examples of ¹⁴C-labeled phenols produced by biolabeling are summarized in Table 11.5. Because of the incorporation of the labeled precursor into multiple metabolic pathways, one of the severe limitations of this approach is the low specific activity of the isolated compounds, which typically is in the nCi/mmol range as opposed to mCi/mmol range in the case of chemical labeling. Although procyanidin B2 was produced quite efficiently by cell suspension cultures of *Crataegus monogyna* (Valls et al. 2007), its specific activity was still 2.7 times lower than procyanidin B2 prepared by chemical synthesis (Viton et al. 2008). However, although the specific activity of biolabeled *trans*-resveratrol was low (4.48 nCi/mmol), it was, nonetheless, possible to carry out a tissue distribution study in mice (Vitrac et al. 2003).

TABLE 11.5
A Selection of ^{14}C-Labeled Phenols Produced By Biolabeling

Compound	Specific Activity	Reference
cis- and *trans-*Resveratrol	4.48–5.22 nCi/mmol	Vitrac et al. (2002)
(+)-Catechin	0.20 mCi/mmol	Valls et al. (2007)
(−)-Epicatechin	0.33 mCi/mmol	Valls et al. (2007)
Procyanidin B2	0.49 mCi/mmol	Valls et al. (2007)
Procyanidin B4	0.13 mCi/mmol	Valls et al. (2007)
Procyanidin B5	0.25 mCi/mmol	Valls et al. (2007)
5-*O*-Caffeoylquinic acid	0.71 mCi/mmol	Valls et al. (2007)
5-*O-trans-p*-Coumaroylquinic acid	0.29 mCi/mmol	Valls et al. (2007)
(+)-Catechin	1.27 nCi/mmol	Deprez et al. (1999)
Procyanidin B3	0.26 nCi/mmol	Deprez et al. (1999)
Procyanidin B2	0.17 nCi/mmol	Deprez et al. (1999)

All the details concerning the chemical and biochemical preparation of radiolabeled phenolics can be found in our sister review (Barron et al. 2012).

11.16.2 PHENYLACETIC ACIDS AND 3′-HYDROXYTYROSOL

The preparation of [2-^{14}C$_1$]-4′-hydroxyphenylacetic acid has been described by Komeno et al. (1982) (Scheme 11.30), en route to the preparation of a radiolabeled β-lactam antibiotic. The radiolabel was introduced at the beginning of the synthesis with ^{14}CO$_2$ generated *in situ* by reaction of 60% perchloric acid with [^{14}C]-barium carbonate. The Grignard reagent prepared from 4-benzyloxy-1-bromoenzene was carboxylated with ^{14}CO$_2$, to give [1-^{14}C$_1$]-4-benzyloxybenzoic acid. [1-^{14}C$_1$]-4-Benzyloxybenzoic acid was reduced to alcohol **91**, chlorinated to compound **92**, which was converted to nitrile **93**. Hydrolysis of **93** gave [2-^{14}C$_1$]-4′-benzyloxyphenylacetic acid, which after deprotection, yielded [2-^{14}C$_1$]-4′-hydroxyphenylacetic acid. [2-^{14}C$_1$]-4′-Hydroxyphenylethanol ([2-^{14}C$_1$]-tyrosol) was prepared from [2-^{14}C$_1$]-4′-hydroxyphenylacetic acid after protection of the phenolic group by acetylation, followed by reduction with LiAlH$_4$ and alkaline hydrolysis of the acetate (Schiefer and Kindl 1971) (Scheme 11.31).

11.17 SUMMARY

In the last 5 years, much progress has been made in the synthesis of sulfated and glucuronidated conjugates of dietary phenolics. The most important advances have concerned the preparation of human metabolites of chlorogenic acids, hydroxycinnamates and hydroxyphenylpropionic acids, as well as of flavanones and isoflavones. The synthesis of γ-valerolactones, the colonic metabolites of tea catechins, has also received a lot of attention, especially from the enantioselective point of view. Noteworthy is now the common use of extremely efficient sulfation (chlorosulfates)

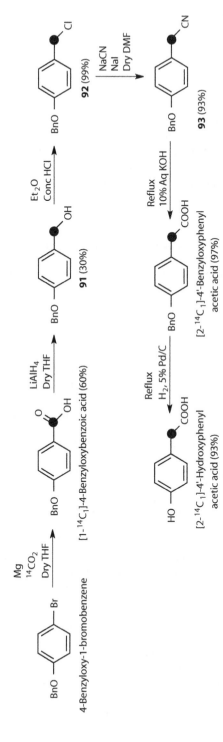

SCHEME 11.30 The synthesis of $[2-^{14}C_1]$-4'-hydroxyphenylacetic acid.

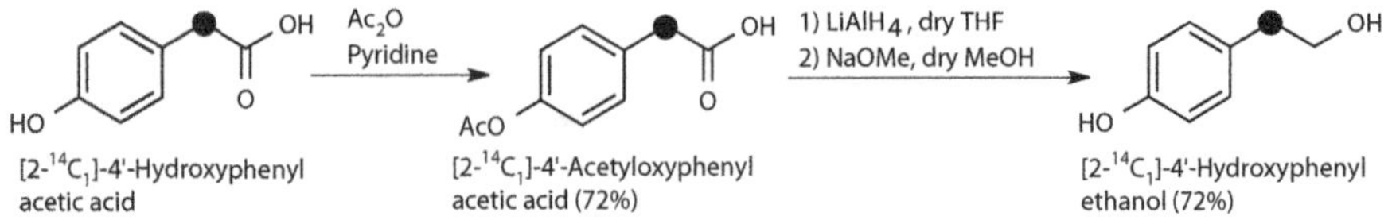

SCHEME 11.31   Synthesis of [2-^{14}C$_1$]-4'-hydroxyphenylethanol.

and glucuronidation (acetimidates) agents. The first mixed sulfated-glucuronidated conjugate was also successfully prepared during this period. Concerning isotopically labeled compounds, there have only been a few studies on the preparation of radiolabeled derivatives, while in contrast, the synthesis of stable isotopes is now common, with the increasing use of the SID-MS analysis.

REFERENCES

Al-Maharik, N. and Botting, N.P. (2006). A facile synthesis of isoflavone 7-*O*-glucuronides. *Tetrahedron Lett.*, **47**, 8703–8706.

Al-Maharik, N. and Botting, N.P. (2008). An efficient method for the glycosylation of isoflavones. *Eur. J. Org. Chem.*, 5622–5629.

Al-Maharik, N. and Botting, N.P. (2010). A versatile synthesis of [2,3,4-^{13}C$_3$]isoflavones. *J. Labelled Compd. Radiopharm.*, **53**, 95–103.

Alo, B.I., Kandil, A., Patil, P.A. et al. (1991). Sequential directed ortho metalation-boronic acid cross-coupling reactions. A general regiospecific route to oxygenated dibenzo[*b,d*] pyran-6-ones related to ellagic acid. *J. Org. Chem.*, **56**, 3763–3768.

Arnaudinaud, V., Nay, B., Verge, S. et al. (2001). Total synthesis of isotopically labelled flavonoids. Part 5: Gram-scale production of ^{13}C-labelled (−)-procyanidin B3. *Tetrahedron Lett.*, **42**, 5669–5671.

Baraldi, P.G., Spalluto, G., Cacciari, B. et al. (1999). Chemical synthesis of [^{13}C]daidzein. *J. Med. Food*, **2**, 99–102.

Barron, D. (2008). Recent advances in the chemical synthesis and biological activity of phenolic metabolites. *Recent Adv. Polyphenol Res.*, **1**, 317–358.

Barron, D., Cren-Olive, C., and Needs, P.W. (2003). Chemical synthesis of flavonoid conjugates. In C. Santos-Buelga and G. Williamson (eds.), *Methods Polyphenol Analysis*. Royal Society of Chemistry Press, Cambridge, pp. 187–213.

Barron, D., Smarrito Menozzi, C. and Viton, F. (2012). (Bio)chemical labelling tools for studying absorption and metabolism of dietary phenols—An overview. *Curr. Org. Chem.*, **16**, 663–690.

Bialonska, D., Kasimsetty, S.G., Khan, S.I. et al. (2009). Urolithins, intestinal microbial metabolites of pomegranate ellagitannins, exhibit potent antioxidant activity in a cell-based assay. *J. Agric. Food Chem.*, **57**, 10181–10186.

Borges, G., Mullen, W., Mullan, A. et al. (2010). Bioavailability of multiple components following acute ingestion of a polyphenol-rich juice drink. *Mol. Nutr. Food Res.*, **54**, S268–S277.

Boumendjel, A., Blanc, M., Williamson, G. et al. (2009). Efficient synthesis of flavanone glucuronides. *J. Agric. Food Chem.*, **57**, 7264–7267.

Caldwell, S.T., Crozier, A., and Hartley, R.C. (2000). Isotopic labeling of quercetin 4'-*O*-β-D-glucoside. *Tetrahedron*, **56**, 4101–4106.

Caldwell, S.T., Petersson, H.M., Farrugia, L.J. et al. (2006). Isotopic labeling of quercetin 3-glucoside. *Tetrahedron*, **62**, 7257–7265.

DeBardeleben, J.F. and Teng, L.C. (1970). Synthesis of carbon-14 labeled chlorogenic acid. *J. Labelled Compds*, **6**, 34–39.

Deprez, S., Mila, I., and Scalbert, A. (1999). Carbon-14 biolabeling of (+)-catechin and proanthocyanidin oligomers in willow tree cuttings. *J. Agric. Food Chem.*, **47**, 4219–4230.

Fumeaux, R., Menozzi-Smarrito, C., Stalmach, A. et al. (2010). First synthesis, characterization, and evidence for the presence of hydroxycinnamic acid sulfate and glucuronide conjugates in human biological fluids as a result of coffee consumption. *Org. Biomol. Chem.*, **8**, 5199–5211.

Fryatt, T. and Botting, N.P. (2005). The synthesis of multiply ^{13}C-labeled plant and mammalian lignans as internal standards for LC-MS and GC-MS analysis. *J. Labelled Compd. Radiopharm.*, **48**, 951–969.

Gabriele, B., Benabdelkamel, H., Plastina, P. et al. (2008). *trans*-Resveratrol-d_4, a molecular tracer of the wild-type phytoalexin; synthesis and spectroscopic properties. *Synthesis*, 2953–2956.

Galland, S., Rakotomanomana, N., Dufour, C. et al. (2008). Synthesis of hydroxycinnamic acid glucuronides and investigation of their affinity for human serum albumin. *Org. Biomol. Chem.*, **6**, 4253–4260.

Ghosal, S., Lal, J., Singh, S.K. et al. (1989). Shilajit. Part 4. Chemistry of two bioactive benzopyrone metabolites. *J. Chem. Res. Synop.*, 350–351.

González-Barrio, R., Truchado, P., Ito, H. et al. (2011). UV and MS identification of urolithins and nasutins, the bioavailable metabolites of ellagitannins and ellagic acid in different mammals. *J. Agric. Food Chem.*, **59**, 1152–1162.

Gonzalez-Manzano, S., Gonzalez-Paramas, A., Santos-Buelga, C. et al. (2009). Preparation and characterization of catechin sulfates, glucuronides, and methyl ethers with metabolic interest. *J. Agric. Food Chem.*, **57**, 1231–1238.

Graefe, E.U. and Veit, M. (1999). Urinary metabolites of flavonoids and hydroxycinnamic acids in humans after application of a crude extract from *Equisetum arvense*. *Phytomedicine*, **6**, 239–246.

Haajanen, K. and Botting, N.P. (2006). Synthesis of multiply ^{13}C-labeled furofuran lignans using ^{13}C-labeled cinnamyl alcohols as building blocks. *Steroids*, **71**, 231–239.

Hakala, U. and Wähälä, K. (2007). Expedient deuterolabeling of polyphenols in ionic liquids-DCl/D_2O under microwave irradiation. *J. Org. Chem.*, **72**, 5817–5819.

Hamada, M., Furuno, A., Nakano, S. et al. (2010). Synthesis of optically pure lactone metabolites of tea catechins. *Synthesis*, 1512–1520.

Hanselaer, R., D'Haenens, L., Martens, M. et al. (1983). *N*-Acylamino acids and peptides. VII. Synthesis of oxygen-sensitive *N*-acylglycines (*N*-caffeoylglycine, *N*-protocatechuoylglycine and *N*-galloylglycine) and a *N*-acyldipeptide (*N*-caffeoylglycyl-L-phenylalanine). *Bull. Soc. Chim. Belg.*, **92**, 1029–1037.

Honohan, T., Hale, R.L., Brown, J.P. et al. (1976). Synthesis and metabolic fate of hesperetin-3-^{14}C. *J. Agric. Food Chem.*, **24**, 906–911.

Jacobs, E. and Metzler, M. (1999). Oxidative metabolism of the mammalian lignans enterolactone and enterodiol by rat, pig, and human liver microsomes. *J. Agric. Food Chem.*, **47**, 1071–1077.

Ji, R., Chen, Z., Corvini, P.F.X. et al. (2005). Synthesis of [^{13}C]- and [^{14}C]-labeled phenolic humus and lignan monomers. *Chemosphere*, **60**, 1169–1181.

Junk, T. and Catallo, W.J. (1997). Hydrogen isotope exchange reactions involving C-H (D, T) bonds. *Chem. Soc. Rev.*, **26**, 401–406.

Kahle, K., Huemmer, W., Kempf, M. et al. (2007). Polyphenols are intensively metabolized in the human gastrointestinal tract after apple juice consumption. *J. Agric. Food Chem.*, **55**, 10605–10614.

Kavvadias, D., Sand, P., Youdim, K.A. et al. (2004). The flavone hispidulin, a benzodiazepine receptor ligand with positive allosteric properties, traverses the blood-brain barrier and exhibits anticonvulsive effects. *Br. J. Pharmacol.*, **142**, 811–820.

Kern, S.M., Bennett, R.N., Mellon, F.A. et al. (2003). Absorption of hydroxycinnamates in humans after high-bran cereal consumption. *J. Agric. Food Chem.*, **51**, 6050–6055.

Khan, M.K., Rakotomanomana, N., Loonis, M. et al. (2010). Chemical synthesis of *Citrus* flavanone glucuronides. *J. Agric. Food Chem.*, **58**, 8437–8443.

Khymenets, O., Joglar, J., Clapes, P. et al. (2006). Biocatalyzed synthesis and structural characterization of monoglucuronides of hydroxytyrosol, tyrosol, homovanillic alcohol, and 3-(4′-hydroxyphenyl)propanol. *Adv. Synth. Catal.*, **348**, 2155–2162.

Kohri, T., Nanjo, F., Suzuki, M. et al. (2001). Synthesis of (−)-[4-^{3}H]epigallocatechin gallate and its metabolic fate in rats after intravenous administration. *J. Agric. Food Chem.*, **49**,1042–1048.

Komeno, T. Nagasaki, T. Katsuyama, Y. et al. (1982). Synthesis of carbon-14-labeled Latamoxef, a potent antibacterial. *J. Labelled Compd. Radiopharm.*, **19**, 981–990.

Kuwajima, H., Takai, Y., Takaishi, K. et al. (1998). Synthesis of ^{13}C-labeled possible intermediates in the biosynthesis of phenylethanoid derivatives, cornoside and rengyosides. *Chem. Pharm. Bull.*, **46**, 581–586.

Lambert, J.D., Rice, J.E., Hong, J. et al. (2005). Synthesis and biological activity of the tea catechin metabolites, M4 and M6 and their methoxy-derivatives. *Bioorg. Med. Chem. Lett.*, **15**, 873–876.

Lappin, G. and Garner, R.C. (2005). The use of accelerator mass spectrometry to obtain early human ADME/PK data. *Expert Opin. Drug Metab. Toxicol.*, **1**, 23–31.

Lappin, G. and Seymour, M. (2010). Addressing metabolite safety during first-in-man studies using ^{14}C-labeled drug and accelerator mass spectrometry. *Bioanalysis*, **2**, 1315–1324.

Lee, D.Y.W., Ji, X.S., and Zhang, X. (2007). Synthesis of tritium-labeled puerarin—a potential antidipsotropic agent. *J. Labelled Compd. Radiopharm.*, **50**, 702–705.

Leppala, E. and Whähälä, K. (2004). Synthesis of [^{2}H$_8$]-enterolactone and [^{2}H$_{10}$]-enterodiol. *J. Labelled Compd. Radiopharm.*, **47**, 25–30.

Lewis, P.T. and Whähälä, K. (1998). Synthesis of the deuterium labeled isoflavone-*O*-glucoside 8,3′,5′-D$_3$-daidzin. *Mol. Online*, **2**, 137–139.

Li, C., Lee, M.J., Sheng, S. et al. (2000). Structural identification of two metabolites of catechins and their kinetics in human urine and blood after tea ingestion. *Chem. Res. Toxicol.*, **13**, 177–184.

Liu, Y., Lien, I.F.F., Ruttgaizer, S. et al. (2004). Synthesis and protection of aryl sulfates using the 2,2,2-trichloroethyl moiety. *Org. Lett.*, **6**, 209–212.

Lucas, R., Alcantara, D., and Morales, J.C.A. (2009). Concise synthesis of glucuronide metabolites of urolithin-B, resveratrol, and hydroxytyrosol. *Carbohydr. Res.*, **344**, 1340–1346.

Menozzi-Smarrito, C., Munari, C., Robert, F. et al. (2008). A novel efficient and versatile route to the synthesis of 5-*O*-feruloylquinic acids. *Org. Biomol. Chem.*, **6**, 986–987.

Menozzi-Smarrito, C., Wong, C.C., Meinl, W. et al. (2011). First synthesis of the potential human metabolites 5-*O*-feruloylquinic acid 4′-*O*-sulphate and 4′-*O*-glucuronide. *J. Agric. Food Chem.*, **59**, 5671–5676.

Meselhy, M.R., Nakamura, N., and Hattori, M. (1997). Biotransformation of (−)-epicatechin 3-*O*-gallate by human intestinal bacteria. *Chem. Pharm. Bull.*, **45**, 888–893.

Momose, T., Tsubaki, T., Iida, T. et al. (1997). An improved synthesis of taurine- and glycine-conjugated bile acids. *Lipids*, **32**, 775–778.

Mullen, W., Borges, G., Lean, M.E.J. et al. (2010). Identification of metabolites in human plasma and urine after consumption of a polyphenol-rich juice drink. *J. Agric. Food Chem.*, **58**, 2586–2595.

Mullen, W., Rouanet, J.-M., Auger, C. et al. (2008). Bioavailabity of [2-^{14}C]quercetin-4′-glucoside in rats. *J. Agric. Food Chem.*, **56**, 12127–12137.

Nakano, S., Hamada, M., Kishimoto, T. et al. (2008). Synthesis of γ-valerolactones as the tea catechin metabolites. *Heterocycles*, **76**, 1001–1005.

Nay, B., Arnaudinaud, V. and Vercauteren, J. (2001). Gram-scale production and applications of optically pure ^{13}C-labeled (+)-catechin and (−)-epicatechin. *Eur. J. Org. Chem.*, 2379–2384.

Nay, B., Arnaudinaud, V. and Vercauteren, J. (2002). Total synthesis of asymmetric flavonoids: the development and applications of ^{13}C-labelling. *C. R. Chim.*, **5**, 577–590.

Needs, P.W. and Kroon, P.A. (2006). Convenient syntheses of metabolically important quercetin glucuronides and sulfates. *Tetrahedron*, **62**, 6862–6868.

Needs, P.W. and Williamson, G. (2001). Syntheses of daidzein-7-yl β-D-glucopyranosiduronic acid and daidzein-4′,7-yl di-β-D-glucopyranosiduronic acid. *Carbohydr. Res.*, **330**, 511–515.

Oldfield, M.F., Chen, L., and Botting, N.P. (2004). Synthesis of [3,4,8-^{13}C$_3$]daidzein. *Tetrahedron*, **60**, 1887–1893.

Oldfield, M.F., Chen, L., and Botting, N.P. (2007). The synthesis of [3,4,1′-^{13}C$_3$]genistein. *J. Labelled Compd. Radiopharm.*, **50**, 1266–1271.

Pelter, A., Ward, R.S., Satyanarayana, P. et al. (1983). Synthesis of lignan lactones by conjugate addition of thioacetal carbanions to butenolide. *J. Chem. Soc. Perkin Trans.* 1, 643–647.

Peng, W. and Yao, Y.F. (2009). Deuterium labeling of theaflavin. *J. Labelled Compd. Radiopharm.*, **52**, 312–315.

Pierre, M.C., Cheze, C., and Vercauteren, J. (1997). Deuterium labeled procyanidin syntheses. *Tetrahedron Lett.*, **38**, 5639–5642.

Rasku, S. and Wähälä, K. (2000a). Synthesis of D$_6$-daidzein. *J. Labelled Compd. Radiopharm.*, **43**, 849–854.

Rasku, S. and Wähälä, K. (2000b). Synthesis of deuterium labeled polyhydroxy flavones and 3-flavonols. *Tetrahedron*, **56**, 913–916.

Rasku, S., Wähälä, K., Koskimies, J. et al. (1999). Synthesis of isoflavonoid deuterium labeled polyphenolic phytoestrogens. *Tetrahedron*, **55**, 3445–3454.

Robbins, R.J. and Schmidt, W.F. (2004). Optimized synthesis of four isotopically labeled (^{13}C-enriched) phenolic acids via a malonic acid condensation. *J. Labelled Compd. Radiopharm.*, **47**, 797–806.

Sakakibara, N., Nakatsubo, T., Suzuki, S. et al. (2007). Metabolic analysis of the cinnamate/monolignol pathway in *Carthamus tinctorius* seeds by a stable-isotope-dilution method. *Org. Biomol. Chem.*, **5**, 802–815.

Schiefer, S. and Kindl, H. (1971). Carbon-14 labeled syntheses of biologically interesting *p*-substituted arylethane compounds. *J. Labelled Compds.*, **7**, 291–297.

Schnabelrauch, M., Wittmann, S., Rahn, K. et al. (2000). New synthetic catecholate-type siderophores based on amino acids and dipeptides. *BioMetals*, **13**, 333–348.

Sharma, A., Joshi, B.P., and Sinha, A.K. (2003). A rapid and efficient microwave-assisted synthesis of substituted 3-phenylpropionic acids from benzaldehydes in minutes. *Chem. Lett.*, **32**, 1186–1187.

Soidinsalo, O. and Wähälä, K. (2004). Synthesis of phytoestrogenic isoflavonoid disulfates. *Steroids*, **69**, 613–616.

Soidinsalo, O. and Wähälä, K. (2006). Synthesis of deuterated isoflavone disulfates. *J. Labelled Compd. Radiopharm.*, **49**, 973–978.

Soidinsalo, O. and Wähälä, K. (2007). Synthesis of daidzein 7-*O*-β-D-glucuronide-4′-*O*-sulfate. *Steroids*, **72**, 851–854.

Stalmach, A., Mullen, W., Barron, D. et al. (2009). Metabolite profiling of hydroxycinnamate derivatives in plasma and urine after the ingestion of coffee by humans: identification of biomarkers of coffee consumption. *Drug Metab. Dispos.*, **37**, 1749–1758.

Sun, W., Cama, L.D., Birzin, E.T. et al. (2006). 6H-Benzo[*c*]chromen-6-one derivatives as selective ERβ agonists. *Bioorg. Med. Chem. Lett.*, **16**, 1468–1472.

Takizawa, Y., Nishimura, H., Morota, T. et al. (2004.) Pharmacokinetics of TJ-8117(Onpi-to), a drug for renal failure (I): Plasma concentration, distribution and excretion of [³H]-(−)-epicatechin 3-*O*-gallate in rats and dogs. *Eur. J. Drug Metab. Pharmacokinet.*, **29**, 91–101.

Uutela, P., Reinila, R., Harju, K. et al. (2009). Analysis of intact glucuronides and sulfates of serotonin, dopamine, and their phase I metabolites in rat brain microdialyzates by liquid chromatography-tandem mass spectrometry. *Anal. Chem.*, **81**, 8417–8425.

Valls, J., Richard, T., Trotin, F. et al. (2007). Carbon-14 biolabeling of flavanols and chlorogenic acids in *Crataegus monogyna* cell suspension cultures. *Food Chem.*, **105**, 879–882.

van Brussel, W. and Van Sumere, C.F. (1978). *N*-Acylamino acids and peptides. VI. A simple synthesis of *N*-acylglycines of the benzoyl and cinnamyl type. *Bull. Soc. Chim. Belg.*, **87**, 791–797.

Viton, F., Landreau, C., Rustidge, D. et al. (2008). First total synthesis of ¹⁴C-labeled procyanidin B2—a milestone toward understanding cocoa polyphenol metabolism. *Eur. J. Org. Chem.*, 6069–6078.

Vitrac, X., Desmouliere, A., Brouillaud, B. et al. (2003). Distribution of [¹⁴C]-*trans*-resveratrol, a cancer chemopreventive polyphenol, in mouse tissues after oral administration. *Life Sci.*, **72**, 2219–2233.

Vitrac, X., Krisa, S., Decendit, A. et al. (2002). Carbon-14 biolabelling of wine polyphenols in *Vitis vinifera* cell suspension cultures. *J. Biotechnol.*, **95**, 49–56.

Whalley, J.L., Oldfield, M.F., and Botting, N.P. (2000). Synthesis of [4-¹³C]-isoflavonoid phytoestrogens. *Tetrahedron*, **56**, 455–460.

Zeng, D., Mi, Q., Sun, H. et al. (2004). A convenient synthesis of ¹⁴C-labeled resveratrol. *J. Labelled Compd. Radiopharm.*, **47**, 167–174.

Zhang, Q. and Botting, N.P. (2004). The synthesis of [2,3,4-¹³C₃]glycitein. *Tetrahedron*, **60**, 12211–12216.

12 Interactions of Flavan-3-ols within Cellular Signaling Pathways

Cesar G. Fraga and Patricia I. Oteiza

CONTENTS

12.1 INTRODUCTION

Plant-derived flavonoids and related phenolic compounds are among the most abundant phytochemicals in the human diet. There is a substantial body of evidence that these compounds, in particular (–)-epicatechin, procyanidins, and related flavan-3-ols, can have beneficial effects on health. This was initially attributed to their antioxidant activity. However, more recent research has shown that they regulate cell signaling pathways through highly specific mechanisms that sometimes, but not always, involve an antioxidant action. This chapter will discuss current evidence on the capacity of the flavan-3-ols to modulate select signaling pathways through redox-dependent and -independent mechanisms. The mechanisms involve free radical scavenging actions, interactions with proteins that generate oxidants, interactions with signaling proteins, and interactions with membrane domains that concentrate signaling molecules. These effects could, in part, explain the health benefits associated with a high consumption of fruit and vegetables.

12.2 CHEMICAL STRUCTURE, BIOAVAILABILITY, AND METABOLISM OF FLAVAN-3-OLS AND PROCYANIDINS

(−)-Epicatechin and procyanidins are flavan-3-ols, a flavonoid subgroup, which in plants are secondary metabolites that have several functions (Crozier et al. 2009). Chemically, flavan-3-ols characterized by a C-ring, which is a saturated heterocycle with a hydroxyl group in position 4, and by the presence of hydroxyl groups in the A- and B-rings. They occur *in planta* as aglycone monomers, as gallic acid derivatives, and procyanidin oligomers and polymers (see Chapter 2). Most naturally occurring flavan-3-ols are stereoisomers in *cis* or *trans* configuration with respect to carbons 2 and 3 [(−)-epicatechin (*cis*) and (+)-catechin (*trans*)]. The relative concentration of each individual flavan-3-ol and procyanidin can vary depending on the plant species (Hammerstone et al. 2000). For example, cocoa (*Theobroma cacao*) synthesizes mostly (−)-epicatechin, (+)-catechin, and procyanidin B-type dimers and larger oligomers of (−)-epicatechin. Changes in isomerization, monomer bonding type, degree of polymerization, or tridimensional structure can have a major impact on the biological actions of these molecules.

The biological actions of flavan-3-ols and procyanidins in animals are largely dependent on their bioavailability. Data from human subjects show that flavan-3-ols and procyanidins are stable during gastric transit (Spencer et al. 2000; Rios et al. 2002), allowing their presence in the gastrointestinal tract at micromolar or higher concentrations. In the small intestine, flavan-3-ols are mostly glucuronidated and methylated (Piskula and Terao 1998; Kuhnle et al. 2000) leaving only trace amounts of native flavan-3-ol, for example, (−)-epicatechin in the mesenteric circulation. Further metabolism, including glucuronidation, O-methylation, and sulfation can occur in the liver and also in other tissues (Piskula and Terao 1998; Donovan et al. 1999, 2001; Abd El Mohsen et al. 2002, Baba et al. 2000b, 2002; Spencer 2003). Once in the blood, monomers are mostly present as conjugated metabolites. Total human plasma concentrations of (−)-epicatechin plus (−)-epicatechin metabolites are in the low micromolar range within 1 h postconsumption of cocoa (Baba et al. 2000b; Rein et al. 2000; Holt et al. 2002; Steinberg et al. 2003; Schroeter et al. 2006). The major metabolite of (−)-epicatechin detected in plasma is 4′-O-methyl-(−)-epicatechin-7-O-glucuronide (Schroeter et al. 2006). Importantly, it was observed that regardless the composition of the food ingested, only plasma oligomers of (−)-epicatechin, but not of (+)-catechin are detected. This indicates that the body distinguishes (−)-epicatechin from (+)-catechin, stressing a high specificity for the interactions of flavan-3-ols with cell components (Baba et al. 2001; Holt et al. 2002; Donovan et al. 2006; Ottaviani et al. 2011). Finally, it is important to stress that the colonic microflora can further metabolize flavan-3-ols and procyanidins by breaking the flavan structure to form simpler phenolics and ring-fission metabolites that can also be absorbed and exert biological actions (Rechner et al. 2002; Unno et al. 2003). More details on the structure, absorption, metabolism, excretion, and sequestration of flavan-3-ols can be found in Chapter 2.

Based on the above-explained flavan-3-ols and procyanidin metabolism, it can be inferred that biological activities in need of high flavan-3-ol and procyanidin concentrations are mostly circumscribed to the gastrointestinal tract. In blood and the vasculature, the potential biological actions should be compatible with nanomolar concentrations of nonmetabolized flavan-3-ols or low micromolar concentration of metabolites, for example, methylated and glucuronidated (–)-epicatechin. Thus, regardless of the limited knowledge on flavonoid tissue availability, it can be speculated that flavan-3-ols are present at no more than low nanomolar concentrations in most tissues participating in very specific interactions.

12.3 FLAVAN-3-OLS AND PROCYANIDINS IN THE REGULATION OF CELL SIGNALING

Extensive evidence has shown that both *in vitro* and *in vivo* models, procyanidin-rich extracts have numerous effects on cell signaling (Nandakumar et al. 2008). Although these studies are of significant value to define the health benefits of flavan-3-ols and procyanidins, they have very limited value to establish mechanisms of action. This is due to the uncertainty on the identity of the molecule/s present in the studied extract that is responsible for the observed effects. We will review the current knowledge on the action of pure or highly purified monomers, usually (–)-epicatechin, their metabolites, and derived procyanidins on cell signaling. We will focus mostly on redox-regulated cell signaling. However, considering the complex net of interconnections among different signaling molecules/cascades, many of the observed effects could also have an impact on redox-independent signaling pathways. As discussed later, the effects of these compounds on redox signaling are in part due to their capacity to scavenge oxidants, to interact with proteins that modulate oxidant production, to interact with signaling proteins, and also to affect the environment of oxidant-generating and/or signaling proteins localized in select membrane domains.

12.3.1 ANTIOXIDANT AND REDOX REGULATION BY FLAVAN-3-OLS AND PROCYANIDINS

Flavan-3-ols and procyanidins are chemically able to prevent free radical-mediated oxidation reactions, and their presence/supplementation has been associated with a decrease of oxidative stress markers in animals and humans. Ubiquitous oxidant species, generally referred to as reactive oxygen species or ROS, can actively participate in the normal regulation of cell signaling, including the triggering of signals involved in adaptive (antioxidant) responses (Forman et al. 2010; Brigelius-Flohe and Flohe 2011). On the other hand, higher and uncontrolled ROS production leads to damage of cell components, which can potentially affect cell physiology. In between, ROS-mediated imbalance of the cellular redox state may not immediately compromise vital cell functions, but could lead in the long term to the onset of disease, for example, chronic inflammation (Reuter et al. 2010).

The capacity to prevent oxidative damage generically categorizes a compound as antioxidant. However, the extrapolation of an antioxidant action *in vitro* to an *in vivo* condition is not trivial. A conceivable approach is to separate "direct antioxidant effects," that is, free radical scavenging and redox active metal sequestration, from "indirect antioxidant effects," for example, regulation of protein synthesis and activities, signaling strategies, etc. This distinction is meaningful when discussing the mechanisms underlying the health effects of flavan-3-ols and procyanidins, because it considers the concentrations (bioavailability) of these compounds to define as antioxidants. In this regard, high concentrations are usually required to act as "direct antioxidant," but "indirect" antioxidant actions can be relevant at significantly lower concentrations (Azzi et al. 2004; Halliwell et al. 2005; Fraga 2007; Dinkova-Kostova and Talalay 2008; Hollman et al. 2011).

Flavan-3-ols and procyanidins are often claimed as direct antioxidants because they affect (i) the occurrence of oxidant-mediated events *in vivo* and *in vitro* (Rice-Evans 2001; Fraga 2007); (ii) the oxidation of different tissues when supplemented to the diet of experimental animals or humans (Fraga et al. 1987, 2005; Koga et al. 1999; Baba et al. 2000a; Toschi et al. 2000; Wang et al. 2000; Frei and Higdon 2003; Orozco et al. 2003; Graziani et al. 2005; Chu et al. 2010); and (iii) the occurrence of pathological situations associated with high oxidant production (Frei and Higdon 2003; Fraga et al. 2005). However, such direct antioxidant actions *in vivo* are likely to be significant in tissues exposed to high concentrations of polyphenols, for example, in the digestive tract where flavan-3-ol and procyanidin concentrations can reach values in the upper micromolar range. The low micromolar concentrations of (−)-epicatechin and its metabolites in blood and the potentially lower (nM) flavonoid concentration in most other tissues do not justify any relevant direct antioxidant action (Galleano et al. 2010). A similar situation is the case for flavan-3-ol- and procyanidin-metal sequestering actions, given the relatively high cell concentrations of many other molecules with high capacity to sequester metals, for example, phosphates, amino acids, albumin, etc. (Frei and Higdon 2003; Fraga 2007; Halliwell 2008; Fraga et al. 2010).

Evidence of indirect "antioxidant" actions of flavan-3-ols and procyanidins includes the regulation of the activity of enzymes that generate oxidants, and local effects at membrane domains that concentrate redox-signaling molecules (Fraga and Oteiza 2011). An example of the regulation of ROS-generating activities, metabolites of (−)-epicatechin were shown to inhibit the activity of NADPH-oxidase in cultures of human umbilical vein endothelial cells (Steffen et al. 2007, 2008). This inhibition seems to be due to the structural similarity between O-methylated (−)-epicatechin metabolites and apocynin, a typical NADPH (reduced nicotinamide adenine dinucleotide phosphate)-oxidase inhibitor. Considering the micromolar concentrations used in cell cultures, this inhibition of NADPH oxidase could be operative *in vivo* for the regulation of redox cell signaling and vascular function.

Another example of the indirect actions of flavan-3-ols and procyanidins is the modulation of oxidant production by interactions with cell membranes, including changes in their physical and chemical characteristics, interactions with receptors present in the membranes, and/or interactions with specialized zones of the membranes, for example, lipid rafts. Experimental evidence of the inhibition of NADPH oxidase activation secondary to the interactions of flavan-3-ols or procyanidins with cell membranes are the modulation of calcium transport and homeostasis (Verstraeten et al. 2008) and the inhibition of the binding of tumor necrosis factor α (TNFα) to TNFα receptor 1 (Erlejman et al. 2008). This last inhibition was inferred by the effects of (–)-epicatechin, catechin, B2 dimer, and hexameric procyanidins, preventing the increase in oxidants occurring upon stimulation with TNFα in intestinal cells. Thus, the interaction with membranes could in part explain the biological actions, including the regulation of cell signaling and antioxidant effects, of large procyanidins that are unlikely being transported inside cells, that is, trimers and larger polymers.

In summary, a direct antioxidant action of flavan-3-ols and procyanidins would be limited to tissue/organs in which relatively high concentrations of these compounds can be achieved. Flavan-3-ols and procyanidins could support indirect antioxidant actions through the regulation of enzymes that generate oxidants, or of membrane-associated events (calcium transport, ligand-receptor binding) that trigger oxidant production. Considering that the resulting steady state level of ROS is central for the regulation of multiple signaling pathways, we will discuss the most relevant redox-regulated signaling cascades that can be modulated by flavan-3-ols and procyanidins and the underlying mechanisms of the modulation.

12.3.2 Modulation of NF-κB by Flavan-3-ols and Procyanidins

NF-κB is a ubiquitous transcription factor that regulates many central events in normal cell function and fate. The regulation of the inflammatory response by NF-κB implies that, when chronically activated, it also underlies the pathophysiology of several inflammation-related diseases (Pahl 1999; Karin and Greten 2005). NF-κB is redox sensitive, and in general, oxidants promote and antioxidants inhibit its activation (Jung et al. 2008). Flavan-3-ols and procyanidins can interfere with NF-κB activation by offsetting changes in cell redox state but also by specific bonding to proteins involved in the NF-κB activation pathway.

(–)-Epicatechin and procyanidins can inhibit NF-κB at different levels in the activation pathway as it is illustrated in Figure 12.1. Once inside the cells, (–)-epicatechin and its metabolites regulate NF-κB activation by: (a) modulating ROS concentration; (b) interacting with different components in the signaling cascade (e.g., IKK, IκB); and (c) interacting with the active NF-κB and preventing its binding to κB sites in the DNA (Mackenzie et al. 2004, 2008, 2009) (Figure 12.1). By contrast, it was observed that (–)-epicatechin, B1 dimer and B2 dimer do not affect the transport of the active NF-κB from the cytosol into the nucleus (Mackenzie et al. 2004, Mackenzie and Oteiza 2006).

As a paradigm of the capacity of flavan-3-ols and dimeric procyanidins to specifically interact with NF-κB, we showed that the B2-dimer inhibits the binding of NF-κB proteins, that is, RelA and p50, to their κB DNA consensus sequence in whole cells, nuclear fractions, and purified chemical systems (Mackenzie et al. 2004, 2008, 2009). The presence of select hydroxyl groups in the B2 dimer is decisive to explain the chemical characteristics of this inhibition. Molecular modeling of the B2 dimer shows a folded structure where ring B stacks onto ring A orienting the hydroxyl groups toward the same edge of the molecule. This mimics the guanine pairs in the κB-DNA sequence that interacts with p50 and RelA. Thus, B2 dimer can establish similar hydrogen bonds to those that the guanine pairs establish with the arginine residues present in the DNA binding region of both p50 (Arg 54 and Arg

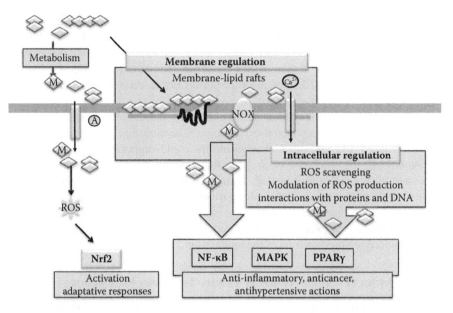

FIGURE 12.1   Redox and nonredox regulation of select signaling pathway by (–)-epicatechin and related procyanidins. (–)-Epicatechin and procyanidin dimers can be absorbed at the gastrointestinal tract, as parent compounds or metabolites and reach the blood and different tissues exerting effects on cell signaling either at the cell membrane or inside the cells. Procyanidins of 3 or more monomer units that are not transported inside the cells can interact with membrane components and either directly affect the activity of signaling proteins, or modify the physical membrane environment, which can have significant impact in the triggering/modulation of signaling. Monomers and metabolites that are transported inside cells can regulate cell signaling through different mechanisms including: (a) the modulation of ROS levels by scavenging oxidants or modulating oxidant production; and (b) the specific interaction with signaling proteins. In addition, select flavan-3-ols have been observed to generate small amounts of ROS that can improve cellular defenses against oxidative stress through the up-regulation of transcription factor Nrf2. All these effects on cell signaling could explain in part the anti-inflammatory and anticarcinogenic actions of these compounds. Single diamonds: Epicatechin monomer; two-linked diamonds: procyanidin diamond; multiple-linked diamonds: procyanidin oligomers; diamonds with "M": metabolite.

56) and RelA (Arg 33 and Arg 35) (Mackenzie et al. 2004). The binding of B2 dimer to NF-κB protein is stereo-specific: dimers B1 and B2, but not A1 and A2, inhibited both NF-κB binding to κB-DNA sites, and NF-κB-dependent gene transcription (Mackenzie et al. 2009). In the B-type dimers, flavan-3-ol monomers are linked through 4β → 8 carbon–carbon bonds, but A-type dimers have an extra 2–O–7 ether bond that does not allow the folding of the dimer necessary for the bonding with RelA and p50 arginines.

We have observed that the binding of (–)-epicatechin and dimers to NF-κB has functional consequences in intestinal and immune cells. Dimeric procyanidins B1 and B2 inhibit NF-κB-dependent IL-2 production in Jurkat T cells triggered by both TNFα and lipopolysaccharide (Mackenzie et al. 2004, 2009). Both (–)-epicatechin and B2 dimer, also inhibit NF-κB in Hodgkin's lymphoma cells characterized by a constitutively activation of this transcription factor. Accordingly, these compounds also inhibited the expression of NF-κB-regulated antiapoptotic and proinflammatory proteins (Mackenzie et al. 2008).

Large procyanidins (three or more monomer units) are not transported inside the cells; thus as parent compounds, they cannot be absorbed at the gastrointestinal tract and reach other tissues. However, they could exert local effect at the intestinal epithelium by interacting with cell membranes and directly regulating the activity of receptors/enzymes or by affecting the membrane biophysical characteristics modifying the environment of receptors and other membrane-associated signaling proteins. As an example of such mechanism, hexameric procyanidins inhibit TNFα-induced NF-κB activation and the expression of the NF-κB-regulated inducible nitric oxide synthase (NOS) expression in intestinal cells (Erlejman et al. 2006). This effect was not observed when exposed to other proinflammatory stimuli, suggesting certain selectivity of the procyanidins for select membrane regions, given that the TNFα receptor is localized in lipid rafts. In summary, *in vitro* data support the capacity of (−)-epicatechin and procyanidins to inhibit NF-κB with high specificity at different levels in the signaling cascade and through different mechanisms of action.

12.3.3 MODULATION OF PROTEIN KINASES AND PHOSPHATASES BY FLAVAN-3-OLS AND PROCYANIDINS

Protein kinases and phosphatases are key molecules in most signaling pathways. Being susceptible to redox regulation, both types of activities can be modulated by the direct or indirect antioxidant actions of flavan-3-ols and procyanidins. In addition, flavan-3-ols and procyanidins can interact with these proteins modulating their activity (Figure 12.1). Dimer B2 inhibits the mitogen activated protein kinase (MAPK) kinase (MEK), MEK phosphorylation, and also the downstream activation of extracellular signal-regulated kinase (ERK), and transcription factors AP-1 and NF-κB, preventing the chemically induced neoplastic transformation of JB6 P + mouse epidermal cells (Kang et al. 2008b). This action could be highly relevant given that the MEK/ERK pathway regulates cell proliferation and high levels of ERK activation are associated with different types of cancer. In monocytes, B2 dimer decreased endotoxin-induced expression of cycloxigenase-2, through the inhibition of the MAPK p38, c-Jun N-terminal kinase (JNK), ERK, and of NF-κB (Zhang et al. 2006). Flavonoids sharing chemical similarities (−)-epicatechin, also inhibit MEK and the upstream kinase Raf potentially through specific molecular interactions; a keto group in the B-ring, and a 3′,4′ catechol group in the C-ring afford an optimum inhibitory action on in vitro MEK activity (Kang et al. 2008a).

Cisplatin-induced ERK1/2 activation was inhibited by (−)-epicatechin in a cochlear organ of Corti-derived cell line, and in a rat model (Lee et al. 2010). Relevant to neurodevelopment and synaptic transmission, (−)-epicatechin and its metabolite 3′-*O*-methyl-(−)-epicatechin activated ERK in neuronal cells, as well as the ERK downstream target cAMP (adenosine 3′ 5′ cyclic monophosphate) response element-binding protein (CREB) (Schroeter et al. 2007). Importantly, these effects depended on the (−)-epicatechin concentration, activating ERK at 100–300 nM but inhibiting ERK in the micromolar range. In neurons, other flavanoids can also regulate the activity of ERK and their downstream kinases through highly specific interactions with receptors, including gamma-aminobutyric acid (GABA) receptors (Adachi et al. 2006), delta-opioid receptors (Panneerselvam et al. 2010), and nicotinic receptors (Lee et al. 2011).

Large procyanidins can also regulate ERK through their interactions with membranes. The secondary bile acid deoxycholic acid triggers the activation of ERK, p-38, and Akt in different cells by changing the lipid raft environment (Fang et al. 2004; Araki et al. 2005). Recent evidence (Da Silva et al. 2011) indicates that hexameric procyanidins can prevent deoxycholate-induced ERK, p-38, and Akt phosphorylation, suggesting the capacity of large procyanidins to interact with specific membrane domains and modulate lipid raft-associated signals. These effects could be relevant in terms of colorectal cancer, given the capacity of ERK to regulate proliferation (Abrams et al. 2010), and of Akt to regulate signals involved in malignant transformation (Chang et al. 2003; Larue and Bellacosa 2005). A grape seed composed mostly of procyanidins of 2–13 subunits induces apoptosis in colorectal cancer cells by both inhibiting Bcl-2 and activating Bax (Shih et al. 2008). Also at the membrane level, (–)-epigallo-catechin-3-O-gallate, but not (–)-epicatechin, inhibits the activation of the epidermal growth factor receptor, which is localized at lipid rafts (Adachi et al. 2007), and that has ERK as one of the downstream activated cascades.

Other kinase modulated by (–)-epicatechin is the calmodulin-dependent kinase II (CaMKII). In cultured, endothelial nitric oxide synthase (eNOS) activation was associated with calcium homeostasis, both independent and dependent on CaMKII, and increased by flavan-3-ols and quercetin through the phosphorylation of eNOS serine residues (Ramirez-Sanchez et al. 2010). Interestingly, these effects were sensitive to the structure of the flavonoid, that is, (–)-epicatechin was more effective than (+)-catechin or quercetin.

12.3.4 MODULATION OF NFR2 AND OTHER SIGNALING PATHWAYS BY FLAVAN-3-OLS AND PROCYANIDINS

In the adaptive response of cells toward oxidative and electrophilic stress, cells activate transcription factor Nrf2 that regulates the expression of antioxidant and drug-metabolizing enzymes (Kensler and Wakabayashi 2010; Kundu and Surh 2010). It has been proposed that select flavonoids can generate small amounts of ROS, which would lead to Nrf2 activation levels that afford an increased antioxidant protective response in cells (Figure 12.1). Nrf2 activation could underlie the cancer-protective effects associated with a high consumption of (–)-epigallocatechin-3-O-gallate-containing foods [reviewed in (Na and Surh 2008). In vitro (–)-epicatechin causes Nrf2 activation in primary cultures of astrocytes and neurons (Bahia et al. 2008). As evidence of the a potential beneficial effect of (–)-epicatechin in the nervous system through Nrf2, (–)-epicatechin supplementation protects mice from Aβ25-35-induced hippocampal toxicity (Cuevas et al. 2009). Stressing a role for Nrf2, (–)-epicatechin also ameliorate stroke-associated brain infarcts and neurologic deficits in wild type mice but not in mice with genetic deficits of Nrf2 (Shah et al. 2010).

(–)-Epicatechin and procyanidins can also regulate several other signaling pathways, including the MAPK, PPARγ (Chen et al. 2006), and the aryl hydrocarbon receptor-mediated signaling (Mukai et al. 2008). In that last study, procyanidn B2 and B5 dimers inhibited the transformation of the aryl hydrocarbon receptor triggered by dioxin. In summary, as indicated for NF-κB, phosphatases and kinases, other signaling cascades have been shown to be modulated by (−)-epicatechin and procyanidins. In most cases, the molecular target/s of this modulation are not yet identified.

12.4 CONCLUSIONS

Flavan-3-ols and procyanidins compounds present in plants that are part of the human diet. These compounds were believed to act mostly providing antioxidant protection by trapping radicals and chelating redox-active metals. However, a significant and increasing body of evidence supports the participation of flavan-3-ols and procyanidins in the regulation of cell signaling. The mechanisms underlying signaling regulation include the capacity of flavan-3-ols and procyanidins to regulate cell oxidant production and antioxidant defenses and hence cell redox state; especially in tissues exposed to high amounts of flavan-3-ols and procyanidins. In addition, specific interactions of flavan-3-ols and procyanidins modulate the activity and biological reactions of cell signaling proteins, and the regulation of membrane-associated cell signaling. All these actions would be limited by the bioavailability of flavan-3-ols in at the target tissue, for example, high micromolar concentration in the digestive tract, low micromolar–high nanomolar concentration in blood, and lower concentrations in other tissues and cells. The protection from cardiac and vascular disease and from cancer associated with a high consumption of fruit and vegetables could be in part explained by the capacity of flavan-3-ols and related procyanidins to modulate proinflammatory and oncogenic signals.

ACKNOWLEDGMENTS

This work was supported by grants from the University of Buenos Aires, UBACyT (001-1111); University of California, Davis; and the CHNR-State of California Vitamin Price Fixing Consumer Settlement Fund. Cesar G. Fraga is principal investigator, and Patricia I. Oteiza is corresponding investigator, CONICET, Argentina.

ABBREVIATIONS

eNOS = endothelial nitric oxide synthase; ERK = extracellular signal-regulated kinase; MAPK = mitogen-activated protein kinase; ROS = reactive oxygen species; TNFα = tumor necrosis factor alpha.

REFERENCES

Abd El Mohsen, M.M., Kuhnle, G., Rechner, A.R. et al. (2002). Uptake and metabolism of epicatechin and its access to the brain after oral ingestion. *Free Radic. Biol. Med.,* **33**, 1693–1702.
Abrams, S.L., Steelman, L.S., Shelton, J.G. et al. (2010). Enhancing therapeutic efficacy by targeting non-oncogene addicted cells with combinations of signal transduction inhibitors and chemotherapy. *Cell Cycle,* **9**, 1839–1846.

Adachi, N., Tomonaga, S., Tachibana, T. et al. (2006). (−)-Epigallocatechin gallate attenuates acute stress responses through GABAergic system in the brain. *Eur. J. Pharmacol.*, **531**, 171–175.

Adachi, S., Nagao, T., Ingolfsson, H. I. et al. (2007). The inhibitory effect of (−)-epigallocatechin gallate on activation of the epidermal growth factor receptor is associated with altered lipid order in HT29 colon cancer cells. *Cancer Res.*, **67**, 6493–5601.

Araki, Y., Katoh, T., Ogawa, A. et al. (2005). Bile acid modulates transepithelial permeability via the generation of reactive oxygen species in the Caco-2 cell line. *Free Radic. Biol. Med.*, **39**, 769–780.

Azzi, A., Davies, K.J. and Kelly, F. (2004). Free radical biology—terminology and critical thinking. *FEBS Lett.*, **558**, 3–6.

Baba, S., Osakabe, N., Natsume, M. et al. (2001). *In vivo* comparison of the bioavailability of (+)-catechin, (−)-epicatechin and their mixture in orally administered rats. *J. Nutr.*, **131**, 2885–2891.

Baba, S., Osakabe, N., Natsume, M. et al. (2002). Absortion and urinary excretion of procyanidin B2 [epicatechin-(4β[s7]-8)-epicatechin] in rats. *Free Radic. Biol. Med.*, **33**, 142–148.

Baba, S., Osakabe, N., Natsume, M. et al. (2000a). Cocoa powder enhances the level of antioxidative activity in rat plasma. *Br. J. Nutr.*, **84**, 673–680.

Baba, S., Osakabe, N., Yasuda, A. et al. (2000b). Bioavailability of (−)-epicatechin upon intake of chocolate and cocoa in human volunteers. *Free Radic. Res.*, **33**, 635–641.

Bahia, P.K., Rattray, M. and Williams, R.J. (2008). Dietary flavonoid (−)-epicatechin stimulates phosphatidylinositol 3-kinase-dependent anti-oxidant response element activity and up-regulates glutathione in cortical astrocytes. *J. Neurochem.*, **106**, 2194–2204.

Brigelius-Flohe, R. and Flohe, L. (2011). Basic principles and emerging concepts in the redox control of transcription factors. *Antioxid. Redox Signal.*, in press.

Chang, F., Lee, J.T., Navolanic, P.M. et al. (2003). Involvement of PI3K/Akt pathway in cell cycle progression, apoptosis, and neoplastic transformation: A target for cancer chemotherapy. *Leukemia*, **17**, 590–603.

Chen, D.M., Cai, X., Kwik-Uribe, C.L. et al. (2006). Inhibitory effects of procyanidin B2 dimer on lipid-laden macrophage formation. *J. Cardiovasc. Pharmacol.*, **48**, 54–70.

Chu, K.O., Chan, K.P., Wang, C.C. et al. (2010). Green tea catechins and their oxidative protection in the rat eye. *J. Agric. Food Chem.*, **58**, 1523–1534.

Crozier, A., Jaganath, I.B. and Clifford, M.N. (2009). Dietary phenolics: Chemistry, bioavailability and effects on health. *Nat. Prod. Rep.*, **26**, 1001–1043.

Cuevas, E., Limon, D., Perez-Severiano, F. et al. (2009). Antioxidant effects of epicatechin on the hippocampal toxicity caused by amyloid-β 25-35 in rats. *Eur. J. Pharmacol.*, **616**, 122–127.

Da Silva, M. et al. (2011). Large procyanidins prevent bile-acid-induced oxidant production and membrane-initiated ERK1/2, p38, and Akt activation in Caco-2 cells. *Free Radic. Biol. Med.*, 2011 Oct 19. [Epub ahead of print].

Dinkova-Kostova, A.T. and Talalay, P. (2008). Direct and indirect antioxidant properties of inducers of cytoprotective proteins. *Mol. Nutr. Food Res.*, **52**, S128–138.

Donovan, J.L., Bell, J.R., Kasim-Karakas, S. et al. (1999). Catechin is present as metabolites in human plasma after consumption of red wine. *J. Nutr.*, **129**, 1662–1668.

Donovan, J.L., Crespy, V., Manach, C. et al. (2001). Catechin is metabolized by both the small intestine and liver of rats. *J. Nutr.*, **131**, 1753–1757.

Donovan, J.L., Crespy, V., Oliveira, M. et al. (2006). (+)-Catechin is more bioavailable than (−)-catechin: Relevance to the bioavailability of catechin from cocoa. *Free Radic. Res.*, **40**, 1029–1034.

Erlejman, A.G., Fraga, C.G., and Oteiza, P.I. (2006). Procyanidins protect Caco-2 cells from bile acid- and oxidant-induced damage. *Free Radic. Biol. Med.*, **41**, 1247–1256.

Erlejman, A.G., Jaggers, G., Fraga, C.G., and Oteiza, P.I. (2008). TNFα-induced NF-κB activation and cell oxidant production are modulated by hexameric procyanidins in Caco-2 cells. *Arch. Biochem. Biophys.*, **476**, 186–195.

Fang, Y., Han, S.I., Mitchell, C. et al. (2004). Bile acids induce mitochondrial ROS, which promote activation of receptor tyrosine kinases and signaling pathways in rat hepatocytes. *Hepatology,* **40**, 961–971.

Forman, H.J., Maiorino, M. and Ursini, F. (2010). Signaling functions of reactive oxygen species. *Biochemistry,* **49**, 835–842.

Fraga, C.G. (2007). Plant polyphenols: How to translate their *in vitro* antioxidant actions to *in vivo* conditions. *IUBMB Life,* **59**, 308–315.

Fraga, C.G., Actis-Goretta, L., Ottaviani, J.I. et al. (2005). Regular consumption of a flavanol-rich chocolate can improve oxidant stress in young soccer players. *Clin. Dev. Immunol.,* **12**, 11–17.

Fraga, C.G., Galleano, M., Verstraeten, S.V. et al. (2010). Basic biochemical mechanisms behind the health benefits of polyphenols. *Mol. Aspects Med.,* **31**, 435–445.

Fraga, C.G., Martino, V.S., Ferraro, G.E. et al. (1987). Flavonoids as antioxidants evaluated by *in vitro* and *in situ* liver chemiluminescence. *Biochem. Pharmacol.,* **36**, 717–720.

Fraga, C.G. and Oteiza, P.I. (2011). Dietary flavonoids: role of (–)-epicatechin and related procyanidins in cell signaling. *Free Radic. Biol. Med.,* **51**, 813–823.

Frei, B. and Higdon, J.V. (2003). Antioxidant activity of tea polyphenols in vivo: Evidence from animal studies. *J. Nutr.,* **133**, 3275S–3284S.

Galleano, M., Verstraeten, S.V., Oteiza, P.I. et al. (2010). Antioxidant actions of flavonoids: Thermodynamic and kinetic analysis. *Arch. Biochem. Biophys.,* **501**, 23–30.

Graziani, G., D'argenio, G., Tuccillo, C. et al. (2005). Apple polyphenol extracts prevent damage to human gastric epithelial cells *in vitro* and to rat gastric mucosa *in vivo*. *Gut,* **54**, 193–200.

Halliwell, B. (2008). Are polyphenols antioxidants or pro-oxidants? What do we learn from cell culture and in vivo studies? *Arch. Biochem. Biophys.,* **476**, 107–112.

Halliwell, B., Rafter, J., and Jenner, A. (2005). Health promotion by flavonoids, tocopherols, tocotrienols, and other phenols: Direct or indirect effects? Antioxidant or not? *Am. J. Clin. Nutr.,* **81**, 268S–276S.

Hammerstone, J.F., Lazarus, S.A., and Schmitz, H.H. (2000). Procyanidin content and variation in some commonly consumed foods. *J. Nutr.,* **130**, 2086S–2090S.

Hollman, P.C., Cassidy, A., Comte, B. et al. (2011). The biological relevance of direct antioxidant effects of polyphenols for cardiovascular health in humans is not established. *J. Nutr.,* **141**, 989S–1009S.

Holt, R.R., Lazarus, S.A., Sullards, M.C. et al. (2002). Procyanidin dimer B2 [epicatechin-(4β-8)-epicatechin] in human plasma after the consumption of a flavanol-rich cocoa. *Am. J. Clin. Nutr.,* **76**, 798–804.

Jung, Y., Kim, H., Min, S.H. et al. (2008). Dynein light chain LC8 negatively regulates NF-κB through the redox-dependent interaction with IkappaBalpha. *J. Biol. Chem.,* **283**, 23863–23871.

Kang, N.J., Lee, K.W., Kwon, J.Y. et al. (2008a). Delphinidin attenuates neoplastic transformation in JB6 Cl41 mouse epidermal cells by blocking Raf/mitogen-activated protein kinase kinase/extracellular signal-regulated kinase signaling. *Cancer Prev. Res.,* **1**, 522–531.

Kang, N.J., Lee, K.W., Lee, D.E. et al. (2008b). Cocoa procyanidins suppress transformation by inhibiting mitogen-activated protein kinase kinase. *J. Biol. Chem.,* **283**, 20664–20673.

Karin, M. and Greten, F.R. (2005). NF-κB: Linking inflammation and immunity to cancer development and progression. *Nat. Rev. Immunol.,* **5**, 749–759.

Kensler, T.W. and Wakabayashi, N. (2010). Nrf2: Friend or foe for chemoprevention? *Carcinogenesis,* **31**, 90–99.

Koga, T., Moro, K., Nakamori, K. et al. (1999). Increase of antioxidative potential of rat plasma by oral administration of proanthocyanidin-rich extract from grape seeds. *J. Agric. Food Chem.,* **47**, 1892–1897.

Kuhnle, G., Spencer, J.P. and Schroeter, H. et al. (2000). Epicatechin and catechin are *O*-methylated and glucuronidated in the small intestine. *Biochem. Biophys. Res. Commun.,* **277**, 507–512.

Kundu, J.K. and Surh, Y.J. (2010). Nrf2-Keap1 signaling as a potential target for chemoprevention of inflammation-associated carcinogenesis. *Pharm. Res.,* **27**, 999–1013.

Larue, L. and Bellacosa, A. (2005). Epithelial-mesenchymal transition in development and cancer: Role of phosphatidylinositol-3′-kinase/AKT pathways. *Oncogene* **24**, 7443–7454.

Lee, B.H., Choi, S.H., Shin, T.J. et al. (2011). Effects of quercetin on α9α10-nicotinic acetylcholine receptor-mediated ion currents. *Eur. J. Pharmacol.,* **650**, 79–85.

Lee, J.S., Kang, S.U., Hwang, H.S. et al. (2010). Epicatechin protects the auditory organ by attenuating cisplatin-induced ototoxicity through inhibition of ERK. *Toxicol Lett.,* **199**, 308–316.

Mackenzie, G.G., Adamo, A.M., Decker, N.P. et al. (2008). Dimeric procyanidin B2 inhibits constitutively active NF-κB in Hodgkin's lymphoma cells independently of the presence of IκB mutations. *Biochem. Pharmacol.,* **75**, 1461–1471.

Mackenzie, G.G., Carrasquedo, F., Delfino, J.M. et al. (2004). Epicatechin, catechin, and dimeric procyanidins inhibit PMA-induced NF-κB activation at multiple steps in Jurkat T cells. *Faseb. J.,* **18**, 167–169.

Mackenzie, G.G., Delfino, J.M., Keen, C.L. et al. (2009). Dimeric procyanidins are inhibitors of NF-κB-DNA binding. *Biochem. Pharmacol.,* **78**, 1252–1262.

Mackenzie, G.G. and Oteiza, P.I. (2006). Modulation of transcription factor NF-κB in Hodgkin's lymphoma cell lines: Effect of (−)-epicatechin. *Free Radic. Res.,* **40**, 1086–1094.

Mukai, R., Fukuda, I., Nishiumi, S. et al. (2008). Cacao polyphenol extract suppresses transformation of an aryl hydrocarbon receptor in C57BL/6 mice. *J. Agric. Food Chem.,* **56**, 10399–10405.

Na, H.K. and Surh, Y.J. (2008). Modulation of Nrf2-mediated antioxidant and detoxifying enzyme induction by the green tea polyphenol EGCG. *Food Chem. Toxicol.,* **46**, 1271–1278.

Nandakumar, V., Singh, T., and Katiyar, S.K. (2008). Multi-targeted prevention and therapy of cancer by proanthocyanidins. *Cancer Lett.,* **269**, 378–387.

Orozco, T.J., Wang, J.F., and Keen, C.L. (2003). Chronic consumption of a flavanol- and procyanindin-rich diet is associated with reduced levels of 8-hydroxy-2′-deoxyguanosine in rat testes. *J. Nutr. Biochem.,* **14**, 104–110.

Ottaviani, J.I., Momma, T.Y., Heiss, C. et al. (2011). The stereochemical configuration of flavanols influences the level and metabolism of flavanols in humans and their biological activity *in vivo. Free Radic. Biol. Med.,* **50**, 237–244.

Pahl, H.L. (1999). Activators and target genes of Rel/NF-κB transcription factors. *Oncogene,* **18**, 6853–6866.

Panneerselvam, M., Tsutsumi, Y.M., Bonds, J.A. et al. (2010). Dark chocolate receptors: Epicatechin-induced cardiac protection is dependent on δ-opioid receptor stimulation. *Am. J. Physiol. Heart. Circ. Physiol.,* **299**, H1604–H1609.

Piskula, M.K. and Terao, J. (1998). Accumulation of (−)-epicatechin metabolites in rat plasma after oral administration and distribution of conjugation enzymes in rat tissues. *J. Nutr.,* **128**, 1172–1178.

Ramirez-Sanchez, I., Maya, L., Ceballos, G. et al. (2010). (−)-Epicatechin activation of endothelial cell endothelial nitric oxide synthase, nitric oxide, and related signaling pathways. *Hypertension,* **55**, 1398–4105.

Rechner, A.R., Kuhnle, G., Bremner, P. et al. (2002). The metabolic fate of dietary polyphenols in humans. *Free Radic. Biol. Med.,* **33**, 220–235.

Rein, D., Lotito, S., Holt, R.R., et al. (2000). Epicatechin in human plasma: *In vivo* determination and effect of chocolate consumption on plasma oxidation status. *J. Nutr.,* **130**, 2109S–2114S.

Reuter, S., Gupta, S.C., Chaturvedi, M.M. et al. (2010). Oxidative stress, inflammation, and cancer: How are they linked? *Free Radic. Biol. Med.,* **49**, 1603–1616.

Rice-Evans, C. (2001). Flavonoid antioxidants. *Curr. Med. Chem.,* **8**, 797–807.

Rios, L.Y., Bennett, R.N., Lazarus, S.A. et al. (2002). Cocoa procyanidins are stable during gastric transit in humans. *Am. J. Clin. Nutr.,* **76**, 1106–1110.

Schroeter, H., Bahia, P., Spencer, J.P. et al. (2007). (–)-Epicatechin stimulates ERK-dependent cyclic AMP response element activity and up-regulates GluR2 in cortical neurons. *J. Neurochem.,* **101**, 1596–1606.

Schroeter, H., Heiss, C., Balzer, J. et al. (2006). (–)-Epicatechin mediates beneficial effects of flavanol-rich cocoa on vascular function in humans. *Proc. Natl. Acad. Sci. USA.,* **103**, 1024–1029.

Shah, Z.A., Li, R.C., Ahmad, A.S. et al. (2010). The flavanol (–)-epicatechin prevents stroke damage through the Nrf2/HO1 pathway. *J. Cereb. Blood Flow Metab.,* **30**, 1951–1961.

Shih, T.C., Hsieh, S.Y., Hsieh, Y.Y. et al. (2008). Aberrant activation of nuclear factor of activated T cell 2 in lamina propria mononuclear cells in ulcerative colitis. *World J. Gastroenterol.,* **14**, 1759–1767.

Spencer, J.P. (2003). Metabolism of tea flavonoids in the gastrointestinal tract. *J. Nutr.,* **133**, 3255S–3261S.

Spencer, J.P.E., Chaudry, F., Pannala, A.S., et al. (2000). Decomposition of cocoa procyanidins in the gastric milieu. *Biochem. Biophys. Res. Commun.,* **272**, 236–241.

Steffen, Y., Gruber, C., Schewe, T., and Sies, H. (2008). Mono-*O*-methylated flavanols and other flavonoids as inhibitors of endothelial NADPH oxidase. *Arch. Biochem. Biophys.,* **469**, 209–219.

Steffen, Y., Schewe, T., and Sies, H. (2007). (–)-Epicatechin elevates nitric oxide in endothelial cells via inhibition of NADPH oxidase. *Biochem. Biophys. Res. Commun.,* **359**, 828–833.

Steinberg, F.M., Bearden, M.M., and Keen, C.L. (2003). Cocoa and chocolate flavonoids: Implications for cardiovascular health. *J. Am. Diet Assoc.,* **103**, 215–223.

Toschi, T.G., Bordoni, A., Hrelia, S. et al. (2000). The protective role of different green tea extracts after oxidative damage is related to their catechin composition. *J. Agric. Food Chem.,* **48**, 3973–3978.

Unno, T., Tamemoto, K., Yayabe, F. et al. (2003). Urinary excretion of 5-(3',4'-dihydroxyphenyl)-γ-valerolactone, a ring-fission metabolite of (–)-epicatechin, in rats and its *in vitro* antioxidant activity. *J. Agric. Food Chem.,* **51**, 6893–6898.

Verstraeten, S.V., Mackenzie, G.G., Oteiza, P.I. et al. (2008). (–)-Epicatechin and related procyanidins modulate intracellular calcium and prevent oxidation in Jurkat T cells. *Free Radic. Res.,* **42**, 864–872.

Wang, J.F., Schramm, D.D., Holt, R.R. et al. (2000). A dose-response effect from chocolate consumption on plasma epicatechin and oxidative damage. *J. Nutr.* **130**, 2115S–2119S.

Zhang, W.Y., Liu, H.Q., Xie, K.Q. et al. (2006). Procyanidin dimer B2 [epicatechin-(4β-8)-epicatechin] suppresses the expression of cyclooxygenase-2 in endotoxin-treated monocytic cells. *Biochem Biophys. Res. Commun.,* **345**, 508–515.

13 Flavonoids and Vascular Function

Ana Rodriguez-Mateos and Jeremy P.E. Spencer

CONTENTS

13.1 INTRODUCTION

Diet has been recognized as an important contributor to cardiovascular disease (CVD). Epidemiological evidence has indicated that the consumption of diets rich in fruit and vegetables lead to a reduction in the risk of CVD (Joshipura et al. 2001; Dauchet et al. 2006; He et al. 2007). The cardioprotective effects of such diets are often attributed to their phytochemical content, in particular, to the flavonoids they contain. Indeed, epidemiological evidence also indicates that there is an association between high dietary intake of flavonoids and the decreased risk of CVD (Hertog et al. 1995; Arts et al. 2005; Buijsse et al. 2006; Mink et al. 2007).

In recent years, there have been extensive *in vitro* and animal model studies designed to support this association, although because of the inherent limitations of such investigations one should interpret results emanating from them with care. For example, many *in vitro* studies have not taken account of the extensive metabolism of flavonoids that occurs *in vivo*, or the very different bioavailability profiles of the various flavonoid subclasses. As such, many *in vitro* studies to date have failed to assess the bioactivity of the relevant human flavonoid metabolites at biological concentrations. In addition to this, studies have not taken account of the differing bioavailability and metabolism of flavonoids within animals and humans, that is, the differing flavonoid metabolomes observed with different mammalian species. Taken together, this means that by far the most effective way of assessing the efficacy of flavonoids toward the human vascular system is in humans using well-powered and well-controlled intervention trials.

To date, there has been a relative lack of well-controlled clinical studies investigating the impact of flavonoids generally on endpoints such as CVD mortality. In contrast, there have been several short-term, small-scale, human intervention studies designed to test the effect of flavonoid-rich foods on several well-characterized CVD risk factors, including vascular function, endothelial dysfunction, hypertension, platelet activity, and platelet clotting time. Despite the limitations highlighted earlier, together the *in vitro*, animal model and human intervention trials with flavonoid rich foods support the notion that flavonoids may improve endothelial function (Cuevas et al. 2000; Heiss et al. 2003, 2005, 2007; Engler et al. 2004; Papamichael et al. 2004; Grassi et al. 2005, 2009; Schroeter et al. 2006; Wang-Polagruto et al. 2006; Widlansky et al. 2007), lower blood pressure (Park et al. 2004; Grassi et al. 2005; Taubert et al. 2007a,b; Erlund et al. 2008; Hooper et al. 2008; Desch et al. 2010), inhibit low density lipoprotein oxidation (Wan et al. 2001; Mathur et al. 2002), inhibit platelet aggregation (Keevil et al. 2000; Rein et al. 2000; Pearson et al. 2002; Erlund et al. 2008), modulate the inflammatory response (Mao et al. 2002; Schramm et al. 2003), alter lipid metabolism (Zern et al. 2005), improve insulin sensitivity (Fuhrman et al. 2005), and decrease vascular cell adhesion molecule expression (Ludwig et al. 2004). This chapter will attempt to highlight the best-controlled and most physiologically relevant evidence supporting the link between flavonoid intake and cardiovascular health. Where possible we will indicate the most potent flavonoids and detail likely mechanisms of action.

13.2 EPIDEMIOLOGICAL EVIDENCE

Epidemiological studies suggest that the daily intake of dietary flavonoids may have cardioprotective effects in humans. However, before firm conclusions can be drawn with regards to their cardiovascular potency, one must take account of the difficulties associated with the accurate assessment of flavonoid intake within the human diet. Precise measurements of intake in human populations have been limited by two principal factors. First, despite best efforts to draw up accurate databases of the flavonoid content of commonly consumed foods, using the United States Department of Agriculture (USDA) flavonoid database and Phenol Explorer (Neveu et al. 2010), it is generally accepted that these are limited because of the inherent variability of flavonoid levels in foods (due to factors such as growing conditions, ripeness, variety, etc.). Second, a lack of suitable analytical methods, capable of measuring biomarkers of flavonoid intake in human biological samples, has further prevented precise analysis of flavonoid intakes in human populations to be made. An additional source of error related to the lack of information regarding the impact of food processing on the flavonoid content of foods, leading very often to an overestimation of the flavonoid content of processed foods. As such, as with other observational studies, analysis of the findings should be made with caution, in that if assessments of intake are flawed, conclusions regarding the association of flavonoid intake with disease made also be so.

A large review of the relationship between flavonoid intake and risk of CVD indicated that 13 of the 15 studied showed a positive correlation between the intake of flavanols, flavonols, flavones, and flavanones and risk of coronary artery disease, with a reduction in mortality of up to 65% (Arts et al. 2005). Furthermore, a prospective study among postmenopausal women indicated that intake of anthocyanins and flavanones was associated with a reduced risk of CVD related mortality (Mink et al. 2007). The Iowa Women's Health study followed 34,489 postmenopausal women for 16 years and estimated the intake of seven flavonoid subclasses using the USDA database. Significant negative correlations were found for: anthocyanin intake and coronary heart disease (CHD), CVD, and total mortality, between flavanones intake and CHD, and between flavones and total mortality. This study also highlighted bran, apple, pear, chocolate, red wine, and strawberries as being the foods most likely to significantly contribute to the reductions in CHD and CVD risk.

The Zupthen Elderly study, which followed 806 elderly men over a 15-year period, found an inverse correlation between flavonoid intake and risk of ischemic heart disease mortality. In this study, cocoa intake, a rich source of flavanols, was highlighted as being associated with a reduction in CVD and total mortality. In agreement with this, an inverse correlation between cocoa intake and blood pressure was also observed (Hertog et al. 1993; Arts et al. 2001; Buijsse et al. 2006). Tea, which is also a rich source of flavanols, has been shown in a meta-analysis to reduce cardiovascular risk by about 11% (intake of three cups per day) (Peters et al. 2001). Furthermore, red wine consumption has also been found to be associated with a reduced risk of CVD (Di Castelnuovo et al. 2002), and a recent systematic review (Hooper et al. 2008) concluded that soy and cocoa flavonoids have a beneficial effect on cardiovascular risk.

As noted in the previous paragraph, many of the foods that are found to be most strongly associated with reductions in CVD risk contain flavanols. In support of this, the study of the Kuna Indians of the San Blas Islands, Panama, is noteworthy because of their very high intake of flavanol-rich cocoa. Observational studies with the island communities and genetically similar communities living in Panama City have indicated that the extremely low risk of CVD and low incidence of hypertension experienced in the island dwellers is likely, in part, to be linked to their very high intake of unprocessed, high flavanol cocoa (Hollenberg et al. 1997). These observations provided compelling evidence that flavonoid-rich foods, such as cocoa, may prevent age-related hypertension and therefore the development of CVD.

Despite these encouraging datasets, there are several epidemiological studies that have concluded no relationship between flavonoid intake and cardiovascular risk (Rimm et al. 1996; Hertog et al. 1997; Sesso et al. 2003; Lin et al. 2007). As mentioned earlier, discrepancies between studies of this nature may be due to inadequacies in the ability of commonly used dietary intake questionnaires and food composition tables to accurately assess flavonoid intake in human populations, or may even result from issues surrounding the nature of the human population studied, that is, well-nourished versus undernourished and/or populations with high baseline flavonoid intake showing no effect (Vita 2005). To avoid such discrepancies, future observational studies must attempt to assess flavonoid intake via the analysis of reliable biomarkers of their intake in plasma, urine or stool samples.

13.3 EFFECTS ON ENDOTHELIAL FUNCTION

As well as being a physical barrier, the vascular endothelium is responsible for the regulation of vessel function through the release of a variety of agents, including vasodilators [i.e., nitric oxide (NO•), prostacyclins], vasoconstrictors (i.e., endothelin-1, prostaglandins) (Stankevicius, et al. 2003), endothelial and smooth muscle cell growth regulators (i.e., fibroblast growth factor, endothelin, transforming growth factor, platelet-derived growth factor, NO•, heparin, heparan sulphate) (Scott-Burden et al. 1994), and factors that influence platelet and leukocyte interactions (i.e., Intercellular adhesion molecule (ICAM), Vascular cell adhesion molecule (VCAM), integrins) (Wittchen 2009). One process thought to disrupt vessel homeostasis and thus underlie CVD pathophysiology is atherosclerosis, a chronic process characterized by the deposition of cholesterol-laden low-density lipoprotein in the walls of the vasculature. These "plaques" obstruct blood flow and, when suitably advanced in terms of their development, rupture triggering the recruitment of platelets and thrombus formation. An important process underlying the development of atherosclerosis is referred to as "endothelial dysfunction" (Ribeiro et al. 2009), a prognostically relevant key event in atherosclerosis, which is characterized by a decreased bioactivity of NO• and an impaired flow-mediated vasodilatation (FMD). Endothelial dysfunction is characterized by a number of physiological changes, including: (1) decreased bioavailability of endothelium-derived vasodilators, particularly NO•; (2) increased bioavailability of endothelial-derived contracting factors (Bonetti et al. 2003); (3) changes in the permeability of the endothelium; (4) amplified adhesion molecule expression on the surface of endothelial cells; (5) reduced secretion of antithrombotic factors and an increased production of procoagulatory factors (McGorisk et al. 1996); and (6) reduced antioxidant and anti-inflammatory capacity of the endothelium (Figure 13.1). Over time, chronic endothelial dysfunction is believed to underlie many features of atherosclerotic pathophysiology, including dysfunction of large vessel homeostasis, a loss of vessel elasticity, and thrombus formation.

To assess endothelial function in humans, one of the methods most widely used is FMD of the brachial artery, using high-resolution ultrasound. Several studies have been performed to test the effect of flavonoid-rich foods, such as cocoa, red wine, grape juice, and tea on endothelial function (Hooper et al. 2008). The majority of these studies have investigated the influence of flavonoid-rich food consumption on endothelial function acutely (i.e., 4–8 h) in healthy subjects or in patients with cardiovascular risk factors. Heiss et al. used FMD to show that the intake of flavanol-rich cocoa, containing 176 mg of flavanols, improves endothelium-dependent vasodilation (as assessed by FMD) via increasing plasma nitric oxide bioavailability in patients with coronary artery disease, hypertension, or diabetes (Heiss et al. 2003). A key feature of this study was the controlled nature of the intervention. The flavanol-rich cocoa drink contained 176 mg of flavanols (monomers and oligomers), whereas the control contained less than 10 mg flavanols but was identical to the flavanol-rich drink in macronutrient and micronutrient contents. Of additional note was the fact that the vascular improvements correlated in time with changes in plasma flavanol metabolites, suggesting a "cause-and-effect" relationship between flavanols and vascular improvements. Flavanol-rich cocoa has also been shown to reverse endothelial dysfunction in smokers (Heiss et al. 2005), something which has also been observed in healthy volunteers after the short-term consumption of flavanol-rich chocolate (Engler et al. 2004).

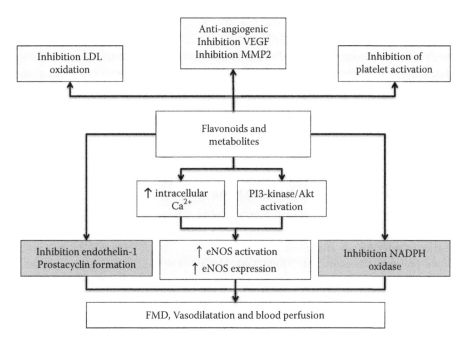

FIGURE 13.1 Overview of the flavonoids and their physiological metabolites on vascular function and blood pressure. Flavonoids may influence vasodilatation and blood perfusion [measured using flow-mediated dilatation (FMD)] via their actions on eNOS expression and activity via their actions on upstream intracellular signaling. Such effects on vascular tone may also be influenced by their inhibition of NADPH oxidase and thus superoxide radical production and/or by the inhibition of vasoconstrictors such as endothelin-1. Concurrently, they have also been shown to inhibit LDL oxidation, platelet activation, and factors that influence vascular wall structure. Together these activities allow flavonoids to maintain nitric oxide bioavailability, reduce blood pressure, prevent vessel ageing (i.e., progression of atherosclerosis), and inhibit thrombus formation.

These effects appear to be directly mediated by flavanols as the ingestion of pure (−)-epicatechin or (−)-epigallocatechin-3-O-gallate has been shown to produce similar vascular effects in healthy volunteers and in patients with CAD (Schroeter et al. 2006; Widlansky et al. 2007). Fisher et al. (2003) provided evidence that such vascular benefits are mediated by the increased generation/reduced clearance of nitric oxide by flavanols. Future studies that go beyond the current single dose, acute intervention paradigm may help to further determine the precise mechanism of action of flavanols *in vivo* species.

An increase in FMD response in healthy volunteers and patients with CVD has also been reported after the consumption of other flavonoid-rich foods, such as black tea, grape juice, and red wine (Stein et al. 1999; Chou et al. 2001; Duffy et al. 2001; Hashimoto et al. 2001; Vita 2003; Papamichael et al. 2004; Whelan et al. 2004). Indeed, the long-term vascular effect of grape juice consumption for 14 days has been reported in patients with CAD (Stein et al. 1999). Furthermore, chronic flavanol-rich cocoa intake has also been shown to increase the FMD response and hyperaemic brachial artery blood flow in hypercholesterolemic postmenopausal women (Wang-Polagruto et al. 2006), and two weeks of flavanol-rich chocolate consumption increases FMD in hypertensive subjects (Grassi et al. 2005). Lastly, long-term tea consumption has been shown to produce a sustained increase in baseline FMD levels and an additional acute-on chronic increase in FMD response (Duffy et al. 2001), something also observed with high flavanol cocoa in healthy smokers and diabetics, respectively (Heiss et al. 2007; Balzer et al. 2008).

13.4　EFFECTS ON BLOOD PRESSURE

Hypertension is a major risk factor for CVD and has been associated with the development of atherosclerosis and thus the progression of the disease. As discussed earlier, observational studies suggest that cocoa intake is inversely related with blood pressure (Buijsse et al. 2006, 2010; McCullough et al. 2006) and in populations where flavonoid intake is high (i.e., the Kuna Indians) hypertension is rare (McCullough et al. 2006). Several clinical intervention studies have been performed to test the effect of flavonoid-rich foods on blood pressure. Here, as with endothelial/vascular function, well-controlled human intervention trials have indicated that high-flavanol foods containing chocolate or cocoa are capable to reduce mild or severe hypertension in healthy subjects and in patients with CVD risk factors (Grassi et al. 2005; Taubert et al. 2007a,b; Hooper et al. 2008; Desch et al. 2010).

Data emanating from randomized clinical trials broadly agree with that collected from epidemiological studies (Murphy et al. 2003; Engler et al. 2004; Fraga et al. 2005; Grassi et al. 2005; Taubert et al. 2007a,b; Crews et al. 2008; Hooper et al. 2008; Muniyappa et al. 2008; Desch et al. 2010). In a meta-analysis of five studies, a decrease of 4.7 mm Hg in systolic BP and 2.8 mm Hg in diastolic BP was observed after consumption of dark chocolate (Taubert et al. 2007b). A more recent meta-analysis of 10 randomized controlled trials comprising 297 subjects (healthy normotensive individuals, prehypertensive subjects, and patients with stage 1 hypertension) reported a decrease of 4.5 mm Hg in systolic and 2.5 mm Hg in diastolic BP after flavanol-rich chocolate or flavanol-rich cocoa consumption for 2–18 weeks (Desch et al. 2010). However, seven of the studies used in the meta-analysis were noted to have used white chocolate as a control for the high-flavanol test intervention. It should be noted here that white chocolate has a very different composition to that of dark chocolate and it is not possible for the volunteers to be blinded. As such, data emanating from such studies need to be treated with caution and perhaps more weight should be afforded to the three studies, which used a low flavanol control matched in appearance, smell, flavor, calories, macro- and micronutrient content, and other potential bioactives such as theobromine and caffeine.

Regarding other flavonoid-rich foods, a meta-analysis by Hooper and colleagues (Hooper et al. 2008), indicated that the chronic intake of red wine, grapes, and black tea did not have significant effects on systolic and diastolic blood pressure (Hertog et al. 1995; Knekt et al. 1996, 2000), whereas a recent prospective study following 156,957 subjects over 14 years indicated that the regular consumption of anthocyanins (predominantly from blueberries and strawberries) reduced hypertension by 8%. Finally, flavones such as apigenin have also been shown to possess hypertension lowering effects (Cassidy et al. 2011). Overall, there is not enough evidence to suggest that flavonoids are capable of lowering blood pressure, mainly because there have not to date been enough well-controlled, longer-term intervention studies investigating the impact of flavonoid consumption on blood pressure. While acute reductions in blood pressure in response to flavonoid intake are of note, physiologically such changes much also be observed following chronic ingestion over a number of months or years to be considered clinically relevant.

13.5 EFFECTS ON PLATELET AGGREGATION

Another significant effect that has been attributed to flavonoids is an ability to inhibit platelet activation and thus the clotting mechanism (Figure 13.1). Platelets play an important role in the early stages of atherosclerosis and coronary thrombosis, and several studies have examined the effects of cocoa ingestion on platelet function. Rein et al. reported an inhibition of platelet activation 2 and 6 h following the consuming a flavanol-rich cocoa drink by healthy subjects (Rein et al. 2000), something also observed after the consumption of a flavanol containing chocolate (Holt et al. 2006). As such, an aspirin-like effect for flavanols has been suggested, although flavanols have been shown to be less effective that aspirin in inhibiting platelet activation (Pearson et al. 2002). Chronic effects of flavanols on platelet function have also been observed after 28 days of supplementation with cocoa flavanols and procyanidins (Murphy et al. 2003). In addition to cocoa, purple grape juice, black tea, coffee, and berries have also been shown to inhibit platelet aggregation following their consumption (Freedman, et al. 2001; Steptoe et al. 2007; Erlund et al. 2008; Natella et al. 2008).

13.6 MECHANISMS OF ACTION

The mechanisms by which flavonoids mediate their vascular effects are not completely understood. However, a growing body of evidence suggests that their mechanism of action is nitric oxide dependent (Figure 13.1). An increase in the nitric oxide concentration due to an increase in nitric oxide synthase (eNOS) activity is considered a likely pathway, although clear evidence of this has been difficult to generate. *In vitro* studies and investigations in animals and humans have shown that flavonoids appear to modulate endothelial nitric oxide (NO) expression and activity (Fitzpatrick et al. 1995; Heiss et al. 2003; Appeldoorn et al. 2009; Schmitt et al. 2009). Although many *in vitro* studies have not used physiologically relevant flavonoid metabolites in cell experiments designed to address mechanisms of action, human studies have proved evidence that vascular and blood pressure improvements correlate with increases in circulating NO and that both of these correlate with the appearance of flavonoids/metabolites in the plasma. Their ability to influence NO levels is likely to be mediated by their ability to modulate endothelial intracellular signaling pathways, such as the PI3-kinase/Akt pathway and intracellular Ca^{2+} levels leading to eNOS phosphorylation and subsequent NO production (Lorenz et al. 2004; Stoclet et al. 2004) (Figure 13.1).

In addition, flavonoids have also been shown to induce an increase on eNOS gene expression, to induce prostacyclin production in endothelial cells and to inhibit endothelin-1 and endothelial NADPH oxidase (Corder et al. 2001; Khan et al. 2002; Wallerath et al. 2002; Steffen et al. 2007, 2008). The latter is of note, as inhibition of NADPH oxidase will reduce superoxide production and thus the possible scavenging of NO by this reactive oxygen species to yield peroxynitrite (Figure 13.1). Such an activity is particularly noteworthy as inhibition of NADPH oxidase may be mediated by bacterial metabolites of flavonoids, which have their origin in the large intestine. Future investigations of such activity are important as they may underpin the long-term beneficial effects of flavonoids seen in intervention studies highlighted earlier. Other mechanisms that have been suggested for the cardioprotective effects of flavonoids are the inhibition of angiogenesis and migration and proliferation of vascular cells via the inhibition of the vascular endothelium growth factor (VEGF) expression and matrix metalloproteinases (MMPs) activation (Stoclet et al. 2004).

13.7 CONCLUSIONS

While the cardiovascular benefits of flavonoids in humans is accumulating, at present there is insufficient evidence to claim a clear and undisputed positive health effect relating to their consumption, particularly with regards to long-term ingestion and health. Epidemiological studies have failed to show conclusive results, although this may be in some cases due to the lack of appropriate nutrient databases and/ or the use of an inappropriately controlled study population. At present, much of the strongest data, particularly with regards to CVD, is based on short-term human studies, where appropriate, well-characterized controls have been utilized. As such, while one might begin to build a case for flavanols as being beneficial in improving vascular function and in preventing CVD, one cannot make a similar case for the remaining flavonoid subgroups, flavonols, flavanones, flavones, and anthocyanins. In future, researchers interested at defining the activity of these flavonoids against CVD risk must also strive to use better defined and controlled human interventions aimed at assessing physiological endpoints linked to disease. Further research is also required regarding the bioavailability of flavonoids, particularly with regards to the effects of food matrices on absorption and the influence on age, gender, and genotype on both absorption and metabolism. These studies are required to help determine the physiological metabolites responsible for activity *in vivo*, as well as to help define adequate biomarkers of flavonoid intake.

At present, while there is a vast literature regarding the potential of flavonoids to improve cardiovascular health, additional long-term, randomized, controlled, dietary intervention trials with appropriate controls are warranted to assess the full and unequivocal role that flavonoids play in preventing CVD. The outcomes of these studies may ultimately be used to make specific dietary recommendations regarding the efficacy of polyphenols in preventing chronic disease risk and to fully validate polyphenols as the new agents against various chronic human diseases.

REFERENCES

Appeldoorn, M.M., Venema, D.P., Peters, T.H. et al. (2009). Some phenolic compounds increase the nitric oxide level in endothelial cells in vitro. *J. Agric. Food Chem.,* **57**, 7693–7699.

Arts, I.C. and Hollman, P.C. (2005). Polyphenols and disease risk in epidemiologic studies. *Am. J. Clin. Nutr.,* **81**, 317S–325S.

Arts, I.C., Jacobs, D.R., Jr., Harnack, L.J, et al. (2001). Dietary catechins in relation to coronary heart disease death among postmenopausal women. *Epidemiology,* **12**, 668–675.

Balzer, J., Rassaf, T., Heiss, C. et al. (2008). Sustained benefits in vascular function through flavanol-containing cocoa in medicated diabetic patients a double-masked, randomized, controlled trial. *J. Am. Coll. Cardiol.,* **51**, 2141–2149.

Bonetti, P.O., Lerman, L.O. and Lerman, A. (2003). Endothelial dysfunction: A marker of atherosclerotic risk. *Arterioscl. Thromb. Vasc. Biol.,* **23**, 168–175.

Buijsse, B., Feskens, E.J., Kok, F.J. et al. (2006). Cocoa intake, blood pressure, and cardiovascular mortality: The Zutphen Elderly Study. *Arch. Intern. Med.,* **166**, 411–417.

Buijsse, B., Weikert, C., Drogan, D. et al. (2010). Chocolate consumption in relation to blood pressure and risk of cardiovascular disease in German adults. *Eur. Heart J.,* **31**, 1616–1623.

Cassidy, A., O'Reilly, E.J., Kay, C. et al. (2011). Habitual intake of flavonoid subclasses and incident hypertension in adults. *Am. J. Clin. Nutr.,* **93**, 338–347.

Chou, E.J., Keevil, J.G., Aeschlimann, S. et al. (2001). Effect of ingestion of purple grape juice on endothelial function in patients with coronary heart disease. *Am. J. Cardiol.,* **88**, 553–555.

Corder, R., Douthwaite, J.A., Lees, D M. et al. (2001). Endothelin-1 synthesis reduced by red wine. *Nature,* **414**, 863–864.

Crews, W.D., Jr., Harrison, D.W., and Wright, J.W. (2008). A double-blind, placebo-controlled, randomized trial of the effects of dark chocolate and cocoa on variables associated with neuropsychological functioning and cardiovascular health: Clinical findings from a sample of healthy, cognitively intact older adults. *Am. J. Clin. Nutr.,* **87**, 872–880.

Cuevas, A.M., Guasch, V., Castillo, O. et al. (2000). A high-fat diet induces and red wine counteracts endothelial dysfunction in human volunteers. *Lipids,* **35**, 143–148.

Dauchet, L., Czernichow, S., Bertrais, S. et al. (2006). Fruits and vegetables intake in the SU.VI.MAX study and systolic blood pressure change. *Arch. Mal. Coeur. Vaiss.,* **99**, 669–673.

Desch, S., Schmidt, J., Kobler, D. et al. (2010). Effect of cocoa products on blood pressure: Systematic review and meta-analysis. *Am. J. Hypertens.,* **23**, 97–103.

Di Castelnuovo, A., Rotondo, S., Iacoviello, L. et al. (2002). Meta-analysis of wine and beer consumption in relation to vascular risk. *Circulation,* **105**, 2836–284.4

Duffy, S.J., Keaney, J.F., Jr., Holbrook, M. et al. (2001). Short- and long-term black tea consumption reverses endothelial dysfunction in patients with coronary artery disease. *Circulation,* **104**, 151–156.

Engler, M.B., Engler, M.M., Chen, C.Y. et al. (2004). Flavonoid-rich dark chocolate improves endothelial function and increases plasma epicatechin concentrations in healthy adults. *J. Am. Coll. Nutr.,* **23**, 197–204.

Erlund, I., Koli, R., Alfthan, G. et al. (2008). Favorable effects of berry consumption on platelet function, blood pressure, and HDL cholesterol. *Am. J. Clin. Nutr.,* **87**, 323–331.

Fisher, N.D., Hughes, M., Gerhard-Herman, M. et al. (2003). Flavanol-rich cocoa induces nitric-oxide-dependent vasodilation in healthy humans. *J. Hypertens.,* **21**, 2281–2286.

Fitzpatrick, D.F., Hirschfield, S.L., Ricci, T. et al. (1995). Endothelium-dependent vasorelaxation caused by various plant extracts. *J. Cardiovasc. Pharmacol.,* **26**, 90–95.

Fraga, C.G., Actis-Goretta, L., Ottaviani, J.I. et al. (2005). Regular consumption of a flavanol-rich chocolate can improve oxidant stress in young soccer players. *Clin. Dev. Immunol.,* **12**, 11–17.

Freedman, J.E., Parker, C., III, Li, L. et al. (2001). Select flavonoids and whole juice from purple grapes inhibit platelet function and enhance nitric oxide release. *Circulation,* **103**, 2792–2798.

Fuhrman, B., Volkova, N., Coleman, R. et al. (2005). Grape powder polyphenols attenuate atherosclerosis development in apolipoprotein E deficient (E0) mice and reduce macrophage atherogenicity. *J. Nutr.,* **135**, 722–728.

Grassi, D., Mulder, T.P., Draijer, R, et al. (2009). Black tea consumption dose-dependently improves flow-mediated dilation in healthy males. *J. Hypertens.,* **27**, 774–781.

Grassi, D., Necozione, S., Lippi, C. et al. (2005). Cocoa reduces blood pressure and insulin resistance and improves endothelium-dependent vasodilation in hypertensives. *Hypertension,* **46**, 398–405.

Hashimoto, M., Kim, S., Eto, M. et al. (2001). Effect of acute intake of red wine on flow-mediated vasodilatation of the brachial artery. *Am. J. Cardiol.,* **88**, 1457–1460, A1459.

He, F.J., Nowson, C.A., Lucas, M. et al. (2007). Increased consumption of fruit and vegetables is related to a reduced risk of coronary heart disease: Meta-analysis of cohort studies. *J. Hum. Hypertens.,* **21**, 717–728.

Heiss, C., Dejam, A., Kleinbongard, P. et al. (2003). Vascular effects of cocoa rich in flavan-3-ols. *JAMA,* **290**, 1030–1031.

Heiss, C., Finis, D., Kleinbongard, P. et al. (2007). Sustained increase in flow-mediated dilation after daily intake of high-flavanol cocoa drink over 1 week. *J. Cardiovasc. Pharmacol.,* **49**, 74-80.

Heiss, C., Kleinbongard, P., Dejam, A. et al. (2005). Acute consumption of flavanol-rich cocoa and the reversal of endothelial dysfunction in smokers. *J. Am. Coll. Cardiol.,* **46**, 1276–1283.

Hertog, M.G., Feskens, E.J., Hollman, P.C. et al. (1993). Dietary antioxidant flavonoids and risk of coronary heart disease: The Zutphen Elderly Study. *Lancet,* **342**, 1007–1011.

Hertog, M.G., Feskens, E.J., and Kromhout, D. (1997). Antioxidant flavonols and coronary heart disease risk. *Lancet,* **349**, 699.

Hertog, M. G., Kromhout, D., Aravanis, C. et al. (1995). Flavonoid intake and long-term risk of coronary heart disease and cancer in the seven countries study. *Arch. Intern. Med.* **155**, 381–386.

Hollenberg, N.K., Martinez, G., McCullough, M. et al. (1997). Aging, acculturation, salt intake, and hypertension in the Kuna of Panama. *Hypertension,* **29**, 171–176.

Holt, R.R., Actis-Goretta, L., Momma, T.Y. et al. (2006). Dietary flavanols and platelet reactivity. *J. Cardiovasc. Pharmacol.,* **47** (Suppl 2), S187–S196; discussion S206–S189.

Hooper, L., Kroon, P.A., Rimm, E.B. et al. (2008). Flavonoids, flavonoid-rich foods, and cardiovascular risk: A meta-analysis of randomized controlled trials. *Am. J. Clin. Nutr.,***88**, 38–50.

Joshipura, K.J., Hu, F.B., Manson, J.E. et al. (2001). The effect of fruit and vegetable intake on risk for coronary heart disease. *Ann. Intern. Med.,* **134**, 1106–1114.

Keevil, J.G., Osman, H.E., Reed, J.D., et al. (2000). Grape juice, but not orange juice or grapefruit juice, inhibits human platelet aggregation. *J. Nutr.,* **130**, 53–56.

Khan, N. Q., Lees, D.M., Douthwaite, J.A. et al. (2002). Comparison of red wine extract and polyphenol constituents on endothelin-1 synthesis by cultured endothelial cells. *Clin. Sci.,* **103** (Suppl 48), 72S–75S.

Knekt, P., Isotupa, S., Rissanen, H. et al. (2000). Quercetin intake and the incidence of cerebrovascular disease. *Eur. J. Clin. Nutr.,* **54**, 415–417.

Knekt, P., Jarvinen, R., Reunanen, A. et al. (1996). Flavonoid intake and coronary mortality in Finland: A cohort study. *BMJ*, **312**, 478–481.

Lin, J., Rexrode, K.M., Hu, F. et al. (2007). Dietary intakes of flavonols and flavones and coronary heart disease in US women. *Am. J. Epidemiol.*, **165**, 1305–1313.

Lorenz, M., Wessler, S., Follmann, E. et al. (2004). A constituent of green tea, epigallocatechin-3-gallate, activates endothelial nitric oxide synthase by a phosphatidylinositol-3-OH-kinase-, cAMP-dependent protein kinase-, and Akt-dependent pathway and leads to endothelial-dependent vasorelaxation. *J. Biol. Chem.*, **279**, 6190–6195.

Ludwig, A., Lorenz, M., Grimbo, N. et al. (2004). The tea flavonoid epigallocatechin-3-gallate reduces cytokine-induced VCAM-1 expression and monocyte adhesion to endothelial cells. *Biochem. Biophys. Res. Commun.*, **316**, 659–665.

Mao, T.K., van de Water, J., Keen, C.L. et al. (2002). Modulation of TNF-alpha secretion in peripheral blood mononuclear cells by cocoa flavanols and procyanidins. *Dev. Immunol.*, **9**, 135–141.

Mathur, S., Devaraj, S., Grundy, S.M. et al. (2002). Cocoa products decrease low density lipoprotein oxidative susceptibility but do not affect biomarkers of inflammation in humans. *J. Nutr.*, **132**, 3663–3667.

McCullough, M.L., Chevaux, K., Jackson, L. et al. (2006). Hypertension, the Kuna, and the epidemiology of flavanols. *J. Cardiovasc. Pharmacol.*, **47** (Suppl 2), S103–1S09; discussion 119–121.

McGorisk, G.M. and Treasure, C.B. (1996). Endothelial dysfunction in coronary heart disease. *Curr. Opin. Cardiol.*, **11**, 341–350.

Mink, P.J., Scrafford, C.G., Barraj, L.M. et al. (2007). Flavonoid intake and cardiovascular disease mortality: A prospective study in postmenopausal women. *Am. J. Clin. Nutr.*, **85**, 895–909.

Muniyappa, R., Hall, G., Kolodziej, T.L. et al. (2008). Cocoa consumption for 2 wk enhances insulin-mediated vasodilatation without improving blood pressure or insulin resistance in essential hypertension. *Am. J. Clin. Nutr.*, **88**, 1685–1696.

Murphy, K.J., Chronopoulos, A.K., Singh, I. et al. (2003). Dietary flavanols and procyanidin oligomers from cocoa (*Theobroma cacao*) inhibit platelet function. *Am. J. Clin. Nutr.*, **77**, 1466–1473.

Natella, F., Nardini, M., Belelli, F. et al. (2008). Effect of coffee drinking on platelets: Inhibition of aggregation and phenols incorporation. *Br. J. Nutr.*, **100**, 1276–1282.

Neveu, V., Perez-Jimenez, J., Vos, F. et al. (2010). Phenol-Explorer: An online comprehensive database on polyphenol contents in foods. *Database (Oxford)*, bap024.

Papamichael, C., Karatzis, E., Karatzi, K. et al. (2004). Red wine's antioxidants counteract acute endothelial dysfunction caused by cigarette smoking in healthy nonsmokers. *Am. Heart J.*, **147**, E5.

Park, Y.K., Kim, J.S. and Kang, M.H. (2004). Concord grape juice supplementation reduces blood pressure in Korean hypertensive men: Double-blind, placebo controlled intervention trial. *Biofactors*, **22**, 145–147.

Pearson, D.A., Paglieroni, T.G., Rein, D. et al. (2002). The effects of flavanol-rich cocoa and aspirin on ex vivo platelet function. *Thromb. Res.*, **106**, 191–197.

Peters, U., Poole, C., and Arab, L. (2001). Does tea affect cardiovascular disease? A meta-analysis. *Am. J. Epidemiol.*, **154**, 495–503.

Rein, D., Paglieroni, T.G., Wun, T. et al. (2000). Cocoa inhibits platelet activation and function. *Am. J. Clin. Nutr.*, **72**, 30–35.

Ribeiro, F., Alves, A.J., Teixeira, M. et al. (2009). Endothelial function and atherosclerosis: Circulatory markers with clinical usefulness. *Rev. Port. Cardiol.*, **28**, 1121–1151.

Rimm, E.B., Katan, M.B., Ascherio, A. et al. (1996). Relation between intake of flavonoids and risk for coronary heart disease in male health professionals. *Ann. Intern. Med.*, **125**, 384–389.

Schmitt, C.A. and Dirsch, V.M. (2009). Modulation of endothelial nitric oxide by plant-derived products. *Nitric Oxide,* 21, 77–91.

Schramm, D.D., Karim, M., Schrader, H.R. et al. (2003). Food effects on the absorption and pharmacokinetics of cocoa flavanols. *Life Sci.,* 73, 857–869.

Schroeter, H., Heiss, C., Balzer, J, et al. (2006). (–)-Epicatechin mediates beneficial effects of flavanol-rich cocoa on vascular function in humans. *Proc. Natl. Acad. Sci. U.S.A.,* 103, 1024–1029.

Scott-Burden, T. and Vanhoutte, P. M. (1994). Regulation of smooth muscle cell growth by endothelium-derived factors. *Texas Heart Inst. J.,* 21, 91–97.

Sesso, H.D., Gaziano, J.M., Liu, S. et al. (2003). Flavonoid intake and the risk of cardiovascular disease in women. *Am. J. Clin. Nutr.,* 77, 1400–1408.

Stankevicius, E., Kevelaitis, E., Vainorius, E. et al. (2003). [Role of nitric oxide and other endothelium-derived factors]. *Medicina,* 39, 333–341.

Steffen, Y., Gruber, C., Schewe, T. et al. (2008). Mono-O-methylated flavanols and other flavonoids as inhibitors of endothelial NADPH oxidase. *Arch. Biochem. Biophys.,* 469, 209–219.

Steffen, Y., Schewe, T., and Sies, H. (2007). (–)-Epicatechin elevates nitric oxide in endothelial cells via inhibition of NADPH oxidase. *Biochem. Biophys. Res. Commun.,* 359, 828–833.

Stein, J.H., Keevil, J.G., Wiebe, D.A. et al. (1999). Purple grape juice improves endothelial function and reduces the susceptibility of LDL cholesterol to oxidation in patients with coronary artery disease. *Circulation,* 100, 1050–1055.

Steptoe, A., Gibson, E.L., Vuononvirta, R. et al. (2007). The effects of chronic tea intake on platelet activation and inflammation: A double-blind placebo controlled trial. *Atherosclerosis,* 193, 277–282.

Stoclet, J.C., Chataigneau, T., Ndiaye, M. et al. (2004). Vascular protection by dietary polyphenols. *Eur. J. Pharmacol.,* 500, 299–313.

Taubert, D., Roesen, R., Lehmann, C. et al. (2007a). Effects of low habitual cocoa intake on blood pressure and bioactive nitric oxide: A randomized controlled trial. *JAMA,* 298, 49–60.

Taubert, D., Roesen, R. and Schomig, E. (2007b). Effect of cocoa and tea intake on blood pressure: A meta-analysis. *Arch. Intern. Med.,* 167, 626–634.

Vita, J.A. (2003). Tea consumption and cardiovascular disease: Effects on endothelial function. *J. Nutr.,* 133, 3293S–3297S.

Vita, J.A. (2005). Polyphenols and cardiovascular disease: Effects on endothelial and platelet function. *Am. J. Clin. Nutr.,* 81, 292S–297S.

Wallerath, T., Deckert, G., Ternes, T. et al. (2002). Resveratrol, a polyphenolic phytoalexin present in red wine, enhances expression and activity of endothelial nitric oxide synthase. *Circulation,* 106, 1652–1658.

Wan, Y., Vinson, J.A., Etherton, T.D. et al. (2001). Effects of cocoa powder and dark chocolate on LDL oxidative susceptibility and prostaglandin concentrations in humans. *Am. J. Clin. Nutr.,* 74, 596–602.

Wang-Polagruto, J.F., Villablanca, A.C., Polagruto, J.A. et al. (2006). Chronic consumption of flavanol-rich cocoa improves endothelial function and decreases vascular cell adhesion molecule in hypercholesterolemic postmenopausal women. *J. Cardiovasc. Pharmacol.,* 47 (Suppl 2), S177–S186; discussion S206–S179.

Whelan, A.P., Sutherland, W.H., McCormick, M.P. et al. (2004). Effects of white and red wine on endothelial function in subjects with coronary artery disease. *Intern. Med. J.* 34, 224–228.

Widlansky, M.E., Hamburg, N.M., Anter, E. et al. (2007). Acute EGCG supplementation reverses endothelial dysfunction in patients with coronary artery disease. *J. Am. Coll. Nutr.,* 26, 95–102.

Wittchen, E.S. (2009). Endothelial signaling in paracellular and transcellular leukocyte transmigration. *Front Biosci.,* **14**, 2522–2545.

Zern, T.L., Wood, R.J., Greene, C. et al. (2005). Grape polyphenols exert a cardioprotective effect in pre- and postmenopausal women by lowering plasma lipids and reducing oxidative stress. *J. Nutr.*, **135**, 1911–1917.

14 Effects of Flavonoids on the Vascular Endothelium: What Is Known and What Is Next?

Antje R. Weseler and Aalt Bast

CONTENTS

14.1 INTRODUCTION

Epidemiological studies suggest that the daily dietary intake of flavonoids benefi-cially affects human cardiovascular health (Arts and Hollman 2005; Hooper et al. 2008). Interestingly, most attention has concentrated on the vascular effects of flavonoids from soy and cocoa (Hooper et al. 2008), and almost all epidemiologic studies regarding flavonoids have originated from the Netherlands, Finland, and the United States (Arts and Hollman 2005). In this chapter, we will focus on the effects of the flavonoid subclasses flavanones, flavones, flavonols, and isoflavones on the vascular endothelium. Information on the effect of flavan-3-ols are provided in Chapter 13. The food items in the diet responsible for the intake of flavonoids are well characterized (USDA 2003; Neveu et al. 2010). However, the explanation for the functional effects of flavonoids on vascular endothelium has changed significantly over time. These changes, rather than being merely the result of scientific dynamics, are rather the consequence of epistemological paradigm shifts with regards to their health effects, as initial concepts (e.g., on radical scavenging properties of the flavonoids) are abandoned, and other more physiological concepts come in investigation.

14.2 INFLUENCE OF MOLECULAR STRUCTURE ON THE ANTIHYPERTENSIVE EFFECTS OF FLAVONOIDS

To illustrate that minor differences in the molecular structure of flavonoids can have a major impact on the mechanism of action and functional influence of flavonoids on the vascular endothelium, four quercetin derivatives will be compared in this section: Quercetin, isoquercitrin (quercetin-3-O-glucoside), rutin (quercetin-3-O-rutinoside), and 7-O-hydroxyethylquercetin-3-O-rutinoside (7-O-hydroxyethylrutin) (Figure 14.1). The main flavonol in the diet is quercetin (Boots et al. 2008), and this polyphenolic compound has been studied extensively for its effect on blood vessel function. In fact, quercetin decreases blood pressure in spontaneously hypertensive rats (SHR) when it is administered p.o. in a daily dose of 10 mg kg^{-1} for 5 weeks (Duarte et al. 2001) and 4 weeks (Romero et al. 2010). Also, in rats fed with a high-fat high-sucrose diet (Yamamoto and Oue 2006), rats treated with L-NAME to make them deficient in nitric oxide (NO) (Duarte et al. 2002), rats with abdominal aortic constriction (Jalili et al. 2006) and in Dahl salt-sensitive rats (Mackraj et al. 2008), quercetin reduces the hypertensive response (Table 14.1). Furthermore, quercetin administered both orally and intravenously has been shown to inhibit the increase in mean arterial pressure by infusion of the vasopressor angiotensin I in rats (Hackl et al. 2002).

Carlstrom et al. (2007) could not confirm earlier reports on the effect of quercetin in SHR when administering quercetin aglycone via the diet (Carlstrom et al. 2007). However, they did obtain a reduction in mean arterial blood pressure in SHR comparable to earlier literature reports when quercetin was given by oral gavage. They concluded that the observed quercetin effects in models of pharmacologically induced hypertension or SHR depend on the mode of delivery. Apparently, a dietary enriched rodent chow with quercetin aglycone may lead to a different result than a daily oral gavage. In contrast to these results, we found that a dietary intervention with quercetin for 5 weeks (in the form of a quercetin enriched tomato paste derived from genetically modified tomatoes) did reduce the increase in blood pressure in SHR (unpublished results from the authors). In the paste, quercetin was mainly present as rutin and isoquercitrin and equaled to 48.8 mg quercetin aglycone equivalents in 100 g tomato

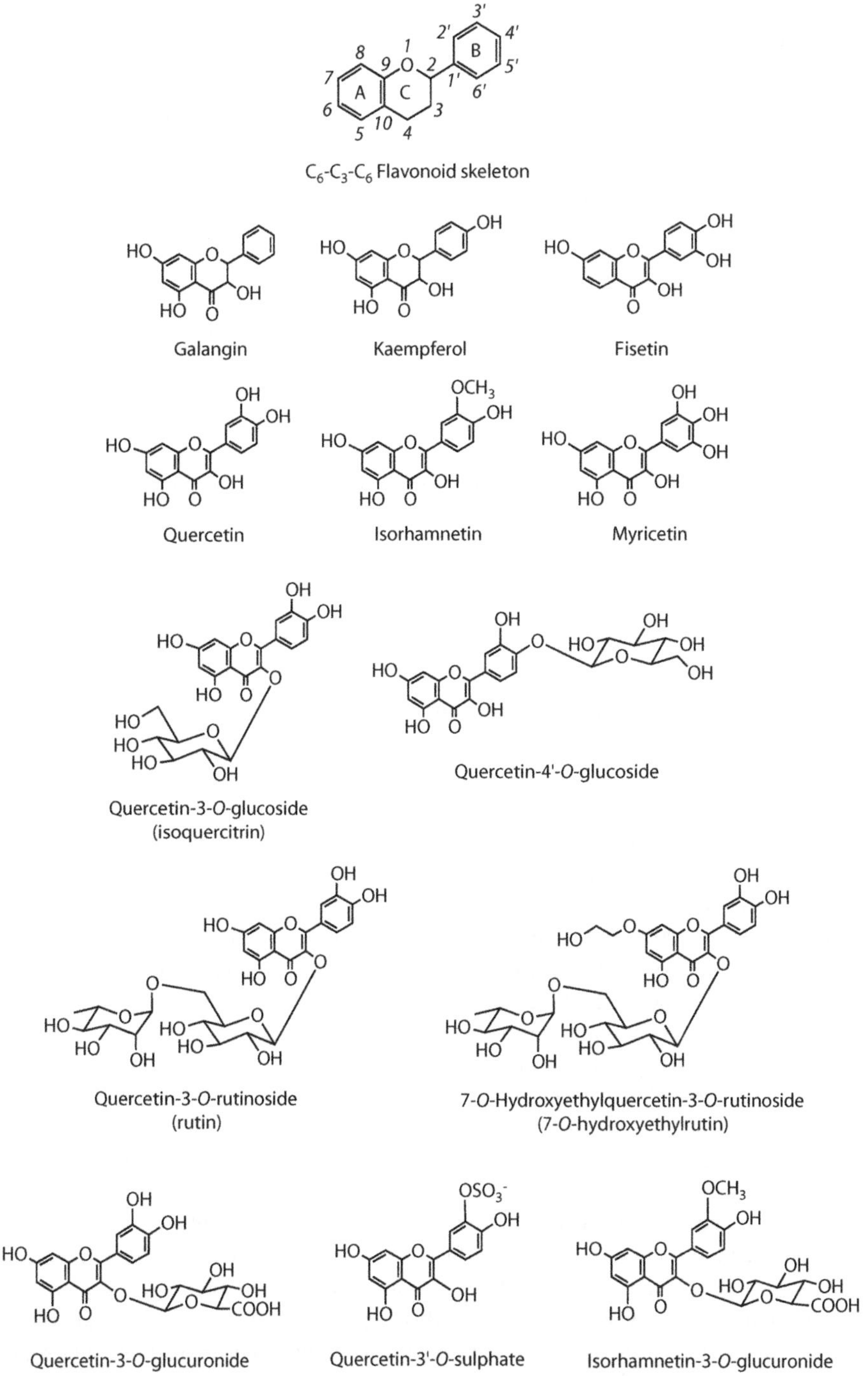

FIGURE 14.1 Structures of the flavonoid skeleton and various flavonols and flavonol metabolites.

TABLE 14.1
Studies on The Effect of Quercetin on The Hypertensive Response in Rats[a]

Study	Quercetin Administration	Effect
Duarte et al. (2001)	5 weeks 10 mg kg^{-1} p.o.	12% reduction mean arterial pressure in SHR, not effective in normotensive WKY rats
Romero et al. (2010)	4 weeks 10 mg kg^{-1} p.o.	14% reduction in mean arterial pressure in SHR
Yamamoto and Oue (2006)	4 weeks diet with 0.02 % to 0.5 %	Dose-dependent decrease in systolic blood pressure in high-fat high sucrose fed rats up to 8%
Duarte et al. (2002)	6 weeks 5 or 10 mg kg^{-1} p.o.	Effective in L-NAME treated rats
Jalili et al. (2006)	Estimated dietary intake of 130 mg kg^{-1} for 3 weeks	Decrease in carotid arterial blood pressure in rats with aortic constriction
Mackraj et al. (2008)	4 weeks with 10 mg kg^{-1} i.p.	Reduced mean arterial pressure in Dahl salt-sensitive rats
Hackl et al. (2002)	88.7 mmol kg^{-1} p.o. or 14.7 mmol kg^{-1} i.v. 45 min before measurement	Reduced mean arterial pressure after infusion of angiotensin I
Carlstrom et al. (2007)	1.5 g kg^{-1} diet for 11 weeks	No effect on arterial blood pressure

[a] SHR: spontaneously hypertensive rat; L-NAME: L-N$_G$-nitroarginine methyl ester; WKY: Wistar Kyoto rats.

paste. The daily amount of quercetin–glycosides received by the rats via the diet was equivalent to approximately 10 mg quercetin aglycone/kg body weight.

In hypertensive humans, quercetin has been shown to reduce blood pressure (Edwards et al. 2007). Supplementation with 730 mg quercetin/day for 28 days resulted in a reduction in mean arterial pressure of 5 mm Hg in a study population of 19 hypertensive subjects. In a study that supplemented quercetin (150 mg/day) for a longer period (42 days) in a larger population, 93 individuals characterized as obese prehypertensive, the systolic blood pressure significantly decreased by 2.9 mm Hg (Egert et al. 2009). However, no blood pressure reductions were observed in prehypertensive individuals. In agreement with this, another study indicated that when healthy volunteers received a high dose of 1000 mg quercetin for 56 days no reduction in systolic blood pressure was observed (Conquer et al. 1998).

Closely related structural analogs of quercetin also reduce blood pressure. For example, the blood pressure lowering effect of the water-soluble quercetin-3-O-glucoside (isoquercitrin, Figure 14.1) in SHR is higher than an equimolar dose of quercetin (Emura et al. 2007). It was suggested that this might be due to a better bioavailability (perhaps due to increased absorption via sugar carriers/transporters) and greater solubilization properties. Indeed, the sugar moiety in flavonoid glycosides is an important determinant in their absorption in rat and man (Hollman et al. 1999; Arts et al. 2004). Apparently, the position of attachment of the sugar (either at the 3- or at the 4′-position in quercetin) does not make much difference for the absorption. At first, the uptake of the glucosides was thought to occur via the sodium-dependent glucose transporting system (Hollman et al. 1999), which was followed by a cytosolic β-glucosidase-mediated hydrolysis of the glucoside. Later, more emphasis was put on lactase phlorizin hydrolase, a β-glycosidase present in the brush border membrane of the small intestinal lumen as a determining factor in the uptake of quercetin–glucosides (Murota and Terao 2003). Quercetin-3-O-glucoside and quercetin-4′-O-glucoside are better absorbed than rutin (Figure 14.1). Quercetin-3-O-glucoside can easily be produced via enzymatic treatment of rutin (Emura et al. 2007), which is found throughout the human diet in many plants such as buckwheat, rhubarb, and asparagus, as well as black tea, onion, and apple.

Early literature reported that rutin may be of value in the prevention of hemorrhages, which occur in malignant hypertension (Hellerstein et al. 1951). The application of rutosides has mainly developed as so-called phlebotonics in the treatment of venous insufficiency. A Cochrane evaluation concluded that there is indication of some efficacy of phlebotonics on oedema but that further clinical trials are needed to substantiate the clinical relevance (Martinez et al. 2005). A standardized mixture of hydroxyethylrutinosides has widely been investigated for its mechanism of action in the protection of capillary fragility. Venous hypertension and reduced venous clearance are important in increasing microvascular permeability. Trapped leukocytes release proteolytic enzymes and reactive oxygen species that damage the capillary basement membrane and this damage may be prevented by hydroxyethylrutinosides. Venoruton is a commercial mixture of hydroxyethylrutinosides in which the major component is 7-O-hydroxyethylrutin (Figure 14.1). We investigated this compound quite extensively for its cardiovascular protection against the toxicity of the anti-tumor agent doxorubicin (van Acker et al. 1995; Abou El Hassan et al. 2003; Bast et al. 2007). Interestingly, intravenous administration of O-hydroxyethylrutinosides to patients showed that the compounds are taken up into the vascular wall in localized endothelial and subendothelial areas (Neumann et al. 1992). This suggests a locally high concentration, which may explain the protective effect of Venoruton on endothelial damage in chronic venous insufficiency (Perrin and Ramelet 2011). This is demonstrated by a Venoruton-induced reduction in biomarkers for endothelial damage, for example, lower levels of circulating endothelial cells in venous blood (Cesarone et al. 2006).

14.3   IS THE VASODILATOR EFFECT OF FLAVONOIDS ENDOTHELIUM DEPENDENT?

There has been quite some debate in the literature on the question to which extent the effect of flavonoids on blood vessel function is endothelium dependent. Initial experiments on the concentration dependent inhibition of quercetin on the noradrenalin-induced contraction of rat aorta strips showed that the effects were not dependent on the endothelium. Moreover, quercetin inhibited the contractions induced by phorbol-12-myristate-13-acetate, which led to the suggestion that the vasodilator effect of quercetin is related to the inhibition of protein kinase C (Duarte et al. 1993a). In agreement with this, the vasodilatory potency of a series of flavonoids in precontracted rat aortic strips correlated with the inhibition of protein kinase C (Duarte et al. 1993b). Herrera et al. (1996) from the same research group suggested that besides inhibition of protein kinase C, the inhibition of cyclic nucleotide phosphodiesterase and Ca^{2+} uptake may be involved. Chan et al. (2000) suggested that flavonols with the characteristic 3-OH group induce an endothelium dependent vasorelaxation in rat aorta, whereas flavones such as luteolin (Figure 14.1), which do not possess this 3-OH moiety, have an endothelium-independent action. Moreover, hydroxyl substitutions in the A ring of the flavonols reduce their vasodilatory efficacy. Thus, flavones and flavonols induce relaxation and impair the contraction caused by the influx of extracellular Ca^{2+} or α_1-adrenoceptor agonist phenylephrine-induced Ca^{2+} release (Chan et al. 2000). In contrast Ajay et al. (2003) emphasized that the inhibition of the phenylephrine induced rat aorta contraction by flavonols, which lack the 3-OH group, may also be endothelium dependent. The concentration might play a role in explaining this difference. At low concentrations, that is, below 30 µM, the relaxant effect of flavonols is endothelium dependent and is then probably mediated by the release of NO• and PGI2 from the endothelium, whereas at higher concentrations, other mechanisms including Ca^{2+} release from intracellular stores may be involved (Ajay et al. 2003).

In rat mesenteric resistance vessels, it has been suggested that the vasodilator action of quercetin and its methylated metabolites is endothelium independent and is the result of a direct effect on the smooth muscle (Perez-Vizcaino et al. 2002). A similar conclusion was drawn in a recent study using porcine coronary arteries (Suri et al. 2010). The endothelium-independent action of quercetin was suggested to occur through enhancing cyclic-GMP-dependent relaxations by a mechanism not involving phosphodiesterase-5.

14.4 HOW ARE FLAVONOIDS TAKEN UP BY VASCULAR CELLS?

Flavonoids undergo extensive first pass metabolism when administered orally. They are rapidly conjugated with among other things glucuronic acid, catalyzed by glucuronosyltransferase. These conjugation processes lead to more water-soluble compounds and consequently a more rapid elimination from the body via the kidneys. Moreover, conjugation with glucuronic acid leads to an altered biological activity of the flavonoids (Day et al. 2000). Indeed, in contrast to quercetin, the metabolites quercetin-3-O-glucuronide, isorhamnetin-3-O-glucuronide, and quercetin-3'-O-sulfate (Figure 14.1) do not have a vasorelaxant effect in thoracic aortic rings (Lodi et al. 2009). However, recently, it has also become clear that local processes determine the availability and uptake of flavonoids. An acute inflammatory response, for example, leads to recruitment of neutrophils to inflamed tissue, where activated neutrophils may excrete β-glucuronidase. It has been suggested that deconjugation catalyzed by β-glucuronidase enhances the local aglycone concentration of the flavonoids, and thus an increase in bioactivity may occur at the site of inflammation. In addition, the modest decrease in pH at the inflammatory site may activate the deconjugation because this environmental change brings the reaction closer to the pH optimum of β-glucuronidase (Bartholome et al. 2010).

Bilitranslocase has been suggested as a plasma membrane flavonoid transporter and is expressed in various rat organs (Passamonti et al. 2009) including the vascular endothelium (Maestro et al. 2010). This transporter has been implicated in the vaso-dilatory action of anthocyanins. It has been purported that the inhibition of the liver clearance of bilirubin, which is transported via bilitranslocase, enhanced the antioxidant action of flavonoids. There is a wide variation in the inhibition of bilitranslocase by flavonoids. Galangin (Figure 14.1) is a noncompetitive inhibitor, while kaempferol and quercetin are mixed-type inhibitors, that is, competitive and noncompetitive. The flavan-3-ol (+)-catechin and the isoflavones genistin (7-O-glucoside of genistein), diadzin (7-O-glucoside of diadzein), daidzein, and puerarin (8-C-glucoside of diadzein) (Figure 14.2) did not inhibit bilitranslocase (Passamonti et al. 2009). The role of bilitranslocase in the uptake of flavonoids in vascular endothelium needs further investigation.

14.5 WHICH ENDOTHELIUM-DERIVED MEDIATORS ARE INFLUENCED BY FLAVONOIDS?

14.5.1 NITRIC OXIDE

We entered the research area of flavonoids via our investigations on the antioxidant action of the phlebotonic Venoruton (*vide supra*). The protection on the capillary fragility was explained by the antioxidant activity of the constituents of the rutoside mixture in Venoruton (Haenen et al. 1993). Interestingly, in the search for effective protectors against the cardiotoxicity of the antitumor agent doxorubicine, the main constituent of Venoruton, 7-O-hydroxyethylrutin (Figure 14.1) appeared to be very effective (van Acker et al. 1995). The protective action of this compound could be explained by its simultaneous antioxidant action and its iron chelating activity, and many new synthetic analogs of 7-O-hydroxyethylrutin were synthesized (van Acker et al. 2001). Clinical trials were conducted with 7-hydroxyethylrutin in an attempt to limit the major drawback of the use of doxorubicin (Bast et al. 2007; Bruynzeel et al. 2007).

The antioxidant actions of flavonoids have been explained on a molecular level (van Acker et al. 1996) and two "antioxidantophore groups" have been described in the flavonols (Heijnen et al. 2001), which took away some confusion that existed in this area. The excellent superoxide radical scavenging activity leant itself to protective vascular effects in that the scavenging of superoxide radicals would reduce the reaction with endothelial NO•, thus limiting the formation of cytotoxic peroxynitrite. Modulation of NO• by flavonoids can occur in additional ways: flavonoids can directly react with NO•, they inhibit NO• formation by inhibiting the inducible form of NO-synthase (iNOS), and they can prevent the expression of iNOS (Paquay et al. 2000). Moreover, flavonoids scavenge peroxynitrite directly (Heijnen et al. 2001). However, despite this, currently the direct antioxidant action of flavonoids is thought to be less important *in vivo* than initially suggested (Hollman et al. 2011).

In various blood vessels, there is ample evidence that polyphenols from very different sources increase endothelium dependent NO• formation (Schini-Kerth et al. 2010). Activation of endothelial NO-synthase (eNOS) can be the result of an increase

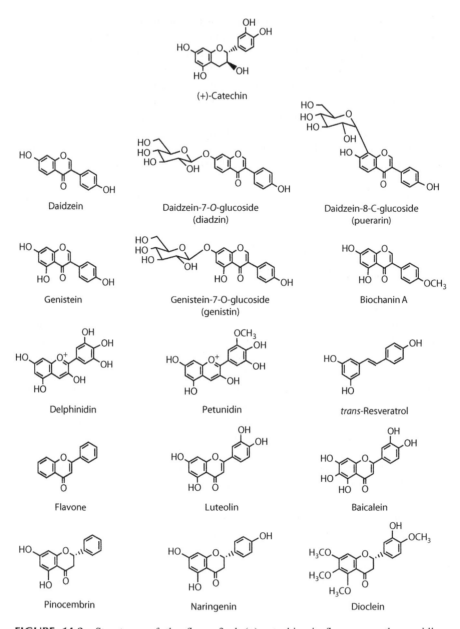

FIGURE 14.2 Structures of the flavan-3-ol (+)-catechin, isoflavones, anthocyanidins, flavones, and flavanones.

in the free cytosolic calcium concentration. However, polyphenols seem to activate NO-synthase primarily via the redox sensitive PI3-kinase/Akt pathway in endothelial cells. This causes phosphorylation of an activator site of eNOS and a dephosphorylation at an inhibitor site of eNOS (Ndiaye et al. 2005). Red wine polyphenols have been shown to increase eNOS expression in endothelial cells both at mRNA and at protein level (Wallerath et al. 2005). In a human endothelial cell line, it appeared that out of 33 phenolic compounds only three gave an increase in NO• level and no clear structure-activity relationship was apparent (Appeldoorn et al. 2009). Arginase competes with NOS for the substrate L-arginine and inhibition of arginase by certain polyphenols has been suggested to increase NO• levels and to improve endothelium-dependent vasorelaxation. For example, cocoa reduces arginase in human erythrocytes (Schnorr et al. 2008); however, other flavonoids have not been investigated yet in this respect.

14.5.2 EDHF

Endothelium-dependent relaxation, which is not sensitive to inhibitors of eNOS and cyclooxygenases, is known as the Endothelium-derived hyperpolarizing factor (EDHF)-mediated relaxation. This relaxation is associated to an endothelium dependent hyperpolarization of vascular smooth muscle cells. EDHF-mediated relaxation and hyperpolarization by red wine polyphenolic compounds has been reported and involves formation of superoxide anions in coronary arteries (Ndiaye et al. 2003). These redox sensitive activations in cellular systems should, however, be regarded with caution, because polyphenols can rapidly generate hydrogen peroxide during incubation in cell culture media (Long et al. 2000).

14.5.3 ENDOTHELIN-1

Of a series of anthocyanidins, delphinidin and petunidin (Figure 14.2) have been shown to inhibit the synthesis of the vasoconstrictor endothelin-1 in bovine cultured endothelial cells (Khan et al. 2002). Some polyphenols have also been shown to exhibit similar inhibitory activity in vascular smooth muscle cells (Ruef et al. 2001) and in human vascular endothelial cells (Nicholson et al. 2010), where their effects on hydrogen peroxide-induced gene expression of the vasoconstrictor endothelin-1 were investigated.

14.6 WHAT IS THE ROLE OF THE INHIBITION OF NADPH OXIDASE, UNCOUPLED NO-SYNTHASE, AND XANTHINE OXIDASE BY FLAVONOIDS?

The indirect antioxidant activity of flavonoids receives growing attention and oxidative stress is known to play a crucial role in endothelial dysfunction. In this respect, vascular enzymes such as nicotinamide adenine dinucleotide phosphate (NADPH) oxidases, uncoupled NO-synthase, and xanthine oxidase are known to be significant sources of vascular superoxide anion radical production. Activation of NADPH oxidases has been found in hypertensive animal models such as rats infused with angiotensin II or spontaneous hypertensive rats (Fukui et al. 1997). Inhibition of these enzymes by flavonoids can be regarded as an indirect antioxidant action and may lead to new treatment possibilities for hypertension (Weseler and Bast 2010). Quercetin and its metabolite 3′-O-methyl quercetin (isorhamnetin) prevented overexpression of p47phox (a subunit of NADPH oxidase) induced by angiotensin II and thus superoxide anion production in rat aorta (Sanchez et al. 2007).

The uncoupling of eNOS due to a shortage of the eNOS cofactor (6R-)5,6,7,8-tetrahydrobiopterin or by oxidative stress also leads to superoxide radical formation (Forstermann and Li 2010), and *trans*-resveratrol (Figure 14.2) has been suggested to reverse eNOS uncoupling. It is known for quite some time that flavonoids are effective inhibitors of xanthine oxidase (Van Hoorn et al. 2002). In this way, flavonoids inhibit the formation of superoxide anion radicals. Remarkably, despite the wide applicability of xanthine oxidase inhibitors (e.g., in hyperuricemia, gout, ulcers, cancer, ischemia, hypertension), only a limited number of drugs with this action profile reached the market. A recent patent survey on xanthine oxidase inhibitors again illustrates the huge interest in this field and emphasizes the interesting activity of flavonoids (Kumar et al. 2011).

14.7 HOW DO FLAVONOIDS MODULATE REDOX SIGNALING?

The important role of reactive oxygen and nitrogen species (ROS) in the physiological regulation of cell activity via redox-sensitive intracellular signaling and gene expression is well established (Droge 2002; Valko et al. 2007). A transient activation of endogenous ROS producing systems (e.g., NADPH oxidases, NO-synthases, cyclooxygenases) initiates redox signaling by subtly increasing ROS levels. This temporary shift in intracellular redox state to more oxidative conditions is rapidly counteracted by endogenous antioxidant systems (e.g., superoxide dismutase, catalase, glutathione peroxidase, thioredoxin) to restore the physiological cellular redox balance. Both, endogenous and environmental circumstances can result in a dysregulation of these processes leading to excessive ROS formation. If such a pro-oxidative state persists, pathological conditions may occur (Droge 2002). Once manifested in the vascular endothelium, persistently high levels of ROS may impair directly and indirectly cellular function eventually leading to endothelial dysfunction, an early manifestation of cardiovascular disease (Heitzer et al. 2001; Bugiardini et al. 2004).

Initially, it was believed that polyphenols worked as powerful antioxidants *in vitro* and *in vivo* due to their ability to donate electrons or hydrogen atoms (Thompson et al. 1976; Haenen et al. 1993; van Acker et al. 1996). However, during the past decades, a shift emerged in understanding of how these compounds exert their biological effects in the body (Figure 14.3). The shift occurred away from the classic, chemically based concept of oxidant–antioxidant interaction, toward an understanding that flavonoids and their metabolites may act as indirect antioxidants by inhibiting ROS producing enzymes such as NADPH- and xanthine oxidases (Van Hoorn et al. 2002; Sanchez et al. 2007). In the meantime, it could also be demonstrated that flavonoids interfere with essential intracellular signaling transmission and gene expression by direct binding to membranes, enzymes, or nuclear receptors as ligands or ligand mimetics (Virgili and Marino 2008). Quercetin, for instance, could rapidly activate the redox sensitive mitogen-activated protein kinases p38, extracellular signaling related kinases-1/2 (Bruynzeel et al. 2007) and c-Jun-N-terminal kinase (JNK) in human coronary endothelial cells (Pasten et al. 2007). As a consequence, gene

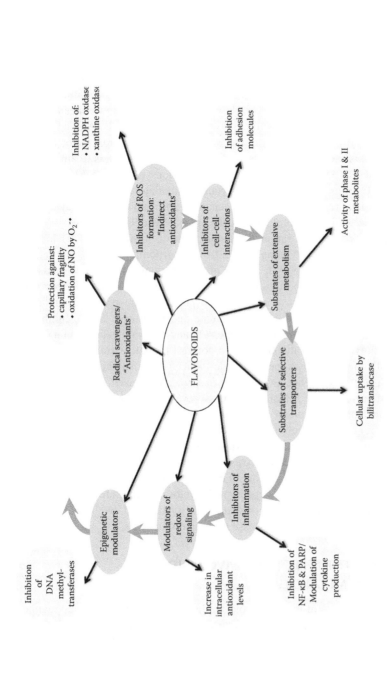

FIGURE 14.3 Changing views on the action of flavonoids in vascular endothelium. At first, radical scavenging was predominantly regarded as the molecular mechanism of action of flavonoids. Over the years, further evidence was provided that flavonoids act as indirect antioxidants, metabolites contribute to and/or modify the effects of the parent compound, exert anti-inflammatory action, and modulate redox signaling and epigenetic traits.

expression of plasminogen activator inhibitor type 1 (PAI-1) was suppressed, which is an essential protein in clot formation. The glucuronide metabolite of quercetin inhibited concentration- and time-dependently angiotensin II-induced activation of JNK and activator protein (AP)-1 mediated gene transcription, thereby preventing hypertrophy of rat smooth muscle cells (Yoshizumi et al. 2002).

The long-term feeding of rats with soy isoflavones led to a marked up-regulated gene expression of eNOS and antioxidant defense enzymes in the vasculature (Mahn et al. 2005). These findings were ascribed to the observation that low concentrations of isoflavones (<1 µM) activate specific kinases in endothelial cells. Among others, this facilitates the translocation of the redox-sensitive transcription factor nuclear factor erythroid 2-related factor (Nrf)-2 to the nucleus where it binds to antioxidant responsive elements (ARE) containing promoter regions (Siow and Mann 2010). Genes with ARE promotor codes for important antioxidant and phase II enzymes including glutathione reductase, heme-oxygenase-1, and NAD(P)H-quinone oxido-reductases-1, which efficiently protect the cells against toxic ROS effects.

The latest research efforts target the modulation of sirtuins, class III histone deacetylases, by polyphenols (Chung et al. 2010). In particular, sirtuin 1 is a central regulator of vascular function and, among other functions, ameliorates oxidative stress levels by deacetylating besides histones a variety of transcription factors (e.g., fork head transcription factors of class O) and proteins [e.g., p53, nuclear factor nuclear factor-kappa B (Nf-κB), eNOS] (Potente and Dimmeler 2008). Inhibition of sirtuin 1 resulted in a down-regulation of eNOS and the manifestation of a senescence-like phenotype of endothelial cells. In contrast to that, sirtuin 1 overexpression successfully prevented hydrogen peroxide-induced endothelial cell damage (Ota et al. 2007). The first evidence exists from *in vitro* and *in vivo* experiments that resveratrol and quercetin can activate sirtuin 1 or reverse its oxidative stress-induced post-translational modification in the vascular endothelium (Chung et al. 2010). Since sirtuin 1 regulates the activity of miscellaneous proteins from distinct signaling pathways, sirtuin 1 activation by certain polyphenols would not only explain their capability to increase the cellular resistance of the vascular endothelium against oxidative stress, but may also shed more light on their anti-inflammatory, antiapoptotic, and angiogenetic mechanisms of action. Therefore, we are sure that in the near future research into the mechanisms of sirtuin 1 activation and its clinical implications for cardiovascular health and disease will rapidly increase.

14.8   WHAT IS THE EFFECT OF FLAVONOIDS ON LIPID METABOLISM?

The evidence that quercetin modulates plasma lipid profiles in a beneficial manner *in vivo* is limited (Perez-Vizcaino and Duarte 2010). Instead, the flavonol proved in animal models to attenuate oxidative modification of low density lipoprotein (LDL) (Hayek et al. 1997) and fatty streak formation (Auger et al. 2005), which are considered as crucial steps in atheroma formation. Interestingly, a recent study indicated that the quercetin-glucuronide, one of the major metabolites of quercetin in human plasma, specifically accumulates in activated macrophages localized in atherosclerotic lesions of human aortas (Kawai et al. 2008). Because of a pronounced β-glucuronidase activity of activated macrophages, quercetin was completely converted into its aglycone and thus became intracellulary available at its side of action. In addition, the methylated quercetin metabolite isorhamnetin contributed effectively to the prevention of damage induced by oxidized LDL in a human endothelial cell line. The unraveled molecular mechanisms of action comprised, among others, attenuation of endothelial NO-synthetase (eNOS) downregulation, lectin-like ox-LDL-receptor-1 upregulation, phosphorylation of p38-mitogen-activated protein kinase (MAPK), and translocation of the inflammatory transcription factor nuclear factor kappa B (Nf-κB) (Bao and Lou 2006).

14.9   HOW DO FLAVONOIDS INFLUENCE PARAOXONASE?

The influence of flavonoids on paraoxonase may be a further putative mechanism by which flavonoids mediate their protective vascular effects as such activity would influence the oxidation of plasma lipoproteins. Paraoxonases (Saponara et al. 2002) are a family of calcium-dependent enzymes consisting of three isoforms: PON 1, 2, and 3 (Draganov et al. 2005). In humans, PON1 and 3 are synthesized in the liver from where they reach the circulation upon secretion in association with HDL particles. In contrast to that, PON2 occurs only intracellularly in many tissues including liver, kidney, brain, testis, and the vascular endothelium (Mackness et al. 1996; Ng et al. 2001). PON received their name from the ability to hydrolyze paraoxon, the active metabolite of the organophosphorus insecticide parathion. In particular, PON1 is characterized by a broad spectrum of substrates comprising various organophosphates, arylesters, lactones, and lipid hydroperoxides (Draganov et al. 2005). PON1 has been shown to prevent LDL (Mackness et al. 1991) and HDL (Aviram et al. 1998) oxidation, thus attenuating an early key event in the development of atherosclerotic plaques. This was substantiated by experiments with PON1$^{-/-}$ knockout mice that proved to be more susceptible to develop atherosclerosis than their wild-type littermates (Shih et al. 1998). In humans, an inverse relationship between PON1 status in serum and the occurrence of coronary heart diseases was found (Camps et al. 2009).

Various cardiovascular drugs and more recently flavonoids, in particular quercetin, have been investigated for their ability to increase PON1 activity and/or expression (Costa et al. 2011). A 6-week dietary supplementation of mice with up to 400 mg quercetin/kg body weight elevated both, liver PON1 RNA expression (twofold) and protein levels (50%) (Boesch-Saadatmandi et al. 2010). On the other hand, young, healthy volunteers supplemented with 50–150 mg quercetin/day did not reveal alterations in plasma PON1 activity, assessed by the substrates paraoxon and phenyl acetate. Interestingly, the authors explained the lack of quercetin-induced PON modulation in humans by the relative low plasma levels of the catechol-O-methyl-transferase-derived quercetin metabolite isorhamnetin (3'-O-methyl-quercetin) (Figure 14.1) compared to mice. In human liver cells, isorhamnetin proved to be a more potent inducer of PON1 than quercetin. The precise molecular mechanism of how quercetin upregulates hepatic PON1 at mRNA and protein level is yet not fully clear (Gouedard et al. 2004). Recent investigations demonstrate involvement of the sterol regulatory element-binding protein (SREBP)-2 (Garige et al. 2010). SREBP2 belongs to a family of membrane-bound transcription factors that regulate lipid homeostasis by binding to sterol-responsive elements (SREs) in the promoter of targeted genes (Horton et al. 2002). Garige et al. showed that quercetin promotes translocation of SREBP-2 from the endoplasmatic reticulum to the nucleus, where it bounds to the SRE-like sequence in the promoter region of the PON1 gene and thereby activated its transcription (Garige et al. 2010).

14.10 WHAT IS THE EFFECT OF FLAVONOIDS ON AMP-ACTIVATED PROTEIN KINASE?

Besides its function as key regulator of cellular energy metabolism, AMP-activated protein kinase (AMPK) has been also suggested to play an essential role in the maintenance of redox balance as well as in the function and structure of blood vessels (Wong et al. 2009; Shirwany and Zou 2010). Activation of AMPK has been shown to reverse diabetes-associated endothelial dysfunction in rats (Suzuki et al. 2008) and increased NO production in human endothelial cells (Morrow et al. 2003). Recently, quercetin, among other polyphenols, has been reported to activate AMPK in different cell types (Hwang et al. 2009). Although appropriate experiments in the vascular endothelium still need to be conducted, we expect that targeting AMPK signaling pathway may be an additional relevant molecular mechanism by which flavonoids evoke their beneficial cardiovascular effects.

14.11 HOW DO FLAVONOIDS MODULATE VASCULAR ION CHANNELS?

The ability of smooth muscle cells to contract determines the resistance of the vasculature and eventually the blood pressure. Numerous ion channels including calcium, potassium, chloride, and stretch-activated cation channels regulate the contractile activity of smooth muscle cells by altering membrane potential. Activation of potassium channels, for instance, results in an increased potassium efflux and thus hyperpolarization of the membrane. Consequently, voltage-gated calcium channels facilitating calcium influx into the cells are closed resulting in vasodilation. Conversely, the closing of potassium channels depolarizes the cell membrane and leads to an activation of voltage gated calcium channels. This triggers calcium release from intracellular stores, thereby eliciting contraction.

Accumulating evidence indicates that various flavonoids exert their cardiovascular effects to a certain extent by the modulation of such ion channels. A recent review article provides an overview on the current knowledge of this topic (Scholz et al. 2010). *In vitro* studies in isolated arterial myocytes revealed that isoflavonoids such as genistein and biochanin A (Figure 14.2), as well as the flavonol myricetin (Figure 14.1), are able to block calcium channels, which might explain their vasorelaxant effects observed in isolated aortic ring experiments. Other flavonoids including fisetin (Figure 14.1), flavone and baicalein (5,6,7-trihydroxflavone) (Figure 14.2), but not quercetin, mediated vasorelaxation at higher concentrations (100 μM) by an inhibition of the intracellular release of calcium from its sarcoplasmatic reticulum stores (Ajay et al. 2003). Interestingly, in smooth muscle cells, the relaxant properties of quercetin cannot be explained by an inhibition of calcium channels, since it has been proven to activate these ion channels (Saponara et al. 2002). However, quercetin has been revealed to inhibit protein kinases including myosin light chain kinase and protein kinase C (Hagiwara et al. 1988). Since these enzymes are crucial in maintaining vasocontraction, it was suggested that their inhibition outweighs the effects of an activated calcium influx and may be responsible for the net relaxant effect of quercetin on vascular tissue (Perez-Vizcaino et al. 2002; Saponara et al. 2002).

The modulation of vascular electrophysiology also comprises the activation of potassium channels. Several flavonoids (see Figures 14.1 and 14.2), including the flavonols kaempferol and quercetin, the flavanones naringenin and dioclein and the isoflavone puerarin, elicited vasorelaxation by activating calcium activated potassium channels (Scholz et al. 2010). Remarkably, the flavanone pinocembrin (Zhu et al. 2007) and the flavone luteolin (Jiang et al. 2005) exerted vasorelaxation by attenuating calcium current at the same time activating potassium channels. In vascular endothelial cells, calcium-activated potassium channels play an essential role in the regulation of NO synthesis. Opening of these channels induces hyperpolarization of the endothelial cell membrane, which in turn initiates a calcium influx and mobilizes NO release (Nilius and Droogmans 2001). Kuhlmann et al. (2005) could demonstrate that quercetin activates these potassium channels and thereby elicits NO and cGMP release in human vascular endothelial cells. The activation of potassium channels and a subsequent release of NO does not only evoke vasorelaxation but might also contribute to the antiangiogenic effects of numerous flavonoids observed *in vitro* and *in vivo* (Kuhlmann et al. 2005). In addition, quercetin has been reported to attenuate the activity of enzymes being associated with cell proliferation like matrix metalloproteinase-2 as well as tyrosine kinases and protein kinase 2 (Agullo et al. 1997).

14.12 HOW DO FLAVONOIDS INHIBIT ENDOTHELIAL INFLAMMATION?

Oxidative stress and low-grade chronic inflammation are major pathophysiological factors contributing to the development of cardiovascular diseases such as hypertension and atherosclerosis. Oxidative stress and inflammation are known to be linked (Weseler and Bast 2010), in that inflammation may lead to oxidative stress, and direct or indirect reduction of oxidative stress in the endothelium will lead to a reduction in inflammation. A good example in this respect is the finding that the oxidative stress elicited by the antitumor agent doxorubicine generates an inflammatory response (Abou El Hassan et al. 2003). Interestingly, the flavonol 7-O-hydroxyethylrutin (Figure 14.1) prevents the doxorubicin-induced adhesion of neutrophils to endothelial cells. It was found that the flavonoid inhibited the doxorubicin-induced expression of the adhesion molecules VCAM and E-selectin. Recently, methylated metabolites of quercetin were also found to inhibit TNFα-induced expression of adhesion molecules in human aortic endothelial cells (Lotito et al. 2011).

The anti-inflammatory action of flavonoids is now widely recognized. A recent study in our laboratory showed the systemic anti-inflammatory action of a single dose of quercetin in sarcoidosis patients (Boots et al. 2011). Also, in the blood of patients (COPD and diabetes mellitus type II) that was challenged *ex vivo* with the inflammatory trigger lipopolysaccharide (LPS), several flavonoids could attenuate this inflammatory response (Weseler et al. 2009). Incubation of porcine coronary arteries with LPS induced NO-synthase and resulted in nitrite production. Quercetin and its metabolites, the 3-glucuronide and the 3'-sulphate, exerted anti-inflammatory effects (Al-Shalmani et al. 2011).

The redox-sensitive transcription factor NF-κB is activated by oxidative stress (van den Berg et al. 2001) and is involved in the production of proinflammatory cytokines (Figure 14.3). The anti-inflammatory action of flavonoids has been ascribed to the inhibition of NF-κB (Lin et al. 2007; Crespo et al. 2008). The nuclear enzyme poly(ADP-ribose) polymerase-1 (PARP-1), which catalyzes the formation of poly(ADP-ribose) (PAR)polymers from its substrate NAD^+ acts as coactivator of NF-κB (Hassa and Hottiger 2002). PARP-1 is activated as a result of DNA damage, and consequently, a rapid depletion of NAD^+ and ATP levels occur. We recently reported (Geraets et al. 2007) that dietary flavonoids are inhibitors of PARP-1 and suggested that by disturbing the coactivator of NF-κB they exert their anti-inflammatory activity. However, no clear structure activity relationship could be discerned in this study.

14.13   CONCLUDING REMARK—WHAT IS NEXT?

What we might learn from the paradigm shifts that are apparently occurring around flavonoids is that the mode of action of these compounds does not follow a simple one-target-one-hit model (Figure 14.3). Rather a multitude of molecular phenomena play a role in explaining the vascular health effects of flavonoids. The challenge for clinical trials to prove the effectiveness of flavonoids is to include a set of biomarkers that adequately takes their pleiotropic actions into account. Moreover, scientists (and for that matter, consumers) should be aware that these paradigm shifts rapidly occur in food science. It may already be predicted that the next wave in vascular flavonoid research will focus on long term effects of these compounds and on the concurrent epigenetic modifications influencing imprinting and silencing of chromosomal domains. This line of research has already begun for the relationship between flavonoids and cancer (Link et al. 2010) and will undoubtedly follow for the subject of this chapter that is of vascular benefits (Duthie 2011).

REFERENCES

Abou El Hassan, M.A., Verheul, H.M., Jorna, A.S. et al. (2003). The new cardioprotector Monohydroxyethylrutoside protects against doxorubicin-induced inflammatory effects in vitro. *Br. J. Cancer,* **89**, 357–362.

Agullo, G., Gamet-Payrastre, L., Manenti, S. et al. (1997). Relationship between flavonoid structure and inhibition of phosphatidylinositol 3-kinase: A comparison with tyrosine kinase and protein kinase C inhibition. *Biochem. Pharmacol.,* **53**, 1649–1657.

Ajay, M., Gilani, A.U., and Mustafa, M.R. (2003). Effects of flavonoids on vascular smooth muscle of the isolated rat thoracic aorta. *Life Sci., 74*, 603–612.

Al-Shalmani, S., Suri, S., Hughes, D.A. et al. (2011). Quercetin and its principal metabolites, but not myricetin, oppose lipopolysaccharide-induced hyporesponsiveness of the porcine isolated coronary artery. *Br. J. Pharmacol. 162*, 1485–1497.

Appeldoorn, M.M., Venema, D.P., Peters, T.H. et al. (2009). Some phenolic compounds increase the nitric oxide level in endothelial cells *in vitro*. *J. Agric. Food Chem., 57*, 7693–7699.

Arts, I.C. and Hollman, P.C. (2005). Polyphenols and disease risk in epidemiologic studies. *Am. J. Clin. Nutr., 81*, 317S–3125S.

Arts, I.C., Sesink, A.L., Faassen-Peters, M. et al. (2004). The type of sugar moiety is a major determinant of the small intestinal uptake and subsequent biliary excretion of dietary quercetin glycosides. *Br. J. Nutr., 91*, 841–847.

Auger, C., Teissedre, P.L., Gerain, P. et al. (2005). Dietary wine phenolics catechin, quercetin, and resveratrol efficiently protect hypercholesterolemic hamsters against aortic fatty streak accumulation. *J. Agric. Food Chem., 53*, 2015–2021.

Aviram, M., Rosenblat, M., Bisgaier, C.L. et al. (1998). Paraoxonase inhibits high-density lipoprotein oxidation and preserves its functions. A possible peroxidative role for paraoxonase. *J. Clin. Investig., 101*, 1581–1590.

Bao, M. and Lou, Y. (2006). Isorhamnetin prevent endothelial cell injuries from oxidized LDL via activation of p38MAPK. *Eur. J. Pharmacol. 547*, 22–30.

Bartholome, R., Haenen, G., Hollman, C.H. et al. (2010). Deconjugation kinetics of glucuronidated phase II flavonoid metabolites by β-glucuronidase from neutrophils. *Drug Metab. Pharmacokin., 25*, 379–387.

Bast, A., Haenen, G.R., Bruynzeel, A.M. et al. (2007). Protection by flavonoids against anthracycline cardiotoxicity: From chemistry to clinical trials. *Cardiovasc. Toxicol., 7*, 154–159.

Boesch-Saadatmandi, C., Egert, S., Schrader, C. et al. (2010). Effect of quercetin on paraoxonase 1 activity—studies in cultured cells, mice and humans. *J. Physiol. Pharmacol., 61*, 99–105.

Boots, A.W., Drent, M., De Boer, V.C. et al. (2011). Quercetin reduces markers of oxidative stress and inflammation in sarcoidosis. *Clin. Nutr., 30*, 506–512.

Boots, A.W., Haenen, G.R., and Bast, A. (2008). Health effects of quercetin: From antioxidant to nutraceutical. *Eur. J. Pharmacol., 585*, 325–337.

Bruynzeel, A.M., Niessen, H.W., Bronzwaer, J.G. et al. (2007). The effect of monohydroxyethylrutoside on doxorubicin-induced cardiotoxicity in patients treated for metastatic cancer in a phase II study. *Br. J. Cancer, 97*, 1084–1089.

Bugiardini, R., Manfrini, O., Pizzi, C. et al. (2004). Endothelial function predicts future development of coronary artery disease: A study of women with chest pain and normal coronary angiograms. *Circulation, 109*, 2518–2523.

Camps, J., Marsillach, J., and Joven, J. (2009). Pharmacological and lifestyle factors modulating serum paraoxonase-1 activity. *Mini Rev. Med. Chem., 9*, 911–920.

Carlstrom, J., Symons, J.D., Wu, T.C. et al. (2007). A quercetin supplemented diet does not prevent cardiovascular complications in spontaneously hypertensive rats. *J. Nutr., 137*, 628–633.

Cesarone, M.R., Belcaro, G., Pellegrini, L. et al. (2006). Circulating endothelial cells in venous blood as a marker of endothelial damage in chronic venous insufficiency: Improvement with venoruton. *J. Cardiovasc. Pharmacol. Therapeut., 11*, 93–98.

Chan, E.C., Pannangpetch, P., and Woodman, O.L. (2000). Relaxation to flavones and flavonols in rat isolated thoracic aorta: Mechanism of action and structure-activity relationships. *J. Cardiovasc. Pharmacol., 35*, 326–333.

Chung, S., Yao, H., Caito, S. et al. (2010). Regulation of SIRT1 in cellular functions: Role of polyphenols. *Arch. Biochem. Biophys., 501*, 79–90.

Conquer, J.A., Maiani, G., Azzini, E. et al. (1998). Supplementation with quercetin markedly increases plasma quercetin concentration without effect on selected risk factors for heart disease in healthy subjects. *J. Nutr.,* **128**, 593–597.

Costa, L.G., Giordano, G., and Furlong, C.E. (2011). Pharmacological and dietary modulators of paraoxonase 1 (PON1) activity and expression: The hunt goes on. *Biochem. Pharmacol.,* **81**, 337–344.

Crespo, I., Garcia-Mediavilla, M.V., Gutierrez, B. et al. (2008). A comparison of the effects of kaempferol and quercetin on cytokine-induced pro-inflammatory status of cultured human endothelial cells. *Br. J. Nutr.,* **100**, 968–976.

Day, A.J., Bao, Y., Morgan, M.R. et al. (2000). Conjugation position of quercetin glucuronides and effect on biological activity. *Free Rad. Biol. Med.,* **29**, 1234–1243.

Draganov, D.I., Teiber, J.F., Speelman, A. et al. (2005). Human paraoxonases (PON1, PON2, and PON3) are lactonases with overlapping and distinct substrate specificities. *J. Lipid Res.,* **46**, 1239–1247.

Droge, W. (2002). Free radicals in the physiological control of cell function. *Physiol. Rev.,* **82**, 47–95.

Duarte, J., Jimenez, R., O'Valle, F. et al. (2002). Protective effects of the flavonoid quercetin in chronic nitric oxide deficient rats. *J. Hyperten.,* **20**, 1843–1854.

Duarte, J., Perez-Palencia, R., Vargas, F. et al. (2001). Antihypertensive effects of the flavonoid quercetin in spontaneously hypertensive rats. *Br. J. Pharmacol.,* **133**, 117–124.

Duarte, J., Perez-Vizcaino, F., Zarzuelo, A. et al. (1993a). Vasodilator effects of quercetin in isolated rat vascular smooth muscle. *Eur. J. Pharmacol.,* **239**, 1–7.

Duarte, J., Perez Vizcaino, F., Utrilla, P. et al. (1993b). Vasodilatory effects of flavonoids in rat aortic smooth muscle. Structure-activity relationships. *General Pharmacol.,* **24**, 857–862.

Duthie, S.J. (2011). Epigenetic modifications and human pathologies: Cancer and CVD. *Proc. Nutr. Soc.,* **70**, 47–56.

Edwards, R.L., Lyon, T., Litwin, S.E. et al. (2007). Quercetin reduces blood pressure in hypertensive subjects. *J. Nutr.,* **137**, 2405–2411.

Egert, S., Bosy-Westphal, A., Seiberl, J. et al. (2009). Quercetin reduces systolic blood pressure and plasma oxidised low-density lipoprotein concentrations in overweight subjects with a high-cardiovascular disease risk phenotype: A double-blinded, placebo-controlled cross-over study. *Br. J. Nutr.,* **102**, 1065–1074.

Emura, K., Yokomizo, A., Toyoshi, T. et al. (2007). Effect of enzymatically modified isoquercitrin in spontaneously hypertensive rats. *J. Nutr. Sci Vitaminol.,* **53**, 68–74.

Forstermann, U. and Li, H. (2010). Therapeutic effect of enhancing endothelial nitric oxide synthase (eNOS) expression and preventing eNOS uncoupling. *Br. J. Pharmacol.,* **164**, 213–233.

Fukui, T., Ishizaka, N., Rajagopalan, S. et al. (1997). p22phox mRNA expression and NADPH oxidase activity are increased in aortas from hypertensive rats. *Circulation Res.,* **80**, 45–51.

Garige, M., Gong, M., Varatharajalu, R. et al. (2010). Quercetin up-regulates paraoxonase 1 gene expression via sterol regulatory element binding protein 2 that translocates from the endoplasmic reticulum to the nucleus where it specifically interacts with sterol responsive element-like sequence in paraoxonase 1 promoter in HuH7 liver cells. *Metabolism: Clin. Expt.* **59**, 1372–1378.

Geraets, L., Moonen, H.J., Brauers, K. et al. (2007). Dietary flavones and flavonols are inhibitors of poly(ADP-ribose)polymerase-1 in pulmonary epithelial cells. *J. Nutr.,* **137**, 2190–2195.

Gouedard, C., Barouki, R., and Morel, Y. (2004). Dietary polyphenols increase paraoxonase 1 gene expression by an aryl hydrocarbon receptor-dependent mechanism. *Mol. Cell. Biol.,* **24**, 5209–5222.

Hackl, L.P., Cuttle, G., Dovichi, S.S., Lima-Landman, M.T., and Nicolau, M. (2002). Inhibition of angiotesin-converting enzyme by quercetin alters the vascular response to brandykinin and angiotensin I. *Pharmacology,* **65**, 182–186.

Haenen, G.R.M.M., Jansen, F.P., and Bast, A. (1993). The antioxidant properties of five O-(β-hydroxyethyl)-rutosides of the flavonoid mixture Venoruton. *Phlebology,* (Suppl1), 10–17.

Hagiwara, M., Inoue, S., Tanaka, T. et al. (1988). Differential effects of flavonoids as inhibitors of tyrosine protein kinases and serine/threonine protein kinases. *Biochem. Pharmacol.,* **37**, 2987–2992.

Hassa, P.O. and Hottiger, M.O. (2002). The functional role of poly(ADP-ribose)polymerase 1 as novel coactivator of NF-kappaB in inflammatory disorders. *Cell. Mol. Life Sci.,* **59**, 1534–1553.

Hayek, T., Fuhrman, B., Vaya, J. et al. (1997). Reduced progression of atherosclerosis in apolipoprotein E–deficient mice following consumption of red wine, or its polyphenols quercetin and catechin, is associated with reduced susceptibility of LDL to oxidation and aggregation. *Arterioscler. Thromb. Vasc. Biol.* **17**, 2744–2752.

Heijnen, C.G., Haenen, G.R., Van Acker, F.A. et al. (2001). Flavonoids as peroxynitrite scavengers: The role of the hydroxyl groups. *Toxicol. In Vitro,* **15**, 3–6.

Heitzer, T., Schlinzig, T., Krohn, K. et al. (2001). Endothelial dysfunction, oxidative stress, and risk of cardiovascular events in patients with coronary artery disease. *Circulation,* **104**, 2673–2678.

Hellerstein, H.K., Orbison, J.L., Rodbard, S., Wilburne, M., and Katz, L.N. 1951. The effect of rutin in experimental malignant hypertension. *Am. Heart J.,* **42**, 271–283.

Herrera, M.D., Zarzuelo, A., Jimenez, J., Marhuenda, E., and Duarte, J. (1996). Effects of flavonoids on rat aortic smooth muscle contractility: Structure-activity relationships. *Gen. Pharmacol.* **27**, 273–277.

Hollman, P.C., Bijsman, M.N., Van Gameren, Y. et al. (1999). The sugar moiety is a major determinant of the absorption of dietary flavonoid glycosides in man. *Free Rad. Res.,* **31**, 569–573.

Hollman, P.C., Cassidy, A., Comte, B. et al. (2011). The biological relevance of direct antioxidant effects of polyphenols for cardiovascular health in humans is not established. *J. Nutr.,* **141**, 989S–1009S.

Hooper, L., Kroon, P.A., Rimm, E.B. et al. (2008). Flavonoids, flavonoid-rich foods, and cardiovascular risk: A meta-analysis of randomized controlled trials. *Am. J. Clin. Nutr.* **88**, 38–50.

Horton, J.D., Goldstein, J.L., and Brown, M.S. (2002). SREBPs: Activators of the complete program of cholesterol and fatty acid synthesis in the liver. *J. Clin. Invest.,* **109**, 1125–1131.

Hwang, J.T., Kwon, D.Y., and Yoon, S.H. (2009). AMP-activated protein kinase: A potential target for the diseases prevention by natural occurring polyphenols. *New Biotechnol.,* **26**, 17–22.

Jalili, T., Carlstrom, J., Kim, S. et al. (2006). Quercetin-supplemented diets lower blood pressure and attenuate cardiac hypertrophy in rats with aortic constriction. *J. Cardiovasc. Pharmacol.,* **47**, 531–541.

Jiang, H., Xia, Q., Wang, X., et al. (2005). Luteolin induces vasorelaxion in rat thoracic aorta via calcium and potassium channels. *Pharmazie* **60**, 444–447.

Kawai, Y., Nishikawa, T., Shiba, Y. et al. (2008). Macrophage as a target of quercetin glucuronides in human atherosclerotic arteries: Implication in the anti-atherosclerotic mechanism of dietary flavonoids. *J. Biol. Chem.,* **283**, 9424–9434.

Khan, N.Q., Lees, D.M., Douthwaite, J.A. et al. (2002). Comparison of red wine extract and polyphenol constituents on endothelin-1 synthesis by cultured endothelial cells. *Clin. Sci.,* **103** (Suppl 48), 72S–75S.

Kuhlmann, C.R., Schaefer, C.A., Kosok, C. et al. (2005). Quercetin-induced induction of the NO/cGMP pathway depends on Ca^{2+}-activated K^+ channel-induced hyperpolarization-mediated Ca^{2+}-entry into cultured human endothelial cells. *Planta Med., 71,* 520–524.

Kumar, R., Darpan, Sharma, S. and Singh, R. (2011). Xanthine oxidase inhibitors: A patent survey. *Expert Opin. Therap. Patents, 21,* 1071–1108.

Lin, R., Liu, J., Gan, W., and Ding, C. (2007). Protective effect of quercetin on the homocysteine-injured human umbilical vein vascular endothelial cell line (ECV304). *Basic Clin. Pharmacol. Toxicol., 101,* 197–202.

Link, A., Balaguer, F., and Goel, A. (2010). Cancer chemoprevention by dietary polyphenols: Promising role for epigenetics. *Biochem. Pharmacol., 80,* 1771–1792.

Lodi, F., Jimenez, R., Moreno, L. et al. (2009). Glucuronidated and sulfated metabolites of the flavonoid quercetin prevent endothelial dysfunction but lack direct vasorelaxant effects in rat aorta. *Atherosclerosis, 204,* 34–39.

Long, L.H., Clement, M.V., and Halliwell, B. (2000). Artifacts in cell culture: Rapid generation of hydrogen peroxide on addition of (–)-epigallocatechin, (–)-epigallocatechin gallate, (+)-catechin, and quercetin to commonly used cell culture media. *Biochem. Biophys. Res. Comm., 273,* 50–53.

Lotito, S.B., Zhang, W.-J., Yang, C.S. et al. (2011). Metabolic conversion of dietary flavonoids alters their anti-inflammatory and antioxidant properties. *Free Rad. Biol. Med., 51,* 454–463.

Mackness, M.I., Arrol, S. and Durrington, P.N. (1991). Paraoxonase prevents accumulation of lipoperoxides in low-density lipoprotein. *FEBS Lett., 286,* 152–154.

Mackness, M.I., Mackness, B., Durrington, P.N. et al. (1996). Paraoxonase: Biochemistry, genetics and relationship to plasma lipoproteins. *Curr. Opin. Lipidol., 7,* 69–76.

Mackraj, I., Govender, T. and Ramesar, S. (2008). The antihypertensive effects of quercetin in a salt-sensitive model of hypertension. *J. Cardiovasc. Pharmacol., 51,* 239–245.

Maestro, A., Terdoslavich, M., Vanzo, A. et al. (2010). Expression of bilitranslocase in the vascular endothelium and its function as a flavonoid transporter. *Cardiovasc. Res., 85,* 175–183.

Mahn, K., Borras, C., Knock, G.A. et al. (2005). Dietary soy isoflavone induced increases in antioxidant and eNOS gene expression lead to improved endothelial function and reduced blood pressure in vivo. *FASEB J., 19,* 1755–1757.

Martinez, M.J., Bonfill, X., Moreno, R.M. et al. (2005). Phlebotonics for venous insufficiency. *Cochrane Database of Syst. Rev.,* CD003229.

Morrow, V.A., Foufelle, F., Connell, J.M. et al. (2003). Direct activation of AMP-activated protein kinase stimulates nitric-oxide synthesis in human aortic endothelial cells. *J. Biol. Chem., 278,* 31629–31639.

Murota, K. and Terao, J. (2003). Antioxidative flavonoid quercetin: Implication of its intestinal absorption and metabolism. *Arch. Biochem. Biophys., 417,* 12–17.

Ndiaye, M., Chataigneau, M., Lobysheva, I. et al. (2005). Red wine polyphenol-induced, endothelium-dependent NO-mediated relaxation is due to the redox-sensitive PI3-kinase/Akt-dependent phosphorylation of endothelial NO-synthase in the isolated porcine coronary artery. *FASEB J., 19,* 455–457.

Ndiaye, M., Chataigneau, T., Andriantsitohaina, R. et al. (2003). Red wine polyphenols cause endothelium-dependent EDHF-mediated relaxations in porcine coronary arteries via a redox-sensitive mechanism. *Biochem. Biophys. Res. Commun., 310,* 371–377.

Neumann, H.A., Carlsson, K., and Brom, G.H. (1992). Uptake and localisation of O-(beta-hydroxyethyl)-rutosides in the venous wall, measured by laser scanning microscopy. *Eur. J. Clin. Pharmacol., 43,* 423–426.

Neveu, V., Perez-Jimenez, J., Vos, F. et al. (2010). Phenol-Explorer: An online comprehensive database on polyphenol contents in foods. Database 2010, doi:10.1093/database/bapO24. Link to database: http://www.phenol-explorer.eu/accessed at 01 Dec 2011.

Ng, C.J., Wadleigh, D.J., Gangopadhyay, A. et al. (2001). Paraoxonase-2 is a ubiquitously expressed protein with antioxidant properties and is capable of preventing cell-mediated oxidative modification of low density lipoprotein. *J. Biol. Chem.,* **276**, 4444–4449.

Nicholson, S.K., Tucker, G.A., and Brameld, J.M. (2010). Physiological concentrations of dietary polyphenols regulate vascular endothelial cell expression of genes important in cardiovascular health. *Br. J. Nutr.,* **103**, 1398–1403.

Nilius, B. and Droogmans, G. (2001). Ion channels and their functional role in vascular endothelium. *Physiol. Rev.,* **81**, 1415–1419.

Ota, H., Akishita, M., Eto, M. et al. (2007). Sirt1 modulates premature senescence-like phenotype in human endothelial cells. *J. Mol. Cell. Cardiol.,* **43**, 571–579.

Paquay, J.B., Haenen, G.R., Stender, G. et al. (2000). Protection against nitric oxide toxicity by tea. *J. Agric. Food Chem.,* 48, 5768–5772.

Passamonti, S., Terdoslavich, M., Franca, R. et al. (2009). Bioavailability of flavonoids: A review of their membrane transport and the function of bilitranslocase in animal and plant organisms. *Curr. Drug Metab.,* **10**, 369–394.

Pasten, C., Olave, N.C., Zhou, L. et al. (2007). Polyphenols downregulate PAI-1 gene expression in cultured human coronary artery endothelial cells: Molecular contributor to cardiovascular protection. *Thromb. Res.,* **121**, 59–65.

Perez-Vizcaino, F. and Duarte, J. (2010). Flavonols and cardiovascular disease. *Mol. Asp. Med.,* **31**, 478–494.

Perez-Vizcaino, F., Ibarra, M., Cogolludo, A.L. et al. (2002). Endothelium-independent vasodilator effects of the flavonoid quercetin and its methylated metabolites in rat conductance and resistance arteries. *J. Pharmacol. Exper. Therapeut.,* **302**, 66–72.

Perrin, M. and Ramelet, A.A. (2011). Pharmacological treatment of primary chronic venous disease: Rationale, results and unanswered questions. *Eur. J. Vasc. Endovasc. Surg.,* **41**, 117–125.

Potente, M. and Dimmeler, S. (2008). Emerging roles of SIRT1 in vascular endothelial homeostasis. *Cell Cycle,* **7**, 2117–2122.

Romero, M., Jimenez, R., Hurtado, B. et al. (2010). Lack of beneficial metabolic effects of quercetin in adult spontaneously hypertensive rats. *Eur. J. Pharmacol.,* **627**, 242–250.

Ruef, J., Moser, M., Kubler, W. et al. (2001). Induction of endothelin-1 expression by oxidative stress in vascular smooth muscle cells. *Cardiovasc. Pathol.* **10**, 311–315.

Sanchez, M., Lodi, F., Vera, R. et al. (2007). Quercetin and isorhamnetin prevent endothelial dysfunction, superoxide production, and overexpression of p47phox induced by angiotensin II in rat aorta. *J. Nutr.,* **137**, 910–915.

Saponara, S., Sgaragli, G., and Fusi, F. (2002). Quercetin as a novel activator of L-type Ca^{2+} channels in rat tail artery smooth muscle cells. *Br. J. Pharmacol.,* **135**, 1819–1827.

Schini-Kerth, V.B., Auger, C., Kim, J.H. et al. (2010). Nutritional improvement of the endothelial control of vascular tone by polyphenols: Role of NO and EDHF. *Eur. J. Physiol.,* **459**, 853–862.

Schnorr, O., Brossette, T., Momma, T.Y. et al. (2008). Cocoa flavanols lower vascular arginase activity in human endothelial cells in vitro and in erythrocytes in vivo. *Arch. Biochem. Biophys.,* **476**, 211–215.

Scholz, E.P., Zitron, E., Katus, H.A. et al. (2010). Cardiovascular ion channels as a molecular target of flavonoids. *Cardiovasc. Ther.,* **28**, e46–e52.

Shih, D.M., Gu, L., Xia, Y.R. et al. (1998). Mice lacking serum paraoxonase are susceptible to organophosphate toxicity and atherosclerosis. *Nature,* **394**, 284–287.

Shirwany, N.A. and Zou, M.H. (2010). AMPK in cardiovascular health and disease. *Acta Pharmacol. Sinica,* **31**, 1075–1084.

Siow, R.C. and Mann, G.E. (2010). Dietary isoflavones and vascular protection: Activation of cellular antioxidant defenses by SERMs or hormesis? *Mol. Asp. Med.,* **31**, 468–477.

Suri, S., Liu, X.H., Rayment, S. et al. (2010). Quercetin and its major metabolites selectively modulate cyclic GMP-dependent relaxations and associated tolerance in pig isolated coronary artery. *Br. J. Pharmacol.,* **159**, 566–575.

Suzuki, K., Uchida, K., Nakanishi, N. et al. (2008). Cilostazol activates AMP-activated protein kinase and restores endothelial function in diabetes. *Am. J. Hyperten.,* **21**, 451–457.

Thompson, M., Williams, C.R., and Elliot, G.E. (1976). Stability of flavonoid complexes of copper (II) and flavonoid antioxidant activity. *Anal. Chim. Acta,* **85**, 375–381.

USDA. (2003). *USDA Database for the Flavonoid Content of Selected Foods* [online]. http://www.nal.usda.gov/fnic/foodcomp/Data/Flav/flav.html last accessed at 01 Dec 2011.

Valko, M., Leibfritz, D., Moncol, J. et al. (2007). Free radicals and antioxidants in normal physiological functions and human disease. *Int. J. Biochem. Cell Biol.,* **39**, 44–84.

van Acker, F.A., Hulshof, J.W., Haenen, G.R. et al. (2001). New synthetic flavonoids as potent protectors against doxorubicin-induced cardiotoxicity. *Free Rad. Biol. Med.* **31**, 31–37.

van Acker, S.A., Kramer, K., Grimbergen, J.A. et al. (1995). Monohydroxyethylrutoside as protector against chronic doxorubicin-induced cardiotoxicity. *Br. J. Pharmacol.,* **115**, 1260–1264.

van Acker, S.A., Van Den Berg, D.J., Tromp, M.N. et al. (1996). Structural aspects of antioxidant activity of flavonoids. *Free Rad. Biol. Med.,* **20**, 331–342.

van Den Berg, R., Haenen, G.R., Van Den Berg, H. et al. (2001).Transcription factor NF-kappaB as a potential biomarker for oxidative stress. *Br. J. Nutr.,* **86** (Suppl 1), S121–S127.

van Hoorn, D.E., Nijveldt, R.J., Van Leeuwen, P.A. et al. (2002). Accurate prediction of xanthine oxidase inhibition based on the structure of flavonoids. *Eur. J. Pharmacol.,* **451**, 111–118.

Virgili, F. and Marino, M. (2008). Regulation of cellular signals from nutritional molecules: A specific role for phytochemicals, beyond antioxidant activity. *Free Rad. Biol. Med.,* **45**, 1205–1216.

Wallerath, T., Li, H., Godtel-Ambrust, U. et al. (2005). A blend of polyphenolic compounds explains the stimulatory effect of red wine on human endothelial NO synthase. *Nitric Oxide,* **12**, 97–104.

Weseler, A.R. and Bast, A. (2010). Oxidative stress and vascular function: Implications for pharmacologic treatments. *Curr. Hyperten. Rep.,* **12**, 154–161.

Weseler, A.R., Geraets, L., Moonen, H.J. et al. (2009). Poly (ADP-ribose) polymerase-1-inhibiting flavonoids attenuate cytokine release in blood from male patients with chronic obstructive pulmonary disease or type 2 diabetes. *J. Nutr.,* **139**, 952–957.

Wong, A.K., Howie, J., Petrie, J.R. et al. (2009). AMP-activated protein kinase pathway: A potential therapeutic target in cardiometabolic disease. *Clin. Sci.,* **116**, 607–620.

Yamamoto, Y. and Oue, E. (2006). Antihypertensive effect of quercetin in rats fed with a high-fat high-sucrose diet. *Biosci., Biotechnol. Biochem.,* **70**, 933–939.

Yoshizumi, M., Tsuchiya, K., Suzaki, Y. et al. (2002). Quercetin glucuronide prevents VSMC hypertrophy by angiotensin II via the inhibition of JNK and AP-1 signaling pathway. *Biochem. Biophys. Res. Commun.,* **293**, 1458–1465.

Zhu, X.M., Fang, L.H., Li, Y.J. et al. (2007). Endothelium-dependent and -independent relaxation induced by pinocembrin in rat aortic rings. *Vasc. Pharmacol.,* **46**, 160–165.

15 Green Tea Flavan-3-ols and Their Role in Protecting against Alzheimer's and Parkinson's Disease Pathophysiology

Orly Weinreb, Tamar Amit, Moussa B.H. Youdim, and Silvia Mandel

CONTENTS

15.1 INTRODUCTION

Polyphenols are found in a variety of beverages that have been manufactured from plant materials, including fruit juices, red wine, and black and green tea. Flavonoids are the largest group of polyphenols and as discussed elsewhere in this book are divided into anthocyanins, flavones, isoflavones, flavan-3-ols, flavans, and flavonols (Butterfield et al. 2002). Flavan-3-ols, sometimes referred to as catechins, account for 30–40% of the dry weight of the leaves of green tea (Yang and Wang 1993; Wang et al. 1994). Green tea belongs to the *Theacease* family one of which is *Camellia sinensis* (Graham 1992), the most widely consumed beverage (aside from water) in Japan, China, and other Asian nations and is presently becoming ever more popular in Western countries. The first historical reports of the medicinal properties of green tea appeared in the 16th century, where extracts were used as therapeutic agents to treat fever, headache, and stomach pains (Sutherland et al. 2006). The intake of green tea is generally regarded as being safe (authorized as such by the U.S. Food and Drug Administration) (Wu and Wei 2002) and has attracted attention for its health benefits in humans, particularly with respect to its potential for preventing and treating cancer, inflammatory diseases, cardiovascular diseases, neurodegenerative diseases, impaired metabolism, diabetes, stroke, and aging (Figure 15.1). For more information on these, please refer to reviews by Hollman et al. (1999), Weisburger and Chung (2002), Mandel et al. (2004b, 2005), Weinreb et al. (2004), Lau et al. (2005), Cabrera et al. (2006), Rossi et al. (2008), and Spencer (2008).

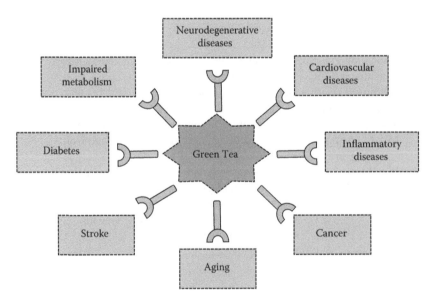

FIGURE 15.1 Schematic illustration of green tea health benefits. Several epidemiological and clinical studies indicated that green tea might help reducing risk of several chronic diseases, particularly with respect to treating cancer, inflammatory diseases, cardiovascular diseases, neurodegenerative diseases, impaired metabolism, diabetes, stroke, and aging.

Green tea contains around four times more free flavanols/catechins compared to black tea (Khokhar and Magnusdottir 2002). These catechins contain phenolic hydroxyl groups on their aromatic rings, which confer both antioxidant and iron chelating properties *in vitro* (van Acker et al. 1996). The principal flavanols found in green tea are (+)-catechin (C), (–)-epicatechin (EC), (+)-gallocatechin (GC), (–)-epigallo-catechin (EGC), (+)-catechin-3-*O*-gallate (CG), (–)-epicatechin-3-*O*-gallate (ECG), (+)-gallocatechin-3-*O*-gallate (GCG), and (–)-epigallocatechin-3-*O*-gallate (EGCG) (see Chapter 2). Among these, EGCG (Figure 15.2) is the major constituent, accounting for more than 10% of the extract dry weight (2.5 g of green tea leaves/200 mL of water), whereas the others are found in lower abundance (Wu and Wei 2002; Lin et al. 2003). Much research has indicated that green tea catechins are potent antioxidants, due to their ability to directly scavenge reactive oxygen species (ROS) and reactive nitrogen species (RNS), to induce endogenous antioxidant enzymes and to bind and chelate excess divalent metal ions, such as copper (Cu^{+2}) and iron (Fe^{+3}) to form inactive complexes [see reviews by Higdon and Frei (2003); Mandel et al. (2005); Perron and Brumaghim (2009); and Weinreb et al. (2009a)]. The relative antioxidant activity of green tea catechins has been found to be EGCG = ECG > EGC > EC (Guo et al. 1999; Nanjo et al. 1999), and they have been found to be more efficient radical scavengers than vitamins E and C (Nanjo et al. 1996; Pannala et al. 1997). This chemical antioxidant nature allows them to scavenge both superoxide and hydroxyl radicals, the 1,1-diphenyl-3-picrylhydrazyl radical, peroxyl radicals, nitric oxide, carbon-center free radicals, singlet oxygen, lipid free radicals, and peroxynitrite (Guo et al. 1996; Pannala et al. 1997; Zhao et al. 2001; Lin et al. 2003).

Several reports have indicated that green tea catechins can access the brain and exert neuroprotective actions following oral intake (Levites et al. 2001; Rezai-Zadeh et al. 2005, 2008). The metabolism of green tea catechins has been studied in animals and humans (Pietta et al. 1998; Li et al. 2000) and have indicated that following oral administration, catechin is absorbed, metabolized and excreted within 24 h (Harada et al. 1999). For example, green tea consumption by healthy individuals leads to the detection of EGCG, EGC, and EC in the plasma in a dose-dependent manner, varying between 0.2% and 2% of the ingested amount, with maximal concentration at 1.4–2.4 h after ingestion (Nakagawa and Miyazawa 1997). In addition, after intake

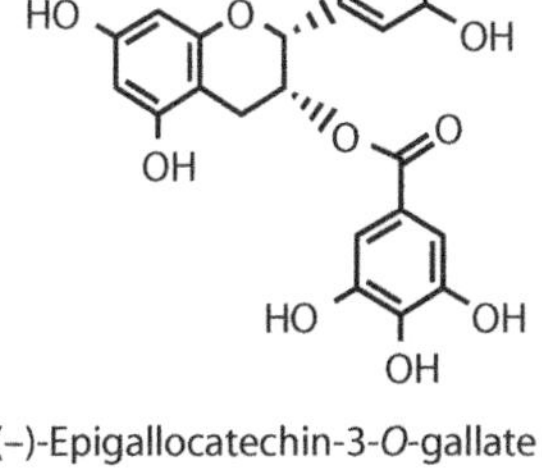

(–)-Epigallocatechin-3-*O*-gallate (EGCG)

FIGURE 15.2 Structure of the green tea flavan-3-ol, (–)-epigallocatechin-3-*O*-gallate (EGCG).

of 1.2 g of green tea solids (dissolved in two cups of warm water), plasma levels of catechins was reported as 46–268 ng/mL (Li et al. 2000). The half-life for EGCG is about 5 h and for EGC and EC between 2.4 and 3.4 h (Yang et al. 1999, 2000). Previously, it was reported that the EC metabolites, epicatechin-O-glucuronide and 3′-O-methylepicatechin-O-glucuronide, were present in the brain of rodents fed epicatechin (Abd El Mohsen et al. 2002). Furthermore, a study with labeled EGCG demonstrated a wide distribution of radioactivity in mouse organs, including brain, after oral administration and a small amount of [^{3}H]EGCG excretion in the urine after direct administration (Suganuma et al. 1998). The bioavailability of EGCG to various brain regions of rats have been demonstrated after oral and intravenous administration and have suggested that EGCG is the most brain bioavailable catechin (Lin et al. 2003, 2007; Chu et al. 2007), whereas the ability of other flavonoids/metabolites to traverse the BBB were dependent on the degree of compound lipophilicity (Youdim et al. 2003).

Even though the question of the precise brain permeability of catechin is still to be resolved, green tea ingestion does appear to inversely correlate with the incidence of brain aging, dementia, and neurodegeneration, including that observed in Alzheimer's disease (AD) and Parkinson's disease (PD) (Checkoway et al. 2002; Kuriyama et al. 2006; Ng et al. 2008; Tan et al. 2008). In accordance with this, green tea extracts and isolated green tea catechins have been tested in *in vitro and in vivo* models of neurodegenerative disease and aging. These have demonstrated their ability to protect and rescue brain neurons against various exogenous damages in a variety of cell culture and animal models of AD and PD. Many of these have been mechanistic studies and have revealed that, in addition to their known antioxidant/metal chelation activities, the catechins can modulate several signal transduction pathways important in cell survival and cell function, which contribute significantly towards overall neuronal and brain viability (see reviews (Mandel et al. 2004b; Spencer 2007; Weinreb et al. 2009a). The aims of this chapter are to compile recent clinical and preclinical studies regarding the beneficial neurotherapeutic properties of green tea and green tea catechins, with regards to neurodegeneration and aged-associated cognitive disorders, particularly AD and PD.

15.2 OXIDATIVE STRESS, IRON, AND NEURODEGENERATION

The etiology of neurodegenerative diseases is not well understood. Nonetheless, over the last two decades, significant evidence has indicated that oxidative stress (OS), increased levels of iron (Zecca et al. 2001, 2004b; Youdim and Buccafusco 2005), increased activity of monoamine oxidase (MAO)-B activity, and reduced levels of antioxidant activity in the brain may be major pathogenic factors in the progression of neurodegenerative pathology (Youdim et al. 1990; Halliwell 1992; Gotz et al. 1994; Jellinger 1999; Berg et al. 2001; Zecca et al. 2001, 2004b; Trushina and McMurray 2007). Free iron can induce OS, because of its interaction with hydrogen peroxide (H_2O_2) in the Fenton reaction, resulting in an increased formation of hydroxyl free radicals. Free radical-related OS causes damage to cellular biomolecules that lead to a critical failure of biological functions, protein modification, misfolding, and aggregation and ultimately cell death (Pratico and Delanty 2000; Halliwell 2001;

Przedborski et al. 2001; Sayre et al. 2001). Indeed, in various neurodegenerative diseases, such as AD, PD, Huntington disease, amyotrophic lateral sclerosis (ALS), and multiple sclerosis (MS), iron has been shown to accumulate at the site of brain lesions, suggesting that it may have a role in neurodegenerative pathology (Riederer et al. 1992; Bartzokis and Tishler 2000; Sayre et al. 2000a; Jenner 2003; Zecca et al. 2004b; Youdim 2008; Altamura and Muckenthaler 2009; Horowitz and Greenamyre 2010). The following sections will discuss the two most prevalent neurodegenerative diseases, AD and PD, with regards to OS and iron accumulation and therapeutic strategies capable of counteracting these using natural iron chelating compounds.

15.2.1 Alzheimer's Disease

AD is the most prevalent neurodegenerative disease in the elderly population, and it has been estimated that about 5% of adults aged above 65 years is affected by this disease (Bullock 2004). Its predominant clinical manifestation is the progressive memory deterioration and other changes in brain function, including disordered behavior and impairment in language, comprehension, and visual-spatial skills (Tsolaki et al. 2001). The neuropathology of AD is characterized by several features, including extracellular deposition of amyloid β peptide (Aβ)-containing plaques in the cerebral cortical regions, accompanied by the presence of intracellular neurofibrillary tangles and a progressive loss of basal forebrain cholinergic neurons leading to reductions in cholinergic markers, such as acetylcholine levels, choline acetyltransferase (ChAT) and muscarinic and nicotinic acetylcholine receptor binding (Selkoe and Schenk 2003; Schliebs 2005). Additionally, there is accumulating evidence that many cytotoxic signals in the AD brain can initiate apoptotic processes, including OS, inflammation, and iron accumulation (Smith et al. 2000; Rogers and Lahiri 2004; Joseph et al. 2005). Iron is significantly concentrated around amyloid senile plaques and neurofibrillary tangles (NFTs), leading to alterations in the pattern of the interaction between iron regulatory proteins (IRPs) and iron responsive elements (IREs) in the 5′ untranslated region (5′UTR) of their related transcripts, and disruption in the sequestration and storage of iron (Lovell et al. 1998; Pinero et al. 2000). In addition, it has been demonstrated that the amount of iron present in the AD neuropil is twice that found in the neuropil of nondemented brains (Pinero et al. 2000).

Further studies have suggested that iron-accumulation could be an important contributor towards oxidative damage in AD pathology and thus, the neurons in AD brains experience high oxidative load (Smith et al. 1997; Castellani et al. 2004, 2007; Honda et al. 2005; Moreira et al. 2005). Indeed, it was found that neurofibrillary tangles and senile plaques contain redox-active transition metals and may exert prooxidant/antioxidant activities, depending on the balance of neuronal antioxidants and reductants (Sayre et al. 2000b). Postmortem analyses of Alzheimer patients' brains have revealed activation of two enzymatic markers of cellular OS: heme oxygenase (HO)-1 (Takeda et al. 2000) and nicotinamide adenine dinucleotide phosphate (NADPH) oxidase (Shimohama et al. 2000). In addition, HO-1 was greatly enhanced in neurons and astrocytes of the hippocampus and cerebral cortex of Alzheimer's subjects, colocalizing to senile plaques and NFTs. A previous study reported that ribosomal RNA acts as a binding site for redox-active iron and thus serves as a redox center within the cytoplasm of vulnerable neurons in AD brain (Honda et al. 2005).

At the biochemical level, iron have been shown to facilitate the aggregation of Aβ and induce aggregation of the major constituent of NFTs, hyperphosphorylated τ (Tau) (Bush 2003; Castellani et al. 2007). It was suggested that the toxicity of Aβ is mediated, at least in part, via redox-active iron. Indeed, neuronal toxicity was significantly attenuated when Aβ was pretreated with the iron chelator DFO, while conversely, the toxicity was restored to original levels following incubation of Aβ with excess free iron (Rottkamp et al. 2001). *In vitro* studies demonstrated that Aβ has high affinity for iron and the iron binding sites are located in the hydrophilic N-terminal part of the peptide (Atwood et al. 2000). In support of this, high levels of iron have been reported in the amyloid plaques of the transgenic mice expressing mutant amyloid precursor protein (APP), Tg2576 mouse model for AD, resembling those seen in the brains of AD patients (Smith et al. 1998).

Another molecular link between iron metabolism and AD pathogenesis was provided by Rogers and coworkers who described the presence of an IRE in the 5′ untranslated region (5′UTR) of the APP transcript with sequence homology to the IREs for ferritin and transferrin (TfR) mRNA (Rogers et al. 2002a; Rogers and Lahiri 2004). Thus, APP translation is selectively responsive to cytoplasmic free iron levels (the labile iron pool) in a pattern that resembles the binding of IRPs to ferritin and TFR mRNA (Klausner et al. 1993). Recently, Duce et al. (2010) have found that APP possesses ferroxidase activity, mediated via a concerned H-ferritin-like active site, which is inhibited specifically by Zn^{2+}. It has been shown, that like ceruloplasmin, APP is capable of oxidizing Fe^{2+} to Fe^{3+} and thus decreasing iron-mediated generation of oxygen radicals. This finding, and the reported iron-export properties of APP from the cytoplasm to the surface (Duce et al. 2010), indicates an important biological activity for a protein that its role is still debatable.

Additionally, higher TfR C2 allele occurrence have been described in AD compared to normal controls, and a possible interaction between C2 alleles and a genetic risk factor for the development of AD, the apolipoprotein (APOE) epsilon 4 allele (Zambenedetti et al. 2003). These observations indicate that iron chelation has the potential to prevent iron-induced ROS, OS, and Aβ peptide aggregation and, therefore, as green tea catechins have been shown to readily chelate transition metals, may be considered a potential therapeutics for the treatment of AD.

15.2.2 Parkinson's Disease

PD is a neurological disease of the elderly, affecting 1–2% of those aged above 60 years, causing motor dysfunctions, such as bradykinesia, resting tremor, rigidity, and postural instability, but also affecting autonomic functions and cognition (Lorincz 2006; Olanow et al. 2009b). Although this disease has long been considered a nongenetic disorder of "sporadic" origin, 5–10% of patients are now known to have monogenic forms (Lesage and Brice 2009). PD is primarily characterized by the progressive, selective, and irreversible loss of dopaminergic neurons in the substantia nigra pars compacta (SNpc) (50–70%) and subsequent decrease of dopamine (DA) concentrations in the striatum (Olanow 1992; Hirsch 1994). The neuropathological hallmark of PD is the formation of eosinophilic Lewy bodies in surviving dopaminergic neurons (Jellinger 1991). The nigral pathology in PD has been shown to be associated with OS, mitochondrial dysfunction, nitric oxide toxic actions, excitotoxicity, inflammation, and defects in the ubiquitin-proteasome system (Youdim et al. 1989; Jenner 1998; Jenner and Olanow 1998; McNaught and Jenner 2001). In addition, it has been shown that iron concentrations are significantly elevated in human Parkinsonian SNpc within the melanized DA neurons (Gotz et al. 2004; Gerlach et al. 2006), as well as in the SNpc of 6-hydroxydopamine (6-OHDA) and N-methyl-4-phenyl-1,2,3,6-tertahydropyridine (MPTP) animal models of PD (Gerlach et al. 2000; Zecca et al. 2004b). In support of this, a number of antioxidants/ iron chelators have been shown to possess neuroprotective activity in animal models of PD. For example, the natural prototype iron chelator/radical scavenger, desferrioxamine (DFO) protects against dopaminergic neurodegeneration induced by 6-OHDA (Ben-Shachar et al. 1991) and MPTP (Lan and Jiang 1997a). It is also noteworthy that iron and ferritin accumulation occurs within the neurons and oligodendrocytes in distinctive regions of the brain with ubiquitin-positive inclusion bodies (Zecca et al. 2004a). Targeted deletion of the gene encoding IRP-2 caused misregulation of iron metabolism and progressive neurodegeneration in mice, as evidenced by axonal degeneration in the CNS and movement disorder characterized by ataxia, bradykinesia, and tremor (LaVaute et al. 2001; Smith et al. 2004). Additional studies have demonstrated that iron misregulation associated with the loss of IRP 2 in the Ireb2(−/−) mice affected DA metabolism in the striatum resulting from the loss of DA and DA-regulating proteins, thus supporting the view that the IRP2−/− genotype may enable neurobiological events associated with PD and aging (LaVaute et al. 2001; Salvatore et al. 2005).

As previously discussed, iron participates in the Fenton chemistry, reacting with H_2O_2 to produce the highly reactive hydroxyl radical. Formation of the hydroxyl radical combined with depletion of endogenous antioxidants, particularly tissue glutathione (GSH), leads to OS (Lan and Jiang 1997b; Han et al. 1999; Bharath et al. 2002). Indeed, in the SN of Parkinsonian brains, there is a drastic depletion of endogenous antioxidants, such as reduced GSH (Riederer et al. 1989; Jenner 1991, 1998; Bharath et al. 2002). Iron also facilitates the decomposition of lipid peroxides to produce highly cytotoxic oxygen-related free radicals, which can cause damage to DNA, lipids, proteins, and ultimately neuronal death (Zecca et al. 2004b). Previous evidence suggests that iron, together with H_2O_2 in the presence of DA, can induce the formation of the neurotoxic 6-OHDA (Jellinger et al. 1995; Linert et al. 1996; Napolitano et al. 2002). Furthermore, 6-OHDA may be also formed from L-dihydroxyphenylalanine (L-DOPA) in the presence of iron and hydrogen peroxide (Maharaj et al. 2005). Iron-dependent OS tends to be tissue specific, owing to differential neuronal cell susceptibility observed in PD (Braak and Del Tredici 2004). Additional OS stems from elevated MAO-B levels, which lead to increased DA production (Youdim and Riederer 2004), increased DA oxidation and further increases in the levels of H_2O_2, creating a vicious circle that leads to progressive levels of neuronal damage. Indeed, postmortem studies reported that MAO-B activity in the human brain is increased in aging and neurodegenerative diseases, in particular PD (Fowler et al. 1980, 2002; Cohen 2000). In addition, increasing lines of evidence suggest that the aggregation of α-synuclein, a major contributor of the neurodegenerative processes occurring in PD pathology, is also associated with iron accumulation (Duda et al. 2000; Halliwell 2001). Together, these observations indicate that compounds capable of chelating metals may have a great therapeutic potential for PD.

15.3 EPIDEMIOLOGICAL AND CLINICAL STUDIES WITH GREEN TEA AND ITS MAJOR CONSTITUENT, EGCG: PD AND AD PREVALENCE

Despite the lack of systematic clinical trials investigating the impact of green tea or catechins on in neurodegenerative disease, accumulating evidence from epidemiological and prospective studies suggest that tea consumption inversely correlates with incidence of dementia, PD, and AD. A previous epidemiological study, aimed at investigating the association between consumption of green tea and cognitive function in elderly Japanese subjects, found that higher consumption of green tea was associated with lower prevalence of cognitive impairment (Kuriyama et al. 2006). A recent cross-sectional study that assessed green tea intake in 2501 participants and 1438 cognitively intact participants aged above 2 years indicated that green tea was inversely associated with cognitive impairment, but not with cognitive decline, possibly due to the small number of green tea drinkers in this cohort (Ng et al. 2008). To add to this, more recently both black, oolong and green tea consumption were also associated with better cognitive performance in community-living Chinese older adults group (Feng et al. 2010).

Regarding clinical trials, the first multicenter, double blind, randomized, placebo-controlled, delayed study to evaluate the safety, tolerability, and efficacy of green tea polyphenols in slowing disease progression in patients with early PD was conducted by the Chinese Parkinson Study Group and supported by The Michael J. Fox Foundation. Four hundred and ten untreated people with PD were enrolled at 32 Chinese Parkinson Study Group sites, and participants were randomized to 0.4 g (102 people), 0.8 g (103 people), or 1.2 g (104 people) of green tea polyphenols daily or placebo (101 people). At 6 months, the placebo group switched to 1.2 g of green tea polyphenols daily for 6 more months. It was postulated that if green tea was capable of exerting a disease-modifying effect, then the improvement seen in the placebo group should not "catch up" with the improvement seen in the group that received treatment from the beginning of the study. To date, the findings have indicated that green tea intake induces a significant improvement in Unified Parkinson's Disease Rating Scale (UPDRS) scores at 6 months for patients in each dosage group, although these improvements were no longer significant at 12 months compared to the placebo patient's group (Chan et al. 2009). The authors concluded that although green tea polyphenols provide a symptomatic benefit in early untreated PD, no obvious disease-modifying effects were evident. Additionally, a current randomized, double-blind efficacy study, sponsored by Charite University, Berlin, Germany, is designed to determine the safety, tolerability, and potential neuroprotective effects of EGCG (given increasing doses, 200–800 mg, of formulated medication over a period of 18 months) in early stage AD patients comedicated with donepezil (estimated study completion date December 2012; U.S. FDA resources; http://clinicaltrials.gov/ct2/show/NCT00951834?term = EGCG&rank = 1).

This clinical evidence demonstrates that green tea and its major component may act as a protective strategy in neurodegenerative diseases, although data to date are relatively limited, emphasizing the need for more well-designed, controlled studies to assess the impact of green tea consumption on disease risk reduction in PD and AD.

15.4 PRECLINICAL ANIMAL AND CELL CULTURE STUDIES WITH GREEN TEA AND GREEN TEA CATECHINS

Although evidence emanating for human studies is limited at present, much animal and *in vitro* model data have been accumulated to indicate the protective effects of green tea catechin against aged-related neurodegenerative diseases. Preclinical animal studies with green tea extract polyphenol fraction or its individual catechin components have shown positive effects on cognitive and behavioral abilities during aging and in neurodegenerative conditions (see reviews: Mandel et al. 2008; Weinreb et al. 2009b). In particular, these studies have enabled investigators to unravel the molecular mechanisms behind the effects of tea catechins on brain function and cognition.

Aging studies have shown that the long-term administration (6 months) of antioxidant supplements, containing β-catechin or a green tea catechin fraction, extended the mean lifespan of senescence-accelerated mice (SAM-P8) (Kumari et al. 1997; Li et al. 2009a) and C57BL/6 J mice (Li et al. 2009b). In old Wistar rats, green tea catechins also prevented spatial learning and memory decline after 7 months of treatment, seemingly due to the reduction of oxidative stress-related damage (Assuncao et al. 2010). Studies using the main green tea catechin constituent, EGCG, have documented an attenuation of cognitive deficits and a decline in antioxidant enzymes and apoptotic parameters in D-galactose-induced mice aging (He et al. 2009) and significant longevity-extending effects in *Caenorhabditis elegans* under stress (Zhang et al. 2009). EGCG has been also shown to improve age-related cognitive decline and protect against cerebral ischemia/reperfusion injuries (Lee et al. 2000; Sutherland et al. 2005).

The positive neuroprotective effects of green tea and EGCG have also been documented in various toxic models of PD and AD. A green tea polyphenol extract, or individual EGCG, prevented striatal DA depletion and SNpc dopaminergic neurons loss when given chronically to mice or rats treated with the Parkinsonism-inducing neurotoxins, MPTP and 6-OHDA (Levites et al. 2001; Chaturvedi et al. 2006; Kim et al. 2010). These findings agree with evidence from another animal study, where EGCG given postdamage was shown to almost completely restore nigrostriatal DA neuron degeneration caused by MPTP (Reznichenko et al. 2010).

In terms of cognition-impaired models, long-term administration of a mixture of green tea catechins (polyphenol E) was demonstrated to improve spatial cognition learning ability in naïve rats and prevent cognitive deficits in intracerebral Aβ-damaged rats (Haque et al. 2006, 2008). In addition, it prevented lipopolysaccharide-mediated neuronal cell death and memory impairment of mice, possibly through inhibition of the elevation of Aβ via the inhibition of β- and γ-secretase (Lee et al. 2009c). Additionally, EGCG reduced cognitive dysfunction in Aβ-injected mice (Lee et al. 2009a) and prevented cerebral amyloidosis and impaired cognition in Alzheimer's transgenic Tg2576 mice (Rezai-Zadeh et al. 2005, 2008). These findings suggest that the green tea flavanol EGCG may itself be a useful agent in preventing Aβ-induced cognitive impairment and brain pathology.

With regards to *in vitro* findings, cell culture studies have demonstrated that flavanol catechin reduces damage induced by hydrogen peroxide, 4-hydroxynonenal, rotenone, and 6-OHDA in primary rat mesencephalic cultures (Mercer et al. 2005). In human neuroblastoma (NB) SH-SY5Y cells and rat pheochromocytoma (PC12) cells, EGCG prevented neuronal cell death caused by 6-OHDA and 1-methyl-4-phenylpyridinium (MPP+) (Levites et al. 2002a; Nie et al. 2002). Furthermore, in a Aβ toxic cell culture model of AD, a green tea extract and EGCG protected primary hippocampal neurons (Choi et al. 2001; Bastianetto et al. 2006) and rat pheochromocytoma (PC12) cells (Mandel et al. 2003), and both catechin and epicatechin were shown to protect cultured rat cortical neurons against Aβ (25–35)-induced neurotoxicity (Ban et al. 2006). Lastly, pure EGCG has been observed to rescue and reduce mortality of NB SH-SY5Y cells induced by a 3 days serum starvation (Reznichenko et al. 2005).

15.5   NEUROPROTECTIVE MECHANISM OF THE GREEN TEA CATECHIN, EGCG

Mechanistic studies aimed at investigating the neuroprotective potential of physiological relevant concentrations of green tea catechins revealed that low micromolar concentrations are responsible for the antiapoptotic/neuroprotective actions, whereas high doses account for the antiproliferative, antiangiogenic, and proapoptotic actions, relevant to cancer management (Levites et al. 2002a; Schroeter et al. 2007). The low neuroprotective doses of EGCG supplemented to naïve human NB SH-SY5Y cells promote the down-regulation of proapoptotic genes, such as mdm2, caspase-1, cyclin-dependent kinase inhibitor p21, and TNF-related apoptosis-inducing ligand, TRAIL, with no significant effect on antiapoptotic genes (Levites et al. 2002a; Weinreb et al. 2003). This biphasic mode of action was shown to be mutual to that of other antioxidants and iron chelators such as ascorbic acid or R-apomorphine (Halliwell 1996; Weinreb et al. 2003).

In addition, an *in vitro*, complementary, large-scale proteomic analysis demonstrated that EGCG induced the differential expression of three major clusters of proteins related to the cytoskeleton (e.g., beta tubulin IV and tropomyosin 3), heat shock (e.g., 14-3-3 gamma, heat shock protein gp96), and metabolic energy (e.g., ATP synthase H+ transporting, mitochondrial F1 complex beta, glucosidase II beta, and nerve vascular growth factor (VGF) inducible precursor) (Weinreb et al. 2007, 2008). By their interaction with more than 100 binding partners, 14-3-3 protein family members modulate the action of proteins that are involved in cell cycle and transcriptional control, signal transduction, intracellular trafficking, regulation of ion channels, and expression of cytoskeletal components (Berg et al. 2003). In this regard, the neuroprotective activity of EGCG, which includes neurite extension, cell body elongation, and up-regulation of the outgrowth associated protein-43 (GAP-43) (Figure 15.3) (Reznichenko et al. 2005), may be associated with the induction of 14-3-3 gamma, interacting with cytoskeletal proteins and cell signaling pathways.

Among the eight catechins found in green tea, there is some evidence that the most abundant constituent of green tea, EGCG, is also the most potent in inducing cellular effects. Hence, EGCG is the focus of many scientific studies and has been associated with most of the newly discovered green tea benefits. In view of the benefits ascribed for EGCG in the epidemiology, clinical, and preclinical studies, in the next sections, we will discuss the underlying neuroprotective mechanism of action of green tea catechins, and of particularly, EGCG (Figure 15.2).

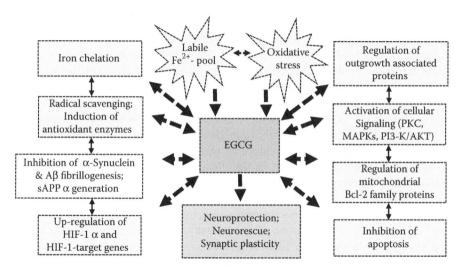

FIGURE 15.3 Simplified scheme depicting basic signaling pathways involved in the neuroprotection of green tea catechins. Green tea catechins are phenolic compounds, and as such, they act as powerful hydrogen-donating radical scavengers of oxygen and nitrogen species and possess the ability to complex transition divalent metal ions, thereby reducing the iron pool and preventing the formation of iron-induced free radicals by Fenton reaction. Chelation of iron by green tea catechins can also interfere with the iron-induced degradation/inactivation of iron regulatory proteins resulting in decrease in the generation of amyloid-beta (Aβ) peptide and α-synuclein fibrils. Green tea catechins may directly disrupt and solubilize Aβ by complexing the high affinity bound iron in the hydrophilic N-terminal part of the peptide. An additional target involving their iron-chelating activity is the inhibition of the iron and oxygen-activated prolyl-4-hydroxylases that regulate hypoxia-inducible factor (HIF)-1 stability, resulting in selective induction of cell survival target genes. Additional triggering pathways include activation of outgrowth associated proteins and regulating intracellular signaling pathways such as, protein kinase C (PKC), mitogen-activated protein kinases (MAPKs), phosphatidylinositide 3'-OH kinase (PI3-K)/Akt, and Bcl-2 family proteins, which in turn leading to inhibition of apoptotic processes and, ultimately, resulted in neuroprotection/neurorescue and neuronal synaptic plasticity. Arrows indicate flow direction.

15.5.1 Antioxidant-Iron Chelating Activities in Redox and Iron Sensitive Cellular Processes

Green tea catechins act as powerful hydrogen-donating radical scavengers of ROS and RNS and possess the ability to chelate transition divalent metal ions (Cu^{2+}, Zn^{2+}, Fe^{2+}), thereby preventing the formation of iron-induced free radicals in *in vitro* and cell/tissue systems (Salah et al. 1995; Nanjo et al. 1996). In brain tissue, green tea and black tea extracts have been shown to strongly inhibit lipid peroxidation chain reactions promoted by iron-ascorbate in homogenates of brain mitochondrial membranes (Levites et al. 2002b). A similar effect was also reported using brain synaptosomes, in which the four major polyphenol catechins of green tea were shown to inhibit iron-induced lipid peroxidation (Guo et al. 1996). In the majority of these studies, EGCG was shown to be more efficient as a radical scavenger (Figure 15.3) than its counterparts ECG, EC, and EGC, which might be attributed to the presence of the trihydroxyl group on the B ring and the gallate moiety esterified at the 3' position in the C ring (Nanjo et al. 1996).

In this respect, a one-month administration of EGCG to 24-month-old Wistar rats augmented the activities of Kreb's cycle enzymes and electron transport chain complexes in brain mitochondria, decreased the expression of hydroxynonenal in aged brain, and up-regulated cellular antioxidant systems (Srividhya et al. 2009). These results support an antioxidant potential for EGCG at the level of mitochondria, and it has been shown that EGCG directly scavenges mitochondrial generated free radicals, as it was found to accumulate in mitochondria and selectively protect rat cerebellar granule neurons from apoptosis induced by various generators of OS (Schroeder et al. 2008). This may partly account for the attenuation of depolarization of the inner mitochondrial membrane potential by a polyphenol-rich green tea extract or EGCG, following oxygen-glucose deprivation in C6 glial culture (Panickar et al. 2009).

It has also been suggested that the neuroprotective effect of green tea polyphenols *in vivo* may involve the regulation of antioxidant protective enzymes (Figure 15.3). EGCG was found to elevate the activity of two major oxygen-radical species metabolizing enzymes, superoxide dismutase (SOD) and catalase in mice striatum (Levites et al. 2001). Oral administration of green tea extracts to young rats or EGCG to aged rats exhibited higher levels of antioxidant enzymes, such as glutathione peroxidase, glutathione reductase, SOD, and catalase in whole brain homogenates (Skrzydlewska et al. 2005; Srividhya et al. 2008) and hippocampal tissue (Assuncao et al. 2010), compared to their respective controls. In addition, EGCG has been shown to inhibit MAO-B activity in C6 astrocyte cells (Mazzio et al. 1998) and adult rat brains (Lin et al. 2010). Moreover, green tea and EGCG have been shown to display *in vitro* effects against the production of inducible nitric oxide synthase (iNOS), an important ROS-generating enzyme and the toxic production of NO (Paquay et al. 2000). In accordance, recent study demonstrated inhibition of iNOS in the SN and striatum of MPTP mice model of PD (Kim et al. 2010).

In peripheral tissues, a number of flavonoids and phenolics have been shown to activate the expression of some stress-response genes, such as phase II drug metabolizing enzymes, glutathione-s-transferase and heme oxygenase-1 (HO-1) at low concentrations, as well as increasing the activity and nuclear binding of the transcription factors Nrf1 and Nrf2 to the antioxidant regulatory element (ARE) sequences contained in their promoters (Owuor and Kong 2002; Mann et al. 2007).

Flavonoids are also potent chelators of transitional metals, such as iron and copper, due to the OH at position 3 of the C ring, the OH at positions 3' and 4' of the B ring or the three OH groups present in the gallol moiety (Nanjo et al. 1996; Kumamoto et al. 2001; Park et al. 2008; Perron et al. 2008). In a recent study examining the differential potency of a series of polyphenols (e.g., EC, vanillic acid, gallic acid, quercetin, myricetin) to prevent DNA damage caused by Fe^{2+} and H_2O_2, it was found that EGCG was the most potent among the 12 phenolic compounds tested, inhibiting more than 90% of the iron-mediated DNA breaks (Perron et al. 2008). By correlating the pKa and IC_{50} values of phenolic compounds for inhibition of Fe_2^+/H_2O_2 -induced neurotoxicity, the same group suggested that the binding of the polyphenols to iron was essential for their antioxidant activity (Perron et al. 2010).

15.5.2 Aggregation of α-Synuclein and Amyloid Precursor Protein

These antioxidant and metal-complexing features of green tea catechins may be of major significance for treatment of PD and AD. The apparent link, discussed earlier, between metal dys-homeostasis and neurodegenerative diseases has given rise to the conception of metal chelation treatment as a therapeutic option. In fact, a number of iron chelators/antioxidants have been shown to possess neuroprotective activity in animal models of neurodegeneration (Kaur et al. 2003; Shachar et al. 2004; Gal et al. 2005; Mandel et al. 2007; Bolognin et al. 2009; Weinreb et al. 2010). Accordingly, the use of nontoxic, naturally occurring neuroprotective and metal-complexing flavonoids with brain access could, in theory, "pull out iron" from those brain areas where it preferentially accumulates and alleviate the brain from free-reactive iron overload by directly influencing aggregation and deposition of Aβ and α-synuclein in brains of AD and PD patients, respectively. Indeed, wine polyphenols (e.g., resveratrol, myricetin, quercetin, kaemferol), curcumin, (+)-catechin, (−)-epicatechin, nordihydroguaiaretic acid, and rosmarinic acid have been shown to inhibit fibril formation, elongation, and destabilization of the formed assemblies of two the major contributors of PD and AD pathology, α-synuclein and Aβ, respectively (Ono et al. 2003; Ono and Yamada 2006). Recently, EGCG was reported to interfere with an early step in the amyloid formation cascade by binding directly to the natively unfolded α-synuclein and Aβ polypeptides, thus inhibiting their fibrillogenesis and redirecting them into an alternative "off pathway" before they become toxic (Figure 15.3) (Ehrnhoefer et al. 2008). The same authors also demonstrated that EGCG converted large, mature α-synuclein and Aβ into nontoxic amorphous monomers or small diffusible oligomers, thus reducing their toxicity in mammalian cell culture (Bieschke et al. 2010). Partial aggregated and oligomerized intracellular Aβ has been documented to be cytotoxic and synaptotoxic in cell culture and impair memory in animal studies (Selkoe 2008). Similarly, a truncated form of α-synuclein showed increased aggregation into large inclusions bodies, increased accumulation of high molecular weight α-synuclein species, and enhanced neurotoxicity in transgenic Drosophila (Periquet et al. 2007). In addition, structure–activity relationship experiments among EGCG-related catechins suggested that the gallate moiety is crucial in the amyloid remodeling process (Bieschke et al. 2010). In support, a recent study demonstrated that gallo-containing polyphenols (e.g., EGCG) promote iron oxidation at a significantly faster rate compared with analogous catechol-containing compounds (e.g., EC) (Perron et al. 2010).

Another outcome of transition metal-complexing molecules, including EGCG, is related to their ability to reduce Aβ formation via translational suppression of its generator, APP, a type I integral membrane protein. This effect is associated with modulation of an iron-responsive element (IRE) located at APP 5′UTR-mRNA (Rogers et al. 2002a, 2002b). In mouse hippocampus, EGCG was shown to down-regulate membrane-associated APP protein level (Levites et al. 2003; Avramovich-Tirosh et al. 2007) and in SH-SY5Y cells EGCG reduced full-length APP, without altering APP mRNA levels, while exogenous iron supplementation reversed its effect (Reznichenko et al. 2006). This suggests a post-transcriptional action, presumably by the mechanism of chelating intracellular iron pools and APP mRNA post-transcriptional inhibition. This is further supported by the observation that EGCG suppressed translation of a luciferase reporter gene driven by the IRE-type II-containing sequences of APP (Reznichenko et al. 2006). Furthermore, it was found that EGCG markedly reduced secreted Aβ levels in the conditioned medium of Chinese hamster ovarian cells, overexpressing "Swedish" mutated APP (CHO/ΔNL) (Reznichenko et al. 2006) and in primary neuronal cells derived from transgenic mice bearing the APP Swedish mutation (Rezai-Zadeh et al. 2005).

15.5.3 Hypoxia Inducible Factor System

An additional level of neuroprotection by iron-complexing compounds involves the stabilization of the transcriptional activator, hypoxia-inducible factor (HIF)-1 and expression of its target protective genes, such as erythropoietin, vascular endothelial growth factor (VEGF), enolase, TfR receptor, and heme oxygenase. HIF-1, a heterodimeric transcription factor, composed of two subunits: HIF-1α, an oxygen-labile protein that becomes stabilized under hypoxic conditions and HIF-1β, which is constantly expressed (Schofield and Ratcliffe 2004). HIF is highly involved in the pathology of a number of diseases associated with tissue hypoxia (reduced oxygen tension) (Giaccia et al. 2003). Within the cells, HIF-1 is under the control of a class of iron-dependent and oxygen-sensor enzymes, HIF prolyl-4-hydroxylases (PHDs) that target the regulatory alpha subunit of HIF-1 for degradation by the proteasome (Schofield and Ratcliffe 2004). The possibility of modulating HIF-1 activity by targeting the free-labile intracellular iron pool by iron chelators and inhibitors of PHDs is gaining recognition, posing HIF-1 as a potential therapeutic target for neurodegenerative diseases (Siddiq et al. 2005, 2009; Kupershmidt et al. 2009; Lee et al. 2009b; Avramovich-Tirosh et al. 2010; Mole 2010; Weinreb et al. 2010). Similar to hypoxia, flavonoids have been demonstrated to induce intracellular accumulation of HIF-1α through activation of enzyme systems and signaling pathways, such as phosphatidylinositide 3′-OH kinase (PI3-K)/Akt and mitogen-activated protein kinase [MAPK, e.g., extracellular signal-regulated protein kinase, (ERK)], as well as by a direct complexing of iron and inactivation of PHDs (Agullo et al. 1997; Wilson and Poellinger 2002; Triantafyllou et al. 2007; Park et al. 2008). For example, the flavonoid containing *Ginkgo biloba* (Ginkgoaceae) extract, EGb 761, has been demonstrated to protect hypoxic PC12 cells and up-regulate HIF-1α protein expression through the activation of the p42/p44 MAPK pathway (Li et al. 2008).

Studies with different cell lines demonstrated that a number of flavonoids including luteolin, fisetin, and quercetin induced HIF-1α levels under normal oxygen pressure and this effect was structurally related to the capability of the molecules to efficiently bind intracellular iron (Jeon et al. 2007; Triantafyllou et al. 2007, 2008). Also, EGCG and ECG were shown to induce HIF-α protein and HIF-1 activity and increase the mRNA expression levels of its prosurvival gene targets glucose transporter-1 (GLUT-1), VEGF, and p21[waf1/cip1], whereas this effect was blocked by iron and ascorbate, indicating that these catechins may activate HIF-1 through the chelation of iron (Figure 15.3) (Zhou et al. 2004; Thomas and Kim 2005). Similar results were reported for quercetin (Park et al. 2008). Applying a neurorescue paradigm in neuronal culture, we have recently found that EGCG decreased mRNA transcript and protein levels of the beta-subunit of prolyl-4-hydroxylase and the protein levels of two molecular chaperones, which are associated with HIF-regulation/degradation, the immunoglobulin-heavy-chain binding protein, and the heat shock protein 90 (HSP90) (Weinreb et al. 2007, 2008). In support, previous finding demonstrated that EGCG directly binds and inhibits HSP90 in cell cultures (Palermo et al. 2005). Inhibition of HSP90 is considered a requirement for the rapid hypoxic stabilization of HIF-1 alpha, which otherwise might be degraded by unspecific pathway (Ibrahim et al. 2005).

An additional link between hypoxia and iron is reflected by the hypoxic-mediated positive regulation of the iron regulatory proteins, IRP1 and IRP2, and consequential transactivation of their target mRNAs, ferritin, and TfR. Interestingly, the free iron-induced proteasomal-mediated degradation of IRP2 involves also activation of a prolyl hydroxylase and is inhibited by iron chelators (Hanson et al. 1999; Hanson and Leibold 1999; Wang and Pantopoulos 2005). It is possible that a common pathway modulates IRP2 and HIF-1 protein abundance. Thus, the reduction in the chelatable iron pool by EGCG may result in the inhibition of prolyl hydroxylases and, consequently, in the concerted activation of both HIF and IRP2. As HIF-1 and IRPs coordinate the expression of a wide array of genes, involved in cellular iron homeostasis, survival, and proliferation (Sharp and Bernaudin 2004), their activation could be of major importance in neurodegenerative diseases.

15.5.4 REGULATION OF CELL SIGNALING

Emerging evidence suggests that the iron chelating and antioxidant activity of green tea catechins cannot be the exclusive mechanisms for their neuroprotective action. Rather, it appears that their ability to interact with and alter signaling pathways may significantly contribute to the cell survival effects (Williams et al. 2004; Spencer 2007). Modulation of cellular survival and signal transduction pathways has significant biological consequences that are important in understanding the various pharmacological and toxicological activities of antioxidant drugs. A number of intracellular signaling pathways have been described to play central functions in EGCG-promoted neuronal protection against a variety of extracellular insults, such as the MAPK (Xia et al. 1995; Singer et al. 1999; Weinreb et al. 2007), protein kinase C (PKC) (Levites et al. 2002a; Cordey et al. 2003; Reznichenko et al. 2005; Weinreb et al. 2008), and PI-3K-Akt pathways (Kaplan and Miller 2000; Koh et al. 2003; Mandel et al. 2003) (Figure 15.3).

The following subsections will outline the ability of green tea catechins, specifically EGCG, to activate various cell signaling pathways and in doing so express neuroprotective effects in the brain, such as up-regulation of cellular antioxidant systems and processes, inhibition of apoptosis, enhancement of mitochondrial function, alteration in APP processing and stimulation of neurite outgrowth.

15.5.4.1 MAPK- and PI-3K-Akt Signaling Systems

MAPK pathways play a critical role in the regulation of cellular processes that are affected in neurodegenerative diseases. As such, their importance as transducers of extracellular stimuli into intracellular activity, via a series of intracellular phosphorylation events is being increasingly recognized. MAPK signaling plays an important role in the attenuation of neuronal death and cellular injury induced by OS (Satoh et al. 2000). Previous *in vitro* studies have demonstrated the potency of EGCG to induce ARE-dependent stress genes, such as phase II drug metabolizing enzymes, glutathione-S-transferase, and heme oxygenase through the activation of the MAPK members, p44/42 ERK 1/2, JNK, and p38 (Chen et al. 2000; Owuor and Kong 2002). Additionally, EGCG has been shown to counteract the decline in ERK1/2 induced by 6-OHDA (Levites et al. 2002a) and to induce phosphorylation of ERK1/2 in serum-deprived SH-SY5Y cells (Weinreb et al. 2007). A previous *in vitro* study indicated that EGCG promoted cell survival by regulating the level and phosphorylation state of the mitochondrial Bcl-2 family proteins through actions on the ERK and Akt signaling pathways (Chung et al. 2003). In rat cortical neurons, epicatechin has been shown to induce the phosphorylation of PI3-K, ERK1/2 and their downstream substrate, cAMP responsive element binding protein (CREB), and subsequent CREB-regulated gene expression, in a concentration-dependent manner (Schroeter et al. 2007). CREB is a transcription factor that binds to the promoter regions of many genes associated with memory and synaptic plasticity, playing a crucial role in long-term potentiation (LTP) and long-term memory formation (Impey et al. 2004; Barco et al. 2006). In agreement with the *in vitro* findings, chronic feeding of two different strains of aged mice with green tea catechins led to an improvement in learning tasks and memory deficits, associated with increased levels of CREB phosphorylation in the hippocampus and increased expressions of brain-derived neurotrophic factor (BDNF) and Bcl-2, two target genes of CREB, having regulatory roles in synaptic plasticity (Li et al. 2009a, 2009b). Similar gene alterations were recently reported in hippocampus of old rats after prolonged administrations of green tea infusion (Assuncao et al. 2010).

15.5.4.2 PKC Signaling Pathway

The neuroprotective activity of EGCG has also been shown to involve the PKC intracellular signaling pathway (Levites et al. 2003; Reznichenko et al. 2005). This pathway plays an essential role in the regulation of cell survival and programmed cell death (Dempsey et al. 2000; Maher 2001; Levites et al. 2002a; Kalfon et al. 2007), and a rapid reduction in neuronal PKC activity is a common consequence of neurodegeneration (Cardell and Wieloch 1993; Busto et al. 1994). Conversely, the induction of PKC activity in neurons by EGCG is thought to be a prerequisite for neuroprotection against several neurotoxins, such as Aβ (Levites et al. 2003), serum withdrawal (Mandel et al. 2003; Reznichenko et al. 2005; Weinreb et al. 2008), and 6-OHDA (Levites et al. 2002a). This is supported by the fact that pharmacological inhibition of PKC activity completely abolishes the protection induced by EGCG (Reznichenko et al. 2005). A more direct demonstration of a possible interaction between green tea catechins and PKC has been provided by the study of Kumazawa et al. (2004), employing solid-state nuclear magnetic resonance and showing that EGCG interacts with the head group region of the phospholipids within lipid bilayers from liposomes and moves freely on the membrane surface (Kajiya et al. 2008). The interaction pattern of EGCG in terms of rotational motion within the lipid bilayers was similar to that described for the phorbol ester 12-O-tetradecanoylphorbol-13-acetate (TPA), a prototype activator of PKC (Saito et al. 1984).

The mechanism by which PKC activation by EGCG results in neuroprotection is the subject of much investigation. The neurorestorative effect of EGCG was demonstrated to involve reduction of the apoptotic markers: cleaved caspase 3, its downstream cleaved substrate poly-ADP-ribose-polymerase (PARP), a nuclear zinc finger DNA-binding protein that detects and binds to DNA strand breaks, and Bad, a member of a group of "BH3 domain only" proteins of the Bcl-2 family (Reznichenko et al. 2005) (Figure 5.3). These events were accompanied by a rapid translocation of the isoform PKCα (particularly important in neuronal growth and differentiation in the brain), to the neuronal membrane compartment (Kim et al. 2004; Reznichenko et al. 2005), an up-regulation of PKCϵ mRNA expression and a concentration-dependent activation of PKCϵ in serum-deprived in SH-SY5Y cells (Weinreb et al. 2008). In support of this, under PKC pathway blockade, EGCG could not overcome the neuronal death, suggesting that this cascade is essential for the neuroprotective and neurorescue effects of EGCG. Recently, we have identified a novel PKC-linked pathway in the neuroprotective mechanism of action of EGCG, which involves a rapid PKC-mediated degradation of the mitochondrial cell death regulator, Bad protein by the ubiquitin proteasome system in NB SH-SY5Y cells (Kalfon et al. 2007). Thus, the newly described role of Bad during the initial response to EGCG-induced cell signaling, may potentially contribute to the elucidation of the EGCG mechanism of neuroprotective action. These findings are supported by animal studies showing that two weeks oral consumption of EGCG prevented the extensive depletion of PKCα and counteracted the robust increase of Bax protein in the striatum and SNPC of mice intoxicated with MPTP (Mandel et al. 2004a).

Established studies of our group have demonstrated that either short- or long-term incubation with EGCG promotes the generation of the nonamyloigogenic, sAPPα, via PKC-dependent activation of α-secretase (Levites et al. 2003; Reznichenko et al. 2006). APP can be processed via alternative pathways: a nonamyloidogenic secretory pathway that includes cleavage of APP to sAPPα by a putative α-secretase within the sequence of the amyloidogenic Aβ peptide, thus precluding the formation of Aβ. The second pathway generates Aβ via the sequential action of β- and γ-secretases (Selkoe 1991). In contrast to Aβ, sAPPα possesses neuroprotective activities against excitotoxic and oxidative insults in various cellular models (Mattson et al. 1997) and promotes neurite outgrowth (Small et al. 1994), synaptogenesis and trophic effects on cerebral neurons in culture (Morimoto et al. 1998). Since sAPPα and Aβ are formed by two mutually exclusive mechanisms, it can be assumed that stimulation of the secretory processing of sAPPα might prevent the formation of the amyloido-genic Aβ. New supportive data emerged from a study conducted in an Alzheimer's transgenic mouse model, showing that EGCG promotes sAPPα generation through activation of α-secretase cleavage (Rezai-Zadeh et al. 2005; Obregon et al. 2006) (Figure 15.3). This was accompanied by a significant reduction in cerebral Aβ levels and β-amyloid plaques. These results were supported *in vitro* by the observation that exogenously added EGCG to primary neurons derived from the above transgenic mice bearing the APP Swedish mutation, markedly reduced secreted Aβ levels in the conditioned medium (Rezai-Zadeh et al. 2005).

Other potential beneficial effects of PKC activation in AD are related to the recent finding showing that overexpression of neuronal PKCε in transgenic mice expressing familial AD-mutant forms of the human APP, decreases Aβ levels and plaque burden and this is accompanied by increased activity of endothelin-converting enzyme (ECE), which degrades Aβ (Choi et al. 2006). Thus, it can be hypothesized that since EGCG has been shown to increase the levels of PKC isoforms α and ε in mice hippocampus and striatum (Levites et al. 2003; Mandel et al. 2004a), EGCG may reduce Aβ levels in AD pathology, both via concomitant stimulation of sAPPα secretion and promotion of Aβ clearance through increased ECE activity. However, although there is a general support for the amyloid hypothesis as a key contributor to neuronal death and dementia in AD, a direct connection between Aβ deposits in senile plaques and neurodegeneration has not been yet established (Robakis 2010).

PKC-dependent pathway activation was also implicated in the regulatory effect of EGCG on DA presynaptic transporters (DAT) internalization (Li et al. 2006). EGCG caused a dose-dependent inhibition of dopamine uptake in DAT-PC12 and decreased membrane-bound DAT, while this effect was abolished by inhibition of PKC. This observation, together with the finding that EGCG inhibited catechol-O-methyltrans-ferase (COMT) activity in rat liver cytosol homogenates (Lu et al. 2003), may be of particular significance for PD patients given that DA and related catecholamine's are physiological substrates of COMT. COMT inhibitors in the clinics dose-dependently inhibit the formation of the major metabolite of levodopa, 3-O-methyldopa, thereby improving its bioavailability in the brain (Deleu et al. 2002).

15.6 CONCLUSIONS AND FUTURE PERSPECTIVES

There are two main aspects that are significantly increasing the view that green tea consumption may enhance brain health. First, the factors and the events that influence the incidence and progression of PD and AD are becoming better defined and understood, and second, experimental data documenting the neuroprotective properties of green tea catechins, both in cell culture and animal studies, are increasing. It is also evident that disorders, such as AD and PD, will require multiple drug therapies to address the varied pathological aspects of the disease (Weinreb et al. 2009c). Therefore, the polypharmacological activities of green tea catechins may be of significance for neuroprotection. Although initially viewed as mere anti-inflammatory and antioxidant compounds, EGCG is now considered a multimodal acting molecule, invoking various cellular neuroprotection and neurorescue mechanisms, including iron-chelation, scavenging of oxygen and nitrogen radical species and activation of PKC signaling and prosurvival genes. Their nontoxic lipophilic nature, and thus presumably brain permeability, is advocated for the "removal of iron" from specific brain areas, where it preferentially accumulates in neurodegenerative diseases. The chelation of reactive free-iron by EGCG and the consequent reduction in APP translation could contribute to decreased $A\beta$ generation and fibrillization, which together with the promotion of the nonamyloidogenic pathway and induction of neurite outgrowth may converge in a slowdown in the process of neuronal loss in AD.

An additional novel approach entails drug combination, providing a practical way to design specific poly-pharmacology. The complex symptomatology of neurodegenerative diseases often necessitates the use of more than one multifunctional drug. Over the years, it has become evident that some combinations do induce a favorable clinical response, not achieved by each of the drugs given alone (Keith et al. 2005). Currently, the choice of drug combinations is based on a trial and error paradigm guided by clinical responses. Understanding the biological principles, by which the combined treatments act, would enable a more "rational" selection of drugs and provide insights into the pathological mechanisms of the neurodegenerative disorders. Indeed, a recent regimen study described a combined treatment of memantine, the first in a novel class of AD medications, and a tea polyphenol, in excitotoxic mouse brain injury (Chen et al. 2008). The findings demonstrated significant neuroprotective effects of the combined treatment, compared with memantine and the tea polyphenol alone, including a reduction in increased synaptosomal ROS and calcium concentration and an attenuation of decreased anion channel ATPase activity and mitochondrial potential, which was accompanied with an improvement in locomotor activity (Chen et al. 2008).

Following this therapeutic strategy rationale, we have recently selected the anti-PD drug, rasagiline (Azilect®), a second generation, irreversible selective inhibitor of MAO-B (Olanow et al. 2009b) and EGCG for an *in vivo* preclinical neurorescue/neurorestorative drug cocktail study, based on preclinical, epidemiological, and human clinical trials, suggestive of a neuroprotective action (Reznichenko et al. 2010). The underlying principle was that the distinct neuroprotective–neurorescue pharmacological profile of rasagiline would complement that of EGCG, thereby restoring the neuronal loss and molecular target damage observed in an animal model of PD. Our results revealed that subliminal doses of rasagiline and EGCG, which individually have no profound protective effect, synergistically restored DA neurons in the SNpc and replenished striatal DA in MPTP-lesioned mice (Reznichenko et al. 2010). A detailed analysis revealed a complementary action of these drugs, differentially acting at MPTP-injured molecules/targets in the SN: induction of BDNF by rasagiline, increased membrane levels of PKCα isoform by EGCG. and a synergistic replenishment of their downstream effecter, the serine/threonine kinase Akt/PKB, suggesting that this kinase might represent one point of convergence of the distinct mechanisms of action of the drug cocktail.

The current vision is to translate these preclinical findings with green tea catechins into the clinic. Thus, future efforts must concentrate on deciphering the cell targets affected by these natural compounds in the context of neurodegenerative diseases pathology for a deeper understanding of their mechanism of action and potential in polypharmacy paradigm.

ACKNOWLEDGMENT

We thank the Rappaport Family Research, Technion-Israel Institute of Technology for their support.

REFERENCES

Abd El Mohsen, M.M., Kuhnle, G., Rechner, A.R. et al. (2002). Uptake and metabolism of epicatechin and its access to the brain after oral ingestion. *Free Radic. Biol. Med.*, **33**, 1693–1702.

Agullo, G., Gamet-Payrastre, L., Manenti, S. et al. (1997). Relationship between flavonoid structure and inhibition of phosphatidylinositol 3-kinase: A comparison with tyrosine kinase and protein kinase C inhibition. *Biochem. Pharmacol.*, **53**, 1649–1657.

Altamura, S., and Muckenthaler, M.U. (2009). Iron toxicity in diseases of aging: Alzheimer's disease, Parkinson's disease and atherosclerosis. *J. Alzheimers Dis.*, **16**, 879–895.

Assuncao, M., Santos-Marques, M.J., Carvalho, F. et al. (2010). Green tea averts age-dependent decline of hippocampal signaling systems related to antioxidant defenses and survival. *Free Radic. Biol. Med.*, **48**, 831–838.

Atwood, C.S., Scarpa, R.C., Huang, X. et al. (2000). Characterization of copper interactions with Alzheimer amyloid beta peptides. identification of an attomolar-affinity copper binding site on amyloid beta1-42. *J. Neurochem.*, **75**, 1219–1233.

Avramovich-Tirosh, Y., Bar-Am, O., Amit, T. et al. (2010). Up-regulation of hypoxia-inducible factor (HIF)-1α and HIF-target genes in cortical neurons by the novel multifunctional iron chelator anti-Alzheimer drug, M30. *Curr. Alzheimer Res.*, **7**, 300–306.

Avramovich-Tirosh, Y., Reznichenko, L., Mit, T. et al. (2007). Neurorescue activity, APP regulation and amyloid-beta peptide reduction by novel multi-functional brain permeable iron-chelating-antioxidants, M-30 and green tea polyphenol, EGCG. *Curr. Alzheimer Res.*, **4**, 403–411.

Ban, J.Y., Jeon, S.Y., Bae, K. et al. (2006). Catechin and epicatechin from *Smilacis chinae* rhizome protect cultured rat cortical neurons against amyloid beta protein$_{(25-35)}$-induced neurotoxicity through inhibition of cytosolic calcium elevation. *Life Sci.*, **79**, 2251–2259.

Barco, A., Bailey, C.H., and Kandel, E.R. (2006). Common molecular mechanisms in explicit and implicit memory. *J. Neurochem.*, **97**, 1520–1533.

Bartzokis, G., and Tishler, T.A. (2000). MRI evaluation of basal ganglia ferritin iron and neurotoxicity in Alzheimer's and Huntingon's disease. *Cell. Mol. Biol.* **46**, 821–833.

Bastianetto, S., Yao, Z.X., Papadopoulos, V. et al. (2006). Neuroprotective effects of green and black teas and their catechin gallate esters against β-amyloid-induced toxicity. *Eur. J. Neurosci.*, **23**, 55–64.

Ben-Shachar, D., Eshel, G., Finberg, J.P. et al. (1991). The iron chelator desferrioxamine (Desferal) retards 6-hydroxydopamine-induced degeneration of nigrostriatal dopamine neurons. *J. Neurochem.,* **56**, 1441–1444.

Berg, D., Gerlach, M., Youdim, M.B.H. et al. (2001). Brain iron pathways and their relevance to Parkinson's disease. *J. Neurochem.,* **79**, 225–236.

Berg, D., Holzmann, C., and Riess, O. (2003). 14-3-3 proteins in the nervous system. *Nat. Rev. Neurosci.,* **4**, 752–762.

Bharath, S., Hsu, M., Kaur, D. et al. (2002). Glutathione, iron and Parkinson's disease. *Biochem. Pharmacol.,* **64**, 1037–1048.

Bieschke, J., Russ, J., Friedrich, R.P., et al. (2010). EGCG remodels mature alpha-synuclein and amyloid-β fibrils and reduces cellular toxicity. *Proc. Natl. Acad. Sci. U.S.A.,* **107**, 7710–7715.

Bolognin, S., Drago, D., Messori, L. et al. (2009). Chelation therapy for neurodegenerative diseases. *Med. Res. Rev.,* **29**, 547–570.

Braak, H., and Del Tredici, K. (2004). Poor and protracted myelination as a contributory factor to neurodegenerative disorders. *Neurobiol. Aging,* **25**, 19–23.

Bullock, R. (2004). Future directions in the treatment of Alzheimer's disease. *Expert Opin. Invest. Drugs,* **13**, 303–314.

Bush, A.I. (2003). The metallobiology of Alzheimer's disease. *Trends Neurosci.,* **26**, 207–214.

Busto, R., Globus, M.Y., Neary, J.T. et al. (1994). Regional alterations of protein kinase C activity following transient cerebral ischemia: Effects of intraischemic brain temperature modulation. *J. Neurochem.,* **63**, 1095–1103.

Butterfield, D., Castegna, A., Pocernich, C. et al. (2002). Nutritional approaches to combat oxidative stress in Alzheimer's disease. *J. Nutr. Biochem.,* **13**, 444.

Cabrera, C., Artacho, R., and Gimenez, R. (2006). Beneficial effects of green tea--a review. *J. Am. Coll. Nutr.,* **25**, 79–99.

Cardell, M., and Wieloch, T. (1993). Time course of the translocation and inhibition of protein kinase C during complete cerebral ischemia in the rat. *J. Neurochem.,* **61**, 1308–1314.

Castellani, R.J., Honda, K., Zhu, X. et al. (2004). Contribution of redox-active iron and copper to oxidative damage in Alzheimer disease. *Ageing Res. Rev.,* **3**, 319–326.

Castellani, R.J., Moreira, P.I., Liu, G. et al. (2007). Iron: The Redox-active center of oxidative stress in Alzheimer disease. *Neurochem. Res.,* **32**, 1640–1645.

Chan, P., Qin, Z., Zheng, Z. et al. (2009). A randomized, double-blind, placebo controlled, delayed start study to assess safety, tolerability and efficacy of green tea polyphenols in Parkinson's disease. *Proceedings of the XVIII WFN World Congress on Parkinson's Disease and Related Disorders,* Miami Beach, FL, USA, vol. 15, p. S145.

Chaturvedi, R.K., Shukla, S., Seth, K. et al. (2006). Neuroprotective and neurorescue effect of black tea extract in 6-hydroxydopamine-lesioned rat model of Parkinson's disease. *Neurobiol. Dis.,* **22**, 421–434.

Checkoway, H., Powers, K., Smith-Weller, T. et al. (2002). Parkinson's disease risks associated with cigarette smoking, alcohol consumption, and caffeine intake. *Am. J. Epidemiol.,* **155**, 732–738.

Chen, C., Yu, R., Owuor, E.D. et al. (2000). Activation of antioxidant-response element (ARE), mitogen-activated protein kinases (MAPKs) and caspases by major green tea polyphenol components during cell survival and death. *Arch. Pharm. Res.,* **23**, 605–612.

Chen, C.M., Lin, J.K., Liu, S.H. et al. (2008). Novel regimen through combination of memantine and tea polyphenol for neuroprotection against brain excitotoxicity. *J. Neurosci. Res.,* **86**, 2696–2704.

Choi, D.S., Wang, D., Yu, G.Q. et al. (2006). PKCepsilon increases endothelin converting enzyme activity and reduces amyloid plaque pathology in transgenic mice. *Proc. Natl. Acad. Sci. U.S.A.,* **103**, 8215–8220.

Choi, Y.T., Jung, C.H., Lee, S.R., et al. (2001). The green tea polyphenol (–)-epigallocatechin gallate attenuates β-amyloid-induced neurotoxicity in cultured hippocampal neurons. *Life Sci.,* **70**, 603–614.

Chu, K.O., Wang, C. C., Chu, C. Y. et al. (2007). Uptake and distribution of catechins in fetal organs following in utero exposure in rats. *Human Reprod.,* **22**, 280–287.

Chung, J.H., Han, J.H., Hwang, E.J. et al. (2003). Dual mechanisms of green tea extract-induced cell survival in human epidermal keratinocytes. *FASEB J.,* **17**, 1913–1915.

Cohen, G. (2000). Oxidative stress, mitochondrial respiration, and Parkinson's disease. *Ann. N.Y. Acad. Sci.,* **899**, 112–120.

Cordey, M., Gundimeda, U., Gopalakrishna, R. et al. (2003). Estrogen activates protein kinase C in neurons: Role in neuroprotection. *J. Neurochem.,* **84**, 1340–1348.

Deleu, D., Northway, M.G., and Hanssens, Y. (2002). Clinical pharmacokinetic and pharmacodynamic properties of drugs used in the treatment of Parkinson's disease. *Clin. Pharmacokinet.,* **41**, 261–309.

Dempsey, E.C., Newton, A.C., Mochly-Rosen, D. et al. (2000). Protein kinase C isozymes and the regulation of diverse cell responses. *Am. J. Physiol. Lung Cel.l Mol. Physiol.,* **279**, L429–L38.

Duce, J.A., Tsatsanis, A., Cater, M.A. et al. (2010). Iron-export ferroxidase activity of beta-amyloid precursor protein is inhibited by zinc in Alzheimer's disease. *Cell,* **142**, 857–867.

Duda, J.E., Lee, V.M., and Trojanowski, J.Q. (2000). Neuropathology of synuclein aggregates. *J. Neurosci. Res.,* **61**, 121–127.

Ehrnhoefer, D.E., Bieschke, J., Boeddrich, A. et al. (2008). EGCG redirects amyloidogenic polypeptides into unstructured, off-pathway oligomers. *Nat. Struct. Mol. Biol.,* **15**, 558–566.

Feng, L., Gwee, X., Kua, E.H. et al. (2010). Cognitive function and tea consumption in community dwelling older Chinese in Singapore. *J. Nutr. Health Aging* **14**, 433–438.

Fowler, C. J., Wiberg, A., Oreland, L. et al. (1980). The effect of age on the activity and molecular properties of human brain monoamine oxidase. *J. Neural. Trans.,* **49**, 1–20.

Fowler, J.S., Logan, J., Wang, G.J. et al. (2002). PET imaging of monoamine oxidase B in peripheral organs in humans. *J. Nucl. Med.,* **43**, 1331–1338.

Gal, S., Zheng, H., Fridkin, M. et al. (2005). Novel multifunctional neuroprotective iron chelator-monoamine oxidase inhibitor drugs for neurodegenerative diseases. In vivo selective brain monoamine oxidase inhibition and prevention of MPTP-induced striatal dopamine depletion. *J. Neurochem.,* **95**, 79–88.

Gerlach, M., Double, K.L., Youdim, M.B H. et al. (2000). Strategies for the protection of dopaminergic neurons against neurotoxicty. *Neurotoxic. Res. Suppl.* **2**, 99–114.

Gerlach, M., Double, K.L., Youdim, M.B. et al. (2006). Potential sources of increased iron in the substantia nigra of parkinsonian patients. *J. Neural. Trans. Suppl.* **70**, 133–142.

Giaccia, A., Siim, B.G., and Johnson, R.S. (2003). HIF-1 as a target for drug development. *Nat. Rev. Drug. Discov.,* **2**, 803–811.

Gotz, M.E., Double, K., Gerlach, M. et al. (2004). The relevance of iron in the pathogenesis of Parkinson's disease. *Ann. NY. Acad. Sci.,* **1012**, 193–208.

Gotz, M.E., Kunig, G., Riederer, P. et al. (1994). Oxidative stress: Free radical production in neural degeneration. *Pharmacol. Ther.,* **63**, 37–122.

Graham, H.N. (1992). Green tea composition, consumption, and polyphenol chemistry. *Prev. Med.,* **21**, 334–350.

Guo, Q., Zhao, B., Li, M. et al. (1996). Studies on protective mechanisms of four components of green tea polyphenols against lipid peroxidation in synaptosomes. *Biochim. Biophys. Acta.,* **1304**, 210–222.

Guo, Q., Zhao, B., Shen, S. et al. (1999). ESR study on the structure-antioxidant activity relationship of tea catechins and their epimers. *Biochim. Biophys. Acta.,* **1427**, 13–23.

Halliwell, B. (1992). Reactive oxygen species and the central nervous system. *J. Neurochem.*, **59**, 1609–1623.

Halliwell, B. (1996). Vitamin C: Antioxidant or pro-oxidant *in vivo*?. *Free Radic. Res.*, **25**, 439–454.

Halliwell, B. (2001). Role of free radicals in the neurodegenerative diseases: Therapeutic implications for antioxidant treatment. *Drugs Aging*, **18**, 685–716.

Han, J., Cheng, F. C., Yang, Z. et al. (1999). Inhibitors of mitochondrial respiration, iron (II), and hydroxyl radical evoke release and extracellular hydrolysis of glutathione in rat striatum and substantia nigra: Potential implications to Parkinson's disease. *J. Neurochem.*, **73**, 1683–1695.

Hanson, E.S., Foot, L.M., and Leibold, E.A. (1999). Hypoxia post-translationally activates iron-regulatory protein 2. *J. Biol. Chem.*, **274**, 5047–5052.

Hanson, E.S., and Leibold, E.A. (1999). Regulation of the iron regulatory proteins by reactive nitrogen and oxygen species. *Gene Expr.*, **7**, 367–376.

Haque, A.M., Hashimoto, M., Katakura, M. et al. (2006). Long-term administration of green tea catechins improves spatial cognition learning ability in rats. *J. Nutr.*, **136**, 1043–1047.

Haque, A.M., Hashimoto, M., Katakura, M. et al. (2008). Green tea catechins prevent cognitive deficits caused by Aβ(1-40) in rats. *J. Nutr. Biochem.*, **19**, 619–626.

Harada, M., Kan, Y., Naoki, H. et al. (1999). Identification of the major antioxidative metabolites in biological fluids of the rat with ingested (+)-catechin and (−)-epicatechin. *Biosci. Biotechnol. Biochem*, **63**, 973–977.

He, M., Zhao, L., Wei, M.J. et al. (2009).. Neuroprotective effects of (−)-epigallocatechin-3-gallate on aging mice induced by D-galactose. *Biol. Pharm. Bull.*, **32**, 55–60.

Higdon, J.V., and Frei, B. (2003). Tea catechins and polyphenols: Health effects, metabolism, and antioxidant functions. *Crit. Rev. Food Sci. Nutr.*, **43**, 89–143.

Hirsch, E.C. (1994). Biochemistry of Parkinson's disease with special reference to the dopaminergic systems. *Mol. Neurobiol.*, **9**, 135–142.

Hollman, P.C., Feskens, E.J., and Katan, M. B. (1999). Tea flavonols in cardiovascular disease and cancer epidemiology. *Proc. Soc. Exp. Biol. Med.*, **220**, 198–202.

Honda, K., Smith, M. A., Zhu, X. et al. (2005). Ribosomal RNA in Alzheimer disease is oxidized by bound redox-active iron. *J. Biol. Chem.*, **280**, 20978–20986.

Horowitz, M.P., and Greenamyre, J. T. (2010). Mitochondrial iron metabolism and its role in neurodegeneration. *J. Alzheimers Dis.*, **20** (Suppl 2), S551–S568.

Ibrahim, N.O., Hahn, T., Franke, C. et al. (2005). Induction of the hypoxia-inducible factor system by low levels of heat shock protein 90 inhibitors. *Cancer Res.*, **65**, 11094–11100.

Impey, S., McCorkle, S.R., Cha-Molstad, H. et al. (2004). Defining the CREB regulon: A genome-wide analysis of transcription factor regulatory regions. *Cell* **119**, 1041–1054.

Jellinger, K.A. (1991). Post mortem studies in Parkinson's disease-is it possible to detect brain areas for specific symptoms? *J. Neural Trans. Suppl.*, **56**, 1–29.

Jellinger K.A. (1999). The role of iron in neurodegeneration: Prospects for pharmacotherapy of Parkinson's disease. *Drugs Aging*, **14**, 115–140.

Jellinger, K., Linert, L., Kienzl, E. et al. (1995). Chemical evidence for 6-hydroxydopamine to be an endogenous toxic factor in the pathogenesis of Parkinson's disease. *J. Neural Trans. Suppl.*, **46**, 297–314.

Jenner, P. (1991). Oxidative stress as a cause of Parkinson's disease. *Acta Neurol. Scand. Suppl.*, **136**, 6–15.

Jenner P. (1998). Oxidative mechanisms in nigral cell death in Parkinson's disease. *Mov. Disord.*, **13**, 24–34.

Jenner P. (2003). Oxidative stress in Parkinson's disease. *Ann. Neurol.* **53** (Suppl 3), S26–S36; discussion S36–S38.

Jenner, P., and Olanow, C.W. (1998). Understanding cell death in Parkinson's disease. *Ann. Neurol.*, **44**, S72–S84.

Jeon, H., Kim, H., Choi, D. et al. (2007). Quercetin activates an angiogenic pathway, hypoxia inducible factor (HIF)-1-vascular endothelial growth factor, by inhibiting HIF-prolyl hydroxylase: A structural analysis of quercetin for inhibiting HIF-prolyl hydroxylase. *Mol. Pharmacol.*, **71**, 1676–1684.

Joseph, J.A., Shukitt-Hale, B., Casadesus, G. et al. (2005). Oxidative stress and inflammation in brain aging: Nutritional considerations. *Neurochem. Res.*, **30**, 927–935.

Kajiya, K., Kumazawa, S., Naito, A. et al. (2008). Solid-state NMR analysis of the orientation and dynamics of epigallocatechin gallate, a green tea polyphenol, incorporated into lipid bilayers. *Magn. Reson. Chem.*, 46, 174–177.

Kalfon, L., Youdim, M.B., and Mandel, S.A. (2007). Green tea polyphenol (−)-epigallocatechin-3-gallate promotes the rapid protein kinase C- and proteasome-mediated degradation of Bad: Implications for neuroprotection. *J. Neurochem.*, **100**, 992–1002.

Kaplan, D.R., and Miller, F.D. (2000). Neurotrophin signal transduction in the nervous system. *Curr. Opin. Neurobiol.*, **10**, 381–391.

Kaur, D., Yantiri, F., Rajagopalan, S. et al. (2003). Genetic or pharmacological iron chelation prevents MPTP-induced neurotoxicity in vivo: A novel therapy for Parkinson's disease. *Neuron*, **37**, 899–909.

Keith, C.T., Borisy, A.A., and Stockwell, B. R. (2005). Multicomponent therapeutics for networked systems. *Nat. Rev. Drug Discov.*, **4**, 71–78.

Khokhar, S., and Magnusdottir, S.G. (2002). Total phenol, catechin, and caffeine contents of teas commonly consumed in the United Kingdom. *J. Agric. Food Chem.*, **50**, 565–570.

Kim, J.S., Kim, J.M., O, J.J., et al. (2010). Inhibition of inducible nitric oxide synthase expression and cell death by (−)-epigallocatechin-3-gallate, a green tea catechin, in the 1-methyl-4-phenyl-1,2,3,6-tetrahydropyridine mouse model of Parkinson's disease. *J. Clin. Neurosci.*, **17**, 1165–1168.

Kim, S.Y., Ahn, B.H., Kim, J. et al. (2004). Phospholipase C, protein kinase C, Ca^{2+}/calmodulin-dependent protein kinase II, and redox state are involved in epigallocatechin gallate-induced phospholipase D activation in human astroglioma cells. *Eur. J. Biochem.*, **271**, 3470–3480.

Klausner, R.D., Rouault, T.A., and Harford, J.B. (1993). Regulating the fate of mRNA: The control of cellular iron metabolism. *Cell*, **72**, 19–28.

Koh, S.H., Kim, S.H., Kwon, H. et al. (2003). Epigallocatechin gallate protects nerve growth factor differentiated PC12 cells from oxidative-radical-stress-induced apoptosis through its effect on phosphoinositide 3-kinase/Akt and glycogen synthase kinase-3. *Mol. Brain Res.*, **118**, 72–81.

Kumamoto, M., Sonda, T., Nagayama, K. et al. (2001). Effects of pH and metal ions on antioxidative activities of catechins. *Biosci. Biotechnol. Biochem.*, **65**, 126–132.

Kumari, M.V., Yoneda, T., and Hiramatsu, M. (1997). Effect of "beta CATECHIN" on the life span of senescence accelerated mice (SAM-P8 strain). *Biochem. Mol. Biol. Int.*, **41**, 1005–1011.

Kumazawa, S., Kajiya, K., Naito, A. et al. (2004). Direct evidence of interaction of a green tea polyphenol, epigallocatechin gallate, with lipid bilayers by solid-state nuclear magnetic resonance. *Biosci. Biotechnol. Biochem.*, **68**, 1743–1747.

Kupershmidt, L., Weinreb, O., Amit, T. et al. (2009). Neuroprotective and neuritogenic activities of novel multimodal iron-chelating drugs in motor-neuron-like NSC-34 cells and transgenic mouse model of amyotrophic lateral sclerosis. *FASEB J.*, **23**, 3766–3779.

Kuriyama, S., Hozawa, A., Ohmori, K. et al. (2006). Green tea consumption and cognitive function: A cross-sectional study from the Tsurugaya Project 1. *Am. J. Clin. Nutr.*, **83**, 355–361.

Lan, J., and Jiang, D.H. (1997a). Desferrioxamine and vitamin E protect against iron and MPTP-induced neurodegeneration in mice. *J. Neural Trans.*, **104**, 469–481.

Lan, J., and Jiang, D.H. (1997b). Excessive iron accumulation in the brain: A possible poten-tial risk of neurodegeneration in Parkinson's disease. *J. Neural. Transm.,* **104**, 649–660.

Lau, F.C., Shukitt-Hale, B., and Joseph, J.A. (2005). The beneficial effects of fruit polyphenols on brain aging. *Neurobiol. Aging,* **26** (Suppl 1), 128–132.

LaVaute, T., Smith, S., Cooperman, S. et al. (2001). Targeted deletion of the gene encoding iron regulatory protein-2 causes misregulation of iron metabolism and neurodegenera-tive disease in mice. *Nature Genet.,* **27**, 209–214.

Lee, D.W., Rajagopalan, S., Siddiq, A. et al. (2009b). Inhibition of prolyl hydroxylase pro-tects against MPTP-induced neurotoxicity: Model for the potential involvement of the hypoxia-inducible factor pathway in Parkinson's disease. *J. Biol. Chem.,* **284**, 29065–29076.

Lee, J.W., Lee, Y.K., Ban, J.O. et al. (2009a). Green tea (−)-epigallocatechin-3-gallate inhibits β-amyloid-induced cognitive dysfunction through modification of secretase activity via inhibition of ERK and NF-kappaB pathways in mice. *J. Nutr.,* **139**, 1987–1993.

Lee, S., Suh, S., and Kim, S. (2000). Protective effects of the green tea polyphenol (−)-epigal-locatechin gallate against hippocampal neuronal damage after transient global ischemia in gerbils. *Neurosci. Lett.,* **287**, 191–194.

Lee, Y.K., Yuk, D.Y., Lee, J.W. et al. (2009c). (−)-Epigallocatechin-3-gallate prevents lipo-polysaccharide-induced elevation of β-amyloid generation and memory deficiency. *Brain Res.,* **1250**, 164–174.

Lesage, S., and Brice, A. (2009). Parkinson's disease: From monogenic forms to genetic sus-ceptibility factors. *Human Mol. Genet.,* **18**, R48–R59.

Levites, Y., Amit, T., and Mandel, S. (2003). Neuroprotection and neurorescue against amyloid beta toxicity and PKC-dependent release of non-amyloidogenic soluble pre-cusor protein by green tea polyphenol (−)-epigallocatechin-3-gallate. *FASEB J.,* **17**, 952–954.

Levites, Y., Amit, T., Youdim, M.B.H. et al. (2002a). Involvement of protein kinase C acti-vation and cell survival/cell cycle genes in green tea polyphenol (−)-epigallocatechin-3-gallate neuroprotective action. *J. Biol. Chem.,* **277**, 30574–30580.

Levites, Y., Weinreb, O., Maor, G. et al. (2001). Green tea polyphenol (−)- epigallocatechin-3-gallate prevents *N*-methyl-4-phenyl-1,2,3,6-tetrahydropyridine-induced dopaminer-gic neurodegeneration. *J. Neurochem.,* **78**, 1073–1082.

Levites, Y., Youdim, M.B.H., Maor, G. et al. (2002b). Attenuation of 6-hydroxydopamine (6-OHDA)-induced nuclear factor-kappaB (NF-kappaB) activation and cell death by tea extracts in neuronal cultures. *Biochem. Pharmacol.,* **63**, 21–29.

Li, C., Lee, M.J., Sheng, S. et al. (2000). Structural identification of two metabolites of cat-echins and their kinetics in human urine and blood after tea ingestion. *Chem. Res. Toxicol.,* **13**, 177–184.

Li, R., Peng, N., Li, X.P., and Le, W.D. (2006). (−)-Epigallocatechin gallate regulates dopa-mine transporter internalization via protein kinase C-dependent pathway. *Brain Res.,* **1097**, 85–89.

Li, Z., Ya, K., Xiao-Mei, W., Lei, Y. et al. (2008). Ginkgolides protect PC12 cells against hypoxia-induced injury by p42/p44 MAPK pathway-dependent upregulation of HIF-1alpha expression and HIF-1DNA-binding activity. *J. Cell. Biochem.,* **103**, 564–875.

Li, Q., Zhao H.F., Zhang Z.F. et al. (2009a). Long-term green tea catechin administration prevents spatial learning and memory impairment in senescence-accelerated mouse prone-8 mice by decreasing Aβ1-42 oligomers and upregulating synaptic plasticity-related proteins in the hippocampus. *Neuroscience,* **163**, 741–749

Li, Q., Zhao, H.F., Zhang, Z.F. et al. (2009b). Long-term administration of green tea cat-echins prevents age-related spatial learning and memory decline in C57BL/6 J mice by regulating hippocampal cyclic amp-response element binding protein signaling cascade. *Neuroscience,* **159**, 1208–1215.

Lin, L.C., Wang, M.N., Tseng, T.Y. et al. (2007). Pharmacokinetics of (−)-epigallocatechin-3-gallate in conscious and freely moving rats and its brain regional distribution. *J. Agric. Food Chem.,* **55**, 1517–1524.

Lin, S.M., Wang, S.W., Ho, S.C. et al. (2010). Protective effect of green tea (-)-epigallocatechin-3-gallate against the monoamine oxidase B enzyme activity increase in adult rat brains. *Nutrition,* **26**, 1195–1200.

Lin, Y.S., Tsai, Y.J., Tsay, J.S. et al. (2003). Factors affecting the levels of tea polyphenols and caffeine in tea leaves. *J. Agric. Food Chem.,* **51**, 1864–1873.

Linert, W., Herlinger, E., Jameson, R. F. et al. (1996). Dopamine, 6-hydroxydopamine, iron, and dioxygen--their mutual interactions and possible implication in the development of Parkinson's disease. *Biochim. Biophys. Acta, 1316,* 160–168.

Lorincz, M. T. 2006. Clinical implications of Parkinson's disease genetics. *Semin. Neurol.,* **26**, 492–498.

Lovell, M.A., Robertson, J.D., Teesdale, W.J. et al. (1998). Copper, iron and zinc in Alzheimer's disease senile plaques. *J. Neurol. Sci.,* **158**, 47–52.

Lu, H., Meng, X., and Yang, C. S. (2003). Enzymology of methylation of tea catechins and inhibition of catechol-*O*-methyltransferase by (−)-epigallocatechin gallate. *Drug Metab. Dispos.,* **31**, 572–579.

Maharaj, H., Sukhdev Maharaj, D., Scheepers, M. et al. (2005). l-DOPA administration enhances 6-hydroxydopamine generation. *Brain Res.,* **1063**, 180–186.

Maher, P. (2001). How protein kinase C activation protects nerve cells from oxidative stress-induced cell death. *J. Neurosci.,* **21**, 2929–2938.

Mandel, S., Maor, G., and Youdim, M.B. (2004a). Iron and α-synuclein in the substantia nigra of MPTP-treated mice: Effect of neuroprotective drugs R-apomorphine and green tea polyphenol (−)-epigallocatechin-3-gallate. *J. Mol. Neurosci.,* **24**, 401–416.

Mandel, S., Reznichenko, L., Amit, T. et al. (2003). Green tea polyphenol (-)-epigallocatechin-3-gallate protects rat PC12 cells from apoptosis induced by serum withdrawal independent of P13-Akt pathway. *Neurotox. Res.,* **5**, 419–424.

Mandel, S., Weinreb, O., Amit, T. et al. (2004b). Cell signaling pathways in the neuroprotective actions of the green tea polyphenol (−)-epigallocatechin-3-gallate: Implications for neurodegenerative diseases. *J. Neurochem.,* **88**, 1555–1569.

Mandel, S., Amit, T., Bar-Am, O. et al. (2007). Iron dysregulation in Alzheimer's disease: Multimodal brain permeable iron chelating drugs, possessing neuroprotective-neurorescue and amyloid precursor protein-processing regulatory activities as therapeutic agents. *Prog. Neurobiol.,* **82**, 348–360.

Mandel, S.A., Amit, T., Weinreb, O. et al. (2008). Simultaneous manipulation of multiple brain targets by green tea catechins: A potential neuroprotective strategy for Alzheimer and Parkinson diseases. *CNS Neurosci. Ther.,* **14**, 352–365.

Mandel, S.A., Avramovich-Tirosh, Y., Reznichenko, L. et al. (2005). Multifunctional activities of green tea catechins in neuroprotection. Modulation of cell survival genes, iron-dependent oxidative stress and PKC signaling pathway. *Neurosignals,* **14**, 46–60.

Mann, G.E., Rowlands, D.J., Li, F.Y. et al. (2007). Activation of endothelial nitric oxide synthase by dietary isoflavones: Role of NO in Nrf2-mediated antioxidant gene expression. *Cardiovasc. Res.,* **75**, 261–274.

Mattson, M.P., Barger, S.W., Furukawa, K. et al. (1997). Cellular signaling roles of TGF beta, TNF-α and β APP in brain injury responses and Alzheimer's disease. *Brain Res. Rev.,* **23**, 47–61.

Mazzio, E.A., Harris, N., and Soliman, K. F. (1998). Food constituents attenuate monoamine oxidase activity and peroxide levels in C6 astrocyte cells. *Planta Medica,* **64**, 603–606.

McNaught, K.S., and Jenner, P. (2001). Proteasomal function is impaired in substantia nigra in Parkinson's disease. *Neurosci. Lett.* **297**, 191–194.

Mercer, L.D., Kelly, B.L., Horne, M.K. et al. (2005). Dietary polyphenols protect dopamine neurons from oxidative insults and apoptosis: Investigations in primary rat mesencephalic cultures. *Biochem. Pharmacol.*, **69**, 339–345.

Mole, D.R. (2010). Iron homeostasis and its interaction with prolyl hydroxylases. *Antioxid. Redox Signal.* **12**, 445–458.

Moreira, P.I., Honda, K., Liu, Q. et al. (2005). Oxidative stress: The old enemy in Alzheimer's disease pathophysiology. *Curr. Alzheimer Res.*, **2**, 403–408.

Morimoto, T., Ohsawa, I., Takamura, C. et al. (1998). Involvement of amyloid precursor protein in functional synapse formation in cultured hippocampal neurons. *J Neurosc. Res.*, **51**, 185–195.

Nakagawa, K., and Miyazawa, T. (1997). Chemiluminescence-high-performance liquid chromatographic determination of tea catechin, (–)-epigallocatechin 3-gallate, at picomole levels in rat and human plasma. *Anal. Biochem.*, **248**, 41–49.

Nanjo, F., Goto, K., Seto, R. et al. (1996). Scavenging effects of tea catechins and their derivatives on 1,1-diphenyl-2-picrylhydrazyl radical. *Free Radic. Biol. Med.*, **21**, 895–902.

Nanjo, F., Mori, M., Goto, K. et al. (1999). Radical scavenging activity of tea catechins and their related compounds. *Biosci. Biotechnol. Biochem.*, **63**, 1621–1623.

Napolitano, M., Centonze, D., Calce, A. et al. (2002). Experimental parkinsonism modulates multiple genes involved in the transduction of dopaminergic signals in the striatum. *Neurobiol. Dis.*, **10**, 387–395.

Ng, T.P., Feng, L., Niti, M. et al. (2008). Tea consumption and cognitive impairment and decline in older Chinese adults. *Am. J. Clin. Nutr.*, **88**, 224–231.

Nie, G., Cao, Y., and Zhao, B. (2002). Protective effects of green tea polyphenols and their major component, (–)-epigallocatechin-3-gallate (EGCG), on 6-hydroxydopamine-induced apoptosis in PC12 cells. *Redox Rep.*, **7**, 171–177.

Obregon, D.F., Rezai-Zadeh, K., Bai, Y. et al. (2006). ADAM 10 activation is required for green tea (–)-epigallocatechin-3-gallate-induced alpha-secretase cleavage of amyloid precursor protein. *J. Biol. Chem.*, **281**, 16419–16427.

Olanow, C.W. (1992). Early therapy for Parkinson's disease. *Eur. Neurol.*, **32**, 30-5.

Olanow, C.W., Rascol, O., Hauser, R. et al. (2009a). A double-blind, delayed-start trial of rasagiline in Parkinson's disease. *N. Engl. J. Med.* **361**, 1268–1278.

Olanow, C.W., Stern, M. B., and Sethi, K. (2009b). The scientific and clinical basis for the treatment of Parkinson disease. *Neurology,* **72**, S1–S136.

Ono, K., and Yamada, M. (2006). Antioxidant compounds have potent anti-fibrillogenic and fibril-destabilizing effects for alpha-synuclein fibrils in vitro. *J. Neurochem.*, **97**, 105–115.

Ono, K., Yoshiike, Y., Takashima, A. et al. (2003). Potent anti-amyloidogenic and fibril-destabilizing effects of polyphenols *in vitro*: Implications for the prevention and therapeutics of Alzheimer's disease. *J. Neurochem.* **87**, 172–181.

Owuor, E.D., and Kong, A.N. (2002). Antioxidants and oxidants regulated signal transduction pathways. *Biochem. Pharmacol.*, **64**, 765–770.

Palermo, C.M., Westlake, C.A., and Gasiewicz, T.A. (2005). Epigallocatechin gallate inhibits aryl hydrocarbon receptor gene transcription through an indirect mechanism involving binding to a 90 kDa heat shock protein. *Biochemistry,* **44**, 5041–5052.

Panickar, K.S., Polansky, M.M., and Anderson, R.A. (2009). Green tea polyphenols attenuate glial swelling and mitochondrial dysfunction following oxygen-glucose deprivation in cultures. *Nutr. Neurosci.*, **12**, 105–113.

Pannala, A.S., Rice-Evans, C.A., Halliwell, B. et al. (1997). Inhibition of peroxynitrite-mediated tyrosine nitration by catechin polyphenols. *Biochem. Biophys. Res. Commun.*, **232**, 164–168.

Paquay, J.B., Haenen, G.R., Stender, G. et al. (2000). Protection against nitric oxide toxicity by tea. *J. Agric. Food Chem.*, **48**, 5768–5772.

Park, S.S., Bae, I., and Lee, Y.J. (2008). Flavonoids-induced accumulation of hypoxia-inducible factor (HIF)-1α/2α is mediated through chelation of iron. *J. Cell. Biochem.*, **103**, 1989–1998.

Periquet, M., Fulga, T., Myllykangas, L. et al. (2007). Aggregated alpha-synuclein mediates dopaminergic neurotoxicity *in vivo*. *J. Neurosci.*, **27**, 338–346.

Perron, N.R., and Brumaghim, J.L. (2009). A review of the antioxidant mechanisms of polyphenol compounds related to iron binding. *Cell Biochem. Biophys.*, **53**, 75–100.

Perron, N.R., Hodges, J.N., Jenkins, M. et al. (2008). Predicting how polyphenol antioxidants prevent DNA damage by binding to iron. *Inorg. Chem.*, **47**, 6153–6161.

Perron, N.R., Wang, H.C., Deguire, S.N. et al. (2010). Kinetics of iron oxidation upon polyphenol binding. *Dalton Trans.*, **39**, 9982–9987.

Pietta, P.G., Simonetti, P., Gardana, C. et al. (1998). Catechin metabolites after intake of green tea infusions. *Biofactors*, **8**, 111–118.

Pinero, D.J., Hu, J., and Connor, J. R. (2000). Alterations in the interaction between iron regulatory proteins and their iron responsive element in normal and Alzheimer's diseased brains. *Cell Mol. Biol.*, **46**, 761–776.

Pratico, D., and Delanty, N. (2000). Oxidative injury in diseases of the central nervous system: Focus on Alzheimer's disease. *Am. J. Med.*, **109**, 577–585.

Przedborski, S., Chen, Q., Vila, M. et al. (2001). Oxidative post-translational modifications of alpha-synuclein in the 1-methyl-4-phenyl-1,2,3,6-tetrahydropyridine (MPTP) mouse model of Parkinson's disease. *J. Neurochem.*, **76**, 637–640.

Rezai-Zadeh, K., Arendash, G.W., Hou, H. et al. (2008). Green tea epigallocatechin-3-gallate (EGCG) reduces β-amyloid mediated cognitive impairment and modulates tau pathology in Alzheimer transgenic mice. *Brain Res.*, **1214**, 177–187.

Rezai-Zadeh, K., Shytle, D., Sun, N. et al. (2005). Green tea epigallocatechin-3-gallate (EGCG) modulates amyloid precursor protein cleavage and reduces cerebral amyloidosis in Alzheimer transgenic mice. *J. Neurosci.*, **25**, 8807–8814.

Reznichenko, L., Amit, T., Youdim, M.B. et al. (2005). Green tea polyphenol (–)-epigallocatechin-3-gallate induces neurorescue of long-term serum-deprived PC12 cells and promotes neurite outgrowth. *J. Neurochem.*, **93**, 1157–1167.

Reznichenko, L., Amit, T., Zheng, H. et al. (2006). Reduction of iron-regulated amyloid precursor protein and β-amyloid peptide by (–)-epigallocatechin-3-gallate in cell cultures: Implications for iron chelation in Alzheimer's disease. *J. Neurochem.*, **97**, 527–536.

Reznichenko, L., Kalfon, L., Amit, T. et al. (2010). Low dosage of Rasagiline and epigallocatechin gallate synergistically restored the nigrostriatal axis in MPTP-induced parkinsonism. *Neurodegen. Dis.*, **7**, 219–231.

Riederer, P., Dirr, A., Goetz, M. et al. (1992). Distribution of iron in different brain regions and subcellular compartments in Parkinson's disease. *Ann. Neurol.*, **32**, S101–S104.

Riederer, P., Sofic, E., Rausch, W. D. et al. (1989). Transition metals, ferritin, glutathione, and ascorbic acid in Parkinsonian brains. *J. Neurochem.*, **52**, 515–520.

Robakis, N.K. (2011). Mechanisms of AD neurodegeneration may be independent of Abeta and its derivatives. *Neurobiol. Aging*, **32**, 372–379.

Rogers, J.T., and Lahiri, D.K. (2004). Metal and inflammatory targets for Alzheimer's disease. *Curr. Drug Targets*, **5**, 535–551.

Rogers, J.T., Randall, J. D., Cahill, C.M. et al. (2002a). An iron-responsive element type II in the 5'-untranslated region of the Alzheimer's amyloid precursor protein transcript. *J. Biol. Chem*, **277**, 45518–45528.

Rogers, J.T., Randall, J.D., Eder, P.S. et al. (2002b). Alzheimer's disease drug discovery targeted to the APP mRNA 5'untranslated region. *J. Mol. Neurosci.*, **19**, 77–82

Rossi, L., Mazzitelli, S., Arciello, M. et al. (2008). Benefits from dietary polyphenols for brain aging and Alzheimer's disease. *Neurochem. Res.*, **33**, 2390–2400.

Rottkamp, C.A., Raina, A.K., Zhu, X. et al. (2001). Redox-active iron mediates amyloid-beta toxicity. *Free Radic. Biol. Med., 30*, 447–450.

Saito, H., Tabeta, R., Kodama, M. et al. (1984). Direct evidence of incorporation of 12-O-[20-2H1]tetradecanoylphorbol-13-acetate into artificial membranes as determined by deuterium magnetic resonance. *Cancer Lett., 22*, 65–69.

Salah, N., Miller, N.J., Paganga, G. et al. (1995). Polyphenolic flavanols as scavengers of aqueous phase radicals and as chain-breaking antioxidants. *Arch. Biochem. Biophys., 322*, 339–346.

Salvatore, M.F., Fisher, B., Surgener, S.P. et al. (2005). Neurochemical investigations of dopamine neuronal systems in iron-regulatory protein 2 (IRP-2) knockout mice. *Brain Res. Mol. Brain Res., 139*, 341–347.

Satoh, T., Nakatsuka, D., Watanabe, Y. et al. (2000). Neuroprotection by MAPK/ERK kinase inhibition with U0126 against oxidative stress in a mouse neuronal cell line and rat primary cultured cortical neurons. *Neurosci. Lett., 288*, 163–166.

Sayre, L.M., Perry, G., Atwood, C.S. et al. (2000a). The role of metals in neurodegenerative diseases. *Cell Mol. Biol., 46*, 731–741.

Sayre, L.M., Perry, G., Harris, P.L. et al. (2000b). *In situ* oxidative catalysis by neurofibrillary tangles and senile plaques in Alzheimer's disease: A central role for bound transition metals. *J. Neurochem., 74*, 270–279.

Sayre, L.M., Smith, M.A., and Perry, G. (2001). Chemistry and biochemistry of oxidative stress in neurodegenerative disease. *Curr. Med. Chem., 8*, 721–738.

Schliebs, R. (2005). Basal forebrain cholinergic dysfunction in Alzheimer's disease--interrelationship with β-amyloid, inflammation and neurotrophin signaling. *Neurochem. Res., 30*, 895–908.

Schofield, C.J., and Ratcliffe, P. (2004). Oxygen sensing by HIF hydroxylases. *Nat. Rev. Mol. Cell Biol., 5*, 343–354.

Schroeder, E. K., Kelsey, N.A., Doyle, J. et al. (2008). Green tea epigallocatechin 3-gallate accumulates in mitochondria and displays a selective anti-apoptotic effect against inducers of mitochondrial oxidative stress in neurons. *Antioxid. Redox Signal. 11*, 469–480.

Schroeter, H., Bahia, P., Spencer, J.P. et al. (2007). (–)-Epicatechin stimulates ERK-dependent cyclic AMP response element activity and up-regulates GluR2 in cortical neurons. *J. Neurochem., 101*, 1596–1606.

Selkoe, D.J. (1991). The molecular pathology of Alzheimer's disease. *Neuron, 6*, 487–498.

Selkoe, D.J. (2008). Soluble oligomers of the amyloid beta-protein impair synaptic plasticity and behavior. *Behav. Brain Res., 192*, 106–113.

Selkoe, D.J., and Schenk, D. (2003). Alzheimer's disease: Molecular understanding predicts amyloid-based therapeutics. *Annu. Rev. Pharmacol. Toxicol., 43*, 545–584.

Shachar, D.B., Kahana, N., Kampel, V. et al. (2004). Neuroprotection by a novel brain permeable iron chelator, VK-28, against 6-hydroxydopamine lesion in rats. *Neuropharmacology, 46*, 254–263.

Sharp, F.R., and Bernaudin, M. (2004). HIF1 and oxygen sensing in the brain. *Nat. Rev. Neurosci., 5*, 437–448.

Shimohama, S., Tanino, H., Kawakami, N. et al. (2000). Activation of NADPH oxidase in Alzheimer's disease brains. *Biochem. Biophys. Res. Commun., 273*, 5–9.

Siddiq, A., Ayoub, I.A., Chavez, J.C. et al. (2005). Hypoxia-inducible factor prolyl 4-hydroxylase inhibition. A target for neuroprotection in the central nervous system. *J. Biol. Chem., 280*, 41732–41743.

Siddiq, A., Aminova, L.R., Troy, C.M. et al. (2009). Selective inhibition of hypoxia-inducible factor (HIF) prolyl-hydroxylase 1 mediates neuroprotection against normoxic oxidative death via HIF- and CREB-independent pathways. *J. Neurosci., 29*, 8828–8838.

Singer, C.A., Figueroa-Masot, X.A., Batchelor, R.H. et al. (1999). The mitogen-activated protein kinase pathway mediates estrogen neuroprotection after glutamate toxicity in primary cortical neurons. *J. Neurosci.,* **19**, 2455–2463.

Skrzydlewska, E., Augustyniak, A., Michalak, K. et al. (2005). Green tea supplementation in rats of different ages mitigates ethanol-induced changes in brain antioxidant abilities. *Alcohol,* **37**, 89–98.

Small, D.H., Nurcombe, V., Reed, G. et al. (1994). A heparin-binding domain in the amyloid protein precursor of Alzheimer's disease is involved in the regulation of neurite outgrowth. *J. Neurosci.,* **14**, 2117–2127.

Smith, M.A., Harris, P.L., Sayre, L.M. et al. (1997). Iron accumulation in Alzheimer disease is a source of redox-generated free radicals. *Proc. Natl. Acad. Sci. U.S.A.,* **94**, 9866–9868.

Smith, M.A., Hirai, K., Hsiao, K. et al. (1998). Amyloid-β deposition in Alzheimer transgenic mice is associated with oxidative stress. *J. Neurochem.,* **70**, 2212–2215.

Smith, M.A., Rottkamp, C.A., Nunomura, A. et al. (2000). Oxidative stress in Alzheimer's disease. *Biochim. Biophys. Act,* **1502**, 139–144.

Smith, S.R., Cooperman, S., Lavaute, T. et al. (2004). Severity of neurodegeneration correlates with compromise of iron metabolism in mice with iron regulatory protein deficiencies. *Ann. N.Y. Acad. Sci.,* **1012**, 65–83.

Spencer, J.P. (2007). The interactions of flavonoids within neuronal signalling pathways. *Genes Nutr.,* **2**, 257–263.

Spencer, J.P. (2008). Flavonoids: Modulators of brain function? *Br. J. Nutr.,* **99**, *E. Suppl.* 1, ES60–ES77.

Srividhya, R., Jyothilakshmi, V., Arulmathi, K. et al. (2008). Attenuation of senescence-induced oxidative exacerbations in aged rat brain by (–)-epigallocatechin-3-gallate. *Int. J. Dev. Neurosci.,* **26**, 217–223.

Srividhya, R., Zarkovic, K., Stroser, M. et al. (2009). Mitochondrial alterations in aging rat brain: Effective role of (–)-epigallocatechin gallate. *Int. J. Dev. Neurosci.,* **27**, 223–231.

Suganuma, M., Okabe, S., Oniyama, M. et al. (1998). Wide distribution of [3H](–)-epigallocatechin gallate, a cancer preventive tea polyphenol, in mouse tissue. *Carcinogenesis,* **19**, 1771–1776.

Sutherland, B.A., Shaw, O.M., Clarkson, A.N. et al. (2005). Neuroprotective effects of (–)-epigallocatechin gallate following hypoxia-ischemia-induced brain damage: Novel mechanisms of action. *FASEB J.,* **19**, 258–260.

Sutherland, B.A., Rahman, R.M., and Appleton, I. (2006). Mechanisms of action of green tea catechins, with a focus on ischemia-induced neurodegeneration. *J. Nutr. Biochem.,* **17**, 291–306.

Takahashi, M., Dore, S., Ferris, C.D. et al. (2000). Amyloid precursor proteins inhibit heme oxygenase activity and augment neurotoxicity in Alzheimer's disease. *Neuron,* **28**, 461–473.

Takeda, A., Perry, G., Abraham, N.G. et al. (2000). Overexpression of heme oxygenase in neuronal cells, the possible interaction with Tau. *J. Biol. Chem.,* **275**, 5395–5399.

Tan, L.C., Koh, W.P., Yuan, J.M. et al. (2008). Differential effects of black versus green tea on risk of Parkinson's disease in the Singapore Chinese Health Study. *Am. J. Epidemiol.,* **167**, 553–560.

Thomas, R., and Kim, M.H. (2005). Epigallocatechin gallate inhibits HIF-1alpha degradation in prostate cancer cells. *Biochem. Biophys. Res. Commun.,* **334**, 543–548.

Triantafyllou, A., Liakos, P., Tsakalof, A. et al. (2007). The flavonoid quercetin induces hypoxia-inducible factor-1alpha (HIF-1α) and inhibits cell proliferation by depleting intracellular iron. *Free Radic. Res.,* **41**, 342–356.

Triantafyllou, A., Mylonis, I., Simos, G. et al. (2008). Flavonoids induce HIF-1alpha but impair its nuclear accumulation and activity. *Free Radic. Biol. Med.,* **44**, 657–670.

Trushina, E., and McMurray, C. T. (2007). Oxidative stress and mitochondrial dysfunction in neurodegenerative diseases. *Neuroscience, 145*, 1233–1248.

Tsolaki, M., Kokarida, K., Iakovidou, V. et al. (2001). Extrapyramidal symptoms and signs in Alzheimer's disease: Prevalence and correlation with the first symptom. *Am. J. Alzheimers Dis. Other Demen., 16*, 268–278.

van Acker, S.A., van den Berg, D.J., Tromp, M.N. et al. (1996). Structural aspects of antioxidant activity of flavonoids. *Free Radic. Biol. Med., 20*, 331–342.

Wang, J., and Pantopoulos, K. (2005). The pathway for IRP2 degradation involving 2-oxoglutarate-dependent oxygenase(s) does not require the E3 ubiquitin ligase activity of pVHL. *Biochim. Biophys. Acta, 1743*, 79–85.

Wang, Z.Y., Huang, M.T., Lou, Y.R. et al. (1994). Inhibitory effects of black tea, green tea, decaffeinated black tea, and decaffeinated green tea on ultraviolet B light-induced skin carcinogenesis in 7,12-dimethylbenz[a]anthracene-initiated SKH-1 mice. *Cancer Res., 54*, 3428–3455.

Weinreb, O., Amit, T., and Youdim, M.B. (2007). A novel approach of proteomics and transcriptomics to study the mechanism of action of the antioxidant-iron chelator green tea polyphenol (–)-epigallocatechin-3-gallate. *Free Radic. Biol. Med., 43*, 546–556.

Weinreb, O., Amit T., and Youdim M.B. (2008). The application of proteomics for studying the neurorescue activity of the polyphenol (–)-epigallocatechin-3-gallate. *Arch. Biochem. Biophys., 476*, 152–160.

Weinreb, O., Amit, T., Mandel, S. et al. (2009a). Neuroprotective molecular mechanisms of (–)-epigallocatechin-3-gallate: A reflective outcome of its antioxidant, iron chelating and neuritogenic properties. *Genes Nutr., 4*, 283–296.

Weinreb, O., Amit, T., Youdim, M.B.H. et al. (2009b). Characterization of the neuroprotective activity of the polyphenol (–)-epigallocatechin-3-gallate in the brain. In H.A.J. McKinley (ed.), *Handbook of Green Tea and Health Research*. Nova Science Publishers, New York, pp. 219–242.

Weinreb, O., Mandel, S., Amit, T. et al. (2004). Neurological mechanisms of green tea polyphenols in Alzheimer's and Parkinson's diseases. *J. Nutr. Biochem., 15*, 506–516.

Weinreb, O., Mandel, S., and Youdim, M.B.H. (2003). CDNA gene expression profile homology of antioxidants and their anti-apoptotic and pro-apoptotic activities in human neuroblastoma cells. *FASEB J., 17*, 935–937.

Weinreb, O., Mandel, S., Bar-Am, O. et al. (2009c). Multifunctional neuroprotective derivatives of Rasagiline as anti-Alzheimer's disease drugs. *Neurotherapeutics, 6*, 163–174.

Weisburger, J.H., and Chung, F.L. (2002). Mechanisms of chronic disease causation by nutritional factors and tobacco products and their prevention by tea polyphenols. *Food Chem. Toxicol., 40*, 1145–1154.

Williams, R.J., Spencer, J.P., and Rice-Evans, C. (2004). Flavonoids: Antioxidants or signalling molecules? *Free Radic. Biol. Med.*, 36, 838–849.

Wilson, W.J., and Poellinger, L. (2002). The dietary flavonoid quercetin modulates HIF-1α activity in endothelial cells. *Biochem. Biophys. Res. Commun.*, 293:446–450.

Wu, C.D., and Wei, G.X. (2002). Tea as a functional food for oral health. *Nutrition, 18*, 443–444.

Xia, Z., Dickens, M., Raingeaud, J. et al. (1995). Opposing effects of ERK and JNK-p38 MAP kinases on apoptosis. *Science, 270*, 1326–1331.

Yang, B., Arai, K., and Kusu, F. (2000). Determination of catechins in human urine subsequent to tea ingestion by high-performance liquid chromatography with electrochemical detection. *Anal. Biochem., 283*, 77–82.

Yang, C.S. and Wang, Z.Y. (1993). Tea and cancer. *J. Natl. Cancer Inst., 85*, 1038–1049.

Yang, C.S., Kim, S., Yang, G.Y. et al. (1999). Inhibition of carcinogenesis by tea: Bioavailability of tea polyphenols and mechanisms of actions. *Proc. Soc. Exp. Biol. Med., 220*, 213–217.

Youdim, K.A., Dobbie, M.S., Kuhnle, G. et al. (2003). Interaction between flavonoids and the blood-brain barrier: *In vitro* studies. *J. Neurochem., **85**, 180–192.

Youdim, M.B. (2008). Brain iron deficiency and excess; cognitive impairment and neurodegeneration with involvement of striatum and hippocampus. *Neurotox. Res., **14**, 45–56.

Youdim, M.B., and Buccafusco, J.J. (2005). Multi-functional drugs for various CNS targets in the treatment of neurodegenerative disorders. *Trends Pharmacol. Sci., **26**, 27–35.

Youdim, M.B.H., Ben-Shachar, D., and Riederer, P. (1989). Is Parkinson's disease a progressive siderosis of substantia nigra resulting in iron and melanin induced neurodegeneration? *Acta Neurol. Scand. Suppl.* **126**, 47–54.

Youdim, M.B.H., Ben-Shachar, D., and Riederer P. (1990). The role of monoamine oxidase, iron-melanin interaction, and intracellular calcium in Parkinson's disease. *J. Neural Trans. Suppl., **32**, 239–248.

Youdim, M.B.H., and Riederer, P. (2004). A review of the mechanisms and role of monoamine oxidase inhibitors in Parkinson's disease. *Neurology, **63**, S32–S35.

Zambenedetti, P., De Bellis, G., Biunno, I. et al. (2003). Transferrin C2 variant does confer a risk for Alzheimer's disease in Caucasians. *J. Alzheimers Dis., **5**, 423–427.

Zecca, L., Gallorini, M., Schunemann, V. et al. (2001). Iron, neuromelanin and ferritin content in the substantia nigra of normal subjects at different ages: Consequences for iron storage and neurodegenerative processes. *J. Neurochem., **76**, 1766–1773.

Zecca, L., Stroppolo, A., Gatti, A. et al. (2004a). The role of iron and copper molecules in the neuronal vulnerability of locus coeruleus and substantia nigra during aging. *Proc. Natl. Acad. Sci. U.S.A., **101**, 9843–9848.

Zecca, L., Youdim, M.B., Riederer, P. et al. (2004b). Iron, brain ageing and neurodegenerative disorders. *Nat. Rev. Neurosci., **5**, 863–873.

Zhang, L., Jie, G., Zhang, J. et al. (2009). Significant longevity-extending effects of EGCG on *Caenorhabditis elegans* under stress. *Free Radic. Biol. Med., **46**, 414–421.

Zhao, B., Guo, Q., and Xin, W. (2001). Free radical scavenging by green tea polyphenols. *Methods Enzymol., **335**, 217–231.

Zhou, Y.D., Kim, Y.P., Li, X.C. et al. (2004). Hypoxia-inducible factor-1 activation by (–)-epicatechin gallate: Potential adverse effects of cancer chemoprevention with high-dose green tea extracts. *J. Nat. Prod., **67**, 2063–2069.

16 Flavonoids and Neuroinflammation

David Vauzour and Katerina Vafeiadou

CONTENTS

16.1 INTRODUCTION

Neuroinflammation is a normal defence mechanism aimed at protecting the central nervous system (CNS) against insults such as infection, injury, or disease. In most cases, it constitutes a beneficial process that resolves on its own once the threat has been eliminated and homeostasis has been restored (Glass et al. 2010). However, sustained neuroinflammatory processes may contribute to the cascade of events leading to the progressive neuronal damage observed in Parkinson's disease and Alzheimer's disease (McGeer and McGeer 2003; Hirsch et al. 2005), and also with neuronal injury associated with stroke (Zheng et al. 2003). In support of this, recent observations suggest that the use of nonsteroidal anti-inflammatory drugs, such as ibuprofen, may delay or even prevent the onset of neurodegenerative disorders, such as Parkinson's disease (Casper et al. 2000; Chen et al. 2003). The majority of existing drug treatments for neurodegenerative disorders does not prevent the underlying degeneration of neurons, and consequently, there is a desire to develop alternative therapies capable of preventing the progressive loss of specific neuronal populations (Legos et al. 2002; Narayan et al. 2002). Since the neuropathology of many neurodegenerative diseases has been linked to increases in brain oxidative stress (Halliwell 2006), historically, strong efforts have been directed at exploring antioxidant strategies to combat neuronal damage.

Recently, there has been intense interest in the neuroprotective effects of a group of plant secondary metabolites known as flavonoids, which are powerful antioxidants *in vitro* (Rice-Evans 2001). Flavonoids are present at high concentrations in a wide variety of fruits, vegetables, and beverages such as red wine and tea. For example, green tea and cocoa are good sources of flavanols such as catechin and epicatechin, citrus fruits are excellent sources of flavanones such as naringenin and hesperetin, and berry fruits have high concentrations of anthocyanins such as cyanidin and malvidin. The latter have been observed to be effective in protecting against both age-related cognitive and motor decline *in vivo* (Joseph et al. 1999; Williams et al. 2008). However, although flavonoids display potent antioxidant capacity *in vitro* (Bastianetto et al. 2000), the precise mechanisms by which they exert their neuroprotective effects *in vivo* is unlikely to be due to this antioxidant activity. This is primarily because flavonoids undergo extensive metabolism in the gastrointestinal tract, liver and colon, which results in the generation of circulating metabolites with reduced antioxidant potential (Hollman and Katan 1997; Kuhnle et al. 2000; Natsume et al. 2004). Therefore, it may be speculated that the classical hydrogen donating antioxidant activity is not the only explanation for the bioactivity of flavonoids *in vivo*. Indeed, it has become evident that flavonoids are more likely to exert their neuroprotective actions by (1) the modulation of intracellular signaling cascades which control neuronal survival, death, and differentiation, (2) affecting gene expression, and (3) interactions with mitochondria (Schroeter et al. 2001; Spencer et al. 2003a).

The aim of the present chapter is to highlight the potential of dietary flavonoids as candidates for neuroprotection in terms of their ability to modulate neuroinflammation in the CNS. For this reason, we provide an outline of the role glial cells play in neuroinflammation and describe the involvement of inflammatory mediators, produced by glia, in the cascade of events leading to neuronal degeneration. In particular, we discuss the involvement of nitric oxide (NO$^\bullet$), interleukin 1-beta (IL-1β) and tumor necrosis factor-alpha (TNF-α) production by glial cells in neuronal death and provide a brief description of the molecular mechanisms involved in cytokine and NO$^\bullet$-induced neuronal death. Then, we highlight the evidence that flavonoids may modulate neuroinflammation by inhibiting the production of these inflammatory agents. Finally, we summarize potential mechanisms of their action, in particular their ability to modulate signaling pathways controlling the activation of glial cells and those determining neuronal apoptosis.

16.2 COMPONENTS OF THE NEUROINFLAMMATORY CASCADE

The immune system plays essential roles in the maintenance of tissue homeostasis and the response to infection and injury. Microglial cells, the primary immune cells in the CNS, have similar actions to that of peripheral macrophages (Kreutzberg 1996). Being immune cells, their primary functions are to promote host defence by destroying invading pathogens, removing potentially deleterious debris, promoting tissue repair, and facilitating tissue homeostasis. Microglial cells also produce factors that influence surrounding astrocytes (another type of glial cell with support functions) and neurons (Glass et al. 2010). However, sustained, uncontrolled activation of microglia may lead to the excess production of a variety of factors that can promote neuronal death, notably, NO•, proinflammatory cytokines (IL-1β, TNF-α) (Gibbons and Dragunow 2006), reactive oxygen species (ROS) (Wang et al. 2006), and glutamate (Takeuchi et al. 2006). The production of NO• by glial cells is mediated by the inducible nitric oxide synthase (iNOS), and excessive NO• may diffuse away from glial cells and induce neuronal cell damage by disrupting correct neuronal mitochondrial electron transport chain function (Stewart and Heales 2003). In particular, NO• may selectively inhibit mitochondrial respiration at cytochrome c oxidase (complex IV), resulting in a disruption of neuronal ATP synthesis and an increased generation of ROS (Moncada and Bolanos 2006). Therefore, the uncontrolled activation of iNOS in glial cells is a critical step in inflammatory-mediated neurodegeneration.

Another important event in activated glia-induced neurotoxicity is the activation of NADPH oxidase (Mander and Brown 2005). NADPH oxidase activation has been suggested to mediate both superoxide ($O_2^{•-}$) production and the release of proinflammatory molecules such as TNF-α, leading to neuronal cell death (Qin et al. 2004). Furthermore, ROS generated by NADPH oxidase may lead to subsequent iNOS induction, suggesting a role for NADPH oxidase in inducible nitric oxide synthase (iNOS) activation (Pawate et al. 2004). In support of its critical role in neurodegeneration, increased NADPH oxidase expression has been reported in the brains of patients with Parkinson's disease (Wu et al. 2003). Lastly, NO• produced in astrocytes and microglia cells may react with $O_2^{•-}$, produced by glial NADPH oxidase (Bal-Price et al. 2002; Abramov et al. 2005), to generate the neurotoxic peroxynitrite (ONOO⁻) (Bal-Price et al. 2002). ONOO⁻ has been observed to inhibit mitochondrial respiration, induce caspase-dependent neuronal apoptosis, and to induce glutamate release resulting in excitotoxicity and neuronal death (Bal-Price et al. 2002, 2003). In addition, NO• may also be detrimental by inducing protein modifications, such as S-nitrosylation and nitration (Zhang et al. 2006). Cytokines produced by activated glial cells also represent major neurotoxic components released by activated glial cells. These may enhance the expression of iNOS and increase NO• production, stimulate the release of other cytokines, and activate neuronal death signaling cascades. For example, cytokines such as TNF-α, IL-1β, and interferon-gamma (INF-γ) induce the expression of low affinity receptor CD23 in glial cells, a protein known to mediate iNOS induction in macrophages (Dugas et al. 1998), resulting in iNOS induction and subsequent increase in NO• production (Hunot et al. 1999). Additionally, glial cytokine production may play a deleterious role in neurodegenerative diseases by binding to specific cell surface receptors expressed in neurons and activate apoptotic pathways. For example, TNF-α binds to the tumor necrosis factor receptor-1 (TNFR1), leading to caspase-8 activation via Fas-associated protein with a death domain (FADD). This leads to subsequent cleavage of caspase-3, which results in neuronal apoptosis (MacEwan 2002; Taylor et al. 2005).

Although substantial evidence suggests that chronic inflammation in the CNS may contribute to neuronal dysfunction and loss, the exact molecular mechanisms involved in inflammatory-induced neuronal death remain unclear. Emerging evidence suggests that regulation of signal transduction pathways, such as the mitogen-activated protein kinase (MAPK) signaling pathway, play an important role in activated glial-induced neuronal death. MAP kinases, which include extracellular signal-regulated kinase (ERK1/2), c-Jun N-terminal kinase (JNK1/2/3), and p38 kinase (p38α/β/γ/δ), are important in the transduction of extracellular signals into cellular responses. When activated, these kinases phosphorylate both cytosolic and nuclear target proteins, resulting in the activation of transcription factors and ultimately the regulation of gene expression (Chang and Karin 2001). Both p38 and JNK have been reported to mediate activated glia-induced neuronal death (Jeohn et al. 2002; Xie et al. 2004). MAPK signaling has been shown to regulate the activation of iNOS and subsequent NO• production and cytokine release in glial cells. For instance, in INF-γ/LPS stimulated glial cells, iNOS and TNF-α expression is regulated by ERK and p38 MAPK cascades (Bhat et al. 1998). Furthermore, JNK1, a specific JNK isoform, modulates iNOS induction in INF-γ/LPS activated astrocytes (Pawate and Bhat 2006), whereas ERK signaling also mediates TNF-α/IL-1β-induced iNOS expression in astrocytes (Marcus et al. 2003). These studies suggest that MAPK plays a pivotal role both in LPS- and cytokine-induced production of proinflammatory molecules (Figure 16.1).

In addition, various transcription factors such as nuclear factor-kappa B (NF-κB), activator protein-1 (AP-1), and the signal transducer, and activator of transcription-1 (STAT-1) have been shown to be involved in pro-inflammatory responses in astrocytes and microglia cells. NF-κB activation has been associated with increased oxidative stress (Flohe et al. 1997), and is considered a good therapeutic target for inflammatory conditions since its activation upon stress rapidly induces gene expression encoding for proinflammatory molecules including cytokines, cytokines receptors, and adhesion molecules (Barnes and Karin 1997). In terms of neuroinflammation, NF-κB activation has been suggested to mediate iNOS induction and thus NO• production in glial cells (Bhat et al. 2002; Davis et al. 2005) and cytokine expression such as TNF-α and IL-1β in activated microglial cells (Jana et al. 2002; Nakajima et al. 2006). NF-κB activation is also involved in the production of cyclooxygenase-2 (COX-2) in activated astrocytes (Dai et al. 2006), a molecule which mediates prostaglandin formation and seems to play a significant role in neuroinflammatory processes (Minghetti 2004). Inappropriate regulation of AP-1 also enhances expression of pro-inflammatory genes such as iNOS, TNF-α, IL-1β, and COX-2 in activated microglial cells (Kang et al. 2004; Bae et al. 2006). Furthermore, STAT-1 signaling is also involved in iNOS expression in activated glial cells (Dell'Albani et al. 2001), and it has also been suggested that suppression of STAT-1 phosphorylation may inhibit the expression of inflammatory molecules in astrocytes (Choi et al. 2005). Interestingly, NF-κB responds to p38 signaling leading to iNOS induction in activated glial cells (Bhat et al. 2002), suggesting that an interplay between signaling pathways, transcription factors, and the production of inflammatory molecules and/or reactive oxygen species is pivotal in determining a neuroinflammatory response in the brain (Figure 16.1).

The modulation of glial signaling cascades and proinflammatory transcription factors as well as cytokines and NO• production may result in suppression of neuroinflammation and ultimately to protection against neurodegeneration. It is plausible that the development of novel therapeutic agents or a cocktail of drugs that target neuroinflammation at various stages may act to reduce neurodegeneration and thus

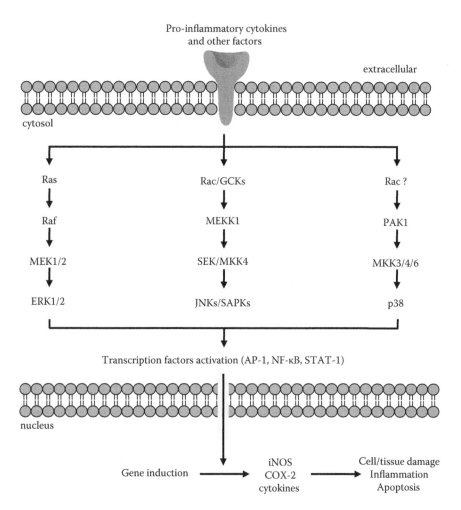

FIGURE 16.1 Potential involvement of MAPK signaling in neuroinflammation. Activation of MAPK signaling cascades may lead to the induction of proinflammatory transcription factors, which in turn lead to an increase in inflammatory molecules expression such as iNOS, cytokines, and COX-2.

delay the progression of neurodegenerative disease. Lately, there has been much interest in the potential neuroprotective potential of a group to dietary phytochemicals known as flavonoids. These dietary components may represent a novel therapeutic strategy to combat neurodegeneration via their reported ability to inhibit oxidative stress-induced neuronal apoptosis (Schroeter et al. 2001; Spencer et al. 2001a).

16.3 BRAIN UPTAKE AND CELLULAR INTERACTIONS

Many studies have reported the bioavailability of flavonoids in the systemic circulation (Manach et al. 2004, 2005; Crozier et al. 2009); however, little is known about their uptake within the brain. To understand whether these phenolic compounds could affect signal transduction pathways and neurohormetic mechanisms, it is crucial to ascertain their presence within the cerebral tissue. In most cases, following oral ingestion and during absorption, flavonoids are extensively metabolized. For example, the majority of flavonoid glycosides and aglycones present in plant-derived foods are extensively conjugated and metabolized during absorption [reviewed in (Spencer et al. 2001b; Spencer 2008)]. In the upper gastrointestinal tract, dietary polyphenols are substrates for a number of enzymes, such as phase I enzymes (hydrolysing and oxidizing) and phase II enzymes (conjugating and detoxifying). During the transfer across the small intestine, and again in the liver, they are transformed into glucuronides, sulphates, and *O*-methylated forms (Spencer et al. 1999; Spencer 2003). Further transformations occur in the colon, where the enzymes of the gut microflora act to metabolize flavonoids to simple phenolics acids, which are absorbed and further metabolized in the liver (Crozier et al. 2009). In addition, flavonoids may undergo at least three types of intracellular metabolism: (1) oxidative metabolism, (2) P450-related metabolism, and (3) Conjugation with thiols, particularly GSH (Spencer et al. 2003b).

16.3.1 Flavonoid Bioavailability to The Brain

The brain is protected by a structurally unique endothelial barrier: the blood–brain barrier (BBB), which differs from other vascular barriers in its physical characteristics including the nature and number of transporters it possesses (Pardridge 2002). The BBB is created by the endothelial cells that form the walls of the capillaries and the combined surface area of these microvessels constitutes by far the largest interface for blood–brain exchange (12 and 18 m^2 for the average human adult) (Abbott et al. 2010). In order for flavonoids to access the brain, they must first cross the physical filter, which controls the entry of xenobiotics into the brain. Flavanones such as hesperetin, naringenin, and their *in vivo* metabolites, along with some dietary anthocyanins, cyanidin-3-*O*-rutinoside, and pelargonidin-3-*O*-glucoside, have been shown to traverse the BBB in relevant *in vitro* and *in situ* models (Youdim et al. 2004). In both cell types, the uptake of hesperetin and naringenin was the greatest, relative to their more polar glucuronidated conjugates, to (−)-epicatechin and its metabolites, and to the dietary anthocyanins and specific phenolic acids derived from colonic biotransformation of flavonoids (Youdim et al. 2004). The uptake of (−)-epicatechin and its *O*-methylated metabolites has been demonstrated in primary cultures of mouse cortical neurons (Spencer et al. 2001a). This uptake of (−)-epicatechin, and indeed its two *O*-methylated metabolites, was low in comparison to that observed in primary mouse astrocytes. A monoglutathionyl adduct of epicatechin was present in both astrocyte lysates and medium following exposure to epicatechin. Similarly, glutathione conjugation was also observed following exposure of quercetin to astrocytes. Although no glutathione adducts of quercetin were observed in neurons, oxidative metabolites were measured in both brain cell types indicating a common pathway for metabolism of quercetin. Consistent with this, the structurally related flavonol, kaempferol, also underwent oxidative metabolism in both neurons and astrocytes (Spencer et al. 2001a). As was observed with

O-methylated metabolites of epicatechin, O-methylated forms of quercetin were also accumulated to a greater amount by both neurons and astrocytes. This accumulation was higher at 2 h than at 18, especially in the case of the 3′-O-methyl metabolite. Astrocytic metabolites, along with parent compounds, may then pass into neurons where they may influence neuronal function. Glial metabolites such as glutathionyl adducts of flavonoids may be important in mediating effects on neurons following their import.

Alternatively, the uptake and metabolism of reactive flavonoids such as quercetin in astrocytes may act to limit their uptake into neurons, thus reducing the potential for neuronal damage (Spencer et al. 2003a; Vafeiadou et al. 2008) (Figure 16.2). The degree of BBB penetration is dependent on compound lipophilicity (Youdim et al. 2003), meaning that less polar O-methylated metabolites may be capable to greater brain uptake than the more polar flavonoid glucuronides. However, evidence exists to suggest that certain drug glucuronides may cross the BBB (Aasmundstad et al. 1995) and exert pharmacological effects (Kroemer and Klotz 1992; Sperker et al. 1997), suggesting that there may be a specific uptake mechanism for glucuronides

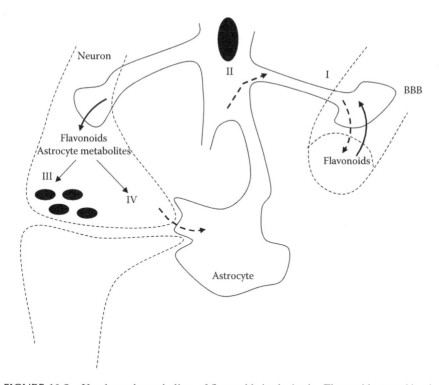

FIGURE 16.2 Uptake and metabolism of flavonoids in the brain. Flavonoids cross blood–brain barrier (I) and enter astrocytes where they are metabolized to form various oxidative metabolites and thiol conjugates (II). Flavonoids and metabolites may then access neurons where they may exert biological function at mitochondria (III) or signaling cascades (IV). Astrocytic metabolism may act to protect neurons from exposure to excessive amounts of potentially reactive and/or toxic polyphenols.

in vivo. Their brain entry may also depend on their interactions with specific efflux transporters expressed in the BBB, such as the P-glycoprotein (Lin and Yamazaki 2003), which appears to be responsible for the differences between naringenin and quercetin flux into the brain *in situ* (Youdim et al. 2004). More recently, a study also evaluated the transmembrane transport of flavonoids across RBE-4 cells (an immortalized cell line of rat cerebral capillary endothelial cells). All of the tested flavonoids (catechin, quercetin, and cyanidin-3-*O*-glucoside) passed across the RBE-4 cells in a time-dependent manner (Faria et al. 2010).

In animals, flavanones have been found to enter the brain following their intravenous administration (Peng et al. 1998), while (–)-epigallocatechin-3-*O*-gallate (EGCG) (Suganuma et al. 1998), (–)-epicatechin (Abd El Mohsen et al. 2002), and anthocyanins (Talavera et al. 2005; El Mohsen et al. 2006) were found in the brain after their oral administration. Furthermore, several anthocyanins have been identified in different regions of the rat (Passamonti et al. 2005) and pig brains (Kalt et al. 2008; Milbury and Kalt 2010) of blueberry fed animals, with 11 intact anthocyanins found in the cortex and cerebellum. A recent investigation also reported anthocyanin presence within rat brains fed with a blackberry anthocyanin-enriched diet for 15 days. A total amount of 0.25 nmol g^{-1} of anthocyanins, including blackberry anthocyanins and peonidin-3-*O*-glucoside, was found in the cerebral tissue of these animals (Talavera et al. 2005). Furthermore, another recent investigation focused on the pharmacokinetic and the ability of the flavonoid constituents to cross the BBB in rats, after oral administration of the *Gingko biloba* extract EGb 761®. A single dose of 600 mg/kg EGb 761 resulted in maximum plasma concentrations of 176, 341, and 183 ng/mL for quercetin, kaempferol, and isorhamnetin/tamarixetin, respectively, and in maximum brain concentrations of 291 ng/g protein for kaempferol and 161 ng/g protein for isorhamnetin/tamarixetin (Rangel-Ordonez et al. 2010). Evidence also exists for the ability of the *O*-methylated flavonoid, tangeretin, to cross the BBB and interact with brain cells (Datla et al. 2001). Concentrations of tangeretin or metabolites derived from it varied, with the brain stem and cerebellum having lowest concentrations (0.17 and 0.27 ng/mg tissue, respectively) while highest concentrations were seen in the hippocampus, striatum, and hypothalamus (2.0, 2.36, and 3.88 ng/mg tissue, respectively) (Datla et al. 2001). Altogether, these results indicate that flavonoids transverse the BBB and are able to localize in the brain, suggesting that they are potentially good candidates for direct neuroprotective and neuromodulatory actions (Figure 16.2).

16.3.2 Flavonoid Interactions with Cellular Signaling

Historically, the bioactivity of flavonoids has been linked with their antioxidant capacity as determined by structurally important features such as a 3′,4′-dihydroxyl catechol group in the B ring, the planarity of the molecule and the presence of 2,3-unsaturation in conjugation with a 4-oxo-function in the C-ring. Although this antioxidant action is reduced markedly *in vivo* because of extensive metabolism, they may be able to prevent the formation of some reactive oxygen species and free radicals via the chelation of transition metal ions such as iron (Levites et al. 2002; Mandel et al. 2004; Mandel et al. 2006). In addition, the increase in endogenous antioxidant enzymes such as catalase and superoxide dismutase by flavonoids in the brain may also contribute to their neuroprotective effects (Levites et al. 2001).

Interestingly, flavonoids have close structural homology with specific inhibitors of cell signaling cascades, such as the PD98059, a MAPK inhibitor, and the LY294002, a phosphatidylinositol-3 kinase (PI3) inhibitor. Indeed, the latter inhibitor was modeled on the structure of quercetin (Vlahos et al. 1994). LY294002 and quercetin fit into the ATP binding pocket of the enzyme, and it appears that the number and substitution of hydroxyl groups on the B ring and the degree of unsaturation of the C2–C3 bond are important determinants of this particular bioactivity. In this regard, quercetin and some of its *in vivo* metabolites have been suggested to inhibit Akt/protein kinase B (PKB) signaling pathways (Spencer et al. 2003a), a mechanism of action consistent with quercetin and its metabolites acting at and inhibiting PI3-kinase activity. Interestingly, the MAPK inhibitor PD98059 has been shown to effectively block iNOS expression and NO$^\bullet$ production in activated glial cells (Bhat et al. 1998), which is highly suggestive that flavonoids may also be capable of exerting anti-inflammatory actions by their inhibition of this signaling pathway (Figure 16.3).

It has recently become clear that flavonoids may exert many effects via their specific interactions within neuronal signaling pathways. For example, Schroeter et al. (2001) demonstrated that epicatechin and its *in vivo* metabolite 3'-*O*-methylepicatechin elicit strong protective effects against oxidized LDL-induced neuronal cell death by inhibiting JNK, c-jun, and caspase-3 activation. In another study by Levites et al. (2002), the neuroprotective effects of EGCG were shown to involve stimulation of the cell survival PKC and modulation of cell survival/cell cycle genes such as Bax, Bad, Bcl-2, Bcl-w, and Bcl-x(L). EGCG has also been shown to protect neurons against oxidative stress-induced injury by activating the (PI3K)/Akt-dependent antiapoptotic pathway and by inhibiting glycogen synthase kinase-3 (GSK-3) activity, which is known to be involved in cell death (Koh et al. 2004). Flavonoids may also exert neuroprotective effects by directly targeting mitochondrial function. For example, flavonoids may prevent mitochondrial damage, attenuate release of cytochrome *c*, DNA fragmentation, and inhibit caspase activation, thus preventing apoptosis (Wang et al. 2001; Smith et al. 2002). Alternatively, flavonoids may protect mitochondrial integrity by controlling calcium homeostasis, as it has been shown for EGCG, which attenuated glutamate induced cytotoxicity by inhibiting calcium influx in neuron cells (Lee et al. 2004).

16.4 INHIBITION OF NEUROINFLAMMATION BY FLAVONOIDS

There is substantial evidence in the literature to suggest that flavonoids may exert neuroprotective effects via interaction with neuronal signaling and synaptic function, as well as modulation of blood flow and neurogenesis (Youdim and Joseph 2001; Schroeter et al. 2002; Mandel et al. 2005; Spencer 2010). More recently, there has been an increasing interest in the potential of flavonoids to exert anti-inflammatory action in the CNS. Indeed, emerging evidence from *in vitro* studies suggests that flavonoids may protect against neurodegenerative diseases by interfering with

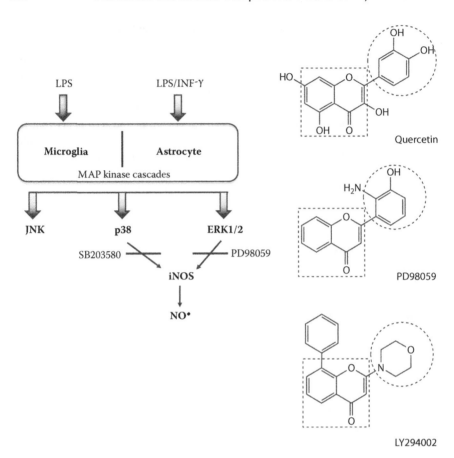

FIGURE 16.3 Structural homology of flavonoids with specific cell signaling cascades inhibitors. Use of specific MAPK inhibitors such as SB203580 and PD98059 inhibits the transcriptional regulation of iNOS in activated glial cells. Interestingly, the structure of PD98059 and other kinase inhibitors have close structural homology to that of flavonoids. It is therefore possible that flavonoids may modulate neuroinflammation by interfering with cell signaling pathways such as MAPK.

inflammatory processes mediated by astrocytes and microglial cells. The individual effects of different flavonoid subclasses will be reviewed below. A summary of the outcomes of studies which investigated the effects of flavonoids on neuroinflammation is presented in Table 16.1.

16.4.1 FLAVONES

Initial studies on the potential anti-inflammatory effects of flavones in the brain focused on wogonin and bacalein, which originate from the root of *Scutellaria baicalensis* Georgi, a plant traditionally used in Chinese medicine for the treatment of inflammatory conditions (Huang et al. 2006). Subsequently, studies included

TABLE 16.1
Summary Outcomes of Studies Investigating The Effects of Dietary Flavonoids on Neuroinflammatory Processes

Flavonoid Subclass	Flavonoid	Summary of Outcomes	Reference
Flavones	Wogonin	Inhibition of NF-κB-mediated iNOS induction, NO production, and inflammatory induced death of C6 glial cells.	Kim et al. (2001)
		Inhibition of LPS-induced TNF-α and IL-1β production, iNOS induction, NF−κB activation, and NO release in microglia and reduction of inflammatory-induced neuronal cell death. Attenuation of inflammatory-induced neuronal death in experimental brain injury models.	Lee et al. (2003)
		Inhibition of LPS-induced NO production in BV2 microglial cells and attenuation of H_2O_2-induced cytotoxicity in SH-SY5Y human neuroblastoma.	Chun et al. (2005)
		Suppression of NF−κB activity and cell motility in MCP-1-stimulated microglia.	Piao et al. (2008)
	Baicalein	Inhibition of iNOS induction, NO production, and apoptosis of LPS-activated BV-2 mouse microglial cells and rat primary microglia cultures. Inhibition of LPS-induced NF−κB activity in BV-2 cells; no effect on caspase-11 activation, IRF-1 induction, or STAT-1 phosphorylation.	Suk et al. (2003)
		Decrease in iNOS protein, mRNA, and promoter activity expression by inhibition of NF-IL6 activity in microglial cells. Weak effects on NF-κB and AP-1 activity. No effect on ERK phosphorylation.	Chen et al. (2004)
		Decrease in LPS-induced dopaminergic neuronal death, inhibition of microglial activation, and subsequent production of LPS-induced TNF-α, NO, and superoxide.	Li et al. (2005)
		Inhibition of hypoxia-induced HIF-1α protein activation and iNOS, COX-2, and VEGF expression by inhibition of ROS and PI 3-kinase/Akt pathway in BV2 microglia. No effect on ERK and p38 phosphorylation.	Hwang et al. (2008)
	Luteolin	Inhibition of LPS-induced iNOS expression, I-κB-alpha degradation, and NO production in BV-2 microglia.	Kim et al. (2006)

(Continued)

TABLE 16.1 (CONTINUED)
Summary Outcomes of Studies Investigating The Effects of Dietary Flavonoids on Neuroinflammatory Processes

Flavonoid Subclass	Flavonoid	Summary of Outcomes	Reference
		Inhibition of IL-1β-induced ROS production, increase in SOD-1 and TRX1 mediators expression in astrocytes. Inhibition of IL-6, IL-8, IP-10, MCP-1, and RANTES release and inflammatory-induced neuronal apoptosis. Modulation of expression of astrocytes specific molecules (glial fibrillary acidic protein, glutamine synthetase, and ceruloplasmin).	Sharma et al. (2007)
		Attenuation of LPS-induced neuronal injury in primary neuron--glia cultures and inhibition of LPS-induced activation of microglia and production of TNF-α, NO, and superoxide in both neuron--glia cultures and microglia-enriched cultures.	Chen et al. (2008)
		Inhibition of LPS-induced IL-6 gene and protein expression via inhibition of JNK signaling and AP-1 activity in primary murine microglia. No effect on NF-κB activity. Reduction in LPS-induced increase in plasma IL-6 and inhibition of LPS-induced IL-6 gene expression in mice hippocampus.	Jang et al. (2008)
	Luteolin	Suppression of IFN-γ-induced CD40 expression, TNF-α and IL-6 production in N9 and primary microglial cells by inhibition of STAT-1 phosphorylation.	Rezai-Zadeh et al. (2008)
		Suppression of proinflammatory and proapoptotic gene expression in LPS-activated BV-2 microglia. Stimulation of genes related to anti-oxidant metabolism, phagocytic uptake, ramification, and chemotaxis. Inhibition of LPS-induced NO secretion and neurotoxicity in 661W photoreceptor cultures.	Dirscherl et al. (2010)
		Inhibition of TNF-α, iNOS and COX-2 induction, TNF-α, NO, and PGE-2 release in LPS-stimulated microglia and reduction of inflammatory-induced neuronal death. Improvement of spatial working memory and restoration of inflammatory markers expression in the hippocampus of aged mice. No effect on spatial working memory or inflammatory markers in young mice.	Jang et al. (2010)

TABLE 16.1 (CONTINUED)
Summary Outcomes of Studies Investigating The Effects of Dietary Flavonoids on Neuroinflammatory Processes

Flavonoid Subclass	Flavonoid	Summary of Outcomes	Reference
	Apigenin	Inhibition of IFN-γ-induced CD40 expression, TNF-α, and IL-6 production in N9 and primary microglia cells by suppression of STAT-1 phosphorylation.	Rezai-Zadeh et al. (2008)
		Inhibition of inflammatory-induced iNOS and COX-2 expression, NO, and PGE2 production by suppression of p38 and JNK phosphorylation in BV-2 microglial cells.	Ha et al. (2008)
	Chrysin	Inhibition of LPS/ IFN-γ-induced NO and TNF-α production in microglial cells and reduction in inflammatory-induced neurotoxicity via decrease in C/EBPdelta protein, gene, and activity levels.	Gresa-Arribas et al. (2010)
	Icariin	Inhibition of LPS-induced NO, PGE-2 and ROS release, TNF-α, IL-6 and IL-1β gene expression, iNOS and COX-2 protein expression in microglia. Suppression of TAK1/IKK/NF-κB, JNK, and p38 MAPK pathways and inhibition of inflammatory-induced neuronal death.	Zeng et al. (2010)
Flavonols	Quercetin	Inhibition on LPS/ IFN-γ-induced iNOS gene expression, NO and TNF-α production in BV-2 microglia. Inhibition of LPS-induced IKK, NF-κB and AP-1 activation, and IFN-γ-induced STAT1 and NF-κB activity. Increase in heme oxygenase-1 expression.	Chen et al. (2005)
		Inhibition of LPS/ IFN-γ-induced iNOS expression and NO production, decrease in ERK, JNK, p38, Akt, Src, Janus kinase-1, Tyk2, STAT-1, and NF-κB activity in BV-2 microglial cells. Inhibition of free radicals release, serine/threonine, and tyrosine phosphatase activity, and disruption of lipid rafts accumulation.	Kao et al. (2010)
	Quercetin	Inhibition of LPS-induced TNF-α and IL-1α gene expression in glial cells and decrease of inflammatory-induced neuronal death in neuronal-microglial co-cultures.	Bureau et al. (2008)

(Continued)

TABLE 16.1 (CONTINUED)
Summary Outcomes of Studies Investigating The Effects of Dietary Flavonoids on Neuroinflammatory Processes

Flavonoid Subclass	Flavonoid	Summary of Outcomes	Reference
		Inhibition of IL-1β-induced ROS production, stimulation of SOD-1 and TRX1 mediators expression in astrocytes. Inhibition of IL-1β-induced release of IL-6, IL-8, IP-10, MCP-1, and RANTES and inhibition of inflammatory-induced neuronal apoptosis. Modulation of astrocytes specific molecules expression (glial fibrillary acidic protein, glutamine synthetase, and ceruloplasmin).	Sharma et al. (2007)
		Improvement in behavioral performance of high-cholesterol-fed old mice by decrease in ROS and protein carbonyl levels and restoring Cu, Zn-SOD activity. Activation of AMPK activity, decrease in iNOS, COX-2, TNF-α, and IL-1β expression via suppression of NF-κB p65 nuclear translocation.	Lu et al. (2010)
	Rutin	Inhibition of trimethyltin-induced hippocampal injury, reduction in microglial activation, and cytokines production in Sprague-Dawley rats.	Koda et al. (2009)
	Fisetin	Inhibition of LPS-induced TNF-α, NO, and PGE-2 production and TNF-α, IL-1β, COX-2, and iNOS gene and protein expression in BV-2 microglia cells and primary microglia cultures. Inhibition of NF-κB activity and p-38 phosphorylation and attenuation of the inflammatory-induced neuronal cell death.	Zheng et al. (2008)
Isoflavones	Daidzein	Inhibition of LPS- and amyloid-beta-induced TNF-α, IL-1, IL-6, and ERβ gene expression and attenuation of inflammatory-induced cell death in astrocytes.	Liu et al. (2009)
	Genistein	Genistein inhibited the hemolysate-induced iNOS and COX-2 gene expression and NF-kB activation in astrocytes.	Lu et al. (2009)
		Inhibition of amyloid-beta-induced TNF-α, IL-1β, COX-2, and iNOS expression by stimulation of PPAR-γ expression in primary astrocytes cultures.	Valles et al. (2010)

TABLE 16.1 (CONTINUED)
Summary Outcomes of Studies Investigating The Effects of Dietary Flavonoids on Neuroinflammatory Processes

Flavonoid Subclass	Flavonoid	Summary of Outcomes	Reference
	Others:		
	Formononetin, daidzein, pratensein, calycosin, and irilone	Attenuation of the LPS-induced decrease in dopamine uptake and the number of dopaminergic neurons in rat neuron--glia cultures and inhibition of LPS-induced TNF-α, NO, and superoxide release in neuron--glia co-cultures and microglia-enriched cultures.	Chen et al. (2008)
	Biochanin A	Attenuation of LPS-induced neuronal death in rat neuron-glia co-cultures and inhibition of LPS-induced of TNF-α, NO and superoxide production in neuron-glia co-cultures and microglia-enriched cultures.	Chen et al. (2007)
	Glycitin, tectoridin, and kakkalide and its bacterial metabolites	Inhibition of LPS-induced TNF-α and NO release in primary cultured microglia and BV2 microglial cell lines. Inhibition of NF-κB, AP-1 and ERK activity by iIrisolidone, a metabolite of kakkalide.	Park et al. (2007)
Flavan-3-ols	Catechin	Inhibition of oxidative-induced cell death and DNA damage in N9 microglial cells and reduction of cell cycle arrest by suppression of NF-κB activation, down-regulation of p53 phosphorylation, and ERK activation.	Huang et al. (2005)
	ECGC	Inhibition of LPS-induced iNOS and TNF-α expression, NO, and TNF-α release in microglia and inhibition of inflammatory-induced neuronal injury in human neuroblastoma cell line (SH-SY5Y) and in primary rat mesencephalic cultures.	Li et al. (2004)
		Inhibition of encephalitogenic T-cells proliferation and TNF-α synthesis in mice, suppression of T cell proliferation by inhibition of NF-κB, neuronal cell death inhibition, and reduction in ROS.	Aktas et al. (2004)
		Improvement in LPS-induced memory deficiency in mice and inhibition in LPS-induced beta-amyloid peptide and secretases levels. Inhibition of LPS-induced iNOS and COX protein expression and prevention of inflammatory-induced neuronal cell death *in vivo*.	Lee et al. (2009)

(Continued)

TABLE 16.1 (CONTINUED)
Summary Outcomes of Studies Investigating The Effects of Dietary Flavonoids on Neuroinflammatory Processes

Flavonoid Subclass	Flavonoid	Summary of Outcomes	Reference
		Inhibition of beta-amyloid peptide-induced apoptosis, iNOS expression, NO, and peroxynitrite production in BV2 microglial cells by increasing glutathione levels.	Kim et al. (2009)
Flavanones	Naringenin	Naringenin inhibited the LPS/IFN-g-induced iNOS expression and NO and TNF-α production in primary glial cells and attenuated inflammatory-induced neuronal death. Naringenin also inhibited LPS/IFN-gamma-induced p38 and STAT-1 phosphorylation in glial cells.	Vafeiadou et al. (2009)
		Naringenin suppressed LPS-induced iNOS and COX-2 expression and NO production in microglial cells.	Chao et al. (2010)
Flavonoid-rich foods	Blueberry extract	Blueberry extract inhibited the LPS-induced gene iNOS and COX-2 expression, NO, TNF-α, IL-1β, and ROS production in BV-2 microglia.	Lau et al. (2007)

Abbreviations: AMPK: AMP-activated protein kinase; AP-1: activator protein-1; C/EBPs: CCAAT/enhancer binding protein; COX-2: cyclooxygenase-2; EGCG: epigallocatechin gallate; ERK: extracellular signal-regulated kinase; HIF-1: hypoxia-inducible factor-1; H_2O_2: hydrogen peroxide; IL: interleukin; IKK: IκB kinase; iNOS: inducible nitric oxide synthase; IRF-1: interferon regulatory factor-1; JNK: c-Jun N-terminal kinase; LPS: lipopolysaccharide; MCP-1: monocyte-chemoattractant protein-1; NF-IL6: nuclear factor IL-6; NF-κB: nuclear factor-kappaB; PGE-2: prostaglandin E; PPAR-γ: peroxisome proliferator-activated receptor gamma; ROS: reactive oxygen species; SOD-1: superoxide dismutase-1; STAT-1: signal transducer and activator of transcription-1; TAK1: transforming growth factor-beta-activated kinase 1; TNF-α: tumor necrosis factor alpha; TRX1: thioredoxin; VEGF: vascular endothelial growth factor.

investigations of the potential neuroprotective effects of other flavones including apigenin and luteolin. The first evidence for wogonin's anti-inflammatory effects in the CNS arises from a study by Kim et al. (2001) in which wogonin inhibited NO• production in LPS-activated in rat glial cells by suppressing iNOS induction and NF-κB activity. Similar anti-inflammatory effects of wogonin and its synthetic derivatives were observed in microglial cell lines (Lee et al. 2003; Chun et al. 2005). In the study by Lee et al. (2003), wogonin treatment suppressed inflammatory cytokines production including TNF-α and IL-1β, and attenuated neuronal cell death in relevant neuron–microglia co-cultures, suggesting that wogonin may exert neuroprotection via inhibition of inflammatory stimulation in the CNS. These observations were supported by data from two experimental brain injury models, in which wogonin attenuated neuronal death via inhibition of hippocampal iNOS and TNF-α induction and down-regulation of microglia activation (Lee et al. 2003). More recently, an additional mechanism of the anti-inflammatory effects of wogonin in the CNS included the suppression of microglial monocyte chemoattractant protein-1 (MCP-1)-induced migration via inhibition of NF-κB activity (Piao et al. 2008).

Bacalein also seems to be able to decrease inflammatory processes in the brain by inhibiting iNOS induction and NO• release resulting to attenuation of inflammatory-induced neuronal death (Suk et al. 2003; Li et al. 2005). Potential mechanisms of bacalein's anti-inflammatory action include modulation of NF-κB activity subsequently resulting in inhibition of downstream inflammatory signaling cascades (Dell'Albani et al. 2001; Suk et al. 2001; Suk et al. 2003). Another target for bacalein's anti-inflammatory effects include the nuclear factor IL-6 (NF-IL6) transcription factor (Chen et al. 2004), which mediates iNOS induction by inflammatory stimulants such as LPS and INF-γ (Dlaska and Weiss 1999). Bacalein may additionally inhibit hypoxia-induced ROS generation and the hypoxia responsive inflammatory genes iNOS and COX-2 in BV2 microglia. Main mechanism includes inhibition of PI 3-kinase/Akt pathway activation (but no effect on ERK and p38 phosphorylation) (Hwang et al. 2008). In agreement with the latter, bacalein treatment had no effect on LPS/INF-γ induced ERK phosphorylation in microglial cells (Chen et al. 2004), suggesting that the PI 3-kinase/Akt pathway may be the main target of bacalein anti-inflammatory action in the brain. However, as discussed previously, MAP kinase signaling appears to be important in the mechanism of activation of glial in response to cytokines (Figure 16.1) and further investigation of their interactions with MAP kinases and how this is related to glial cells activation are warranted.

More recent studies are also suggestive of a potential anti-inflammatory action of luteolin in the CNS, a flavone usually occurring as glycosylated conjugate in celery, green pepper, perilla leaf, and camomile tea. Luteolin attenuated IL-6, TNF-α, COX-2, and iNOS gene expression and proinflammatory cytokine production in activated microglia and inhibited the cytokine-induced neuronal death in cultured neurons (Chen et al. 2008a; Jang et al. 2008, 2010; Kim et al. 2006). In *in vivo* models, luteolin ingestion reduced inflammatory cytokine gene expression in mice hippocampus and subsequently improved working memory in aged mice (Jang et al. 2010). In the first published genome-wide search by Dirscherl et al. (2010), luteolin triggered global changes in the microglial transcriptome by suppressing pro-inflammatory marker expression and inducing anti-oxidant genes expression in activated microglia. Reported cellular mechanisms of luteolin anti-inflammatory action include inhibition of JNK, AP-1 (Jang et al. 2008) and STAT-1 phosphorylation and CD-40 expression (Rezai-Zadeh et al. 2008), suggesting that luteolin may act as modulator of cell signaling, transcription pathways, and immunomodulatory molecules. Limited evidence for other flavones anti-inflammatory effects in the CNS also exists for apigenin, chrysin, and icariin (Ha et al. 2008; Gresa-Arribas et al. 2010; Zeng et al. 2010).

16.4.2 Flavonols

There is limited evidence on the anti-inflammatory effects of flavonols in the brain. Initial investigations showed that the main flavonol quercetin might inhibit inflammatory processes by attenuating NO• production and iNOS gene expression in microglia. However, in this study, quercetin's circulating metabolite, quercetin-3'-O-sulphate (Day et al. 2001), failed to demonstrate any anti-inflammatory actions (Chen et al. 2005). These findings were subsequently supported by other studies in which quercetin was not only also successful in attenuating NO• production and iNOS gene expression (Kao et al. 2010) but also inhibited inflammatory cytokines production and reduced neuroinflammatory-induced neuronal cell death in neuronal–glial co-cultures (Sharma et al. 2007; Bureau et al. 2008). The reported potential underlying cellular mechanisms that mediate quercetin's anti-inflammatory effects include down-regulation of tyrosine kinase and MAPK, predominantly ERK, JNK, p38, and Akt, suppression of transcription factors involved in inflammatory signaling such as NF-κB, AP-1, and STAT-1 and disruption of membrane lipid raft accumulation, which is of crucial importance to cell signaling (Day et al. 2001; Kao et al. 2010). However, the concentrations of quercetin needed to express a significant inhibitory effect were much higher than those expected *in vivo*. Furthermore, like most flavonoids, the aglycone form of quercetin is present in the plasma at very low concentrations, if at all (Day et al. 2001; Mullen et al. 2006), suggesting that future *in vitro* investigations should concentrate on the actions its circulating metabolite forms, rather than the aglycone itself. Although quercetin-3′-*O*-sulphate failed to exert any effect on NO• production and iNOS expression, it is still possible that it may exert neuroprotective biological effects through modulation of neuronal signaling cascades, as it has been previously reported for other flavonoid subclasses (Spencer et al. 2003a).

In the only reported animal investigation, oral administration of quercetin led to an improvement in the behavioural performance of high-cholesterol-fed old mice in both a step-through test and the Morris water maze task. This was attributed to anti-inflammatory actions of quercetin in the brain of high-cholesterol-fed old mice including among others the reduction of ROS, restoration of antioxidant enzymes activity, down-regulation of iNOS and COX-2, and cytokines expression (Lu et al. 2010). In this study, quercetin's anti-inflammatory effects were attributed to its interaction with AMP-activated protein kinase (AMPK), suggesting that AMPK may be a potential target to enhance the resistance of neuronal cells to inflammatory-induced neurodegeneration. Anti-inflammatory action in the brain has also been reported in the literature for quercetin rutinoside form, known as rutin (Koda et al. 2009) and fisetin, a flavonol mainly found in strawberries (Zheng et al. 2008). Since fisetin has previously been shown to induce cAMP response element-binding protein (CREB) phosphorylation in rat hippocampus and to enhance memory (Maher et al. 2006), it could be argued that fisetin could be another flavonol with a potential neuroprotective potential.

16.4.3 ISOFLAVONES

There is increasing evidence from cell culture studies to suggest that isoflavones have the potential to exert neuroprotective effects via suppression of microglia activation, cytokines, NO•, and superoxide production (Chen et al. 2007, 2008b; Jin et al. 2008; Liu et al. 2009; Yuan et al. 2009; Valles et al. 2010). This is of particular interest since isoflavones have structural homology to oestrogen and display binding affinity for oestrogen receptors, in particular, ERβ, which is expressed in glial cells (Kuiper et al. 1998). Isoflavones may therefore induce oestrogen receptor activation and subsequent modulation of oestrogen receptor sensitive signaling pathways and gene expression involved in inflammatory cascades (Baker et al. 2004). Thus, it is plausible that isoflavones may mimic the actions of oestrogen in the brain and modulate inflammation-mediated neurodegeneration. However, isoflavones undergo extensive metabolism in the gut, and it is therefore highly likely that it is the metabolites rather than the glucosides that reach tissues and exert biological effects at a cellular level (Clarke et al. 2002; Shelnutt et al. 2002). For example, in the study by Park et al. (2007), the colonic isoflavone metabolites irisolidone, tectorigenin, and glycitein were more potent than their parent glycosylated forms in inhibiting LPS-induced release of TNF-α, IL-1β, and NO• in primary cultured microglia and BV2 microglial cell lines. Reported mechanisms of isoflavones anti-inflammatory action in the CNS include inhibition of NF-κB signaling pathway (Lu et al. 2009) and activation of peroxisome proliferator activated receptors (PPARs) (Valles et al. 2010), which in turn lead to the suppression of inflammatory processes.

16.4.4 FLAVAN-3-OLS

Data on the potential anti-inflammatory effects of flavan-3-ols exists for (+)-catechin and EGCG (Huang et al. 2005). Catechin has been shown to increase microglial cells survival following exposure to the oxidative agent *tert*-butylhydroperoxide (tBHP) by decreasing tBHP-induced hydroxyl radical (•OH) generation and diminishing DNA damage via inhibition of NF-κB, down-regulation of p53 activity, and activation of ERK signaling pathway. However, the concentrations used in this study (0.13–2.0 mM) far exceed the circulating concentrations achieved following ingestion, and therefore, it is difficult to accurately assess the relevance of these results to the *in vivo* situation. EGCG, the main flavanol present in green tea, has been shown to inhibit LPS-induced release of NO• and TNF-α in microglia by down-regulating iNOS and TNF-α gene expression and subsequently inhibit inflammatory-induced neuronal injury (Li et al. 2004). EGCG suppressed brain inflammatory processes in orally treated mice by inhibiting the proliferation of encephalitogenic T-cells and the synthesis of TNF-α, further suggesting its potential of reducing neuroinflammatory processes and neuronal damage (Aktas et al. 2004). Furthermore, EGCG induced cell-cycle arrest in human myelin specific CD4$^+$ T-cells, which are known to mediate myelin sheath destruction in multiple sclerosis (Mouzaki et al. 2004). Main mechanism of action included the blockage of 20S/26S proteasome complex resulting in an intracellular accumulation of IκB-α and subsequent inhibition of NF-κB. In the same study, EGCG exerted antioxidant and neuroprotective effects by blocking ROS formation. In other studies, EGCG recovered LPS-induced memory impairment and reduced beta-amyloid peptide generation as well as apoptotic cell death in mice models (Lee et al. 2009) and inhibited the beta-amyloid-induced oxidative and nitrosative cell death via enhancing the intracellular antioxidant pool (Kim et al. 2009). EGCG can penetrate the BBB and localize in the brain (Suganuma et al. 1998), and therefore, it could prove a potent candidate agent for alleviating inflammatory induced neurodegeneration.

16.4.5 FLAVANONES

There is currently very limited information on citrus flavanones anti-inflammatory effects in the CNS. Naringenin was reported to inhibit iNOS expression and NO•

production in both primary glial cells (Vafeiadou et al. 2009) and migrolia cell line (Chao et al. 2010) and effectively inhibit inflammatory-induced neuronal cell death in relevant neuronal glial cocultures. Possible reported mechanisms of naringenin's anti-inflammatory action in the brain include inhibition of p38 MAPK phosphorylation and downstream signal transducer and activator of transcription-1 (STAT-1) (Vafeiadou et al. 2009). There is also evidence from the same study to suggest that hesperetin may also inhibit neuroinflammation by inhibition of cytokines production; nevertheless, hesperetin seems to have weaker anti-inflammatory potential than naringenin (Vafeiadou et al. 2009). Since naringenin has been previously found to localize in the brain following its oral ingestion (Youdim et al. 2003), it may hold the most potential for exerting antineuroinflammatory actions *in vivo* and thus in combating neurodegenerative disease. More studies on animal models are warranted to confirm these findings.

16.4.6 ANTHOCYANINS

There is very limited evidence on the effects of anthocyanins on neuroinflammation. In one study, the anthocyanidins, cyanidin and pelargonidin, failed to exert anti-inflammatory effects in a primary glial cell culture model (Vafeiadou et al. 2009). Nevertheless, since anthocyanins are complex components that undergo extensive metabolism following ingestion (Kay et al. 2005), the biological effects of the parent anthocyanins and their metabolic products, which will ultimately reach the brain, warrant further investigation.

16.4.7 OTHER FLAVONOID-RICH FOODS

To our knowledge, there is only one study in the literature that has investigated the effects of flavonoids rich foods on neuroinflammation. In this study, a blueberry extract inhibited the inflammatory stimulation of microglial cells via suppression of iNOS and COX-2 gene and protein expression, suggesting that flavonoid-rich fruits may have the potential of attenuating inflammatory conditions in the CNS (Lau et al. 2007). More research is definitely needed to investigate the anti-inflammatory effects of flavonoid-rich foods and their precise underlying mechanism of action.

16.5 SUMMARY

In the recent years, there is an increasing amount of emerging data from mainly cellular *in vitro* studies that supports the notion that dietary flavonoids exert neuroprotective effects via suppression of neuroinflammation mediated by both microglial cells and astrocytes, leading to inhibition of neuronal apoptosis triggered by inflammatory molecules and neurotoxins. Reported cellular mechanisms of flavonoids anti-inflammatory action include inhibition of inflammatory cytokines and chemokines production such as TNF-α, interleukins and COX-2, suppression of iNOS expression and NO$^\bullet$ release, inhibition of NADPH oxidase and subsequent ROS release, and increase in intracellular antioxidant enzyme pool. These effects appear to be mediated by the ability of flavonoids to modulate signaling cascades and transcription factors that promote inflammatory gene expression including MAPK, NF-κB, STAT-1, and AP-1.

There is evidence from animal and *in vitro* studies suggesting that flavonoids may transverse the BBB and localize in the brain; thus, flavonoids may have the potential to be considered as neuromodulatory dietary agents. Nevertheless, the evidence for flavonoids anti-inflammatory effects in the CNS stems mainly from *in vitro* studies in which in some cases non physiological concentrations of flavonoids were used. In the future, the investigations of the cellular mechanism of flavonoids action should focus on characterized flavonoids circulating metabolites and forms which have the potential to penetrate the BBB and should employ flavonoids concentrations that correspond to those found in plasma or key tissues. Furthermore, animal studies are needed to investigate whether the reported anti-inflammatory activity of flavonoids can lead to decrease neurodegeneration and improved cognitive performance. Finally, in order for flavonoids to be established as potential neuroprotective agents, sophisticated clinical supplementation studies are also required, which when linked with the data from cellular and animal studies, would potentially enable public health recommendations for flavonoids intake and inflammatory-induced neurodegeneration.

REFERENCES

Aasmundstad, T.A., Morland, J., and Paulsen, R.E. (1995). Distribution of morphine 6-glucuronide and morphine across the blood-brain barrier in awake, freely moving rats investigated by *in vivo* microdialysis sampling. *J. Pharmacol. Exp. Ther.,* **275**, 435–441.

Abbott, N.J., Patabendige, A.A., Dolman, D.E. et al. (2010). Structure and function of the blood-brain barrier. *Neurobiol. Dis.,* **37**, 13–25.

Abd El Mohsen, M.M., Kuhnle, G., Rechner, A.R., et al. (2002). Uptake and metabolism of epicatechin and its access to the brain after oral ingestion. *Free Radic. Biol. Med.,* **33**, 1693–1702.

Abramov, A.Y., Jacobson, J., Wientjes, F., et al. (2005). Expression and modulation of an NADPH oxidase in mammalian astrocytes. *J. Neurosci.,* **25**, 9176–9184.

Aktas, O., Prozorovski, T., Smorodchenko, A. et al. (2004). Green tea epigallocatechin-3-gallate mediates T cellular NF-kappa B inhibition and exerts neuroprotection in autoimmune encephalomyelitis. *J. Immunol.,* **173**, 5794–5800.

Bae, E.A., Kim, E.J., Park, J.S., et al. (2006). Ginsenosides Rg3 and Rh2 inhibit the activation of AP-1 and protein kinase A pathway in lipopolysaccharide/interferon-gamma-stimulated BV-2 microglial cells. *Planta Med.,* **72**, 627–633.

Baker, A.E., Brautigam, V.M., and Watters, J.J. (2004). Estrogen modulates microglial inflammatory mediator production via interactions with estrogen receptor beta. *Endocrinology,* **145**, 5021–5032.

Bal-Price, A., Matthias, A., and Brown, G.C. (2002). Stimulation of the NADPH oxidase in activated rat microglia removes nitric oxide but induces peroxynitrite production. *J. Neurochem.,* **80**, 73–80.

Barnes, P.J. and Karin, M. (1997). Nuclear factor-kappaB: a pivotal transcription factor in chronic inflammatory diseases. *N. Engl. J. Med.,* **336**, 1066–1071.

Bastianetto, S., Zheng, W.H., and Quirion, R. (2000). Neuroprotective abilities of resveratrol and other red wine constituents against nitric oxide-related toxicity in cultured hippocampal neurons. *Br. J. Pharmacol.,* **131**, 711–720.

Bhat, N.R., Feinstein, D.L., Shen, Q., and Bhat, A.N. (2002). p38 MAPK-mediated transcriptional activation of inducible nitric-oxide synthase in glial cells. Roles of nuclear factors, nuclear factor kappa B, cAMP response element-binding protein, CCAAT/enhancer-binding protein-beta, and activating transcription factor-2. *J. Biol. Chem.*, **277**, 29584–29592.

Bhat, N.R., Zhang, P., Lee, J.C., and Hogan, E.L. (1998). Extracellular signal-regulated kinase and p38 subgroups of mitogen-activated protein kinases regulate inducible nitric oxide synthase and tumor necrosis factor-alpha gene expression in endotoxin-stimulated primary glial cultures. *J. Neurosci.*, **18**, 1633–1641.

Brown, G.C. and Bal-Price, A. (2003). Inflammatory neurodegeneration mediated by nitric oxide, glutamate, and mitochondria. *Mol. Neurobiol.*, **27**, 325–355.

Bureau, G., Longpre, F., and Martinoli, M.G. (2008). Resveratrol and quercetin, two natural polyphenols, reduce apoptotic neuronal cell death induced by neuroinflammation. *J Neurosci Res*, **86**, 403–410.

Casper, D., Yaparpalvi, U., Rempel, N. et al. (2000). Ibuprofen protects dopaminergic neurons against glutamate toxicity *in vitro*. *Neurosci. Lett.*, **289**, 201–204.

Chang, L. and Karin, M. (2001). Mammalian MAP kinase signalling cascades. *Nature*, **410**, 37–40.

Chao, C.L., Weng, C.S., Chang, N.C. et al. (2010). Naringenin more effectively inhibits inducible nitric oxide synthase and cyclooxygenase-2 expression in macrophages than in microglia. *Nutr Res*, **30**, 858–864.

Chen, C.J., Raung, S.L., Liao, S.L., and Chen, S.Y. (2004). Inhibition of inducible nitric oxide synthase expression by baicalein in endotoxin/cytokine-stimulated microglia. *Biochem. Pharmacol.*, **67**, 957–965.

Chen, H., Zhang, S.M., Hernan, M.A. et al. (2003). Nonsteroidal anti-inflammatory drugs and the risk of Parkinson disease. *Arch. Neurol.*, **60**, 1059–1064.

Chen, H.Q., Jin, Z.Y., and Li, G.H. (2007). Biochanin A protects dopaminergic neurons against lipopolysaccharide-induced damage through inhibition of microglia activation and proinflammatory factors generation. *Neurosci Lett.*, **417**, 112–117.

Chen, H.Q., Jin, Z.Y., Wang, X.J. et al. (2008a). Luteolin protects dopaminergic neurons from inflammation-induced injury through inhibition of microglial activation. *Neurosci. Lett.*, **448**, 175–179.

Chen, H.Q., Wang, X.J., Jin, Z.Y. et al. (2008b). Protective effect of isoflavones from Trifolium pratense on dopaminergic neurons. *Neurosci Res*, **62**, 123–130.

Chen, J.C., Ho, F.M., Pei-Dawn Lee, C., et al. (2005). Inhibition of iNOS gene expression by quercetin is mediated by the inhibition of IkappaB kinase, nuclear factor-kappa B and STAT1, and depends on heme oxygenase-1 induction in mouse BV-2 microglia. *Eur. J. Pharmacol.*, **521**, 9–20.

Choi, W.H., Ji, K A., Jeon, S.B, et al. (2005). Anti-inflammatory roles of retinoic acid in rat brain astrocytes: Suppression of interferon-gamma-induced JAK/STAT phosphorylation. *Biochem. Biophys. Res. Commun.*, **329**, 125–131.

Chun, W., Lee, H.J., Kong, P.J. et al. (2005). Synthetic wogonin derivatives suppress lipopolysaccharide-induced nitric oxide production and hydrogen peroxide-induced cytotoxicity. *Arch. Pharm. Res.*, **28**, 216–219.

Clarke, D.B., Lloyd, A.S., Botting, N.P. et al. (2002). Measurement of intact sulfate and glucuronide phytoestrogen conjugates in human urine using isotope dilution liquid chromatography-tandem mass spectrometry with [^{13}C(3)]isoflavone internal standards. *Anal. Biochem.*, **309**, 158–172.

Crozier, A., Jaganath, I.B., and Clifford, M.N. (2009). Dietary phenolics: chemistry, bioavailability and effects on health. *Nat. Prod. Rep.*, **26**, 1001–1043.

Dai, Y.Q., Jin, D.Z., Zhu, X.Z., and Lei, D.L. (2006). Triptolide inhibits COX-2 expression via NF-kappa B pathway in astrocytes. *Neurosci. Res.*, **55**, 154–160.

Datla, K.P., Christidou, M., Widmer, W.W. et al. (2001). Tissue distribution and neuroprotective effects of citrus flavonoid tangeretin in a rat model of Parkinson's disease. *Neuroreport*, 12, 3871–3875.

Davis, R.L., Sanchez, A.C., Lindley, D.J. et al. (2005). Effects of mechanistically distinct NF-kappaB inhibitors on glial inducible nitric-oxide synthase expression. *Nitric Oxide*, 12, 200–209.

Day, A.J., Mellon, F., Barron, D., et al. (2001). Human metabolism of dietary flavonoids: identification of plasma metabolites of quercetin. *Free Radic. Res.*, 35, 941–952.

Dell'Albani, P., Santangelo, R., Torrisi, L., et al. (2001). JAK/STAT signaling pathway mediates cytokine-induced iNOS expression in primary astroglial cell cultures. *J. Neurosci. Res.*, 65, 417–424.

Dirscherl, K., Karlstetter, M., Ebert, S., et al. (2010). Luteolin triggers global changes in the microglial transcriptome leading to a unique anti-inflammatory and neuroprotective phenotype. *J. Neuroinflamm.*, 7, 3.

Dlaska, M. and Weiss, G. (1999). Central role of transcription factor NF-IL6 for cytokine and iron-mediated regulation of murine inducible nitric oxide synthase expression. *J. Immunol.*, 162, 6171–6177.

Dugas, N., Palacios-Calender, M., Dugas, B., et al. (1998). Regulation by endogenous INTERLEUKIN-10 of the expression of nitric oxide synthase induced after ligation of CD23 in human macrophages. *Cytokine*, 10, 680–689.

El Mohsen, M.A., Marks, J., Kuhnle, G., et al. (2006). Absorption, tissue distribution and excretion of pelargonidin and its metabolites following oral administration to rats. *Br. J. Nutr.*, 95, 51–58.

Faria, A., Pestana, D., Teixeira, D., et al. (2010). Flavonoid transport across RBE4 cells: A blood-brain barrier model. *Cell Mol. Biol. Lett.*, 15, 234–241.

Flohe, L., Brigelius-Flohe, R., Saliou, C. et al. (1997). Redox regulation of NF-kappa B activation. *Free Radic. Biol. Med.*, 22, 1115–1126.

Gibbons, H.M. and Dragunow, M. (2006). Microglia induce neural cell death via a proximity-dependent mechanism involving nitric oxide. *Brain Res.*, 1084, 1–15.

Glass, C.K., Saijo, K., Winner, B. et al. (2010). Mechanisms underlying inflammation in neurodegeneration. *Cell*, 140, 918–934.

Gresa-Arribas, N., Serratosa, J., Saura, J. et al. (2010). Inhibition of CCAAT/enhancer binding protein delta expression by chrysin in microglial cells results in anti-inflammatory and neuroprotective effects. *J. Neurochem.*, 115, 526–536.

Ha, S.K., Lee, P., Park, J.A. et al. (2008). Apigenin inhibits the production of NO and PGE2 in microglia and inhibits neuronal cell death in a middle cerebral artery occlusion-induced focal ischemia mice model. *Neurochem. Int.*, 52, 878–886.

Halliwell, B. (2006). Oxidative stress and neurodegeneration: where are we now? *J. Neurochem.*, 97, 1634–1658.

Hirsch, E.C., Hunot, S., and Hartmann, A. (2005). Neuroinflammatory processes in Parkinson's disease. *Parkinsonism Relat. Disord.*, 11, *Suppl. 1*, S9–S15.

Hollman, P.C. and Katan, M.B. (1997). Absorption, metabolism and health effects of dietary flavonoids in man. *Biomed. Pharmacother.*, 51, 305–310.

Huang, Q., Wu, L.J., Tashiro, S, et al. (2005). (+)-Catechin, an ingredient of green tea, protects murine microglia from oxidative stress-induced DNA damage and cell cycle arrest. *J. Pharmacol. Sci.*, 98, 16–24.

Huang, W.H., Lee, A.R., and Yang, C.H. (2006). Antioxidative and anti-inflammatory activities of polyhydroxyflavonoids of *Scutellaria baicalensis* GEORGI. *Biosci. Biotechnol. Biochem.*, 70, 2371–2380.

Hunot, S., Dugas, N., Faucheux, B. et al. (1999). FcepsilonRII/CD23 is expressed in Parkinson's disease and induces, *in vitro*, production of nitric oxide and tumor necrosis factor-alpha in glial cells. *J. Neurosci.*, 19, 3440–3447.

Hwang, K.Y., Oh, Y.T., Yoon, H. et al. (2008). Baicalein suppresses hypoxia-induced HIF-1alpha protein accumulation and activation through inhibition of reactive oxygen species and PI 3-kinase/Akt pathway in BV2 murine microglial cells. *Neurosci. Lett.*, **444**, 264–269.

Jana, M., Dasgupta, S., Liu, X., and Pahan, K. (2002). Regulation of tumor necrosis factor-alpha expression by CD40 ligation in BV-2 microglial cells. *J. Neurochem.*, **80**, 197–206.

Jang, S., Dilger, R.N., and Johnson, R.W. (2010). Luteolin inhibits microglia and alters hippocampal-dependent spatial working memory in aged mice. *J. Nutr.*, **140**, 1892–1898.

Jang, S., Kelley, K.W., and Johnson, R.W. (2008). Luteolin reduces IL-6 production in microglia by inhibiting JNK phosphorylation and activation of AP-1. *Proc. Natl. Acad. Sci. U.S.A.*, **105**, 7534–7539.

Jeohn, G.H., Cooper, C.L., Wilson, B. et al. (2002). p38 MAP kinase is involved in lipopolysaccharide-induced dopaminergic neuronal cell death in rat mesencephalic neuron-glia cultures. *Ann. NY. Acad. Sci.*, **962**, 332–346.

Jin, G.H., Ha, S.K., Park, H.M. et al. (2008). Synthesis of azaisoflavones and their inhibitory activities of NO production in activated microglia. *Bioorg. Med. Chem. Lett.*, **18**, 4092–4094.

Joseph, J.A., Shukitt-Hale, B., Denisova, N.A. et al. (1999). Reversals of age-related declines in neuronal signal transduction, cognitive, and motor behavioral deficits with blueberry, spinach, or strawberry dietary supplementation. *J. Neurosci.*, **19**, 8114–8121.

Kalt, W., Blumberg, J.B., McDonald, J.E. et al. (2008). Identification of anthocyanins in the liver, eye, and brain of blueberry-fed pigs. *J. Agric. Food Chem.*, **56**, 705–712.

Kang, G., Kong, P. J., Yuh, Y.J. et al. (2004). Curcumin suppresses lipopolysaccharide-induced cyclooxygenase-2 expression by inhibiting activator protein 1 and nuclear factor kappab bindings in BV2 microglial cells. *J. Pharmacol. Sci.*, **94**, 325–328.

Kao, T.K., Ou, Y.C., Raung, S.L, et al. (2010). Inhibition of nitric oxide production by quercetin in endotoxin/cytokine-stimulated microglia. *Life Sci.*, **86**, 315–321.

Kay, C.D., Mazza, G.J., and Holub, B. J. (2005). Anthocyanins exist in the circulation primarily as metabolites in adult men. *J. Nutr.*, **135**, 2582–2588.

Kim, C.Y., Lee, C., Park, G.H., and Jang J.H. (2009). Neuroprotective effect of epigallocatechin-3-gallate against beta-amyloid-induced oxidative and nitrosative cell death via augmentation of antioxidant defense capacity. *Arch. Pharm. Res.*, **32**, 869–881.

Kim, H., Kim, Y. S., Kim, S.Y., and Suk, K. (2001). The plant flavonoid wogonin suppresses death of activated C6 rat glial cells by inhibiting nitric oxide production. *Neurosci. Lett.*, **309**, 67–71.

Kim, J.S., Lee, H., Lee, M.H., et al. (2006). Luteolin inhibits LPS-stimulated inducible nitric oxide synthase expression in BV-2 microglial cells. *Planta Med.*, **72**, 65–68.

Koda, T., Kuroda, Y., and Imai, H. (2009) Rutin supplementation in the diet has protective effects against toxicant-induced hippocampal injury by suppression of microglial activation and pro-inflammatory cytokines: protective effect of rutin against toxicant-induced hippocampal injury. *Cell. Mol. Neurobiol.*, **29**, 523–531.

Koh, S.H., Kim, S.H., Kwon, H. et al. (2004). Phosphatidylinositol-3 kinase/Akt and GSK-3 mediated cytoprotective effect of epigallocatechin gallate on oxidative stress-injured neuronal-differentiated N18D3 cells. *Neurotoxicology*, **25**, 793–802.

Kreutzberg, G.W. (1996). Microglia: a sensor for pathological events in the CNS. *Trends Neurosci.*, **19**, 312–318.

Kroemer, H.K. and Klotz, U. (1992). Glucuronidation of drugs. A re-evaluation of the pharmacological significance of the conjugates and modulating factors. *Clin. Pharmacokinet.*, **23**, 292–310.

Kuhnle, G., Spencer, J.P., Schroeter, H., et al. (2000). Epicatechin and catechin are O-methylated and glucuronidated in the small intestine. *Biochem. Biophys. Res. Commun.*, **277**, 507–512.

Kuiper, G.G., Lemmen, J.G., Carlsson, B. et al. (1998). Interaction of estrogenic chemicals and phytoestrogens with estrogen receptor β. *Endocrinology,* **139**, 4252–4263.

Lau, F.C., Bielinski, D.F., and Joseph, J.A. (2007). Inhibitory effects of blueberry extract on the production of inflammatory mediators in lipopolysaccharide-activated BV2 microglia. *J. Neurosci. Res.,* **85**, 1010–1017.

Lee, H., Kim, Y.O., Kim, H. et al. (2003). Flavonoid wogonin from medicinal herb is neuroprotective by inhibiting inflammatory activation of microglia. *FASEB J.,* **17**, 1943–1944.

Lee, J.H., Song, D.K., Jung, C.H. et al. (2004). (–)-Epigallocatechin gallate attenuates glutamate-induced cytotoxicity via intracellular Ca modulation in PC12 cells. *Clin. Exp. Pharmacol. Physiol.,* **31**, 530–536.

Lee, Y.K., Yuk, D.Y., Lee, J.W. et al. (2009). (–)-Epigallocatechin-3-gallate prevents lipopolysaccharide-induced elevation of β-amyloid generation and memory deficiency. *Brain Res.,* **1250**, 164–174.

Legos, J.J., Tuma, R.F., and Barone, F.C. (2002). Pharmacological interventions for stroke: failures and future. *Expert Opin. Investig. Drugs,* **11**, 603–614.

Levites, Y., Amit, T., Youdim, M.B. et al. (2002). Involvement of protein kinase C activation and cell survival/ cell cycle genes in green tea polyphenol (–)-epigallocatechin 3-gallate neuroprotective action. *J. Biol. Chem.,* **277**, 30574–30580.

Levites, Y., Weinreb, O., Maor, G. et al. (2001). Green tea polyphenol (–)-epigallocatechin-3-gallate prevents N-methyl-4-phenyl-1,2,3,6-tetrahydropyridine-induced dopaminergic neurodegeneration. *J. Neurochem.,* **78**, 1073–1082.

Li, F.Q., Wang, T., Pei, Z. et al. (2005). Inhibition of microglial activation by the herbal flavonoid baicalein attenuates inflammation-mediated degeneration of dopaminergic neurons. *J. Neural. Transm.,* **112**, 331–347.

Lin, J.H. and Yamazaki, M. (2003). Role of P-glycoprotein in pharmacokinetics: clinical implications. *Clinical Pharmacokin.,* **42**, 59–98.

Liu, M.H., Lin, Y.S., Sheu, S. Y. et al. (2009). Anti-inflammatory effects of daidzein on primary astroglial cell culture. *Nutr. Neurosci.,* **12**, 123–134.

Li, R., Huang, Y.G., Fang, D. et al. (2004). (–)-Epigallocatechin gallate inhibits lipopolysaccharide-induced microglial activation and protects against inflammation-mediated dopaminergic neuronal injury. *J. Neurosci. Res.,* **78**, 723–731.

Lu, H., Shi, J.X., Zhang, D.M. et al. (2009). Inhibition of hemolysate-induced iNOS and COX-2 expression by genistein through suppression of NF-small ka, CyrillicB activation in primary astrocytes. *J. Neurol. Sci.,* **278**, 91–95.

Lu, J., Wu, D.M., Zheng, Y.L. et al. (2010). Quercetin activates AMP-activated protein kinase by reducing PP2C expression protecting old mouse brain against high cholesterol-induced neurotoxicity. *J. Pathol.,* **222**, 199–212.

MacEwan, D.J. (2002). TNF receptor subtype signalling: differences and cellular consequences. *Cell Signal.,* **14**, 477–492.

Maher, P., Akaishi, T., and Abe, K. (2006). Flavonoid fisetin promotes ERK-dependent long-term potentiation and enhances memory. *Proc. Nat. Acad. Sci. U.S.A.,* **103**, 16568–16573.

Manach, C., Scalbert, A., Morand, C. et al. (2004). Polyphenols: food sources and bioavailability. *Am. J. Clin. Nutr.,* **79**, 727–747.

Manach, C., Williamson, G., Morand, C. et al. (2005). Bioavailability and bioefficacy of polyphenols in humans. I. Review of 97 bioavailability studies. *Am. J. Clin. Nutr.,* **81**, 230S–242S.

Mandel, S., Amit, T., Reznichenko, L. et al. (2006). Green tea catechins as brain-permeable, natural iron chelators-antioxidants for the treatment of neurodegenerative disorders. *Mol. Nutr. Food Res.,* **50**, 229–234.

Mandel, S., Maor, G., and Youdim, M.B. (2004). Iron and alpha-synuclein in the substantia nigra of MPTP-treated mice: effect of neuroprotective drugs R-apomorphine and green tea polyphenol (–)-epigallocatechin-3-gallate. *J. Mol. Neurosci.,* **24**, 401–416.

Mandel, S.A., Avramovich-Tirosh, Y., Reznichenko, L. et al. (2005). Multifunctional activities of green tea catechins in neuroprotection. Modulation of cell survival genes, iron-dependent oxidative stress and PKC signaling pathway. *Neurosignals,* **14**, 46–60.

Mander, P. and Brown, G.C. (2005). Activation of microglial NADPH oxidase is synergistic with glial iNOS expression in inducing neuronal death: a dual-key mechanism of inflammatory neurodegeneration. *J Neuroinflammation,* **2**, 20.

Marcus, J.S., Karackattu, S.L., Fleegal, M.A. et al. (2003). Cytokine-stimulated inducible nitric oxide synthase expression in astroglia: role of Erk mitogen-activated protein kinase and NF-kappaB. *Glia,* **41**, 152–160.

McGeer, E.G. and McGeer, P.L. (2003). Inflammatory processes in Alzheimer's disease. *Progr. Neuro-Psychopharmacol. Biol. Psych.,* **27**, 741–749.

Milbury, P.E. and Kalt, W. (2010). Xenobiotic metabolism and berry flavonoid transport across the blood-brain barrier. *J. Agric. Food Chem.,* **58**, 3950–3956.

Minghetti, L. (2004). Cyclooxygenase-2 (COX-2) in inflammatory and degenerative brain diseases. *J. Neuropathol. Exp. Neurol.,* **63**, 901–910.

Moncada, S. and Bolanos, J.P. (2006). Nitric oxide, cell bioenergetics and neurodegeneration. *J. Neurochem.,* **97**, 1676–1689.

Mouzaki, A., Tselios, T., Papathanassopoulos, P. et al. (2004). Immunotherapy for multiple sclerosis: basic insights for new clinical strategies. *Curr. Neurovasc. Res.,* **1**, 325–340.

Mullen, W., Edwards, C.A., and Crozier, A. (2006). Absorption, excretion and metabolite profiling of methyl-, glucuronyl-, glucosyl- and sulpho-conjugates of quercetin in human plasma and urine after ingestion of onions. *Br. J. Nutr.,* **96**, 107–116.

Nakajima, K., Matsushita, Y., Tohyama, Y. et al. (2006). Differential suppression of endotoxin-inducible inflammatory cytokines by nuclear factor kappa B (NFkappaB) inhibitor in rat microglia. *Neurosci. Lett.,* **401**, 199–202.

Narayan, R.K., Michel, M.E., Ansell, B., et al. (2002). Clinical trials in head injury. *J. Neurotrauma,* **19**, 503–557.

Natsume, M., Osakabe, N., Yasuda, A, et al. (2004). *In vitro* antioxidative activity of (−)-epicatechin glucuronide metabolites present in human and rat plasma. *Free Radic. Res.,* **38**, 1341–1348.

Pardridge, W.M. (2002). Drug and gene targeting to the brain with molecular Trojan horses. *Nat. Rev. Drug Discov.,* **1**, 131–139.

Park, J.S., Woo, M.S., Kim, D.H. et al. (2007). Anti-inflammatory mechanisms of isoflavone metabolites in lipopolysaccharide-stimulated microglial cells. *J. Pharmacol. Exp. Ther.,* **320**, 1237–1245.

Passamonti, S., Vrhovsek, U., Vanzo, A. et al. (2005). Fast access of some grape pigments to the brain. *J. Agric. Food Chem.,* **53**, 7029–7034.

Pawate, S. and Bhat, N.R. (2006). C-Jun N-terminal kinase (JNK) regulation of iNOS expression in glial cells: predominant role of JNK1 isoform. *Antioxid. Redox Signal,* **8**, 903–909.

Pawate, S., Shen, Q., Fan, F. et al. (2004). Redox regulation of glial inflammatory response to lipopolysaccharide and interferongamma. *J. Neurosci. Res.,* **77**, 540–551.

Peng, H.W., Cheng, F.C., Huang, Y.T. et al. (1998). Determination of naringenin and its glucuronide conjugate in rat plasma and brain tissue by high-performance liquid chromatography. *J. Chromatogr. B,* **714**, 369–374.

Piao, H.Z., Choi, I.Y., Park, J.S. et al. (2008). Wogonin inhibits microglial cell migration via suppression of nuclear factor-kappa B activity. *Int. Immunopharmacol.,* **8**, 1658–1662.

Qin, L., Liu, Y., Wang, T. et al. (2004). NADPH oxidase mediates lipopolysaccharide-induced neurotoxicity and proinflammatory gene expression in activated microglia. *J. Biol. Chem.,* **279**, 1415–1421.

Rangel-Ordonez, L., Noldner, M., Schubert-Zsilavecz, M. et al. (2010). Plasma levels and distribution of flavonoids in rat brain after single and repeated doses of standardized Ginkgo biloba extract EGb 761(R). *Planta Med.,* **76**, 1683–1690.

Rezai-Zadeh, K., Ehrhart, J., Bai, Y. et al. (2008). Apigenin and luteolin modulate microglial activation via inhibition of STAT1-induced CD40 expression. *J. Neuroinflammation,* **5**, 41.

Rice-Evans, C. (2001). Flavonoid antioxidants. *Curr. Med. Chem.,* **8**, 797–07.

Schroeter, H., Boyd, C., Spencer, J.P. et al. (2002). MAPK signaling in neurodegeneration: influences of flavonoids and of nitric oxide. *Neurobiol. Aging,* **23**, 861–880.

Schroeter, H., Spencer, J.P., Rice-Evans, C. et al. (2001). Flavonoids protect neurons from oxidized low-density-lipoprotein-induced apoptosis involving c-Jun N-terminal kinase (JNK), c-Jun and caspase-3. *Biochem. J.,* **358**, 547–557.

Sharma, V., Mishra, M., Ghosh, S, et al. (2007). Modulation of interleukin-1beta mediated inflammatory response in human astrocytes by flavonoids: implications in neuroprotection. *Brain Res. Bull.,* **73**, 55–63.

Shelnutt, S.R., Cimino, C.O., Wiggins, P.A. et al. (2002). Pharmacokinetics of the glucuronide and sulfate conjugates of genistein and daidzein in men and women after consumption of a soy beverage. *Am. J. Clin. Nutr.,* **76**, 588–594.

Smith, J.V., Burdick, A.J., Golik, P. et al. (2002). Anti-apoptotic properties of Ginkgo biloba extract EGb 761 in differentiated PC12 cells. *Cell Mol. Biol.,* **48**, 699–707.

Spencer, J.P. (2003). Metabolism of tea flavonoids in the gastrointestinal tract. *J. Nutr.,* **133**, 3255S–3261S.

Spencer, J.P. (2008). Food for thought: the role of dietary flavonoids in enhancing human memory, learning and neuro-cognitive performance. *Proc. Nutr. Soc.,* **67**, 238–252.

Spencer, J.P., Chowrimootoo, G., Choudhury, R, et al. (1999). The small intestine can both absorb and glucuronidate luminal flavonoids. *FEBS Lett.,* **458**, 224–230.

Spencer, J.P., Rice-Evans, C. and Williams, R.J. (2003a). Modulation of pro-survival Akt/protein kinase B and ERK1/2 signaling cascades by quercetin and its in vivo metabolites underlie their action on neuronal viability. *J. Biol. Chem.,* **278**, 34783–34793.

Spencer, J.P., Schroeter, H., Crossthwaithe, A.J. et al. (2001a). Contrasting influences of glucuronidation and O-methylation of epicatechin on hydrogen peroxide-induced cell death in neurons and fibroblasts. *Free Radic. Biol. Med.,* **31**, 1139–1146.

Spencer, J.P., Schroeter, H., Rechner, A.R. et al. (2001b). Bioavailability of flavan-3-ols and procyanidins: gastrointestinal tract influences and their relevance to bioactive forms in vivo. *Antioxid. Redox Signal.,* **3**, 1023–1039.

Spencer, J.P.E. (2010). The impact of fruit flavonoids on memory and cognition. *Br. J. Nutr.,* **104** (Suppl. 3), S40–S47.

Spencer, J.P.E., Kuhnle, G.G., Williams, R.J. et al. (2003b). Intracellular metabolism and bioactivity of quercetin and its *in vivo* metabolites. *Biochem. J.,* **372**, 173–181.

Sperker, B., Backman, J.T., and Kroemer, H.K. (1997). The role of β-glucuronidase in drug disposition and drug targeting in humans. *Clin. Pharmacokinet.,* **33**, 18–31.

Stewart, V.C. and Heales, S.J. (2003). Nitric oxide-induced mitochondrial dysfunction: implications for neurodegeneration. *Free Radic. Biol. Med.,* **34**, 287–303.

Suganuma, M., Okabe, S., Oniyama, M, et al. (1998). Wide distribution of [H-3](−)-epigallocatechin gallate, a cancer preventive tea polyphenol, in mouse tissue. *Carcinogenesis,* **19**, 1771–1776.

Suk, K., Chang, I., Kim, Y.H, et al. (2001). Interferon gamma (IFNgamma) and tumor necrosis factor alpha synergism in ME-180 cervical cancer cell apoptosis and necrosis. IFNgamma inhibits cytoprotective NF-kappa B through STAT1/IRF-1 pathways. *J. Biol. Chem.,* **276**, 13153–13159.

Suk, K., Lee, H., Kang, S. S. et al. (2003). Flavonoid baicalein attenuates activation-induced cell death of brain microglia. *J. Pharmacol. Exp. Ther.,* **305**, 638–645.

Takeuchi, H., Jin, S., Wang, J. et al. (2006). Tumor necrosis factor-α induces neurotoxicity via glutamate release from hemichannels of activated microglia in an autocrine manner. *J. Biol. Chem.*, **281**, 21362–21368.

Talavera, S., Felgines, C., Texier, O. et al. (2005). Anthocyanin metabolism in rats and their distribution to digestive area, kidney, and brain. *J. Agric. Food Chem.*, **53**, 3902–3908.

Taylor, D.L., Jones, F., Kubota, E.S. et al. (2005). Stimulation of microglial metabotropic glutamate receptor mGlu2 triggers tumor necrosis factor alpha-induced neurotoxicity in concert with microglial-derived Fas ligand. *J. Neurosci.*, **25**, 2952–2964.

Vafeiadou, K., Vauzour, D., Lee, H.Y. et al. (2009). The citrus flavanone naringenin inhibits inflammatory signalling in glial cells and protects against neuroinflammatory injury. *Arch. Biochem. Biophys.*, **484**, 100–109.

Vafeiadou, K., Vauzour, D., Rodriguez-Mateos, A. et al. (2008). Glial metabolism of quercetin reduces its neurotoxic potential. *Arch. Biochem. Biophys.*, **478**, 195–200.

Valles, S.L., Dolz-Gaiton, P., Gambini, J. et al. (2010). Estradiol or genistein prevent Alzheimer's disease-associated inflammation correlating with an increase PPAR gamma expression in cultured astrocytes. *Brain Res.*, **1312**, 138–144.

Vlahos, C.J., Matter, W.F., Hui, K.Y. et al. (1994). A specific inhibitor of phosphatidylinositol 3-kinase, 2-(4-morpholinyl)-8-phenyl-4H-1-benzopyran-4-one (LY294002). *J. Biol. Chem.*, **269**, 5241–5248.

Wang, C.N., Chi, C.W., Lin, Y.L. et al. (2001). The neuroprotective effects of phytoestrogens on amyloid beta protein-induced toxicity are mediated by abrogating the activation of caspase cascade in rat cortical neurons. *J. Biol. Chem.*, **276**, 5287–5295.

Wang, J.Y., Wen, L.L., Huang, Y.N. et al. (2006). Dual effects of antioxidants in neurodegeneration: direct neuroprotection against oxidative stress and indirect protection via suppression of glia-mediated inflammation. *Curr. Pharm. Des.*, **12**, 3521–3533.

Williams, C.M., El Mohsen, M.A., Vauzour, D. et al. (2008). Blueberry-induced changes in spatial working memory correlate with changes in hippocampal CREB phosphorylation and brain-derived neurotrophic factor (BDNF) levels. *Free Radic. Biol. Med.*, **45**, 295–305.

Wu, D.C., Teismann, P., Tieu, K., et al. (2003). NADPH oxidase mediates oxidative stress in the 1-methyl-4-phenyl-1,2,3,6-tetrahydropyridine model of Parkinson's disease. *Proc. Natl. Acad. Sci. U.S.A.*, **100**, 6145–6150.

Xie, Z., Smith, C.J. and Van Eldik, L.J. (2004). Activated glia induce neuron death via MAP kinase signaling pathways involving JNK and p38. *Glia*, **45**, 170–179.

Youdim, K.A., Dobbie, M.S., Kuhnle, G., et al. (2003). Interaction between flavonoids and the blood-brain barrier: *in vitro* studies. *J. Neurochem.*, **85**, 180–192.

Youdim, K.A. and Joseph, J.A. (2001). A possible emerging role of phytochemicals in improving age-related neurological dysfunctions: a multiplicity of effects. *Free Radic Biol. Med.*, **30**, 583–594.

Youdim, K.A., Spencer, J.P., Schroeter, H. et al. (2002). Dietary flavonoids as potential neuroprotectants. *Biol. Chem.*, **383**, 503–519.

Youdim, K.A., Qaiser, M.Z., Begley, D. J. et al. (2004). Flavonoid permeability across an in situ model of the blood-brain barrier. *Free Radic. Biol. Med.*, **36**, 592–604.

Yuan, D., Xie, Y.Y., Bai, X. et al. (2009). Inhibitory activity of isoflavones of *Pueraria* flowers on nitric oxide production from lipopolysaccharide-activated primary rat microglia. *J. Asian Nat. Prod. Res.*, **11**, 471–481.

Zeng, K.W., Fu, H., Liu, G.X. et al. (2010). Icariin attenuates lipopolysaccharide-induced microglial activation and resultant death of neurons by inhibiting TAK1/IKK/NF-kappaB and JNK/p38 MAPK pathways. *Int. Immunopharmacol.*, **10**, 668–678.

Zhang, L., Dawson, V.L., and Dawson, T.M. (2006). Role of nitric oxide in Parkinson's disease. *Pharmacol. Ther.*, **109**, 33–41.

Zheng, L.T., Ock, J., Kwon, B.M. et al. (2008). Suppressive effects of flavonoid fisetin on lipo-polysaccharide-induced microglial activation and neurotoxicity. *Int. Immunopharmacol.,* **8**, 484–494.

Zheng, Z., Lee, J.E., and Yenari, M.A. (2003). Stroke: molecular mechanisms and potential targets for treatment. *Curr. Mol. Med.,* **3**, 361–372.

17 Effects of Flavonoids on Cognitive Performance

Shibu M. Poulose and Barbara Shukitt-Hale

CONTENTS

17.1   INTRODUCTION

It is rare to see a day pass without hearing or reading media reports on how the population is aging and about the associated risks of the aging population in industrialized countries. We have begun to see a dramatic increase in age-associated diseases such as cancer and cardiovascular diseases. However, the most devastating diseases are the ones that involve the nervous system, such as Alzheimer's disease (AD), Parkinson's disease (PD), dementia, and others. Aging greatly affects the cognitive performance of an individual as seen in the brain's ability to acquire and process knowledge, which includes remembering, thinking, problem solving, and judging. Although scientists have long been interested in the aging process, this interest has greatly accelerated, not only since the elderly form an ever-increasing percentage of the population, but also because they utilize a significant proportion of the national medical expenditures. Animal studies indicate that the loss of memory as a function of age can occur from alterations in either the hippocampus or the striatum. The hippocampus mediates allocentric spatial navigation or place learning, whereas the striatum mediates egocentric spatial orientation or response/cue learning (Devan et al. 1996; Oliveira et al. 1997). However, overall deficits in cognitive performance occur primarily in secondary memory systems and are reflected in the retrieval of newly acquired information. Based on human studies (West 1993; Muir 1997) and using animal models, we and others (Bartus 1990; Ingram et al. 1994; Shukitt-Hale et al. 1998) have reported that the impairments in retrieval are attributed to deficits in such encoding processes as motivation, attention, processing depth, and organizational skills.

Numerous contemporary studies indicate the potential health benefits of plant-derived secondary metabolites, collectively known as phytochemicals, such as flavonoids, carotenoids, coumarins, terpenoids, essential oils, tannins, and limonoids, in ameliorating the risks of age-related diseases. These compounds are often synthesized by plants in response to various environmental stresses like diseases, insects and pests, climate, and ultraviolet radiation (Dixon and Paiva 1995; Winkel-Shirley 2002). However, various classes of these phytochemicals possess human health-promoting properties such as antioxidant, antiallergic, anti-inflammatory, antiviral, antiproliferative, and anticarcinogenic effects (Spencer 2009, 2010; Poulose et al. 2005, 2006; Vauzour et al. 2008). Over the past decades, studies to determine the health-promoting benefits of phytochemicals have primarily been focused on examining their roles in reducing risk factors associated with cancer and heart disease. Limited research has been directed towards their roles in reducing age-related decrements in brain functions such as learning and memory, and in overall cognitive functions, which are likely to have a negative impact on the quality of life.

17.2 FLAVONOIDS: FORMS, SOURCES, AND BIOAVAILABILITY IN THE BRAIN

Flavonoids constitute the most abundant phytonutrients in the human diet and are structurally classified into six major groups including: (i) flavonols (isorhamnetin, kaempferol, myricetin, quercetin), (ii) flavones (apigenin, luteolin), (iii) flavanones (eriodictyol, hesperetin, naringenin), (iv) flavan-3-ols [(+)-catechin, (−)-epicatechin, (−)-epicatechin-3-O-gallate, (−)-epigallocatechin, (−)-epigallocatechin-3-O-gallate, (+)-gallocatechin], (v) theaflavins (theaflavin, theaflavin-3-O-gallate, theaflavin-3″-O-gallate, theaflavin-3,3″-O-digallate, thearubigins), and (vi) anthocyanidins (cyanidin, delphinidin, malvidin, pelargonidin, peonidin, petunidin) (Crozier et al. 2006; USDA Database 2007). Flavonoids possess a common skeleton of the flavan nucleus, and differences between various classes are due to changes in ring C (presence of a double bond, a 3-hydroxy group, and/or a 4-oxo group) and in the number and position of hydroxyl and methoxyl groups in rings A and B (Crozier et al. 2006; USDA Database 2007). The most abundant flavonoids in the diet are flavan-3-ols (catechins and proanthocyanidins), anthocyanins (anthocyanidins with a glycoside moiety), and their oxidation products, and these account for about two thirds of our daily total intake (Scalbert et al. 2000). The major dietary sources from which these compounds are obtained include fruits, vegetables, and beverages (e.g., fruit juice, wine, tea, coffee, chocolate, and beer), with an average total daily intake of approximately 1 g (Manach et al. 2004). Variability in flavonoid composition in plants occurs because of cultivar variation, environmental conditions or growing location, agricultural practices, processing and storage conditions, and preparation methods (Amiot et al. 1995; Hakkinen et al. 2000; Van der Sluis et al. 2001). Most of the flavonoids in food are present in glycosylated forms except for the flavan-3-ols, which are typically present either in free forms or as gallic acid esters (e.g., in tea) (Crozier et al. 2006; USDA Database 2007).

Regarding potential benefits on brain health, it is important to determine the form and amount of each flavonoid that enters the brain to induce the desired effect. After oral ingestion, flavonoids are extensively conjugated and metabolized during absorption in the small intestine and then again in the liver (Yodim et al. 2004a). The intact form of flavonoid and respective metabolites derived from flavonoid biotransformation in the gastrointestinal tract and in the liver are the forms that enter the circulation and ultimately reach the brain (Abbott 2002; Ross and Kasum 2002). However, the entry of flavonoids into the brain is further complicated because of the blood–brain barrier (BBB), which controls the composition of extracellular fluid in the central nervous system, preventing entry of all but the smallest molecules (Abbott 2002; Yodim et al. 2004b). The BBB is formed by the endothelium of brain microvessels under the inductive influence of astrocytes, and transporters such as multidrug resistance-associated proteins (MRPs) assist in the crossover of substrates across the blood–brain interface (Suganuma et al. 1998; Singh et al. 2008). Furthermore, the size of the brain makes tissue availability limited, in particular, when attempting to localize polyphenolics in specific structures (Yodim et al. 2003, 2004b). Therefore, correlating *in vitro* effects with clinical effects is problematic because of poor absorption of naturally occurring flavonoid glycosides and extensive conjugation of the flavonoid aglycones after absorption. Even though the presence of most of these flavonoids inside the brain is highly difficult to detect, due to the extremely low levels as well as conjugated forms, some of these compounds have been reported to cross the BBB and localize within brain structures. These include (−)-epigallocatechin-3-*O*-gallate (Suganuma et al. 1998; Abd El Mohsen et al. 2002), a major component found in green and black tea as well as wine; naringenin (Peng et al. 1998), hesperetin (Tsai and Chen 2000), quercetin (Youdim et al. 2003), and tangeretin (Datla et al. 2001). In addition, synthetic flavonoid-like compound, Emd 49209, has been localized both in the adult rat brain and developing fetal rat brain (Schroder-vander Elst et al. 1998). The studies also indicate that methylated flavonoids cross the BBB more readily than their phenolic counterparts (Youdim et al. 2004a).

There are conflicting reports on the presence of anthocyanins inside brain tissues. While one study indicated detectable levels of anthocyanins in the hippocampus of rats fed blueberries for 8 weeks (Shukitt-Hale et al. 2008a), another study was not able to establish the presence of raspberry anthocyanins and ellagitannins in the rat brain following berry consumption (Borges et al. 2007). One hour after administration of one dose of raspberry juice, anthocyanins and ellagitannins were transiently detected in the plasma, but were absent in the brain. However, in an extended study using pigs, berry polyphenols were found to accumulate in several tissues, including eye and brain, following four weeks of supplementation (Kalt et al. 2008). Interestingly, in plasma, the levels of anthocyanins were undetectable, and the data also suggest that the levels of anthocyanins in neural tissue were not in equilibrium with plasma levels. Despite the lack of strong evidence to suggest polyphenolics are able to penetrate into the brain, there is support to show that they are able to promote a variety of neuroprotective actions to curtail deficits associated with aging.

17.3 FLAVONOIDS AND BRIAN HEALTH

Most of the studies on the health benefits of flavonoids are related to cancer and cardiovascular disease prevention; the role of flavonoids on brain health, particularly on how they affect cognitive performance and specific neuronal function, has received relatively less attention. There are quite a few reports that flavonoids significantly suppress the activation of nuclear factor kappa B (NF-kB) and activator protein 1 (AP-1) as well as mitogen activated protein kinase (MAPK) pathways in activated microglia, resulting in an attenuation of the production of inflammatory molecules (Chen et al. 1996; Luterman et al. 2000; Bodles et al. 2004). NF-kB and AP-1 are important transcription factors, and MAPK-extracellular signal regulated kinases 1 and 2 (ERK1/2), p38, and c-Jun-N-terminal kinase (c-JNK) pathways are involved in regulating inflammatory gene expression in neurons (McGeer and McGeer 1995; Joseph et al. 2000). Microglial cells, an important type of brain cell, protect the neurons from oxidative stress and inflammation (Kreutzberg 1996; Rozovsky et al. 1998). However, under a highly activated state caused by extensive oxidative or inflammatory insults, these microglial cells produce inflammatory molecules such as cytokines, superoxide, and nitric oxide. Inflammatory cytokines such as interleukins (IL) disrupt hippocampal-dependent cognitive processing during infection (McGeer and McGeer 1995; Rabin et al. 1998; Bickford et al. 1999; Joseph et al. 2000, 2005). Specifically, IL-1β inhibits memory consolidation, making it more difficult to learn new tasks. Elevated hippocampal levels of IL-1β are associated with impaired synaptic plasticity in the CA1 region of the hippocampus and deficits in long-term potentiation (LTP), which is thought to be a biological substrate for learning and memory (Vereker et al. 2001). The inflammatory cytokine IL-6 may also play a role in modulating cognition during neurodegenerative disease and aging. The severity of dementia in Alzheimer's disease (AD) has been linked to high levels of IL-6 in plaques (Jang and Johnson 2010). Therefore, inhibition of these molecules, as well as reduction in the activation of microglia, has been shown to be beneficial in mitigating neurodegenerative disorders (Jang and Johnson 2010).

Quercetin, a major dietary flavonol, has been shown to suppress lipopolysaccharide (LPS)-induced nitrous oxide (NO) production and inducible nitric oxide synthase gene transcription by blocking activation of an inhibitor of kappa B (IkB) kinase, NF-kB, and AP-1 (Chen et al. 2005). Luteolin, another flavonoid, has been shown to attenuate IL-1β, tumor necrosis factor-α (TNF-α), NO, and prostaglandin E2 (PGE2), and affect the microglial transcriptome leading to an anti-inflammatory, antioxidative, and neuroprotective phenotype (Jang et al. 2008). Similarly, apigenin and luteolin suppressed microglial TNF-α and IL-6 production stimulated by interferon-gamma in the presence of CD40 ligation (Rezai-Zadeh et al. 2008; Dirscherl et al. 2010). Fisetin, another flavonoid, has shown attenuated TNF-α and NO production (Zheng et al. 2008). It has also been shown to suppress nuclear translocation of NF-kB and phosphorylation of p30-MAP kinase in LPS-stimulated BV-2 microglia cells. Genistein, a soy isoflavone, inhibited activation of microglia and production of TNF-α, NO, and superoxides, in rat mesencephalic neuron–glia cultures stressed with LPS (Wang et al. 2005). Similar effects were also noticed for another flavonoid, wogonin (6,7-dihydroxy-8-methoxyflavone) (Lee et al. 2003).

Along with genistein, other flavonoids such as resveratrol and quercetin have been shown to protect dopaminergic neurons, loss of which has been implicated in Parkinson's disease (Wang et al. 2005; Bureau et al. 2008). Pharmacological effects of flavonoids also include free-radical scavenging; anti-inflammatory, antispasmodic, antiallergic, and antiviral effects; as metal chelators, exhibiting interaction with key enzymes; regulating intrinsic calcium, cell signaling, and gene expression; reduction in capillary fragility; inhibiting platelet aggregation; and enhancing neurogenesis (Shukitt-Hale et al. 2008a; Joseph et al. 2009; Spencer 2009, 2010; Del Rio et al. 2010).

17.4 FLAVONOIDS IN AGEING AND COGNITION

There is an overwhelming amount of data suggesting the pathophysiological processes that lead to losses in neuronal function and eventual cognitive decline in aging. These include, but are not limited to, sensitivity to oxidative stress, inflammation, loss of endogenous antioxidant systems, altered receptor sensitivity and membrane alterations, loss of calcium homeostasis, and loss of proteosome/autophagy function, which leads to loss of housekeeping in neurons and eventual neuronal death (Shukitt-Hale et al. 2008a; Joseph et al. 2009; Spencer 2009, 2010; Del Rio et al. 2010; Poulose et al. 2011a). However, flavonoids present in fruits and vegetables can counter insults on the brain and could be potential candidates for the development of neuroprotective therapies. Numerous studies indicate the beneficial effects of flavonoid supplementation on brain health, particularly in reducing oxidative stress, neuroinflammation, and cognitive decline in animal models. For example, when aged mice (22 and 24 months old) were fed with luteolin for 8–12 weeks, their brains showed an inhibition of microglial over-activation and improved hippocampal-dependent spatial and working memory (Jang et al. 2010). We have reported that feeding aged rats with flavonoid-rich fruits and vegetables such as blueberry, strawberry, or spinach extracts exhibited improved learning and better memory retention than nonsupplemented animals (Joseph et al. 1998, 1999). In another study, feeding mice with apigenin and quercetin, as well as soy isoflavones, reversed the age- and LPS-induced retention deficits in passive avoidance and elevated plus-maze tasks (Wang et al. 2002; Patil et al. 2003). Quercetin, as well as onion extracts, showed protection against ischemic neuronal damage in the gerbil hippocampus (Hwang et al. 2009). In Alzheimer's disease and amyotrophic lateral sclerosis (ALS) animal models, ginkgo biloba, grape seed extracts, and green tea catechins attenuated the accumulation of beta amyloid plaques, suppressed morphologic and functional regression of the brain, as well as blocked spatial memory deficits (Stackman et al. 2003; Haque et al. 2006; Xu et al. 2006; Wang et al. 2009).

Epidemiological and observational studies also support the beneficial effects of flavonoids on cognitive deficits in old persons or demented patients (Gillette et al. 2007; Letenneur et al. 2007; Commenges et al. 2000; Krikorian et al. 2010a,b). Letenneur et al. (2007) reported an inverse relationship between flavonoid intake and cognitive decline in a so-called PAQUID study, in which a cluster sample of 1640 subjects in the southwestern regions of Gironde and Dordogne in France were followed up for about 5 and 10 years. In this study, an average intake of 14.3 ± 5.85 mg/day of dietary flavonoid resulted in less cognitive decline in subjects aged 65 years or older. Patients who were given a flavonoid-rich diet showed better cognitive performance at baseline as well as a better evolution of performance over time. Similarly, Commenges et al. (2000), in a cohort study of 1367 subjects above 65 years of age, also reported a decreased risk of dementia with increased flavonoid-rich diet consumption. Consumption of fruit and vegetable juices high in flavonoid polyphenolics, at least three times per week, is thought to delay the onset of AD, particularly in ApoE4 carriers (Dai et al. 2006). In another community-based case–control cohort study by Morris et al. (2002) with 194 AD patients and 1790 non-AD patients, risk of AD was significantly lowered with a high intake of the Mediterranean diet, consisting of high amounts of fruits, vegetables, cereals, and fish, mild to moderate amounts of alcohol, and low amounts of red meat and dairy products. The Mediterranean diet has also been shown to reduce the incidence of dementia as well as mortality in patients with AD (Scarmeas et al. 2006a,b, 2007).

A recent human study by Krikorian et al. (2010a) involving nine older adults, where the subjects were supplemented with blueberry (BB) juice for 12 weeks, showed improved memory with respect to paired associate learning and word list recall when compared to that of demographically matched subjects who consumed a placebo beverage. The study also indicated reduced depressive symptoms and lower glucose levels. In another human study conducted by the same group, 12 older adults with mild cognitive impairment but not dementia were supplemented with Concord grape juice for 12 weeks, in a randomized, placebo-controlled, double blind trial (Krikorian et al. 2010b). The results indicated significant improvement in a measure of verbal learning and a nonsignificant trend toward enhancement of verbal and spatial recall. These studies provide substantial evidence for the beneficial effects of flavonoids in improving human cognitive functions. However, it has been argued that the subjects in most of these studies were close to the onset of dementia (aged 65 years and more), when the vulnerability for oxidative stress is very high and degeneration of neurons is already initiated (Singh et al. 2008). Therefore, the relative long-term effects of these flavonoids on cognitive performance could be underestimated and needs to be investigated beginning with younger subjects and over a longer period of time.

17.5 EFFECTS OF BERRY FLAVONOIDS ON COGNITIVE HEALTH AND RELATED NEURONAL SIGNALING

Berries, such as blueberry (BB), strawberry (SB), raspberry, blackberry, cranberry, and acai berry, are known to possess an array of phytochemicals and to be a predominantly rich source of flavonoids. The major flavonoids in the popular berries of the United States have been summarized in Table 17.1. Our laboratory has previously shown that feeding with blueberry, strawberry, or spinach extract (1–2% of the diet) was effective in reversing age-related decrements in cognitive or neuronal function in Fischer 344 rats (Joseph et al. 1998, 1999). However, only the BB-supplemented group exhibited improved performance on tests of motor function, particularly balance and coordination, rod walking and the accelerating rotarod,

TABLE 17.1

Major Types of Flavonoids in the Popular Berry Fruits and their Health Effects on Cognitive Performance

Berry Fruit	Flavonoids	Types	mg/100g	Specific Brain Health Effect
Blueberries wild	Anthocyanidins	Cyanidin	42	Improved memory and cognition (human and animals),
		Delphidin	93	↓ neuroinflammation (NF-kB, cytokines, p-CREB, OX-6, COX 1&2, TNF-α,
		Malvidin	103	↑ GST, SOD.
		Peonidin	23	↑ Synaptic plasticity via ↑ ERK1/2, IGF-1, IGF-1R, P38-MAPK, JNKs
		Petunidin	58	Reducing the sensitivity of MAChR to ROS
Blueberries cultivated	Anthocyanidins	Cyanidin	17	Improved working memory in both striatum and hippocampus of rats
		Delphidin	47	Improved memory and cognition (human and animals),
		Malvidin	61	↓ neuroinflammation via NF-kB, cytokines, p-CREB, OX-6, COX 1&2, TNF-α
		Peonidin	12	Microglial modulation
		Petunidin	27	↑ Synaptic plasticity via ↑ ERK1/2, IGF-1, IGF-1R, P38-MAPK, JNKs
	Flavan-3-ols	Catechin	37	Reducing the sensitivity of MAChR to ROS
		Epicatechin	14	
Cranberries	Anthocyanidins	Cyanidin	41	Improved memory in humans, reducing ROS
		Delphidin	8	
		Peonidin	42	
	Flavonol	Quercetin	15	
Strawberries	Anthocyanidins	Cyanidin	2	Improved memory and cognition in animals, primarily affecting hippocampus, reducing ROS,
		Pelargonidin	32	↓ neuroinflammation (NF-kB, cytokines IL-1B, p-CREB, OX-6, COX 1&2, TNF-α microglial
	Flavan-3-ols	Catechin	7	modulation
Raspberries	Anthocyanidins	Cyanidin	102	
		Delphidin	13	
		Malvidin	3	
		Peonidin	5	
	Flavonol	Quercetin	113	

(Continued)

TABLE 17.1 (CONTINUED)
Major Types of Flavonoids in the Popular Berry Fruits and their Health Effects on Cognitive Performance

Berry Fruit	Flavonoids	Types	mg/100g	Specific Brain Health Effect
Acai berries	Anthocyanidins	Procyanidin Dimers	1086	↓ NO production,
		Procyanidin trimers	2016	↓ iNOS,
	Flavan-3-ols	Catechins	67	↓ inflammation ((NF-kB, cytokines IL-1B)
		Procatechuic acid	630	alteration of pain perception
				↓ Oxidative stress
Blackberries	Anthocyanidins	Cyanidin	90	↑ motor and cognition in rats,
	Flavan-3-ols	Catechin	37	↑ short-term memory in rats
				↓ oxidative stress
Bilberry	Anthocyanidins	Cyanidin	113	Protection of retinal ganglion cells in the eye
		Delphidin	162	Inhibition of lipid peroxidation in forebrain of rats
		Malvidin	54	Neuroprotection via antioxidant mechanisms
		Peonidin	51	
		Petunidin	51	
Cowberries	Anthocyanidins	Cyanidin	44	
Elderberries	Flavonol	Quercetin	13	
	Anthocyanidins	Cyanidin	758	
	Flavonol	Quercetin	42	

whereas none of the other supplemented groups differed from control on these tasks. In another study, results showed that short-term (8 weeks) BB supplementation could prevent the onset of age-related deficits in cognitive behavior, improving memory performance in the radial arm water maze (Casadesus et al. 2004). Furthermore, the improvement in spatial memory was correlated to improved hippocampal plasticity parameters such as hippocampal neurogenesis, extracellular receptor kinase activation, and insulin-like growth factor-1 (IGF-1), and IGF-1 receptor (IGF-1R) levels (Casadesus et al. 2004).

Subsequent studies from our laboratory and others have also indicated an improvement in several neuronal signaling indices, such as muscarinic acetylcholine receptor (MAChR) sensitivity, reduced microglial activity as measured by immunoreactivity to major histocompatibility complex class II molecules (MHC-II OX-6), and glial fibrillary acidic protein (GFAP), and reduced inducible nitric oxide synthase (iNOS), using BB- and SB-supplemented animals, as well as cell culture models (Joseph et al. 2003a,b, 2005, 2006, 2009, 2010; Shukitt-Hale et al. 2008; Malin et al. 2010). BB diet also reversed age-related "dysregulation" in Ca^{2+} buffering capacity and protected the brain against kainic acid-induced neurotoxicity (Duffy et al. 2008; Shukitt-Hale et al. 2008a,b). Examinations of ROS production in brain tissue induced by hydrogen peroxide and assayed using DCF;2',7'-dichlorofluorescein diacetate indicated that the striata obtained from all of the supplemented groups exhibited significantly lower ROS levels than the controls (Crivello et al. 2007).

Evidences indicate a significant role of flavonoids in effecting neuronal signaling, which is directly correlated with the cognitive performance of an individual. For example, several reports indicate that protein kinase C (PKC) activity is important in memory formation, particularly spatial memory (Serrano et al. 1994; Micheau and Riedel 1999), and that treatment with PKC inhibitors impairs memory formation (Micheau and Riedel 1999). Colombo et al. (1997) showed that young rats with the best performance in spatial memory also had the highest PKCγ localization on the membranes of the hippocampal neurons, while PKCβ$_2$ was localized in the cytosol. Similarly, calcium-dependent PKCs play a critical role in the conversion of short- to long-term memory, and MAP kinases are critical in long-term memory formation through the activation of cyclic-AMP response element-binding (CREB) proteins (Serrano et al. 1994; Micheau and Riedel 1999; Colombo et al. 1997; Brewer et al. 2009). MAP kinases also play a key role in synaptic plasticity via modulation of extracellular signal regulated kinases (ERK) 1/2, and the Jun kinases (JNKs) (Colombo et al. 1997; Brewer et al. 2009). In rat primary hippocampal neurons, BB fractions rich in proanthocyanidins attenuated the deleterious effects of oxidative/ inflammatory stress (induced by 6-hydroxydopamine, amyloid beta, or LPS) through increased activation of P38, MAPK, and by modulating JNK and ERKs (Brewer et al. 2009; Joseph et al. 2010). Furthermore, in BV-2 mouse microglial cells, purified BB extracts suppressed the LPS-induced increases in iNOS and NF-kB, indicating a secondary protection on neurons by microglial cells (Lau et al. 2007). This suggests that flavonoids act differently in two key types of brain cells involved in memory and cognition; in glial cells, these flavonoids could potentially play a protective role by activating redox mechanisms and housekeeping, whereas in neurons, they could potentially activate prosurvival signals.

Substantiating the improvements in cognitive effects of blueberry flavonoids to the molecular events in neurons, Williams et al. (2008) showed that improvements in spatial memory of aged rats, fed with 2% (w/w) blueberry diets for 12 weeks, were correlated to the activation of CREB and increases in the levels of brain-derived neurotrophic factor (BDNF) in the hippocampus. Activation of CREB has been widely considered to improve the memory deficits and cognitive declines observed in Alzheimer's patients (Tully et al. 2003), and BDNF has been implicated in synaptic plasticity and long-term memory (Bramham and Messaoudi 2005). The study also showed increased phosphorylation of ERK1/2, cytoskeletal-associated protein (Arc/Arg31), and phospho inositol 3 kinase/Akt, and modulation in mammalian target of rapamycin (mTOR). Increase in the activity of these signals is not only related to neuronal plasticity and memory formation, but also to the synthesis of new mRNA and proteins (Williams et al. 2008).

Similarly, the striatal muscarinic receptor (MAChR) plays a key role in a variety of functions including memory, amyloid precursor protein (APP) processing, and vascular functioning (Rossner et al. 1998; Elhusseiny et al. 1999). The sensitivity of MAChR decreases with oxidative stress and aging, as well as in the exacerbated conditions of AD and vascular dementia (VaD), leading to cognitive, behavioral, and neuronal aberrations (Joseph et al. 2002; DeSarno et al. 2003). Experiments from our laboratory using COS-7 cells transfected with the M1 receptor, among five MAChRs, showed increased sensitivity to oxidative stress as a function of loss of Ca^{2+} buffering following oxotremorine-induced depolarization (Joseph et al. 2006). However, blueberry extracts effectively reduced the sensitivity of the M1 receptors, thereby reducing the susceptibility to oxidative stress (Joseph et al. 2006). Similarly, the loss of calcium buffering has been consistently linked to aging (Herman et al. 1998), and such losses significantly affect the functioning and viability of the cell (Vannucci et al. 2001), ultimately resulting in motor- and memory-function decrements in senescent rats (Huidobro et al. 1993). However, in vitro, the losses in Ca^{2+} buffering significantly recovered when BV-2 microglial and primary hippocampal cells were pretreated with blueberry, strawberry, and acai berry extracts (Joseph et al. 2010; Poulose et al. 2011b).

Numerous animal studies indicate profound differences were observed with respect to different aging factors such as morphology, electrophysiology, and receptor sensitivity for the different regions of the brain such as hippocampus, cerebellum, and striatum (Joseph et al. 1996; Hartmann et al. 1996; Nyakas et al. 1997; Kaufmann et al. 2001). Experimental results indicated that mechanistically the protective effects of the BB may be derived from increasing MAPK signaling and via attenuation of pCREB and PKCγ (Joseph et al. 2003; Shukitt-Hale et al. 2008a; Vuong et al. 2010). It has also been observed that transgenic mice with mutations for amyloid precursor protein and presenilin-1 (PS-1), a model for Alzheimer's disease, did not show behavioral deficits in Y-maze performance when supplemented with BB diet for 11 months when compared to control animals (Joseph et al. 2003). Moreover, the supplemented mice showed enhancements in several signaling molecules such as PKC-α, ERK, and GTPase, which are associated with cognitive function. In other studies involving rat brain, BB supplementation was effective in antagonizing the H_2O_2-induced age-related changes in neuronal signaling, such as decrements in striatal synaptosomal calcium buffering and behavior, by both anti-inflammatory and antioxidant activities (Vuong et al. 2010). Isolated striatal slices from the brains of rats whose diets had been supplemented with BB extracts showed increased oxotremorine-enhanced dopamine release, which is essential for maintaining the functional integrity of the striatal dopaminergic system and has a major impact on certain behavioral parameters (Shukitt-Hale et al. 2007; Joseph et al. 1996; Bickford et al. 1999, 2000).

Blueberries were also shown to prevent and/or reverse age-related declines in cerebellar noradrenergic receptor function, whereas strawberries or blueberries exhibited reversal of age-induced declines in β-adrenergic receptor function in cerebellar Purkinje neurons (Bickford et al. 2000; Andres-Lacueva et al. 2005; Shukitt-Hale et al. 2007). Similarly, rats given a strawberry-supplemented diet were shown to have enhanced striatal muscarinic receptor sensitivity, which was directly correlated with the reversal of cognitive behavioral deficits (Shukitt-Hale et al. 2006). Therefore, it can be argued that the blueberry flavonoids affect both short- and long-term memory formation not only by affecting the major neuronal signaling mechanisms such as those that involve cAMP dependent kinase, calcium-calmodulin kinases, PKCs, and MAPKs, but also by the synthesis of new mRNA and proteins, and by the covalent modification of preexisting proteins.

While most of the studies on brain health with berry effects were done using blueberries and strawberries, a few studies have also shown the beneficial effects of other berries. For example, in another study from our laboratory, when 19-month-old Fischer 344 rats were given a diet supplemented with 2% blackberry for 8 weeks, they showed improved motor performance on three tasks which rely on balance and coordination: the accelerating rotarod, wire suspension, and the small plank walk (Shukitt-Hale et al. 2009). Results for the Morris water maze also showed that the blackberry-fed rats had significantly greater working, or short-term, memory performance than the control rats. In a separate study, rats given intraperitoneal injections of 200 mg/kg/day bilberry anthocyanins for 5 days had significantly more triiodothyronine (T3) in their brains than rats given only the solvent (26% alcohol). Since T3 can enter the brain by a specific transport in the capillaries, it has been postulated that anthocyanins could mediate T3 transport at the capillary level (Saija et al. 1990; Ramirez et al. 2005). The report also indicates superior memory, better vision, and better control of sensory input among the bilberry-treated animals when compared with that of control-fed animals.

17.6 SUMMARY

Overall, the evidence points toward the convincing effects of flavonoids in reducing or reversing the age-related cognitive decline in animals as well as in humans. The effects of flavonoids related to cognitive performance are summarized in Figure 17.1. Most of the effects are exerted via reducing inflammation and oxidative stress in key parts of the memory formation, such as hippocampus, cortex, or striatum, along with the modification of microglial-derived signaling mechanisms. However, new evidences are emerging indicating that the flavonoid effects in brain extend beyond their antioxidant and anti-inflammatory properties, to restoration of neuronal

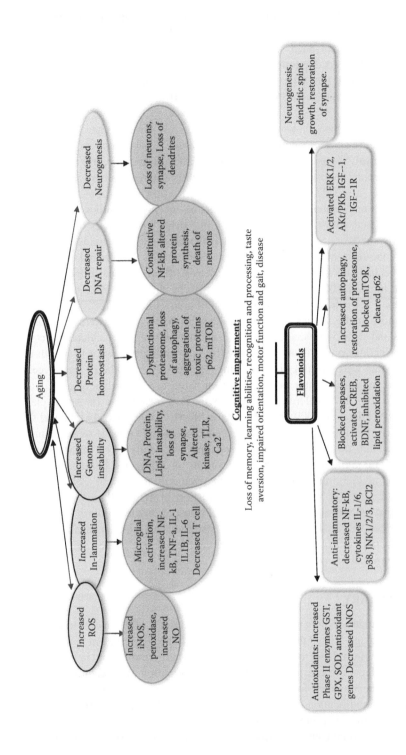

FIGURE 17.1　Schematic representation of the possible effects of aging leading to decline in cognitive performance and known effects of flavonoids in improving the cognitive decrements.

housekeeping, neurogenesis, and gene regulations. Considering the consistent reduction of neurodegenerative effects in animal models of diseases such as AD, PD, and ALS, and the improvement of cognitive performance in recent human studies involving elderly individuals, flavonoid-rich diets represent an intriguing alternative or adjunctive therapy for the prevention and treatment of age-related brain dysfunction. Berries loaded with very high amounts of flavonoids are proven to be an effective option in improving motor and cognitive performances in aged animals. As sensitivity to oxidative stress and inflammation continue to rise in an ever-increasing aging population, nutritional interventions, particularly those involving berries and colorful fruits and vegetables, can play a pivotal role in maintaining cognitive, motor, and neuronal functions, and possibly reverse decrements in these parameters in senescence.

REFERENCES

Abbott, N.J. (2002). Astrocyte-endothelial interactions and blood-brain barrier permeability. *J. Anatomy*, **200**, 629–638.

Abd El Mohsen, M.M., Kuhnle,G., Rechner, A.R. et al. (2002). Uptake and metabolism of epicatechin and its access to the brain after oral ingestion. *Free Rad. Biol. Med.*, **33**, 1693–1702.

Amiot, M.J., Tacchini, M., Aubert, S.Y. et al. (1995). Influence of cultivar, maturity stage and storage conditions on phenolic composition and enzymatic browning in pear fruits. *J. Agric. Food Chem.*, **43**, 1132–1137.

Andres-Lacueva, C., Shukitt-Hale, B., Galli, R.L. et al. (2005). Anthocyanins in aged blueberry-fed rats are found centrally and may enhance memory. *Nutr.Neurosci.*, **8**, 111–120.

Bartus, R.T. (1990). Drugs to treat age-related neurodegenerative problems. The final frontier of medical science? *J. Am. Geriatr. Soc.*, **38**, 680–695.

Bickford, P.C., Shukitt-Hale, B., and Joseph, J. (1999). Effects of aging on cerebellar noradrenergic function and motor learning: Nutritional interventions. *Mech. Ageing Develop.*, **111**, 141–154.

Bickford, P.C., Gould, T., Briederick, L. et al. (2000). Antioxidant-rich diets improve cerebellar physiology and motor learning in aged rats. *Brain Res.*, **866**, 211–217.

Bodles, A.M. and Barger, S.W. (2004). Cytokines and the aging brain—what we don't know might help us. *Trends Neurosci.*, **27**, 621–626.

Borges, G., Roowi, S., Rouanet, J.M. et al. (2007). The bioavailability of raspberry anthocyanins and ellagitannins in rats. *Mol. Nutr. Food Res.*, **51**, 714–725.

Bramham, C.R. and Messaoudi, E. (2005). BDNF function in adult synaptic plasticity: The synaptic consolidation hypothesis. *Progr. Neurobiol.*, **76**, 99–125.

Brewer, G.J., Torricelli, J.R., Lindsey, A.L. et al. (2010). Age-related toxicity of amyloid-beta associated with increased pERK and pCREB in primary hippocampal neurons: Reversal by blueberry extract. *J. Nutr. Biochem.*, **21**, 991–998.

Bureau, G., Longpre, F. and Martinoli, M.G. (2008). Resveratrol and quercetin, two natural polyphenols, reduce apoptotic neuronal cell death induced by neuroinflammation. *J. Neurosci. Res.*, **86**, 403–410.

Casadesus, G., Shukitt-Hale, B., Stellwagen, H. M. et al. (2004). Modulation of hippocampal plasticity and cognitive behavior by short-term blueberry supplementation in aged rats. *Nutr. Neurosci.*, **7**, 309–316.

Chen, S., Frederickson, R.C., and Brunden, K.R. (1996). Neuroglial-mediated immunoinflammatory responses in Alzheimer's disease: Complement activation and therapeutic approaches. *Neurobiol. Aging*, **17**, 781–787.

Chen, J.C., Ho, F.M., Pei-Dawn Lee, C. et al. (2005). Inhibition of iNOS gene expression by quercetin is mediated by the inhibition of IkappaB kinase, nuclear factor-kappa B and STAT1, and depends on heme oxygenase-1 induction in mouse BV-2 microglia. *Eur. J. Pharmacol.,* **521**, 9–20.

Colombo, P.J., Wetsel, W.C. and Gallagher, M. (1997). Spatial memory is related to hippocampal subcellular concentrations of calcium-dependent protein kinase C isoforms in young and aged rats. *Proc. Natl. Acad. Sci., U.S.A,* **94**, 14195–14199.

Commenges, D., Scotet, V., Renaud, S. et al. (2000). Intake of flavonoids and risk of dementia. *Eur. J. Epidemiol.,* **16**, 357–363.

Crivello, N.A., Rosenberg, I.H., Shukitt-Hale, B.I (2007). Aging modifies brain region-specific vulnerability to experimental oxidative stress induced by low dose hydrogen peroxide. *Age,* **29**, 191–203.

Dai, Q., Borenstein, A.R., Wu, Y. et al. (2006). Fruit and vegetable juices and Alzheimer's disease: The Kame Project. *Am. J. Med.,* **119**, 751–759.

Datla K.P., Christidou, M., Widmer, W.W. et al. (2001). Tissue distribution and neuroprotective effects of citrus flavonoid tangeretin in a rat model of Parkinson's disease. *Neuroreport,* **12**, 3871–3875.

Del Rio, D., Borges, G., and Crozier, A. (2010). Berry flavonoids and phenolics: Bioavailability and evidence of protective effects. *Br. J. Nutr.,* **3**, S67–S90.

DeSarno, P., Shestopal, S.A., King, T.D. et al. (2003). Muscarinic receptor activation protects cells from apoptotic effects of DNA damage, oxidative stress and mitochondrial inhibition. *J. Biol. Chem.,* **278**, 11086–11093.

Devan, B.D., Goad, E.H., and Petri, H.L. (1996). Dissociation of hippocampal and striatal contributions to spatial navigation in the water maze. *Neurobiol. Learn. Mem.,* **66**, 305–323.

Dirscherl, K., Karlstetter, M., Ebert, S. et al. (2010). Luteolin triggers global changes in the microglial transcriptome leading to a unique anti-inflammatory and neuroprotective phenotype. *J. Neuroinflamm.,* **14**, 7, 3.

Dixon, R.A. and Paiva, N.L. (1995). Stress-induced henylpropanoid metabolism. *Plant Cell,* **7**, 1085–1097.

Duffy, K.B., Spangler, E L., Devan, B.D. et al. (2008). A blueberry-enriched diet provides cellular protection against oxidative stress and reduces a kainate-induced learning impairment in rats. *Neurobiol. Aging,* **29**, 1680–1689.

Elhusseiny, A., Cohen, Z., Olivier, A. et al. (1999). Functional acetylcholine muscarinic receptor subtypes in human brain microcirculation: Identification and cellular localization. *J. Cerebr. Blood Flow Metab.,* **19**, 794–802.

Engelhart, M.J., Geerlings, M.I., Ruitenberg, A. et al. (2002). Dietary intake of antioxidants and risk of Alzheimer disease. *JAMA,* **287**, 3223–3232.

Gillette Guyonnet, S., Abellan Van Kan, G., Andrieu, S. et al. (2007). IANA task force on nutrition and cognitive decline with aging. *J. Nutr. Health Aging,* **11**, 132–152.

Hakkinen, S.H., Karenlampi. S.O., Mykkanen, H.M. et al. (2000). Influence of domestic processing and storage on flavonoid contents in berries. *J. Agric. Food Chem.,* **48**, 2960–2965.

Haque, A.M., Hashimoto, M., Katakura, M. et al. (2006). Long-term administration of green tea catechins improves spatial cognition learning ability in rats. *J. Nutr.,* **136**, 1043–1047.

Hartmann, H., Velbinger, K., Eckert, A. et al. (1996). Region-specific downregulation of free intracellular calcium in the aged rat brain. *Neurobiol. Aging,* **17**, 557–563.

He, F.J., Nowson, C.A., and MacGregor, G A. (2006). Fruit and vegetable consumption and stroke: Meta-analysis of cohort studies. *Lancet,* **367**, 320–326.

Herman, J.P., Chen, K.C., Booze, R. et al. (1998). Upregulation of αDCa^{2+} channel subunit mRNA expression in the hippocampus of aged 344 rats. *Neurobiol. Aging,* **19**, 581–587.

Huidobro, A., Blanco, P., Villalba, M. et al. (1993). Age-related changes in calcium homeostatic mechanisms in synaptosomes in relation with working memory deficiency. *Neurobiol. Aging*, **14**, 479–486.

Hwang, I.K., Lee, C.H., Yoo, K.Y. et al. (2009). Neuroprotective effects of onion extract and quercetin against ischemic neuronal damage in the gerbil hippocampus. *J. Med. Food*, **12**, 990–995.

Ingram, D.K., Spangler, E.L., Iijima S. et al. (1994). New pharmacological strategies for cognitive enhancement using a rat model of age-related memory impairment. *Ann. NY Acad. Sci.*, **717**, 16–32.

Jang, S., Kelley, K.W., and Johnson, R.W. (2008). Luteolin reduces IL-6 production in microglia by inhibiting JNK phosphorylation and activation of AP-1. *Proc. Natl. Acad. Sci. U.S.A.*, **105**, 7534–7539.

Jang, S. and Johnson, R.W. (2010). Can consuming flavonoids restore old microglia to their youthful state? *Ann. Rev. Nutr.*, **68**, 719–728.

Jang, S., Dilger, R.N., and Johnson, R.W. (2010). Dietary luteolin inhibits microglia and improves hippocampal-dependent spatial working memory in aged mice. *J. Nutr.*, **140**, 1892–1898.

Joseph, J.A., Villalobos-Molina, R., Denisova, N. et al. (1996). Age differences in sensitivity to H2O2-or NO-induced reductions in K+-evoked dopamine release from superfused striatal slices: Reversals by PBN or Trolox. *Free Radic. Biol. Med.*, **20**, 821–830.

Joseph, J.A., Shukitt-Hale, B., Denisova, N.A. et al. (1998). Long-term dietary strawberry, spinach, or vitamin E supplementation retards the onset of age-related neuronal signal-transduction and cognitive behavioral deficits. *J. Neurosci.*, **18**, 8047–8055.

Joseph, J.A., Shukitt-Hale, B., Denisova, N.A. et al. (1999). Reversals of age-related declines in neuronal signal transduction, cognitive, and motor behavioral deficits with blueberry, spinach, or strawberry dietary supplementation. *J. Neurosci.*, **19**, 8114–8121.

Joseph, J.A., Shukitt-Hale, B., McEwen, J. et al. (2000). CNS-induced deficits of heavy particle irradiation in space: The aging connection. *Adv. Space Res.*, **25**, 2057–2064.

Joseph, J.A., Fisher, D.R., and Strain, J. (2002). Muscarinic receptor subtype determines vulnerability to oxidative stress in COS-7 cells. *Free Radic. Biol. Med.*, **32**, 153–161.

Joseph, J.A., Denisova, N.A., Arendash, G. et al. (2003). Blueberry supplementation enhances signaling and prevents behavioral deficits in an Alzheimer disease model. *Nutr. Neurosci.*, **6**, 153–162.

Joseph, J.A., Shukitt-Hale, B., and Casadesus, G. (2005a). Reversing the deleterious effects of aging on neuronal communication and behavior: Beneficial properties of fruit polyphenolic compounds. *Am. J. Clin. Nutr.*, **81**, 313S–316S.

Joseph, J.A., Shukitt-Hale, B., Casadesus, G. et al. (2005b). Oxidative stress and inflammation in brain aging: Nutritional considerations. *Neurochem. Res.*, **30**, 927–935.

Joseph, J.A., Fisher, D.R., Carey, A.N. et al. (2006). Dopamine-induced stress signaling in COS-7 cells transfected with selectively vulnerable muscarinic receptor subtypes is partially mediated via the i3 loop and antagonized by blueberry extract. *J. Alzheimer Dis.*, **10**, 423–437.

Joseph, J.A., Shukitt-Hale, B., and Willis, L.M. (2009a). Grape juice, berries, and walnuts affect brain aging and behavior. *J. Nutr.*, **139**, 1813S–1817S.

Joseph, J. A., Cole, G., Head, E. et al. (2009b). Nutrition, brain aging, and neurodegeneration. *J. Neurosci.*, **29**, 12795–12801.

Joseph, J.A., Shukitt-Hale, B., Brewer, G.J. et al. (2010). Differential protection among fractionated blueberry polyphenolic families against DA-, Aβ(42)- and LPS-induced decrements in $Ca^{(2+)}$ buffering in primary hippocampal cells. *J. Agric. Food Chem.*, **58**, 8196–8204.

Kalt, W., Blumberg, J.B., McDonald, J.E. et al. (2008). Identification of anthocyanins in the liver, eye, and brain of blueberry-fed pigs. *J. Agric. Food Chem.*, **56**, 705–712.

Kaufmann, J.L., Bickford, P.C., and Taglialatela, G. (2001). Oxidative stress dependent up-regulation of Bcl-2 expression in the central nervous system of aged Fisher 344 rats. *J. Neurochem.*, **76**, 1099–1108.

Kreutzberg, G.W. (1996). Microglia: A sensor for pathological events in the CNS. *Trends Neurosci.*, **19**, 312–318.

Krikorian, R., Shidler, M.D., Nash, T.A. et al. (2010a). Blueberry supplementation improves memory in older adults. *J. Agric. Food Chem.* 58, 3996–4000.

Krikorian, R., Nash, T.A., Shidler, M.D. et al. (2010b). Concord grape juice supplementation improves memory function in older adults with mild cognitive impairment. *Br. J. Nutr.*, **103**, 730–734.

Lau, F.C., Bielinski, D.F. and Joseph, J.A. (2007). Inhibitory effects of blueberry extract on the production of inflammatory mediators in lipopolysaccharide-activated BV2 microglia. *J. Neurosci. Res.*, 85, 1010–1017.

Laurin, D., Masaki, K.H., Foley, D.J. et al. (2004). Midlife dietary intake of antioxidants and risk of late-life incident dementia: The Honolulu-Asia Aging Study. *Am. J. Epidemiol.*, **159**, 959–967.

Lee, H., Kim, Y.O., Kim, H. et al. (2003). Flavonoid wogonin from medicinal herb is neuroprotective by inhibiting inflammatory activation of microglia. *FASEB J.*, **17**, 1943–1944.

Letenneur, L., Proust-Lima, C., Le Gouge, A. et al. (2007). Flavonoid intake and cognitive decline over a 10-year period. *Am. J. Epidemiol.*, **165**, 1364–1371.

Luterman, J.D., Haroutunian, V., Yemul, S. et al. (2000). Cytokine gene expression as a function of the clinical progression of Alzheimer disease dementia, *Arch. Neurol.*, **57**, 1153–1160.

Malin, D.H., Lee, D.R., Goyarzu, P. et al. (2011). Short-term blueberry-enriched diet prevents and reverses object recognition memory loss in aging rats. *Nutrition*, **27**, 338–342.

Manach, C., Scalbert, A., Morand, C. et al. (2004). Polyphenols: Food sources and bioavailability. *Am. J. Clin. Nutr.*, **79**, 727–747.

McGeer, P.L. and McGeer, E.G. (1995). The inflammatory response system of the brain: Implications for therapy of Alzheimer and other neurodegenerative diseases. *Brain Res. Rev.*, **21**, 195–218.

Micheau, J. and Riedel, G. (1999). Protein kinases: Which one is the memory molecule? *Cell. Mol. Life Sci.*, **55**, 534–548.

Morris, M.C., Evans, D.A., Bienias, J.L. et al. (2002). Dietary intake of antioxidant nutrients and the risk of incident Alzheimer disease in a biracial community study. *JAMA*, **287**, 3230–3237.

Muir, J.L. (1997). Acetylcholine, aging, and Alzheimer's disease. *Pharmacol. Biochem. Behav.*, **56**, 687–696.

Nyakas, C., Oosterink, B.J., Keijser, J. et al. (1997). Selective decline of 5-HT1A receptor binding sites in rat cortex, hippocampus and cholinergic basal forebrain nuclei during aging. *J. Chem. Neuroanat.*, **13**, 53–61.

Oliveira, M.G., Bueno, O.F., Pomarico A.C. et al. (1997). Strategies used by hippocampal- and caudate putamen-lesioned rats in a learning task. *Neurobiol. Learn. Mem.*, **68**, 32–41.

Patil, C.S., Singh, V.P., Satyanarayan, P.S. et al. (2003). Protective effect of flavonoids against aging and lipopolysaccharide-induced cognitive impairment in mice. *Pharmacology*, **69**, 59–67.

Peng, H.W., Cheng, F.C., Huang, Y.T. et al. (1998). Determination of naringenin and its glucoronide conjugate in rat plasma and brain tissue by high-performance liquid chromatography. *J. Chromatogr.*, **714**, 369–374.

Poulose, S.M., Harris, E.D., and Patil, B.S. (2005). Citrus limonoids induce apoptosis in human neuroblastoma cells and have radical scavenging activity. *J. Nutr.*, **135**, 870–877.

Poulose, S.M., Harris, E.D., and Patil, B.S. (2006). Antiproliferative effects of citrus limonoids against human neuroblastoma and colonic adenocarcinoma cells. *Nutr. Cancer*, **56**, 103–112.

Poulose, S.M., Bielinski, D.B., Carrihill-Knoll, K. et al. (2011). Exposure to 16O particle irradiation causes age-like decrements in rats through increased oxidative stress, inflammation and loss of autophagy. *Rad. Res.*, **176**, 761–769.

Poulose, S.M., Fisher, D.R., Larson, J. et al. (2011). Anthocyanin rich acai fruit pulp fractions (Euterpe oleracea mart) attenuates inflammatory stress signalling in mouse brain BV-2 microglial cells. *J. Agric. Food Chem.* (In press)

Rabin, B.M., Joseph, J.A., and Erat, S. (1998). Effects of exposure to different types of radiation on behaviors mediated by peripheral or central systems. *Adv. Space Res.*, **22**, 217–225.

Ramirez, M.R., Izquierdo, I., do Carmo Bassols Raseira, M. et al. (2005). Effect of lyophilised Vaccinium berries on memory, anxiety and locomotion in adult rats. *Pharmacol.Res.*, **52**, 457–462.

Rezai-Zadeh, K., Ehrhart, J., Bai, Y. et al. (2008). Apigenin and luteolin modulate microglial activation via inhibition of STAT1-induced CD40 expression. *J. Neuroinflamm.*, **25**, 41.

Ross, J.A. and Kasum, C.M. (2002). Dietary flavonoids: Bioavailability, metabolic effects, and safety. *Ann. Rev. Nutr.*, **22**, 19–34.

Rossner, S., Ueberham, U., Schliebs, R. et al. (1998). The regulation of amyloid precursor protein metabolism by cholinergic mechanisms and neurotrophin receptor signaling, *Prog. Neurobiol.*, **56**, 541–569.

Rozovsky, I., Finch, C.E., and Morgan, T.E. (1998). Age-related activation of microglia and astrocytes: *In vitro* studies show persistent phenotypes of aging, increased proliferation, and resistance to down-regulation. *Neurobiol. Aging*, **19**, 97–103.

Saija, A., Princi, P., D'Amico, N. et al. (1990). Effect of *Vaccinium myrtillus* anthocyanins on triiodothyronine transport into brain in the rat. *Pharmacol. Res.*, **3**, 59–60.

Scalbert, A., Déprez, S., Mila, I. et al. (2000). Proanthocyanidins and human health: Systemic effects and local effects in the gut. *Biofactors*, **13**, 115–120.

Scarmeas, N., *Stern*, Y., Tang, M.X. et al. (2006a). Mediterranean diet and risk for Alzheimer's disease. *Ann. Neurol.*, **59**, 912–921.

Scarmeas, N., Stern, Y., Mayeux, R. et al. (2006b). Mediterranean diet, Alzheimer disease, and vascular mediation. *Arch. Neurol.*, **63**, 1709–1717.

Scarmeas, N., Luchsinger, J.A., Mayeux, R. et al. (2007). Mediterranean diet and Alzheimer disease mortality. *Neurology*, **69**, 1084–1093.

Schröder-van der Elst, J.P., van der Heide, D., Rokos, H. et al. (1998). Synthetic flavonoids cross the placenta in the rat and are found in fetal brain. *Am. J. Physiol.*, **274**, E253–256.

Serrano, P.A., Beniston, D.S., Oxonian, M.G. et al. (1994). Differential effects of protein kinase inhibitors and activators on memory formation in the 2-day-old chick. *Behav. Neural Biol.*, **61**, 60–72.

Shukitt-Hale, B., Mouzakis, G., and Joseph, J.A. (1998). Psychomotor and spatial memory performance in aging male Fischer 344 rats. *Exp. Gerontol.*, **33**, 615–624.

Shukitt-Hale, B., Cheng V, Bielinski D. et al. (2006). Differential brain regional specificity to blueberry and strawberry polyphenols in improved motor and cognitive function in aged rats. *Abstract 88.15/LL24 in Society for Neuroscience*. Atlanta, GA.

Shukitt-Hale, B., Carey, A.N., Jenkins, D. et al. (2007). Beneficial effects of fruit extracts on neuronal function and behavior in a rodent model of accelerated aging. *Neurobiol. Aging*, **28**, 1187–1194.

Shukitt-Hale, B., Lau, F.C., and Joseph, J.A. (2008a). Berry fruit supplementation and the aging brain. *J. Agric. Food Chem.*, **56**, 636–641.

Shukitt-Hale, B., Lau, F. C., Carey, A.N. et al. (2008b). Blueberry polyphenols attenuate kainic acid-induced decrements in cognition and alter inflammatory gene expression in rat hippocampus. *Nutr. Neurosci.*, **11**, 172–182.

Shukitt-Hale, B., Cheng, V., and Joseph, J A. (2009). Effects of blackberries on motor and cognitive function in aged rats. *Nutr. Neurosci.*, **12**, 135–140.

Singh, M., Arseneault, M., Sanderson T. et al. (2008). Challenges for research on polyphenols from foods in Alzheimer's disease: Bioavailability, metabolism, and cellular and molecular mechanisms. *J. Agric. Food Chem.*, **56**, 4855–4873.

Spencer, J.P. (2009). Flavonoids and brain health: Multiple effects underpinned by common mechanisms. *Genes Nutr.*, **4**, 243–250.

Spencer, J.P. (2010). The impact of fruit flavonoids on memory and cognition. *Br. J. Nutr.* **104**, 40–47.

Stackman, R.W., Eckenstein. F., Frei, B. et al. (2003). Prevention of age-related spatial memory deficits in a transgenic mouse model of Alzheimer's disease by chronic *Ginkgo biloba* treatment. *Exp. Neurol.*, **184**, 510–520.

Suganuma, M., Okabe, S., Oniyama, M. et al. (1998). Wide distribution of [³H](−)-epigallocatechin gallate, a cancer preventive tea polyphenol, in mouse tissue. *Carcinogenesis* **19**, 1771–1776.

Tsai, T.H. and Chen, Y.F. (2000). Determination of unbound hesperetin in rat blood and brain by microdialysis coupled to microbore liquid chromatography. *J. Food Drug Anal.*, **8**, 331–336.

Tully, T., Bourtchouladze, R., Scott, R. et al. (2003). Targeting the CREB pathway for memory enhancers. *Nat. Rev. Drug Discov.*, **2**, 267–277.

USDA database for the flavonoid content of selected foods, 2007, Release 2.1

Van der Sluis, A.A., Dekker, M., de Jager, A. et al. (2001). Activity and concentration of polyphenolic antioxidants in apple: Effect of cultivar, harvest year, and storage conditions. *J. Agric. Food Chem.*, **49**, 3606–3613.

Vannucci, R.C., Brucklacher, R.M., and Vannucci, S.J. (2001). Intracellular calcium accumulation during the evolution of hypoxicischemic brain damage in the immature rat. *Develop. Brain Res.*, **126**, 117–120.

Vauzour, D., Vafeiadou, K., Rodriguez-Mateos, A. et al. (2008). The neuroprotective potential of flavonoids: A multiplicity of effects. *Genes Nutr.* **3**, 115–126.

Vereker, E., O'Donnell, E., Lynch, A. et al. (2001). Evidence that interleukin-1beta and reactive oxygen species production play a pivotal role in stress-induced impairment of LTP in the rat dentate gyrus. *Eur. J. Neurosci.*, **14**, 1809–1819.

Vuong, T., Matar, C., Ramassamy, C. et al. (2010). Biotransformed blueberry juice protects neurons from hydrogen peroxide-induced oxidative stress and mitogen-activated protein kinase pathway alterations. *Br. J. Nutr.*, **104**, 656–663.

Wang, M.J., Lin, W.W., Chen, H.L. et al. (2002). Silymarin protects dopaminergic neurons against lipopolysaccharide-induced neurotoxicity by inhibiting microglia activation. *Eur. J. Neurosci.*, **16**, 2103–2121.

Wang, X., Chen, S., Ma, G. et al. (2005). Genistein protects dopaminergic neurons by inhibiting microglial activation. *Neuroreport*, **16**, 267–270.

Wang, Y.J., Thomas, P., Zhong, J.H. et al. (2009). Consumption of grape seed extract prevents amyloid-β deposition and attenuates inflammation in brain of an Alzheimer's disease mouse. *Neurotox. Res.*, **15**, 3–14.

West, M.J. (1993). Regionally specific loss of neurons in the aging human hippocampus. *Neurobiol. Aging*, **14**, 287–293.

Williams, C. M., El Mohsen, M.A., Vauzour, D. et al. (2008). Blueberry-induced changes in spatial working memory correlate with changes in hippocampal CREB phosphorylation and brain-derived neurotrophic factor (BDNF) levels. *Free Radic. Biol. Med.*, **45**, 295–305.

Winkel-Shirley B. (2002). Biosynthesis of flavonoids and effects of stress. *Curr. Opin. Plant Biol.*, **5**, 218–223.

Xu, Z., Chen, S., Li, X. et al. (2006). Neuroprotective effects of (–)-epigallocatechin-3-gallate in a transgenic mouse model of amyotrophic lateral sclerosis. *Neurochem. Res.* **31**, 1263–1269.

Youdim, K.A., Dobbie, M.S., Kuhnle, G. et al. (2003). Interaction between flavonoids and the blood-brain barrier: In vitro studies. *J. Neurochem.*, **85**, 180–192.

Youdim, K.A., Shukitt-Hale, B. and Joseph, J.A. (2004a). Flavonoids and the brain: Interactions at the blood-brain barrier and their physiological effects on the central nervous system. *Free Radic. Biol. Med.*, **37**, 1683–1693.

Youdim, K., Qaiser, M.Z., Begley, D. et al. (2004b). Flavonoid permeability across an *in situ* model of the blood brain barrier. *Free Radic. Biol. Med.*, **36**, 592–604.

Zheng, L.T., Ock, J., Kwon, B.M. et al. (2008). Suppressive effects of flavonoid fisetin on lipopolysaccharide-induced microglial activation and neurotoxicity. *Int. Immunopharmacol.*, **8**, 484–494.

18 Flavonoids and Oral Cancer

Thomas Walle

CONTENTS

18.1 ORAL CANCER

Tobacco smoking is the most important cause of human oral squamous cell carcinoma (SCC) (Schmidt et al. 2004), although betel chewing is an important contributor in some countries (Awang 1988). The presence of tobacco-related carcinogen-DNA adducts, such as adducts with polycyclic aromatic hydrocarbons, has greatly strengthened the link between the development of oral cancer and smoking (Franceschi et al. 1990; Hsu et al. 1997; Hecht 2002; Schmidt et al. 2004). Two cytochrome P450 (CYP) enzymes, that is, CYP1A1 and CYP1B1, are mainly involved in the bioactivation of benzo[*a*]pyrene, resulting in the most common smoking-related cellular DNA adducts (Nebert et al. 2004; Wen and Walle 2005; Chi et al. 2009). DNA adducts may lead to cancer by causing mutation in genes essential for key functions, including apoptosis, proliferation, and differentiation. Significantly increased oral cancer risks are also found in heavy drinkers (Franceschi et al. 1990), presumably by facilitating absorption of the tobacco carcinogens.

Among dietary factors affecting oral cancer, fruits and vegetables exert the greatest protective impact. These observations have formed the basis for many epidemiological studies attempting to pinpoint the nature of the protective dietary effects (Winn et al. 1984; Gridley et al. 1990; Tavani et al. 2001; Sanchez et al. 2003; Llewellyn et al. 2004; De Stefani et al. 2005; Boeing et al. 2006; Maserejian et al. 2006; Pavia et al. 2006; Garavello et al. 2009; Lucenteforte et al. 2009; Petti and Scully 2009; Sandoval et al. 2009).

Protective effects of dietary components can be provided by inactivation of carcinogens, prevention of covalent binding of carcinogens to DNA, or, maybe most important, interaction with many of the signal transduction events affected by essential mutations. Such dietary components may reach their target sites systemically after gastrointestinal absorption or locally in the oral cavity. The latter possibility was studied in a bioengineered human gingival epithelial tissue construct. Both the carcinogen benzo[*a*]pyrene and several potential chemoprotectors, including the flavonoids quercetin and 5,7-dimethoxyflavone and the polyphenol resveratrol, penetrated and accumulated in the basal cells of this highly complex tissue (Walle et al. 2006).

18.2 PREVENTION BY FLAVONOIDS AND THEIR GLYCOSIDES

Most dietary flavonoids are present in fruits and vegetables as glycosides with various sugar moieties. Although some of these glycosides may be actively transported and absorbed intact in the intestine (Walgren et al. 2000), most must first undergo hydrolytic cleavage to their aglycones before intestinal absorption can occur. The hydrolysis could take place in the lumen by the broad-specific β-glucosidase (BSBG) or by the lactase phloridzin hydrolase (LPH) (Day et al. 1998, 2000; Walle 2004). The situation in the oral cavity may be similar, although our knowledge at this potential absorption site is limited. However, several flavonoid glycosides have been shown to be hydrolyzed by the saliva, including quercetin 4′-glucoside and genistein 7-glucoside (Walle et al. 2005). The hydrolysis by saliva appeared to be partially bacterial in nature, as evidenced by inhibition by antibacterial washes of the mouth prior to saliva sampling (Walle et al. 2005). Oral streptococci have been shown to hydrolyze the glycoside rutin to its quercetin aglycone (Parisis and Pritchard 1983). Of particular interest was the finding of large interindividual variation in the rate of salivary hydrolysis of genistin to genistein (Figure 18.1) (Walle et al. 2005) in a small population of young, healthy human subjects. The nature of this more than 30-fold variability in genistin hydrolysis rate is not known but may be due to genetic variation in the expression of BSBG, LPH, or in the oral bacterial flora. This may be of great importance with respect to the chemoprotective properties of the soy isoflavonoids as well as of other flavonoids. The ability of intact flavonoid glycosides to inhibit cancer

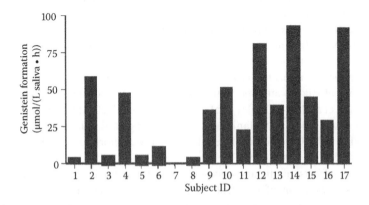

FIGURE 18.1 Genistin hydrolysis to genistein in the saliva from a group of young, healthy adults. Saliva was incubated for 1 h with 25 μM genistin. (Walle, T. et al., *J. Nutr.*, **135**, 48–52, 2005. With permission.)

cell proliferation, that is, without prior hydrolysis, was investigated using the human oral squamous carcinoma SCC-9 cell as a model (Browning et al. 2005). The cellular uptake and intracellular hydrolysis of four glycosides, that is, genistin, spiraeoside, diosmin, and quercitrin, were assessed as well. Genistin had a low potency (minimum effective concentration, MEC ≈ 100 μM) compared with that of its aglycone genistein (MEC ≈ 30 μM) (Figure 18.2a). Although 100 μM is a fairly high concentration, this concentration can easily be obtained in the oral cavity, where dietary chemicals reach their highest levels. The effect of genistin appeared to be due to rapid uptake by the SCC-9 cells and hydrolysis to genistein (Figure 18.2b).

Spiraeoside (quercetin-4′-O-glucoside) behaved similarly to genistin with an MEC of 200 μM, compared to an MEC of 30 μM for quercetin. Also in this case, the glycoside showed hydrolysis to quercetin, presumably explaining its antiproliferative effect. Diosmin, the 7-O-rutinoside of diosmetin, interestingly, was more potent than diosmetin (MEC 10 μM versus 20 μM) (Figure 18.3). SCC-9 cell uptake of diosmin and its hydrolysis to diosmetin was difficult to determine, as both compounds were unstable. Semiquantitative measurements indicated an approximately 10-fold higher uptake of the aglycone. This appeared to indicate that the glycoside diosmin was a more potent inhibitor of the cancer cell proliferation than the aglycone diosmetin. Ciolino et al. (1998) reported that diosmin and diosmetin both were potent arylhydrocarbon receptor (AhR) agonists, resulting in cytochrome P450 1A1 induction. AhR agonists have been demonstrated to inhibit the growth of prostate cancer LNCaP cells (Morrow et al. 2004). The SCC-9 cell growth inhibition may thus have been mediated by a similar mechanism. Diosmin being more potent than diosmetin is contrary to the common assumption that dietary flavonoid glycosides lack biological activity. This may thus not always be true. However, quercitrin, quercetin-3-O-rhamnoside, had no effect on SCC-9 cell proliferation, at least up to a concentration of 200 μM. This is in agreement with the lack of evidence of either uptake or hydrolysis of quercitrin. It is clear that more

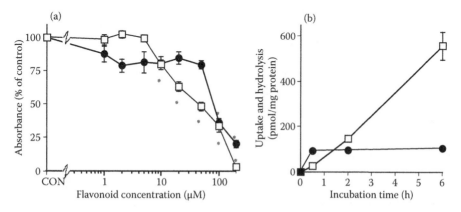

FIGURE 18.2 (a) Antiproliferative effects of genistin (●) and genistein (□) in SCC-9 cells (mean ± SEM; n = 12). *Lower than control, $P < 0.05$. (b) Uptake of genistin (●) by SCC-9 cells and hydrolysis to genistein (□) ($n = 24$). (Browning, A.M., Walle, U.K., and Walle, T., *J. Pharm. Pharmacol.,* **57**, 1037–1041, 2005. With permission.)

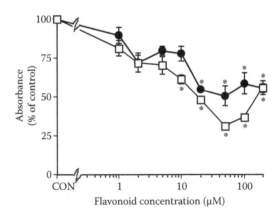

FIGURE 18.3 Antiproliferative effects of diosmin (□) and diosmetin (●) in SCC-9 cells (mean ± SEM; $n = 12$–18). *Lower than control, $P < 0.05$. (Browning, A.M., Walle, U.K., and Walle, T., *J. Pharm. Pharmacol.*, **57**, 1037–1041, 2005. With permission.)

studies of the effects of flavonoid glycosides and their aglycones on oral cancer cell proliferation are needed.

18.3 PREVENTION BY BERRIES, A WHOLE FOOD APPROACH

The cancer preventive activities of flavonoids are due to a large variety of biological mechanisms, a fact that might be beneficial to attain maximum protection against cancer progression. Thus, a mixture of flavonoids, as would be found in most foods, may be advantageous, although there is very limited information on this topic, at least for oral cancer. Despite this, it seems likely that a mixed-food approach could be most beneficial in the prevention of oral cancer development. The study of tea components is one example of such an approach, as tea comprises a wide variety of chemicals with different biological properties. However, because of limited oral bioavailability of the tea catechins, tea does not seem like the diet of choice, even though some effectiveness in oral cancer has been shown (Li et al. 2002). A better choice would appear to be a berry approach, particularly focusing on black raspberries, which several groups at Ohio State University have been pursuing and refining. Improved modes of administration have been utilized to maximize the delivery of the polyphenols present within the berries. Research on black raspberries started with berry extracts administered in the hamster cheek pouch model (Casto et al. 2002; Han et al. 2005). Here, there was a significant inhibition of dimethylbenzanthracene-induced tumors in the oral cavity after 10 weeks of administration of the berry extract. These effects appeared to be due to multiple mechanisms, reflecting the content of a variety of cancer preventive chemicals present in the extract, for example, anthocyanins, a class of flavonoids, and ellagitannins (Stoner 2009), and a variety of other chemicals (Del Rio et al. 2010). Similar effects of the berry extract were seen in human oral SCC cells (Rodrigo et al. 2006). To increase the effectiveness of the berries in human experiments, formulation work has been undertaken using mucoadhesive gels (Mallery et al. 2007) for local delivery of the berry extracts. Similarly, millicylindrical implants (Desai et al. 2010) and nanoparticles (Holpuch et al. 2010) are novel delivery systems designed for targeted delivery of cancer preventive mixtures of flavonoids and other polyphenols.

18.4 PREVENTION BY METHYLATED FLAVONES

As has been seen frequently in the literature both for flavonoids (Walle 2004) and many other polyphenolic dietary products, not least the tea flavonoids (Yang et al. 2009), their bioavailability, that is, their ability to enter the central circulation from their sites of absorption and therefore become available for tissue uptake and biological effects, is highly limited. This limits their potential usefulness as effective chemoprotective and/or therapeutic agents. The reasons for this are twofold, one being that they are too polar to penetrate the lipid cellular barriers and the other, most likely more important, being that they serve as acceptors for conjugating enzymes, which are most abundant in the gastrointestinal barriers, thus, converting flavonoids to highly polar and in most cases biologically inactive conjugates (glucuronides and sulfates). The key chemical feature for this to occur is the presence of unprotected hydroxyl groups, which serve as highly effective receptors for both sulfate groups, from sulfotransferases, and glucuronic acid groups, from UDP-glucuronic acid transferases. In this respect, it should be pointed out that oxidative metabolism, the most common pathway for degradation of foreign chemicals, is rather ineffective for flavonoids. To protect the flavonoids from metabolic degradation through conjugating enzymes, the effect of O-methylation on these conjugative activities was determined, employing chrysin (5,7-dihydroxyflavone) and apigenin (5,7,4′-trihydroxyflavone) and their O-methylated analogs as model substrates. The effects are shown in Figure 18.4, demonstrating very rapid metabolism of the two unmethylated flavones with almost no metabolism of the corresponding O-methylated analogs (Wen and Walle 2006). Figure 18.5 shows the rapid absorption of the methylated flavonoids in

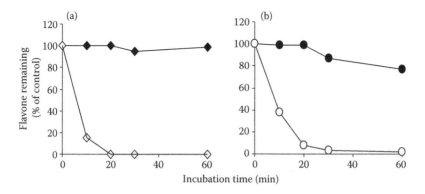

FIGURE 18.4 Time-dependent metabolism of unmethylated and methylated flavonoids (5 μM) in pooled human liver S9 fraction with cofactors. (a) 5,7-dimethoxyflavone (filled symbols) and chrysin (open symbols); (b) 5,7,4′-trimethoxyflavone (filled symbols) and apigenin (open symbols). (Wen, X. and Walle, T., *Xenobiotica*, **36**, 387–397, 2006. With permission.)

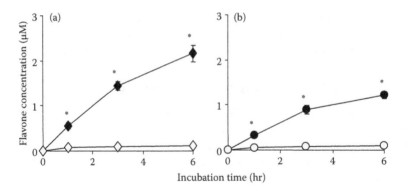

FIGURE 18.5 Caco-2 cell transport of unmethylated and methylated flavonoids. Flavonoids (5–10 µM) in transport buffer were added on the apical side of the Transwell chamber, and samples were taken from the basolateral side. (a) 5,7-dimethoxyflavone (filled symbols) and chrysin (open symbols); (b) 5,7,4'-trimethoxyflavone (filled symbols) and apigenin (open symbols). (Wen, X. and Walle, T., *Xenobiotica*, **36**, 387–397, 2006. With permission.)

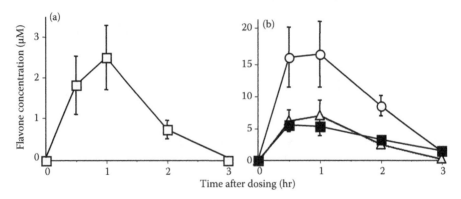

FIGURE 18.6 Plasma and tissue levels of 5,7-dimethoxyflavone and chrysin after oral administration of 5 mg/kg in rats. (a) Plasma 5,7-dimethoxyflavone (no chrysin could be detected); (b) Tissue 5,7-dimethoxyflavone in liver (O), lung (■) and kidney (Δ) (no chrysin could be detected); Mean ± SEM; five animals at each time-point. (Walle, T. et al., *Biochem. Pharmacol.*, **73**, 1288–1296, 2007. With permission.)

the Caco-2 cell human absorption model when compared to the very poor absorption of the unmethylated analogs (Wen and Walle 2006). Based on these observations, it was anticipated that the methylated analogs might be absorbed more effectively *in vivo* as well. This was demonstrated in a separate study in the rat, in which 5,7-dimethoxyflavone (5,7-DMF) and chrysin were administered simultaneously by gavage at 5 mg/kg (Walle et al. 2007). 5,7-DMF was clearly detected in plasma with a peak concentration of 2.5 µM at 1 h. after the dose (Figure 18.6a). In sharp contrast, the unmethylated chrysin was not detectable in plasma at any time. 5,7-DMF was also clearly detected in liver, lung, and kidney tissue with concentrations in the liver exceeding those in the plasma by as much as sevenfold, that is, 16.5 ± 5 µM at 1 h after the dose (Figure 18.6b). Chrysin was not detected in any of these tissues.

We also examined the effects of these flavonoids on oral cancer cell proliferation (Walle et al. 2007). 5,7-DMF, interestingly, was considerably more potent than chrysin, with an IC_{50} of 8 μM compared to 108 μM for chrysin (Figure 18.7a). Similarly, 5,7,4'-trimethoxyflavone (5,7,4'-TMF) had a much lower IC_{50} (5 μM) compared with apigenin (40 μM) (Figure 18.7b). To determine whether the growth inhibitory effect of the two pairs of flavones was accompanied by cell cycle arrest, SCC-9 cells were treated with increasing concentrations of the compounds for 48 h. Asynchronized cells were harvested, fixed in ethanol, stained with propidium iodide, and analyzed for cell cycle distribution by flow cytometry (Walle et al. 2007). Figure 18.8a shows that 5,7-DMF caused a notable dose-dependent increase in the G1 phase with a concomitant rather prominent decrease in the S phase and no change in the G2/M phase populations. This was distinctly different from the very modest and opposite effects of chrysin on the cell cycle distribution (Figure 18.8b). Figure 18.8c shows the corresponding effects of 5,7,4'-TMF with a prominent dose-dependent increase in the G1 phase, as with 5,7-DMF, and a marked decrease in the S phase and not much change in the G2/M phase populations. Figure 18.8d shows the effects of apigenin, which qualitatively is similar to the effects of chrysin. Thus, not only were the methylated flavones distinctly more potent inhibitors of oral cancer cell proliferation than their unmethylated analogs, but their mechanisms of action, as expressed in their effects on cell cycle progression, were dramatically different. These relationships should be further explored, as they might be critically important for prevention as well as treatment of oral cancers.

O-Methylation has been shown more recently to enhance the metabolic stability as well as anticancer activity of additional flavonoids in SCC-9 cells (Walle and Walle 2007). These include flavonoid inhibitors of cytochrome P450 1B1, an important carcinogen-activating enzyme in the oral cavity, and may thus have a protective

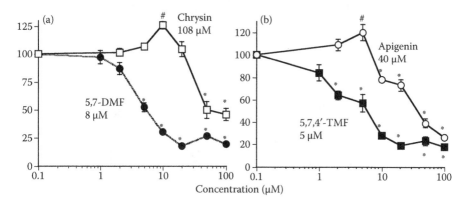

FIGURE 18.7 Effect of the methylated flavones: (a) 5,7-dimethoxyflavone and (b) 5,7,4'-trimethoxyflavone compared to their unmethylated analogs chrysin and apigenin on SCC-9 cell proliferation. Cell proliferation (as % of control) was measured as BrdU incorporation into cellular DNA after 24 h exposure of the cells to the flavones. Mean ± SEM; $n = 10$; *Lower than control, $P < 0.05$; #Higher than control, $P < 0.05$. (Walle, T. et al., *Biochem. Pharmacol.*, **73**, 1288–1296, 2007. With permission.)

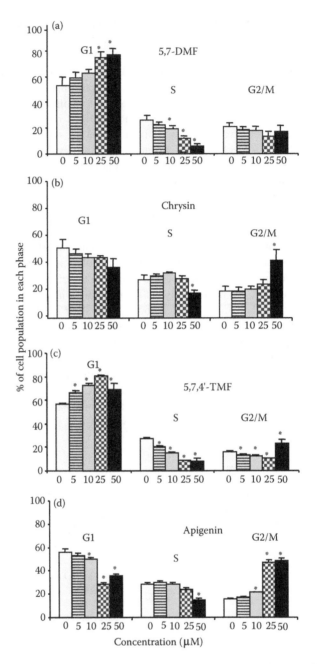

FIGURE 18.8 Effect of 5,7-dimethoxyflavone (a) compared to chrysin (b) and 5,7,4′-trime-thoxyflavone (c) compared to apigenin (d) on SCC-9 cell cycle progression (Walle, T. et al., *Biochem. Pharmacol.*, **73**, 1288–1296, 2007. With permission.). The cells were exposed to 0–50 μM flavone for 48 h. The percentage of cells in G1, S, and G2/M phase was measured by flow cytometry after propidium iodide staining. Mean of three experiments with duplicate samples. *Significantly different from control, $P < 0.05$ or better.

effect in particular against chemicals such as benzo[a]pyrene present in cigarette smoke. In another study in HL-60 leukemia cells, the methylated flavonoids also displayed improved bioavailability and inhibition of cell proliferation by arresting the cells in the G1 phase. However, in these cells, the methylated flavonoids showed different responses with regard to apoptosis and proteasomal activities (Landis-Piwowar et al. 2008). In a very recent study, 5,7,3′,4′,5′-pentamethoxyflavone was a more potent inhibitor of gastrointestinal adenomas in the Apc^{Min} mouse than the partially methylated analog or especially the nonmethylated apigenin (Cai et al. 2009). More extensive studies into the mechanisms by which methylated flavonoids exert effects in oral and other cancer cells are most certainly needed.

ACKNOWLEDGMENT

U. Kristina Walle is greatly acknowledged for her help with preparation of this manuscript.

REFERENCES

Awang, M.N. (1988). Fate of betel nut chemical constituents following nut treatment prior to chewing and its relation to oral precancerous and cancerous lesion. *Dental J. Malaysia,* **10**, 33–37.

Boeing, H., Dietrich, T., Hoffmann, K. et al. (2006). Intake of fruits and vegetables and risk of cancer of the upper aero-digestive tract: The prospective EPIC-study. *Cancer Causes Control,* **17**, 957–969.

Browning, A.M., Walle, U.K. and Walle, T. (2005). Flavonoid glycosides inhibit oral cancer cell proliferation—role of cellular uptake and hydrolysis to the aglycones. *J. Pharm. Pharmacol.,* **57**, 1037–1041.

Cai, H., Sale, S., Schmid, R. et al. (2009). Flavones as colorectal cancer chemopreventive agents—phenyl-*O*-methylation enhances efficacy. *Cancer Prev. Res.,* **2**, 743–750.

Casto, B.C., Kresty, L.A., Kraly, C.L. et al. (2002). Chemoprevention of oral cancer by black raspberries. *Anticancer Res.,* **22**, 4005–4015.

Chi, A.C., Appleton, K., Henriod, J.B. et al. (2009). Differential induction of CYP1A1 and CYP1B1 by benzo[a]pyrene in oral squamous cell carcinoma cell lines and by tobacco smoking in oral mucosa. *Oral Oncol.,* **45**, 980–985.

Ciolino, H.P., Wang, T.T.Y., and Yeh, G.C. (1998). Diosmin and diosmetin are agonists of the aryl hydrocarbon receptor that differentially affect cytochrome P450 1A1 activity. *Cancer Res.,* **58**, 2754–2760.

Day, A.J., Cañada, F.J., Diaz, J.C. et al. (2000). Dietary flavonoid and isoflavone glycosides are hydrolysed by the lactase site of lactase phlorizin hydrolase. *FEBS Lett.,* **468**, 166–170.

Day, A.J., Dupont, M.S., Ridley, S. et al. (1998). Deglycosylation of flavonoid and isoflavonoid glycosides by human small intestine and liver β-glucosidase activity. *FEBS Lett.,* **436**, 71–75.

De Stefani, E., Boffetta, P., Ronco, A.L. et al. (2005). Dietary patterns and risk of cancer of the oral cavity and pharynx in Uruguay. *Nutr. Cancer,* **51**, 132–139.

Del Rio, D., Borges, G., and Crozier, A. (2010). Berry flavonoids and phenolics: Bioavailability and evidence of protective effects. *Br. J. Nutr.,* **104**, S67–S90.

Desai, K.G.H., Olsen, K.F., Mallery, S.R. et al. (2010). Formulation and in vitro-in vivo evaluation of black raspberry extract-loaded PLGA/PLA injectable millicylindrical implants for sustained derivery of chemopreventive anthocyanins. *Pharm. Res.,* **27**, 628–643.

Franceschi, S., Talamini, R., Barra, S. et al. (1990). Smoking and drinking in relation to cancers of the oral cavity, pharynx, larynx, and esophagus in northern Italy. *Cancer Res.,* **50**, 6502–6507.

Garavello, W., Lucenteforte, E., Bosetti, C. et al. (2009). The role of foods and nutrients on oral and pharyngeal cancer risk. *Minerva Stomatol.,* **58**, 25–34.

Gridley, G., Mclaughlin, J.K., Block, G. et al. (1990). Diet and oral and pharyngeal cancer among blacks. *Nutr. Cancer,* **14**, 219–225.

Han, C., Ding, H., Casto, B. et al. (2005). Inhibition of the growth of premalignant and malignant human oral cell lines by extracts and components of black raspberries. *Nutr. Cancer,* **51**, 207–217.

Hecht, S.S. (2002). Cigarette smoking and lung cancer: Chemical mechanisms and approaches to prevention. *Lancet Oncol.,* **3**, 461–469.

Holpuch, A.S., Hummel, G.J., Tong, M. et al. (2010). Nanoparticles for local drug delivery to the oral mucosa: Proof of principle studies. *Pharm. Res.,* **27**, 1224–1236.

Hsu, T.M., Zhang, Y.J. and Santella, R.M. (1997 Immunoperoxidase quantitation of 4-aminobiphenyl- and polycyclic aromatic hydrocarbon-DNA adducts in exfoliated oral and urothelial cells of smokers and nonsmokers. *Cancer Epidemiol. Biomarkers,* **6**, 193–199.

Landis-Piwowar, K.R., Milacic, V., and Dou, Q.P. (2008). Relationship between the methylation status of dietary flavonoids and their growth-inhibitory and apoptosis-inducing activities in human cancer cells. *J. Cellular Biochem.,* **105**, 514–523.

Li, N., Chen, X., Liao, J. et al. (2002). Inhibition of 7,12-dimethylbenz[a]anthracene (DMBA).-induced oral carcinogenesis in hamsters by tea and curcumin. *Carcinogenesis,* **23**, 1307–1313.

Llewellyn, C.D., Linklater, K., Bell, J. et al. (2004). An analysis of risk factors for oral cancer in young people: A case-control study. *Oral Oncol.,* **40**, 304–313.

Lucenteforte, E., Garavello, W., Bosetti, C. et al. (2009). Dietary factors and oral and pharyngeal cancer risk. *Oral Oncol.,* 45, 461–467.

Mallery, S.R., Stoner, G.D., Larsen, P.F. et al. (2007). Formulation and in-vitro and in-vivo evaluation of a mucoadhesive gel containing freeze dried black raspberries: Implications for oral cancer chemoprevention. *Pharm. Res.,* **24**, 728–737.

Maserejian, N.N., Giovannucci, E., Rosner, B. et al. (2006). Prospective study of fruits and vegetables and risk of oral premalignant lesions in men. *Am. J. Epidemiol.,* **164**, 556–566.

Morrow, D., Qin, C., Smith, R.I. et al. (2004). Aryl hydrocarbon receptor-mediated inhibition of LNCaP prostate cancer cell growth and hormone-induced transactivation. *J. Steroid Biochem. Mol. Biol.,* **88**, 27–36.

Nebert, D.W., Dalton, T.P., Okey, A.B. et al. (2004). Role of aryl hydrocarbon receptor-mediated induction of the CYP1 enzymes in environmental toxicology and cancer. *J. Biol. Chem.,* **279**, 23847–23850.

Parisis, D.M. and Pritchard, E.T. (1983). Activation of rutin by human oral bacterial isolates to the carcinogen-mutagen quercetin. *Arch. Oral Biol.,* **28**, 583–590.

Pavia, M., Pileggi, C., Nobile, C.G. et al. (2006). Association between fruit and vegetable consumption and oral cancer: A meta-analysis of observational studies. *Am. J. Clin. Nutr.,* **83**, 1126–1134.

Petti, S. and Scully, C. (2009). Polyphenols, oral health and disease: A review. *J. Dentistry,* **37**, 413–423.

Rodrigo, K.A., Rawal, Y., Renner, R.J. et al. (2006). Suppression of the tumorigenic phenotype in human oral squamous cell carcinoma cells by an ethanol extract derived from freeze-dried black raspberries. *Nutr. Cancer,* **54**, 58–68.

Sanchez, M.J., Martinez, C., Nieto, A. et al. (2003). Oral and oropharyngeal cancer in Spain: Influence of dietary patterns. *Eur. J. Cancer Prev.,* **12**, 49–56.

Sandoval, M., Font, R., Manos, M. et al. (2009). The role of vegetable and fruit consumption and other habits on survival following the diagnosis of oral cancer: A prospective study. *Int. J. Oral Maxillof. Surg.*, **38**, 31–39.

Schmidt, B.L., Dierks, E.J., Homer, L. et al. (2004). Tobacco smoking history and presentation of oral squamous cell carcinoma. *J. Oral Maxillofac. Surg.*, **62**, 1055–1058.

Stoner, G.D. (2009). Foodstuffs for preventing cancer: The preclinical and clinical development of berries. *Cancer Prev. Res.*, **2**, 187–194.

Tavani, A., Gallus, S., La Vecchia, C. et al. (2001). Diet and risk of oral and pharyngeal cancer. An Italian case-control study. *Eur. J. Cancer Prev.*, **10**, 191–195.

Walgren, R.A., Lin, J.-T., Kinne, R.K.-H. et al. (2000). Cellular uptake of dietary flavonoid quercetin 4′-β-glucoside by sodium-dependent glucose transporter SGLT1. *J. Pharmacol. Exp. Ther.*, **294**, 837–843.

Walle, T. (2004). Absorption and metabolism of flavonoids. *Free Radic. Biol. Med.*, **36**, 829–837.

Walle, T., Browning, A.M., Steed, L. S. et al. (2005). Flavonoid glucosides are hydrolyzed and thus activated in the oral cavity in humans. *J. Nutr.*, **135**, 48–52.

Walle, T., Ta, N., Kawamori, T. et al. (2007). Cancer chemopreventive properties of orally bioavailable flavonoids—methylated versus unmethylated flavones. *Biochem. Pharmacol.*, **73**, 1288–1296.

Walle, T., Walle, U.K., Sedmera, D. et al. (2006). Benzo[*a*]pyrene-induced oral carcinogenesis and chemoprevention—studies in bioengineered human tissue. *Drug Metab. Dispos.*, **34**, 346–350.

Walle, U.K. and Walle, T. (2007). Bioavailable flavonoids: Cytochrome P450-mediated metabolism of methoxyflavones. *Drug Metab. Dispos.*, **35**, 1985–1989.

Wen, X. and Walle, T. (2005). Preferential induction of CYP1B1 by benzo[*a*]pyrene in human oral epithelial cells: Impact on DNA adduct formation and prevention by polyphenols. *Carcinogenesis*, **26**, 1774–1781.

Wen, X. and Walle, T. (2006). Methylation protects dietary flavonoids from rapid hepatic metabolism. *Xenobiotica*, **36**, 387–397.

Winn, D.M., Ziegler, R.G., Pickle, L.W. et al. (1984). Diet in the etiology of oral and pharyngeal cancer among women from the southern United States. *Cancer Res.*, **44**, 1216–1222.

Yang, C.S., Wang, X., Lu, G. et al. (2009). Cancer prevention by tea: Animal studies, molecular mechanisms and human relevance. *Nat. Rev. Cancer*, **9**, 429–439.

19 Flavonoids and Cancer— Effects on DNA Damage

Piyawan Sitthiphong, Annett Klinder,
Johanna W. Lampe, and Ian Rowland

CONTENTS

19.1 INTRODUCTION

Plants produce a diverse range of secondary metabolites that function in the cellular communication within and among plants, in reproduction and as mediators of responses to biotic and abiotic stresses. Such compounds include polyphenols, most notably the flavonoids. In human cell cultures and in animal models, these compounds have been shown to exert wide range of biological effects, suggesting they may have cancer preventing properties, including free radical scavenging, reducing DNA damage, inducing detoxifying enzymes, and modulating cell signaling pathways involved in apoptosis and cell cycle control (Naczk and Shahidi 2006; Watson and Preedy 2010). In this chapter, we review the evidence for effects of flavonoids on cancer risk in humans with a focus on prevention of DNA damage.

19.2 FLAVONOIDS AND CANCER RISK: EPIDEMIOLOGIC STUDIES

Early epidemiologic studies suggested that diets rich in fruit and vegetables are associated with reduced risk of some cancers (Negri et al. 1991). Case–control studies in particular and some cohort studies have provided strong evidence that plant foods decrease risk for several cancers of epithelial origin, including stomach, esophagus, colon, and lung (Steinmetz and Potter 1996). In more recent large prospective studies, the inverse association has weakened, and based on a systematic review, a report from the World Cancer Research Fund and American Institute for Cancer Research (2007) categorized the strength of the evidence as "probable." The report examined fruit and nonstarchy vegetable intake in relation to the major categories of cancer. Based on the totality of the cohort data available, the report indicated that fruit and nonstarchy vegetable intake probably protects against cancers of the mouth, larynx, pharynx, oesophagus, and stomach, and that fruits also protect against lung cancer. The report also indicated that there was limited evidence that fruits may also protect against the following cancers: nasopharynx; pancreas; liver; and colorectum.

Flavonoids are widely distributed in fruits and nonstarchy vegetables and beverages of plant origin. Flavonoids in six major subgroups—flavonols, flavones, anthocyanidins, catechins, flavanones, and isoflavones—are relatively common in human diets (Aherne and O'Brien 2002), and consequently, some of the cancer-protective effects of fruits and vegetables have been hypothesized to be due to flavonoid content. Several research groups (Hertog et al. 1992; Hertog et al. 1993; Sampson et al. 2002) and the United States Department of Agriculture have generated flavonoid databases to use with various dietary assessment approaches, including self-administered food frequency questionnaires and structured interviews.

Previously, paralleling the findings for fruits and vegetables, results of case–control studies suggested that intake of total flavonoids, flavonoid subgroups, or individual flavonoids were associated with reduced risk of lung, gastric, colorectal, breast, ovarian, and endometrial cancers, and non-Hodgkin lymphoma (from Wang et al. 2009). However, based on the results of more recent cohort studies, these relationships are weaker. Case–control studies, where cases are selected after cancer diagnosis, are subject to dietary recall bias (i.e., participants' reporting of diet may differ depending whether they are cases or controls) and effect of the long latency of cancer development; thus, these data need to be interpreted with caution (Kristal and Lampe 2011). Prospective cohort studies select a group of healthy individuals for long-term study, characterize their diet and life-style at recruitment, and follow them to determine who is diagnosed with cancer. Cohort studies are not subject to bias by disease status; however, they may be subject to residual confounding and need to be very large in order ultimately to provide sufficient numbers of cancer cases.

Several cohort studies have examined the relationship between overall cancer incidence and intake of select flavonoids or flavonoid-rich foods. Flavonoid intake was not associated with the all-cause cancer or mortality from all-cause cancer in the Zutphen Elderly Study, a cohort study of men in the Netherlands (Hertog et al. 1994). In the Iowa Women's Study, women in the highest quintile of total flavonoid intake had a 12% lower incidence of any cancer compared to women in the lowest quintile in age and energy adjusted models; however, this association was no longer observed after multivariable adjustment (Cutler et al. 2008). Similar results were observed for flavanones, flavonols, flavan-3-ols, proanthocyanidins, and total proanthocyanidins. After multivariable adjustment, only isoflavone intake was associated with a modestly reduced incidence of any cancer, with women in the highest quintile of intake having a 7% lower incidence compared to those in the lowest quintile (HR = 0.93; 95% CI: 0.86–1.00, p for trend across quintiles = 0.03). Wang et al. (2009) also reported no association between intake of total flavonoids or individual flavones or flavonols and total cancer in a study of 3234 incident cancer cases among women in the Women's Health Study in the United States.

Given that the mechanisms of action of flavonoids may differentially influence risk of certain cancers based on tissue-type, etc., other studies have examined the association between intake of individual or specific groups of flavonoids and specific cancers. In the Zutphen Elderly Study cohort, no association was found between individual flavonoid intake and the incidence of specific cancers in Dutch men (Hertog et al. 1994; Arts et al. 2002). In contrast, an inverse association was reported between total flavonoids, or individual or some subgroup of flavonoids and risk of lung cancer (Hirvonen et al. 2001; Knekt et al. 2002; Mursu et al. 2008), pancreatic cancer (Bobe et al. 2008), and prostate cancer (Knekt et al. 2002) in three cohorts in Finland. For example, men with higher quercetin intakes had a lower lung cancer incidence (RR = 0.42; 95% CI 0.25, 0.72; $P = 0.001$), and men with higher myricetin intakes had a lower prostate cancer risk (0.43; 0.22, 0.86; $P = 0.002$) (Knekt et al. 2002). Among U.S. cohorts, high flavonoid intake was associated with lower risk of ovarian cancer in the Nurses' Health Study I (Gates et al. 2007) and the California Teachers Study (Chang et al. 2007); lung cancer (Cutler et al. 2008), and rectal cancer (Arts et al. 2002) in the Iowa Women's Health Study; and pancreatic cancer among current smokers in the Multiethnic Cohort Study (Nöthlings et al. 2007). However, these findings are not consistent across cohorts, and no association between flavonoid intake and risk of cancers at many other sites has been reported in other U.S. cohort studies (Arts et al. 2002; Adebamowo et al. 2005; Lin et al. 2006; Wang et al. 2009).

The associations between intake of flavonoid-rich foods and beverages and cancer risk have also been examined in cohort studies; however, few of these have reported significant associations. For example, in several large cohort studies, intake of onions and apples, good sources of quercetin, were not associated with risk of specific cancers (Knekt et al. 1997; Arts et al. 2001; Lin et al. 2006; Gates et al. 2007; Bobe et al. 2008; Cutler et al. 2008) or total cancer (Wang et al. 2009), although apple intake was associated with lower lung cancer risk in the Finnish Mobile Clinic Health Examination Survey study (Knekt et al. 1997). Similarly, an inverse association between intake of soy foods, rich sources of isoflavones, and breast cancer risk has been reported in some cohort studies of populations that consume soy (Wu et al. 2002; Yamamoto et al. 2003), but not in others (Hiriyama et al. 1992; Key et al. 1999; Nishio et al. 2007).

The usual differences in study designs and participant characteristics may explain some of the inconsistencies in findings across cohort studies of flavonoid intake and cancer risk; however, factors related specifically to flavonoid exposure itself may also contribute to the inconsistency. Flavonoid intakes vary greatly among different populations and the food sources of flavonoids vary too (Zamora-Ros et al. 2010). For example, isoflavone intake contributed considerably to the total flavonoid intake in Japanese (Arai et al. 2000), whereas it was a minor component of the total flavonoid intake of a Spanish cohort (Zamora-Ros et al. 2010). Thus, comparison of findings across populations is difficult. Further, accurately assessing flavonoid exposure in population-based studies is also challenging. Many food frequency questionnaires used for these studies were not designed to measure flavonoid intakes specifically.

In addition, amounts of flavonoids in plant foods can differ by species, variety, growing conditions, food preparation and processing (Aherne and O'Brien 2002), and systemic exposure in humans can differ because of genetic variation and gut microbial community differences (Lampe and Chang 2007).

Overall, although several epidemiologic cohort studies suggest that flavonoid-rich plant foods may play a role in chemoprevention of some cancers, the findings to date are not consistent and the studies, in their totality, do not support a strong protective role of flavonoids against cancer. Nevertheless, there is extensive evidence from studies in mammalian cell cultures and laboratory animals that indicates that flavonoids can have preventive effects on carcinogenesis via a wide range of potential mechanisms. These include their capacity to quench reactive oxygen species and reactive nitrogen species such as superoxide radical, hydroxyl radical, peroxyl radical, and peroxynitrite and to modulate the activity of various detoxifying enzymes, including lipoxygenase, cycloxygenase, and xanthine oxidase (Gibellini et al. 2011). They have been shown to have effects at various stages of the cancer process including induction of cell cycle arrest, inhibition of cell adhesion and invasion, and enhancing immune function. Flavonoids also have the ability to modulate cell signal transduction pathways related to cellular proliferation, apoptosis, and angiogenesis (Williams et al. 2004). There is growing evidence that some flavonoids can exert epigenetic effects in cancer cells by modulating DNA methylation and influencing chromatin remodeling, for example, epigallocatechin gallate is a potent and efficacious *in vitro* inhibitor of DNA methyltransferase and resveratrol activates SIRT-1, a histone deacetylase that has an important role in apoptosis (Gibellini et al. 2011).

Of the various chemopreventative activities of flavonoids summarized earlier, the most extensively investigated both *in vitro* and, importantly, in human intervention trials has been protection against free radical damage to DNA, and this is the subject of the remainder of this review.

19.3   FREE RADICALS AND DNA DAMAGE IN CANCER

Cancer is a disease in which uncontrolled cellular growth occurs as a result of alteration or damage to the genetic material. For example, colorectal cancer (CRC), one of the most common cancers of the Western world is characterized by a multistep pathway involving sequential mutations and deletions of key oncogenes and tumor suppressor genes such as *APC*, *K-ras*, *p53* in colonic epithelial cells (Fearon and Vogelstein 1990). These changes result in evasion of cell signaling pathways and cell cycle control points, leading to hyperproliferation, avoidance of apoptosis, and eventual invasion and metastasis. Key causes of genetic damage are free radicals derived from endogenous and exogenous sources. Endogenous free radicals are produced as metabolic byproducts, while exogenous sources of radicals include tobacco smoke, infectious agents, radiation, industrial chemicals, certain drugs, and carcinogens in food. Important radicals that impact upon physiological functions are reactive oxygen species (ROS) such as superoxide anions (O_2^-) and hydroxyl radicals (OH•), which cause oxidative damage to DNA, RNA, proteins, and lipids (Azad et al. 2008).

Reaction of ROS with DNA results in the formation of oxidized DNA bases, apurinic/apyrimidinic sites (AP sites) or baseless sugars, or DNA strand breaks (Friedberg et al. 2006). In particular, 8-oxo-7,8-dihydro-2′-deoxyguanosine (8-OHdG) is one of the most easily formed oxidative DNA lesions. Moreover, 8-OHdG can react with compounds such as peroxynitrate to produce further mutagenic lesions (Friedberg et al. 2006). In addition, ROS also react with phospholipids in the cell membrane and form lipid peroxidation products such as 4-hydroxynonenal, which in turn react with the DNA and lead to the formation of cyclic-DNA adducts (Knasmüller 2009). Other abundant lesions of oxidative DNA damage, which are also highly mutagenic, result from GC to TA transversions (Barnes and Lindahl 2004).

A defense system to neutralize the toxicity of free radicals includes detoxifying enzymes such as superoxide dismutase (SOD), catalase, glutathione peroxidase, and glutathione reductase. Additionally, high molecular weight antioxidants such as albumin and ferritin and low molecular weight antioxidants such as glutathione (GSH) are involved in detoxification. An overproduction of free radicals in the body can cause oxidative stress because of an imbalance between free radical generation and the natural defense system.

However, there are also dietary sources of antioxidants, especially from fruits and vegetables, and these compounds may protect cells from DNA damage. These include vitamin C, vitamin E, carotenoids, and polyphenolic compounds such as glucosinolates, flavonoids, tannins, and lignans (Naczk and Shahidi 2006; Watson and Preedy 2010). The level of each of the phytochemicals varies between plant species and depends on the growing method, the cultivar, the ripening process, storage conditions, and the cooking process. It is considered that hydroxyl groups of aromatic rings in plant polyphenols can trap the free radicals and prevent them from reacting with cellular proteins and DNA.

19.4 MEASUREMENT OF DNA DAMAGE

Several human nucleated cells types have been used to measure DNA damage. DNA damage can be determined in local organs either by taking biopsies from, for example, prostate or colon or by isolating exfoliated epithelium cells such as buccal, bladder, or intestinal cells. Systemic DNA damage is usually measured in white blood cells, as they can be easily isolated from blood samples. In addition, they are believed to reflect DNA damage in other body tissues. Studies of DNA damage can be performed *in vivo*, *ex vivo*, as well as *in vitro*. Analysis is complicated by the fact that there are many factors that can have an effect during the experiment such as DNA repair and the method used for assessing DNA damage.

The most frequently applied assay to detect strand breaks and related events is the alkaline single cell gel electrophoresis (Comet) assay (Singh et al. 1991), which has largely replaced the technique of alkaline elution. The method is applicable for a wide range of cell types and for *in vivo* and *in vitro* studies. It involves embedding the cell suspension of lymphocytes or treated cell cultures in agarose on a microscope slide, lysing the cells to liberate the DNA, and then treating under alkaline conditions causing the DNA to unwind from sites of strand breakage. The slides are then subjected to electrophoresis and DNA fragments induced by genotoxic agents migrate to the anode to form a comet "tail." The DNA is stained with a fluorescent dye and the proportion of damaged DNA in the tail assessed by image analysis. A modification to the procedure in which enzymes are added, which convert oxidized bases to strand breaks, allows oxidative DNA damage to be quantified. Specifically, the enzyme endonuclease III (EndoIII), which specifically nicks DNA at sites of oxidized pyrimidines or formamidopyrimidine DNA glycosylase (FPG), which recognizes 8-oxoGua and other oxidized purines, is added after lysis (Collins 2011).

Additionally, the production of DNA adducts such as 8-OHdG or lipid peroxidation products in plasma or urine, as measured by analytical methods, is also often used as a marker for systemic DNA damage.

19.5 EFFECT OF FLAVONOIDS ON DNA DAMAGE

The impact of flavonoids on DNA damage has been investigated in both *in vitro* studies and human intervention trials. Most of the *in vitro* studies use purified flavonoids and so provide direct evidence for their potential to reduce DNA damage. The disadvantage to these studies is that they do not take into account bioavailability, or the extensive metabolism that flavonoids undergo by mammalian enzymes and gut microbial activities *in vivo*, both of which are likely to alter the bioactivity of the phytochemicals. Evidence for the effectiveness of flavonoids in human trials is deduced from interventions with vegetables and fruits rich in flavonoids such as celery, Brussels sprouts, spinach, red cabbage, red onion, broccoli, kiwi fruit, grape, berries, and citrus fruits. All of these foods contain flavonoid compounds, which have been shown to decrease the level of DNA damage *in vitro* (Duthie et al. 1997b; Aherne and O'Brien 2000; Wilms et al. 2005; Yeh et al. 2005).

19.5.1 *In Vitro* Evidence

The *in vitro* evidence of the effects of flavonoids on DNA damage is summarized in Table 19.1, and in this context, the flavonol quercetin is the most well-studied compound. Interestingly, when incubated with HeLa cells, HepG cells, or lymphocytes, quercetin (and a related flavonoid myricetin) actually induced DNA damage, although only at relatively high dose levels (100 μM or greater; Duthie et al. 1997b; Johnson and Loo 2000). However, several studies have shown that quercetin and myricetin at lower doses are effective in preventing DNA damage induced in human lymphocytes, or lymphocyte cell lines, by H_2O_2 (Duthie et al. 1997a; Johnson and Loo 2000; Wilms et al. 2005). In addition, Melidou et al. (2005) investigating the ability of a range of flavonoids to protect Jurkat cells (a T lymphocyte cell line) from DNA damage induced by H_2O_2 found that the most potent inhibitors of single strand breaks included several flavonols (quercetin, galangin, myricetin, fisetin) and flavones (7,8-dihydroxyflavone, baicalein, and luteolin). In contrast, flavanones (naringenin, taxifloin) and flavan-3-ols (catechin) showed little or no protective activity. The number of flavonoids studied enabled some structure activity relationships to be elucidated: the presence of dihydroxy groups in the orthoposition of ring A or B was associated with strong DNA-protecting capacities as was a

TABLE 19.1

Effect of Flavonoids on DNA Damage *in vitro*[a]

Reference	Cell Type/Cell Line	Test Compounds	Study Design	Result
Duthie (1997b)	Caco-2 HeLa HepG2 lymphocytes	Quercetin (Q) Myricetin (M) Silymarin (S)	Cells incubated with flavonoids 0–1000 μM) DNA damage (strand breaks and oxidized pyrimidines) assessed by Comet assay	↑ DNA damage in HepG2, HeLa, lymphocyte when incubate with (Q) ↑ DNA damage in HeLa when incubate with (M) ↑ DNA damage in HeLa when incubate with (S)
Aherne (2000)	Caco-2	Quercetin (Q) Rutin (R)	Cells preincubated with flavonoids 24 h before exposure to *tert*-BOOH. DNA damage assessed by Comet assay	↓ DNA damage by Q and R, Q was more protective
Lazze (2003)	Rat smooth muscle Rat hepatoma cells	Anthocyanins: Delphinidin (DP) Delphinidin-3-*O*-glucoside Delphinidin-3-*O*-rutinoside Cyanidin Cyanidin-3-*O*-glucoside Cyanidin-3-*O*-rutinoside	Cells preincubated with anthocyanins (100 μM) for 2 h. Cells washed then treated with *tert*-BOOH. DNA damage and oxidative damage assessed by Comet assay	↓ r TBH induced DNA damage by most anthocyanins, particular DP in both cell lines. No effect on oxidative DNA damage
Duthie (1997a)	Lymphocytes	Quercetin (Q) Myricetin (M) Silymarin (S)	Pretreated cell for 30 min before exposure to 200 μM H_2O_2. DNA damage/ oxidized DNA assessed by Comet assay	↓ H_2O_2 induced DNA damage (Q) ↓ H_2O_2 induced DNA damage (M) S no effect ↓ H_2O_2 induced oxidation of pyrimidines (Q)
Wilms et al. (2005)	Human lymphocytes	Quercetin	(1) Cells pretreated with quercetin (1–100 μM) 1 h then exposed to H_2O_2 (25 μM) (2) Cells pretreated with quercetin then exposed to benzo(a)pyrene (1 uM)	(1) ↓ DNA damage, dose related ($p < 0.01$) (2) ↓ Benzo(a)pyrene diol epoxide DNA adducts at Quercetin conc. of 10 – 100 μM ($p < 0.05$)

(Continued)

TABLE 19.1 (CONTINUED)
Effect of Flavonoids on DNA Damage *in vitro*[a]

Reference	Cell Type/Cell Line	Test Compounds	Study Design	Result
Johnson and Loo (2000)	Jurkat (T lymphocyte cell line)	Quercetin (−)-Epicatechin-3-O-gallate (EGCG)	(1) Preincubate with 10uM quercetin or EGCG for 30 min, then expose to 25 µM H_2O_2 at 4°C or 100 µM SIN-1* for 30 min (2) Cell incubate with quercetin/ EGCG 100 µM, 20 min	(1) ↓ oxidative DNA damage induced by H_2O_2 and SIN-1 (2) ↑ DNA damage induced by quercetin or EGCG
Melidou et al. (2005)	Jurkat (T lymphocyte cell line)	Quercetin, galangin, kaempferol, morin, myricetin, fisetin, 7,8-dihydroxyflavone, baicalein, apigenin, luteolin, naringenin, taxifolin, catechin	Cells exposed to flavonoids (0.001–100 µM; 15 min) then H_2O_2 (by glucose oxidase; 10 min)	Reduction in H_2O_2-induced DNA damage observed with quercetin, galangin, myricetin, fisetin, 7,8-dihydroxyflavone, baicalein, and luteolin. Naringenin, taxifolin, morin, kaempferol catechin showed weak or no effects
Gill et al. (2004)	HT 29	Cruciferous and leguminous sprouts extract (CLS)	(1) Cells incubated with CLS 1 and 24 hr (2) Cells preincubated with CLS 1 and 24 hr then treated with 75 µM H_2O_2	(1) ↔ 100 µL/mL of CLS did not increase DNA damage ↑ 200 µL/mL of CLS increased DNA damage (2) ↓ DNA damage
Zhu et al. (2001)	Lymphocytes	Raw, cooked, autolyzed Brussels sprout, sinigrin (glucosinolate)	Cells preincubated with aqueous extract of powder of raw, cook, autolyzed Brussels sprout, sinigrin for 60 min at 37°C then exposed to 100 µM H_2O_2	↓ Cooked, autolyzed and sinigrin inhibited DNA damage induced by H_2O_2, Raw Brussel sprout had no DNA protective effect ↔ cook, autolyzed and sinigrin did not reduce EndoIII,vFPG site

[a] *tert* BOOH = *tert*-butyl-hydroperoxide; SIN-1 = 3-morpholinosydnonimine (peroxynitrite generator)

hydroxyl group at position 3 of the C ring together with an oxo group at position 4. The presence of a double bond at position of C_3 or C_4 of the C ring is also related to protective ability. However, the protection against DNA damage by flavonoids also depended on the ability of flavonoids to penetrate the cell membrane and access the interior of cells as well as the ability of flavonoids to remove loosely bound redox-active iron from intracellular locations. As a result of this, the prevention of the formation of hydroxyl radical at sites of redox-active iron in cells might be much more effective than hydroxyl radical scavenging itself. However, scavenging of ROS also plays a role in the antioxidant activity of flavonoids. Johnson and Loo (2000) compared the antioxidant activity of quercetin and (−)-epigallocatechin-3-*O*-gallate (EGCG). They found that EGCG was a stronger scavenger of 2,2-diphenyl-1-picrylhydrazyl (DPPH) than quercetin. However, quercetin was better than EGCG in preventing DNA damage in lymphocytes caused by H_2O_2. The reason is probably that the content of quercetin in cells is higher because quercetin has a higher ability to penetrate the cell membrane than EGCG.

19.5.2 EVIDENCE FROM HUMAN DIETARY TRIALS

In vivo studies in human subjects have focused on using fruits and vegetables containing flavonoids and other phytochemicals, rather than flavonoid supplements, and so the evidence for a role of the compounds is indirect. Nevertheless, consumption of fruit and vegetables rich in flavonoids have been shown to be protective against DNA damage, especially oxidative damage in several human trials (Table 19.2). For instance, Park et al. (2003) found that drinking purple grape juice daily for 8 weeks could reduce endogenous DNA damage in lymphocytes and also decreased reactive oxygen species release, which was positively correlated to the level of DNA damage. This suggests that a reduction of free radical release may partially reduce cellular DNA damage. Also, Weisel et al. (2006) demonstrated that the intake of mixed fruit juice with a high content of flavonoids reduced oxidative DNA damage. This might result from direct antioxidant effects as well as from enhancing the cellular antioxidant system since during intervention an increase of the glutathione status was found. In a recent well-controlled study (Kaspar et al. 2011), male volunteers ($n = 12$) were fed either 150 g/day yellow potatoes (high in phenolic acids and carotenoids) or purple potatoes (high in anthocyanins) for 6 weeks. A control group was fed a similar amount of white potatoes. Plasma 8-OHdG was lower in the yellow potato group (30.3 ± 2.4 μg/L) and purple potato group (26.0 ± 1.5 μg/L) ($P < 0.05$) than in the controls (38.0 ± 2.3 μg/L; $P < 0.05$) although there were no changes in total antioxidant activity in plasma.

Consumption of fruit rich in flavonoids has also been shown to enhance cellular resistance to ROS as shown in decreased of DNA damage after exposure to ROS-releasing agents. Collins et al. (2003) found that consumption of kiwi fruit reduced DNA damage in lymphocytes after challenge with H_2O_2, indicating that antioxidant status in lymphocytes was enhanced (Table 19.2). Moreover, this protection also occurs with endogenous oxidation as a decrease of oxidized bases in lymphocytes was found. The authors suggested that this protection may result from an increased rate of DNA repair since a significant increase in DNA repair was seen after kiwi

TABLE 19.2

Dietary Interventions with Flavonoids or Flavonoid-Containing Foods—Impact on DNA Damage and Oxidative DNA Damage

Reference	Type of Fruits and Vegetables	Study Design	Effect on DNA Damage	P	N	Control
Smith et al. (1999)	Fruit powder capsule (850 mg powder /capsule): pineapple, papaya, apple, orange cranberries, peaches, Vegetable powder capsule (750 mg powder/capsule): carrot parsley, beets, broccoli kale, cabbage, spinach tomato.	2 Fruit capsules + 2 vegetable capsules/day for 80 days	↓ 67% in DNA damage post- vs pretreatment	$P < 0.0001$	20	No concurrent control
Park et al. (2003)	Purple grape juice	480 mL juice/day for 8 weeks. DNA damage in lymphocytes assessed by Comet assay	↓ 20% in DNA damage, post- vs pretreatment	$P < 0.001$	67	No concurrent control
Pool-Zobel et al. (1997)	Tomato juice, carrot powder, spinach powder	Weeks 1–2: depletion Weeks 3–4: 330 mL/day tomato juice Weeks 5–6: 330 mL/day carrot juice Weeks 7–8: 10 g/day dried spinach powder DNA damage/oxidized damage in lymphocytes assessed by Comet assay +/− Endo III	↓ in endogenous SB for tomato juice and carrot ↓ Oxidized pyrimidines for carrot	$P > 0.05$ post- vs. pretreatment $P < 0.05$	23	No concurrent control
Moller et al. (2004)	Gp1 blackcurrant juice (600 mg/L anthocyanin) Also contained Vit. C (140 mg/day) Gp 2 .anthocyanin concentrate drink (600 mg/L anthocyanin) No Vit. C Gp 3 Control drink (low flavonoids diet) Amounts consumed 475–1000 mL/day depending on body wt	Parallel study of 3 weeks. Oxidative DNA damage in lymphocytes assessed by Comet assay + Endo III and Fpg	↔ No sig. change in level of damage with treatment ↑ Fpg sensitive sites in blackcurrant juice group.		57	Placebo controlled

Reference	Intervention	Methods	Results	P value	N	Notes
Thompson et al. (2005)	Wide range of F and V including lettuce, cabbage broccoli, radish melon, blueberry, apple, tomato, citrus fruits, carrot, grape, celery	Group 1 = 3.6 serving F and V/day for 2 weeks. Group 2 = 12.1 serving F and V/day for 2 weeks	↓ lymphocyte 8OHdG and urinary F2-isoprostanes in high V and F consumption	$P < 0.01$	64	High vs. low F and V comparison
Briviba et al. (2008)	Mixed fruit and vegetable.	Weeks 1–4: All volunteers consumed 2 servings F and V/day. 1 serving =100 g F and V or 200 mL juice. Week 5–8: Group 1 = 2 serving/day; Group 2 = 5 serving/day; Group 3 = 8 serving/day. Oxidative DNA damage in lymphocytes assessed by Comet assay + Endo III and Fpg. Lipid peroxidation assessed by plasma F-2 isoprostanes and malondiadehyde	No significant difference between groups in SB, oxidative DNA damage, plasma F-2 isoprostanes or malondiadehyde		64	Placebo controlled
Carmen Ramirez-Tortosa et al. (2004)	Dessert made of concentrated juice of grape, cherry, blackberry, blackcurrant, raspberry (224 mg/kg anthocyanins)	22 elderly subjects given 200 g juice dessert/day for 2 weeks. 8 control subjects. Blood samples taken pre- and postintervention. DNA damage in lymphocytes assessed by Comet assay. Serum assayed for TEAC and FRAP total antioxidant capacity	No significant change in DNA strand breaks or serum total antioxidant capacity		22	Placebo controlled
Weisel et al. (2006)	Mixed fruit juice: red grape, blackberry, sour cherry, blackcurrant and chokeberry	Weeks 1–2: run in; Weeks 3–6: juice intake; Weeks 7–9: wash out. Consume 700 mL of juice daily	↓ DNA oxidative damage during juice intake	$P < 0.005$		Placebo controlled
	Control – polyphenol-depleted juice	Oxidative DNA damage in lymphocytes assessed by Comet assay + FPG. Lipid peroxidation assessed by urinary F-2 isoprostanes	↑ total GSH. No effect on F2-isoprostanes	$P < 0.005$	9 control 18 treatment	

(Continued)

TABLE 19.2 (CONTINUED)

Dietary Interventions with Flavonoids or Flavonoid-Containing Foods—Impact on DNA Damage and Oxidative DNA Damage

Reference	Type of Fruits and Vegetables	Study Design	Effect on DNA Damage	P	N	Control
Bub et al. (2003)	Juice A: aronia, blueberries, boysenberries (rich in anthocyanins, mainly cyanidin glycosides). 330 mL/day. Juice B: green tea, apricot, lime (rich in flavan-3-ols mainly EGCG), 330mL/day	Cross over study, Weeks 1–2: run in Weeks 3–4: juice A or B Weeks 5–6: wash out Weeks 7–8: juice B or A Weeks 9–10: wash out. Oxidative DNA damage in lymphocytes assessed by Comet assay + EndoIII ; Lipid peroxidation assessed by plasma malondialdehyde	Juice consumption had no effect on single strand breaks but decreased oxidative DNA damage. ↓ plasma malondialdehyde	$P < 0.05$	27	No concurrent control
Riso et al. (2005)	Blood orange juice (OJ) 600 mL/day	Cross over study Weeks 1–3: No OJ or OJ Weeks: 4–6 wash out Weeks 7–9: OJ or No OJ Resistance to H_2O_2 induced DNA damage in lymphocytes assessed by Comet assay. Malondialdehyde in plasma	↓ DNA damage after H_2O_2 challenge ex vivo No effect on plasma malondialdehyde	$P < 0.05$	16	Control group without OJ
Kang et al. (2004)	'Green vegetable' drink 240 mL/day containing: Angleica Keiskei extract, Kale extract, turmeric, Rooibos tea extract	2 week dietary restriction of F and V intake, then 8 week intervention with vegetable drink. DNA damage in lymphocytes assessed by Comet assay	↓ DNA damage	$P = 0.00$	19	No concurrent control
Haegele et al. (2000)	Wide range of F and V including lettuce, cabbage broccoli, radish, apple melon, blueberry, cranberry, apple, tomato, citrus fruits, carrot, grape, celery, artichoke, strawberry spinach, soybean, garlic and onion	High F and V: 12 servings/day for 2 weeks Low F and V: 3.8 servings/day for 2 weeks	Inverse correlation between plasma carotenoid and lymphocyte 8-OHdG and urinary F-2 isoprostanes	$P \leq 0.006$	37	High vs. low F and V comparison

Reference	Intervention	Design	Findings	P value	N	Control
Thompson et al. (1999)	Wide range of F and V including lettuce, cabbage broccoli, radish melon, blueberry, apple, tomato, citrus fruits, carrot, grape, celery, artichoke, and strawberry	Baseline (3 days) 5–8 portions F and V/day Intervention (14 days 12 portions F and V/day Oxidative damage assessed by 8-OHdG in lymphocytes and malondialdehyde, 8-OHdG, F2-isoprostanes in urine.	↓ lymphocyte 8-OHdG ↓ urinary 8-OHdG ↓ urinary F-2 isoprostanes	$P = 0.113$ $P = 0.108$ $P = 0.004$	28	No concurrent control
Moller et al. (2003)	Group 1: 600 g/day Mixed F and V (apple, pear, orange juice, broccoli, carrots, onion, tomato. Group 2: Antioxidant vitamins/minerals (inc Vit C, VitE, β-carotene; no flavonoids) equiv to 600 g F and V in a tablet Group 3: Control, low flavonoid diet (no F and V, tea, chocolate, alcohol) placebo tablet	Parallel study, n = 12–16/group Oxidative DNA damage in lymphocytes assessed by Comet assay + Endo III and FPG; 24h urinary excretion of 8-oxodG	No effect of F and V or supplement on any marker of DNA damage in lymphocytes or urine. No effect on sensitivity to H2O2 damage		24 days	12–16/group
Hoelzl et al. (2008)	Brussels sprouts	5 days—restricted diet period 6 days—300 g/day of Brussels sprout. DNA damage in lymphocytes assessed by Comet assay	↔ no significant change in DNA damage induced ex vivo by TRP-p-2 ↓ oxidize purine 45%(FPG) ↓ oxidize pyrimidine (Endo III) ↓ DNA damage after H2O2 challenge	$P < 0.01$	8	No concurrent control
Kim et al. (2003)	Polyphenol rich F and V diet: red cabbage, red chicory, onion, red lettuce, mushroom, apple, kidney bean, black rice and grape juice	Day 0–6: phenol depleted diet Day 7–9: wash out Day 10–15: phenol-rich diet DNA damage in lymphocytes assessed by Comet assay	↔ no significant change in DNA damage		19	No concurrent control
Collins et al. (2003)	Kiwi fruit	Subjects given in random order 1, 2, or 3 kiwi fruits/day for 3 weeks With 2 week wash out period between each treatment	↓ DNA strand breaks after H2O2 challenge ↓ oxidized pyrimidine and purine in all groups	$P < 0.0001$ $P < 0.001$	14	

(Continued)

TABLE 19.2 (CONTINUED)

Dietary Interventions with Flavonoids or Flavonoid-Containing Foods—Impact on DNA Damage and Oxidative DNA Damage

Reference	Type of Fruits and Vegetables	Study Design	Effect on DNA Damage	P	N	Control
Gill et al. (2007)	Watercress, raw, 85 g/day	DNA damage in lymphocytes assessed by Comet assay +/– EndoIII and FPG for measurement of oxidative damage. Double blind, cross over study control, 8 week habitual diet; washout 7 week; watercress 8 week. DNA damage in lymphocytes assessed by Comet assay	↓ basal DNA damage ↓ basal+ oxidative purine DNA damage ↓ DNA damage after H2O2 challenge	$P = 0.03$ $P = 0.002$ $P = 0.07$	60	Control group no supplement
Kaspar et al. (2011)	White potato (WP), yellow potato (YP; higher conc. of phenolic acids + carotenoids), purple potato (PP; higher conc. of phenolic acids + anthocyanins)	150 g/day potatoes for 6 weeks, n = 12 males/group	↓ plasma 8-OHdG in YP and PP groups.	$P < 0.05$	12	Placebo controlled
Wilms et al. (2005)	Blueberry /apple juice containing 18 mg quercetin	Consumed daily for 4 weeks, n = 8 DNA damage assessed by Comet assay	Non sig ↓ *ex vivo* H2O2-induced DNA damage in lymphocytes and benzopyrene adducts	$P = 0.07$	8	No concurrent control
Gill et al. (2004)	Sprouting stage of broccoli, radish, alfalfa, clover	Parallel study: 115 g per day for 2 week control no supplement. DNA damage assessed by Comet assay	↓ DNA strand breaks after H2O2 challenge No effect on endogenous DNA Damage	$P < 0.05$	10/group	Control group no supplement

fruit intake. Other studies that showed protection against oxidative DNA damage in lymphocytes with high flavonoid foods include that of Riso et al. (2005) using blood orange juice, Bub et al. (2003) using juices high in anthocyanins or EGCG, Weisel et al. (2006) also with an anthocyanin rich fruit juice mixture, and Smith et al. (1999) who tested fruit and vegetable powders in capsule form.

Some studies with high flavonoid juices, however, showed no protective effects on lymphocyte DNA damage, for example, Carmen Ramirez-Tortosa et al. (2004) who gave a grape and berry juice mixture containing 224 mg/kg anthocyanins, to elderly subjects for 2 weeks and Moller et al. (2004) who used a blackcurrant juice with 600 mg/l anthocyanins. Moller et al. (2004) ascribed the lack of effect to the very low DNA damage level in the subjects at baseline. Other factors that may influence the results of *in vivo* studies include the time period of consumption of the foods. In the study by Bub et al. (2003), fruit juice intake at an early period of intervention had no significant effect on DNA damage, while after a period of 5 weeks, a reduction in DNA damage was found. This result suggests that flavonoids may not only be acting via the mechanism of ROS scavenging but also by exerting an influence on the antioxidant defence system by modulating the activity of phase I and phase II enzymes that play a role in preventing DNA damage.

Three studies that used cruciferous vegetables, Gill et al. 2004 (broccoli and radish sprouts), Hoelzl et al. 2008 (Brussels sprout), and Gill et al. 2007 (watercress) all reported decreases in oxidative DNA damage in lymphocytes, although of course flavonoids are not the main phytochemical in the *Cruciferae*. Other studies have used mixtures of fresh fruits and vegetables, which contain various types of phytochemicals, including carotenoids and glucosinolates, as well as flavonoids. The results from these studies are inconsistent (Table 19.2). For example, Thompson et al. (1999) demonstrated that increasing fruit and vegetable intake from 5.8 serving/day to 12.0 serving/day over 14 days decreased the level of 8-OHdG in DNA from lymphocytes and in urine. Notably, the reduction was greater in individuals who had lower average levels of plasma α-carotene at preintervention (56 ng/mL) than individuals with higher plasma α-carotene (148 ng/mL). The same group of Thompson et al. (2005) found that consumption of mixed fruits and vegetables in high amounts (12.1 servings/day) resulted in a decrease of DNA oxidation in lymphocytes and reduced lipid oxidation products excreted in urine, while DNA oxidation was unchanged in the group with low intake of fruit and vegetables. Another study evaluated the effect of a phenol-depleted and a phenol-rich diet on blood oxidative stress including analysis of antioxidant enzyme activity (superoxide dismutase, SOD) and DNA damage (Kim et al. 2003). It found that the erythrocyte SOD activity was slightly decreased during the phenol-depleted diet period while at day 6 of phenol-rich diet erythrocyte SOD activity was significant increased. However, there was no reduction in lymphocyte DNA damage. Negative results for mixed fruit and vegetables on oxidative DNA damage in lymphocytes were also reported by Brivba et al. (2008) who compared 2, 5, and 8 servings of plant foods per day, Moller et al. (2003) who gave their subjects 600 g per day mixed fruit and vegetables, and Kim et al. (2003) who used an intervention of high polyphenol foods including red cabbage, grape juice, and apples. It is perhaps not surprising that interventions with plant foods give both positive and negative results for protection against DNA damage given the diversity of flavonoids and other phytochemicals in different fruits and vegetables and also the day-to-day variation in phytochemical levels reported in certain plants (Gill et al. 2007; Jin et al. 2009). Furthermore, the studies vary in duration, characteristics of volunteers and methods used, and background flavonoid intake, all of which may have an impact on response.

19.6 SUMMARY

There is extensive evidence from *in vitro* studies and laboratory animal models that plant flavonoids can have preventive effects on carcinogenesis via a wide range of potential mechanisms. Their antioxidant properties, which encompass the capacity to quench reactive oxygen species and modulate intracellular detoxification enzymes that inactivate free radicals, have received considerable attention because of the role oxidative damage plays in the process of carcinogenesis. Overall, human intervention trials with flavonoid-rich fruits and vegetables, in particular, fruit and fruit juices with high anthocyanin content, indicate a capacity to decrease significantly oxidative damage to DNA, suggesting beneficial effects on cancer risk. However, there are inconsistencies in the evidence base, possibly because of the diversity and amounts of flavonoids and other phytochemicals in different fruits and vegetables, as well as differences in study design.

Epidemiologic cohort studies suggest that flavonoid-rich plant foods may play a role in chemoprevention of some cancers, although the findings to date are not consistent and the studies, in their totality, do not support a strong protective role of flavonoids against cancer. As with the DNA damage studies, differences in study designs and participant characteristics may explain some of the inconsistencies, but also factors related specifically to flavonoid exposure itself may also contribute. For example, flavonoid intakes vary greatly among different populations making comparisons of findings across populations difficult, and accurately assessing flavonoid exposure in population-based studies is very challenging.

ACKNOWLEDGMENTS

Piyawan Sitthiphong acknowledges receipt of a scholarship from the Royal Thai Government. Annett Klinder and Ian Rowland are grateful to the Alpro Foundation for financial support. Johanna Lampe is supported in part by United States National Cancer Institute grant R56 CA070913 and the Fred Hutchinson Cancer Research Center.

REFERENCES

Adebamowo, C.A., Cho, E., Sampson, L. et al. (2005). Dietary flavonols and flavonol-rich foods intake and the risk of breast cancer. *Int. J. Cancer,* **114,** 628–633.

Aherne, S.A. and O'Brien, N.M. (2000). Mechanism of rotection by the flavonoids, quercetin and rutin, against *tert*-butylhydroperoxide- and menadione-induced DNA single strand breaks in Caco-2 cells. *Free Radic. Biol. Med.,* **29,** 507–514.

Arai, Y., Watanabe, S., Kimira, M. et al. (2000). Dietary intakes of flavonols, flavones and iso-flavones by Japanese women and the inverse correlation between quercetin intake and plasma LDL cholesterol concentration. *J. Nutr.,* **130,** 2243–2250.

Arts, I.C., Hollman, P.C., Bueno De Mesquita, H.B. et al. (2001). Dietary catechins and epithelial cancer incidence: The Zutphen Elderly Study. *Int. J. Cancer,* **92,** 298–302.

Arts, I.C., Jacobs, D.R., Jr., Gross, M. et al. (2002). Dietary catechins and cancer incidence among postmenopausal women: The Iowa Women's Health Study (United States). *Cancer Causes Control,* **13,** 373–382.

Azad, N., Rojanasakul, Y. and Vallyathan, V. (2008). Inflammation and lung cancer: Roles of reactive oxygen/nitrogen species. *J. Toxicol. Environ. Health, Pt. B Crit. Rev.,* **11,** 1–15.

Barnes, D.E. and Lindahl, T. (2004). Repair and genetic consequences of endogenous DNA base damage in mammalian cells. *Annu. Rev. Genet.,* **38,** 445–476.

Bobe, G., Weinstein, S. J., Albanes, D. et al. (2008). Flavonoid intake and risk of pancreatic cancer in male smokers (Finland). *Cancer Epidem. Biomar.,* **17,** 553–562.

Briviba, K., Bub, A., Moseneder, J. et al. (2008). No differences in DNA damage and antioxidant capacity between intervention groups of healthy, nonsmoking men receiving 2, 5, or 8 servings/day of vegetables and fruit. *Nutr. Cancer,* **60,** 164–170.

Bub, A., Watzl, B., Blockhaus, M. et al. (2003). Fruit juice consumption modulates antioxidative status, immune status and DNA damage. *J. Nutr. Biochem.,* **14,** 90–98.

Carmen Ramirez-Tortosa, M., Garcia-Alonso, J., Luisa Vidal-Guevara, M. et al. (2004). Oxidative stress status in an institutionalised elderly group after the intake of a phenolic-rich dessert. *Br. J. Nutr.,* **91,** 943–950.

Chang, E.T., Lee, V.S., Canchola, A.J. et al. (2007). Diet and risk of ovarian cancer in the California Teachers Study Cohort. *Am. J. Epidemiol.,* **165,** 802–813.

Collins, A.R. (2011). The use of bacterial repair endonucleases in the Comet assay. *Methods Mol. Biol.,* **691,** 137–147.

Collins, A.R., Harrington, V., Drew, J. et al. (2003). Nutritional modulation of DNA repair in a human intervention study. *Carcinogenesis,* **24,** 511–515.

Cutler, G.J., Nettleton, J.A., Ross, J.A. et al. (2008). Dietary flavonoid intake and risk of cancer in postmenopausal women: The Iowa Women's Health Study. *Int. J. Cancer,* **123,** 664–671.

Duthie, S.J., Collins, A.R., Duthie, G.G. et al. (1997a). Quercetin and myricetin protect against hydrogen peroxide-induced DNA damage (strand breaks and oxidised pyrimidines) in human lymphocytes. *Mutat. Res.-Genet. Toxicol. Environ. Mutag.,* **393,** 223–231.

Duthie, S.J., Johnson, W. and Dobson, V.L. (1997b). The effect of dietary flavonoids on DNA damage (strand breaks and oxidised pyrimidines) and growth in human cells. *Mutat. Res.-Genet. Toxicol. Environ. Mutag.,* **390,** 141–151.

Fearon, E.R. and Vogelstein, B. (1990). A genetic model for colorectal tumorigenesis. *Cell,* **61,** 759–767.

Friedberg, E.C., Aguilera, A., Gellert, M. et al. (2006). DNA repair: From molecular mechanism to human disease. *DNA Repair,* **5,** 986–996.

Gates, M.A., Tworoger, S.S., Hecht, J.L. et al. (2007). A prospective study of dietary flavonoid intake and incidence of epithelial ovarian cancer. *Int. J. Cancer,* **121,** 2225–2232.

Gibellini, L., Pinti, M., Nasi, M. et al. (2011). Quercetin and cancer chemoprevention. *Evid. Based Complement Alternat. Med.,* **2011,** 591356.

Gill, C.I., Haldar, S., Porter, S. et al. (2004). The effect of cruciferous and leguminous sprouts on genotoxicity, *in vitro* and *in vivo. Cancer Epidem. Biomar.,* **13,** 1199–205.

Gill, C.I.R., Haldar, S., Boyd, L. A. et al. (2007). Watercress supplementation in diet reduces lymphocyte DNA damage and alters blood antioxidant status in healthy adults. *Am. J. Clin. Nutr.,* **85,** 504–510.

Haegele, A.D., Gillette, C., O'Neill, C. et al. (2000). Plasma xanthophyll carotenoids correlate inversely with indices of oxidative DNA damage and lipid peroxidation. *Cancer Epidem. Biomar.,* **9,** 421–425.

Hertog, M.G., Feskens, E.J., Hollman, P.C. et al. (1994). Dietary flavonoids and cancer risk in the Zutphen Elderly Study. *Nutr. Cancer,* **22,** 175–184.

Hertog, M.G.L., Hollman, P.C.H. and Katan, M.B. (1992). Content of potentially anticarcinogenic flavonoids of 28 vegetables and 9 fruits commonly consumed in the Netherlands.
J. Agric. Food Chem., **40,** 2379–2383.

Hertog, M.G.L., Hollman, P.C.H. and van de Putte, B. (1993). Content of potentially anticarcinogenic flavonoids of tea infusions, wines, and fruit juices. *J. Agric. Food Chem.,* **41,**
1242–1246.

Hirayama, T. (1992). Life-style and cancer: From epidemiological evidence to public behavior change to mortality reduction of target cancers. *J. Natl. Cancer Inst. Monogr.,*
65–74.

Hirvonen, T., Virtamo, J., Korhonen, P. et al. (2001). Flavonol and flavone intake and the risk
of cancer in male smokers (Finland). *Cancer Causes Control,* **12,** 789–796.

Hoelzl, C., Glatt, H., Meinl, W. et al. (2008). Consumption of Brussels sprouts protects periph-
eral human lymphocytes against 2-amino-1-methyl-6-phenylimidazo 4,5-B pyridine
(Phlp) and oxidative DNA-damage: Results of a controlled human intervention trial.
Mol. Nutr. Food Res., **52,** 330–341.

Jin, J., Koroleva, O.A., Gibson, T. et al. (2009). Analysis of phytochemical composition
and chemoprotective capacity of Rocket (*Eruca sativa* and *Diplotaxis tenuifolia*)
leafy salad following cultivation in different environments. *J. Agric. Food Chem.,* **57,**
5227–5234.

Johnson, M.K. , Loo, G. (2000). Effects of epigallocatechin gallate and quercetin on oxidative
damage to cellular DNA. *Mutat. Res./DNA Repair,* **459,** 211–218.

Kang, M.H., Park, Y.K., Kim, H.Y. et al. (2004). Green vegetable drink consumption protects
peripheral lymphocyte DNA damage in Korean smokers. *BioFactors,* **22,** 245–247.

Kaspar, K.L., Park, J.S., Brown, C.R. et al. (2011). Pigmented potato consumption alters oxidative stress and inflammatory damage in men. *J. Nutr.,* **141,** 108–111.

Key, T.J., Sharp, G.B., Appleby, P.N. et al. (1999). Soya foods and breast cancer risk: A prospective study in Hiroshima and Nagasaki, Japan. *Br. J. Cancer,* **81,** 1248–1256.

Kim, H.Y., Kim, O.H. and Sung, M.K. (2003). Effects of phenol-depleted and phenol-rich diets
on blood markers of oxidative stress, and urinary excretion of quercetin and kaempferol
in healthy volunteers. *J. Am. Coll. Nutr.,* **22,** 217–223.

Knasmüller, S. (2009). *Chemoprevention of Cancer and DNA Damage by Dietary Factors,*
Weinheim, Wiley-VCH.

Knekt, P., Jarvinen, R., Seppanen, R. et al. (1997). Dietary flavonoids and the risk of lung
cancer and other malignant neoplasms. *Am. J. Epidemiol.,* **146,** 223–230.

Knekt, P., Kumpulainen, J., Jarvinen, R. et al. (2002). Flavonoid intake and risk of chronic
diseases. *Am. J. Clin. Nutr.,* **76,** 560–568.

Kristal, A.R. and Lampe, J.W. (2011). Prioritization of diet and cancer manuscripts: A brief
primer. *Cancer Epidem. Biomar.,* **20,** 725–726.

Lazze, M.C., Pizzala, R., Savio, M. et al. (2003). Anthocyanins protect against DNA damage
Induced by *tert*-butyl-hydroperoxide in rat smooth muscle and hepatoma cells. *Mutat.
Res.-Genet. Toxicol. Environ. Mutag.,* **535,** 103–115.

Lin, J., Zhang, S.M., Wu, K. et al. (2006). Flavonoid intake and colorectal cancer risk in men
and women. *Am. J. Epidemiol.,* **164,** 644–651.

Melidou, M., Riganakos, K. and Galaris, D. (2005). Protection against nuclear DNA damage
offered by flavonoids in cells exposed to hydrogen peroxide: The role of iron chelation.
Free Radic. Biol. Med., **39,** 1591–1600.

Moller, P., Loft, S., Alfthan, G. et al. (2004). Oxidative DNA damage in circulating mononuclear blood cells after ingestion of blackcurrant juice or anthocyanin-rich drink. *Mutat.
Res.-Fundam. Mol. Mech. Mutag.,* **551,** 119–126.

Moller, P., Vogel, U., Pedersen, A. et al. (2003). No effect of 600 grams fruit and vegetables
per day on oxidative DNA damage and repair in healthy nonsmokers. *Cancer Epidem.
Biomar.,* **12,** 1016–1022.

Mursu, J., Nurmi, T., Tuomainen, T.P. et al. (2008). Intake of flavonoids and risk of cancer in Finnish men: The Kuopio Ischaemic Heart Disease Risk Factor Study. *Int. J. Cancer,* **123,** 660–663.

Naczk, M. and Shahidi, F. (2006). Phenolics in cereals, fruits and vegetables: Occurrence, extraction and nalysis. *J. Pharm. Biomed. Anal.,* **41,** 1523–1542.

Negri, E., Lavecchia, C., Franceschi, S. et al. (1991). Vegetable and fruit consumption and cancer risk. *Int. J. Cancer,* **48,** 350–354.

Nishio, K., Niwa, Y., Toyoshima, H. et al. (2007). Consumption of soy foods and the risk of breast cancer: Findings from the Japan Collaborative Cohort (Jacc) Study. *Cancer Causes Control,* **18,** 801–808.

Nothlings, U., Murphy, S.P., Wilkens, L. R. et al. (2007). Flavonols and pancreatic cancer risk: The Multiethnic Cohort Study. *Am. J. Epidemiol.,* **166,** 924–931.

Park, Y.K., Park, E., Kim, J.S. et al. (2003). Daily grape juice consumption reduces oxidative DNA damage and plasma free radical levels in healthy Koreans. *Mutat. Res.-Fundam. Mol. Mech. Mutag.,* **529,** 77–86.

PoolZobel, B.L., Bub, A., Muller, H. et al. (1997). Consumption of vegetables reduces genetic damage in humans: First results of a human intervention trial with carotenoid-rich foods. *Carcinogenesis,* **18,** 1847–1850.

Riso, P., Visioli, F., Gardana, C. et al. (2005). Effects of blood orange juice intake on antioxidant bioavailability and on different markers related to oxidative stress. *J. Agric. Food Chem.,* **53,** 941–947.

Sampson, L., Rimm, E., Hollman, P. C. et al. (2002). Flavonol and flavone intakes in US health professionals. *J. Am. Diet. Assoc.,* **102,** 1414–1420.

Singh, N.P., McCoy, M.T., Tice, R.R. et al. (1988). A simple technique for quantitation of low-levels of DNA damage in individual cells. *Exp. Cell Res.,* **175,** 184–191.

Smith, M.J., Inserra, P.F., Watson, R.R. et al. (1999). Supplementation with fruit and vegetable extracts may decrease DNA damage in the peripheral lymphocytes of an elderly population. *Nutr. Res.,* **19,** 1507–1518.

Steinmetz, K.A. and Potter, J.D. (1991). Vegetables, fruit, and cancer 2 mechanisms. *Cancer Causes Control,* **2,** 427–442.

Thompson, H.J., Heimendinger, J., Gillette, C. et al. (2005). *In vivo* investigation of changes in biomarkers of oxidative stress induced by plant food-rich diets. *J. Agric. Food Chem.,* **53,** 6126–6132.

Thompson, H.J., Heimendinger, J., Haegele, A. et al. (1999). Effect of increased vegetable and fruit consumption on markers of oxidative cellular damage. *Carcinogenesis,* **20,** 2261–2266.

Wang, L., Lee, I.M., Zhang, S.M. et al. (2009). Dietary intake of selected flavonols, flavones, and flavonoid-rich foods and risk of cancer in middle-aged and older women. *Am. J. Clin. Nutr.,* **89,** 905–912.

Watson, R.R. and Preedy, V.R. (2010). *Bioactive Foods in Promoting Health. Fruits and Vegetables,* Amsterdam; Boston, Academic Press.

Weisel, T., Baum, M., Eisenbrand, G. et al. (2006). An anthocyanin/polyphenolic-rich fruit juice reduces oxidative DNA damage and increases glutathione level in healthy probands. *Biotechnol. J.,* **1,** 388–397.

Williams, R.J., Spencer, J.P., and Rice-Evans, C. (2004). Flavonoids: Antioxidants or signalling molecules? *Free Radic. Biol. Med.,* **36,** 838–849.

Wilms, L.C., Hollman, P.C.H., Boots, A.W. et al. (2005). Protection by quercetin and quercetin-rich fruit juice against induction of oxidative DNA damage and formation of Bpde-DNA adducts in human lymphocytes. *Mutat. Res.,* **582,** 155–162.

World Cancer Research Fund/American Institute for Cancer Research. (2007). *Food, Nutrition, Physical Activity, and the Prevention of Cancer: A Global Perspective.* Washington, DC, AICR.

Wu, A.H., Wan, P., Hankin, J. et al. (2002). Adolescent and adult soy intake and risk of breast cancer in Asian-Americans. *Carcinogenesis,* **23,** 1491–1496.

Yamamoto, S., Sobue, T., Kobayashi, M. et al. (2003). Soy, isoflavones, and breast cancer risk in Japan. *J. Natl. Cancer Inst.,* **95,** 906–913.

Yeh, S.L., Wang, W.Y., Huang, C.H. et al. (2005). Pro-oxidative effect of β-carotene and the interaction with flavonoids on UVa-induced DNA strand breaks in mouse fibroblast C3h10t1/2 cells. *J. Nutr. Biochem.,* **16,** 729–735.

Zamora-Ros, R., Andres-Lacueva, C., Lamuela-Raventos, R.M. et al. (2010). Estimation of dietary sources and flavonoid intake in a Spanish adult population (Epic-Spain). *J. Am. Diet. Assoc.,* **110,** 390–398.

Zhu, C.Y. and Loft, S. (2001). Effects of Brussels sprouts extracts on hydrogen peroxide-induced DNA strand breaks in human lymphocytes. *Food Chem. Toxicol.,* **39,** 1191–1197.

Index

Printed and bound by CPI Group (UK) Ltd, Croydon, CR0 4YY

21/10/2024

01777103-0010